S0-BVO-340

Systems Analysis and Simulation in Ecology

VOLUME II

Contributors

Michael Brylinsky
Hal Caswell
George I. Child
A. Ben Clymer
Theodore C. Foin, Jr.
Gilberto C. Gallopín
Herman E. Koenig
Orie L. Loucks
Howard T. Odum
Bernard C. Patten
G. J. Paulik
James A. Resh
Quentin E. Ross
Saul B. Saila
Herman Henry Shugart, Jr.
Richard B. Williams

SYSTEMS ANALYSIS AND SIMULATION IN ECOLOGY

Edited by BERNARD C. PATTEN
Department of Zoology
University of Georgia
Athens, Georgia

VOLUME II

ACADEMIC PRESS New York and London 1972

ACADEMIC PRESS, INC.
111 Fifth Avenue, New York, New York 10003

United Kingdom Edition published by
ACADEMIC PRESS, INC. (LONDON) LTD.
24/28 Oval Road, London NW1

LIBRARY OF CONGRESS CATALOG CARD NUMBER: 76-127695

PRINTED IN THE UNITED STATES OF AMERICA

Contents

List of Contributors

Numbers in parentheses indicate the pages on which the authors' contributions begin.

MICHAEL BRYLINSKY,* Department of Zoology, University of Georgia, Athens, Georgia (81)

HAL CASWELL, Department of Zoology, Michigan State University, East Lansing, Michigan (4)

GEORGE I. CHILD,† Institute of Ecology, University of Georgia, Athens, Georgia (103)

A. BEN CLYMER, Ohio Department of Health, Columbus, Ohio (533)

THEODORE C. FOIN, JR., Environmental Systems Group, Institute of Ecology, University of California, Davis, California (476)

GILBERTO C. GALLOPÍN,‡ Section of Ecology and Systematics, Division of Biological Sciences, Cornell University, Ithaca, New York (241)

HERMAN E. KOENIG, Department of Electrical Engineering and Systems Science, Michigan State University, East Lansing, Michigan (4)

ORIE L. LOUCKS, Institute for Environmental Studies, University of Wisconsin, Madison, Wisconsin (419)

HOWARD T. ODUM, Environmental Engineering Sciences, University of Florida, Gainsville, Florida (139)

* Present address: Department of Biology, Dalhousie University, Halifax, Nova Scotia, Canada.

† Present address: University of Guam, Agana, Guam.

‡ Present address: Fundacion Bariloche, Rivadavia 986 (7°), Buenos Aires, Argentina.

BERNARD C. PATTEN, Department of Zoology and Institute of Ecology, University of Georgia, Athens, Georgia (284)

G. J. PAULIK,* Center for Quantitative Science in Forestry, Fisheries and Wildlife, University of Washington, Seattle, Washington (373)

JAMES A. RESH, Department of Electrical Engineering and Systems Science, Michigan State University, East Lansing, Michigan (4)

QUENTIN E. ROSS,† Department of Zoology, Michigan State University, East Lansing, Michigan (4)

SAUL B. SAILA, Graduate School of Oceanography, University of Rhode Island, Kingston, Rhode Island (331)

HERMAN HENRY SHUGART, JR.,‡ Institute of Ecology, University of Georgia, Athens, Georgia (103, 284)

RICHARD B. WILLIAMS, Bureau of Commercial Fisheries, Center for Estuaries and Menhaden Research, Beaufort, North Carolina (213)

*Deceased.

†Present address: Department of Biology, University of Notre Dame, South Bend, Indiana.

‡Present address: Ecological Sciences Division, Oak Ridge National Laboratory, Oak Ridge, Tennessee.

Preface

This volume concludes the original concept for *Systems Analysis and Simulation in Ecology*, and at the same time initiates a continuing series under the same title. The original idea, in 1968, was to draw together a collection of systems ecology articles as a convenient benchmark to the state of this emerging new field and as a stimulus to broader interest. These purpose will continue to motivate the series in highlighting, from time to time, accomplishments, trends, and prospects.

Present-day systems ecology is exploratory, engaged in a search for central themes, problems, and methods. Volumes I and II both reflect this in the tentative and prospective nature of just about every chapter. The systems approach has not yet contributed new fundamental knowledge about ecology. It has not been responsible for important discoveries either of fact or generalization. What is in progress is a gradual paradigm shift, paralleling a similar turn to systems in other areas of science.

Still, as a minimum achievement, systems ecology has shown that ecological systems can be modeled usefully. Whether or not a model culminates in a successful simulation or route to systems analysis, the heuristic value of the modeling exercise by itself justifies the effort. Individual researchers and small groups of collaborators are learning the coordinating power of models and the utility of the modeling process in clarifying thinking, identifying data needs, and determining research priorities. Large investigations, such as the "biome" programs of the U.S. International Biological Program are finding in addition that a modeling focus helps to provide thematic and administrative coherence and thus serves a program-organizing function. Modeling, as structured thought, has been established as competitive with other known modes of conceiving and elucidating scientific problems. Neither does it preclude, as many think, the subjective, qualitative intuition of the experienced investigator, but rather provides an operational medium for drawing out and expressing insights. As a result, new avenues and approaches to ecological research are opening up (Parts I, II, and III), and specific problems of many kinds have already been addressed and solved rationally and effectively (Part IV).

There are too many cross currents of activity at the present time to be very clear about trends. Four key problems, however, would seem inevitably to lead systems ecology in the direction of theory.

First, a clear distinction between modeling behavior (dynamics) and modeling the mechanisms that produce behavior has not been arrived at. Mechanisms are nonlinear, but behavior, particularly the nominal or small perturbation behavior of adapted systems, tends to be linear. Models based on nonlinear representations of mechanisms have not been too successful in simulating behavior. The mathematical properties of nonlinear systems are just not consistent with the dynamics of real systems. The problems are compounded when complex patterns of interaction are involved (e.g., Chapter 5). A trend to linear modeling will probably occur as experience teaches that linear theory suffices for a large class of problems of interest to the ecologist. This will open the way for excursions away from simulation, which now dominates, and into linear systems analysis, much of the theory for which already exists. When this occurs, systems ecology will assume greater formality as it becomes aligned more and more with system theory. Chapter 1, particularly, and also Chapter 2 of Volume I illustrate some of the possibilities.

A system is a collection of components linked in interaction, and so a model must also be a set of interacting components. The homomorphic basis of modeling (Chapter 1, Volume I) dictates a many-to-one mapping of the parts and connections of the real system into those of the model. This brings up the old problems of taxonomy in a new form. There are many ways to cross-classify components of natural systems, particularly where function is concerned. When one way is picked for a particular model much of the richness of the real system, as the ecologist intuitively experiences it through mental cross-classifying, is lost. The problem referred to is *aggregation*—the process of selecting model components and adequate state variables to represent them. It is a basic problem that will continue to give difficulty to systems ecologists for a very long time. Resolution will probably be through abstract modeling theory, of the kind being developed by, e.g., Ziegler[1] and others. Meanwhile, the ecologist will have to rely on experience and judgment in structuring his systems models and interpreting results. The need for abstraction

[1] Ziegler, B. P. (1971), *U. Mich. Tech. Reps.* 032960–11–T, 032960–15–T.

at so fundamental a level will further draw systems ecology in a formalizing direction.

Equally fundamental and equally problematic is model *validation*—the process of gaining confidence that a model aptly represents its real counterpart system for the purposes at hand. The current concept of validation is to control trajectories in the model's state space to conform to empirical trajectories of the real system. At base, this amounts to curve fitting through parameter adjustments, and is a very weak criterion. Needed is some form of validation theory that will guarantee output for certain classes and ranges of input and also guarantee correspondence of a model's internal structure to that of the real system. Otherwise, properties of a model revealed by systems analysis may not conform to properties of the prototype. The validation problem will also cause systems ecology to look to theory.

Finally, the *connectivity* of systems will become an increasingly important subject. Connectivity refers to the degree and pattern of component interaction in the internal structure of systems. Chapters 1 through 6 are all concerned with various aspects of this subject, either directly or indirectly. It is not necessary to know systems theory to appreciate the significance of the connectivity idea. For this reason, connectivity may well provide an immediate bridge between systems ecology and conventional ecology. Like diversity, connectivity lends itself to various forms of measurement and expression (one can easily foresee proliferation of "connectivity indices") that can readily involve the field ecologist. Large-system connectivity has already been examined preliminarily for a possible relationship to stability, and the concept of "critical connectance" for stability expounded.[2] The critical level above which linear systems went unstable was found to be about 13%; coincidentally, most large scale total ecosystem models have been in the range of 10–20% connected. The possible realtionship of easily-measured connectivity to not-so-easily-measured stability should serve to involve the systems ecologist in stability theory and related subjects of systems science.

As to prospects, my own personal belief is that, in systems ecology, we have not even begun to imagine where the systems approach may lead. There are exciting times ahead for those who will let themselves be captivated by this elusive something we call "system." Much original thinking is going to be required to elaborate this concept in an ecologically meaningful way—and much hard work and frustration. The philosophical

[2] Gardner, M. R., and Ashby, W. R. (1970), *Nature* **228**, 784; and Somorjai, R. L., and Goswami, D. N. (1972), *Nature* **236**, 466.

basis is already present in the General Systems Theory of von Bertalanffy, Ashby, and others, and a methodology exists in the various formalisms of systems science as evolved for physical systems. Effective joining of these disciplines with the subject matter and methods of traditional ecology will require revisions that extend into the classroom. *Systems Analysis and Simulation in Ecology*, it is hoped, may help in this effort.

The present volume differs from the outline projected in Volume I mainly in the addition of Part I, Introduction to Systems Science, whose sole chapter was uncertain until after the first volume went to press. Part II presents example applications of systems analysis methods to ecosystems. Part III is devoted to new theory, and Part IV to applications in a variety of human interest areas. Taken together, both books represent pretty well the current scope of systems ecology in premise, theory, methods, and applications. If there is a significant omission, it probably is the absence of treatment of a large-scale total ecosystem model, several examples of which now exist in various stages of development. This will be rectified in a subsequent volume devoted to such a model, the theory, description, and analysis of which grew out of bounds in the writing, originally intended for this volume.

Until then, it is a distinct pleasure to offer the present representations of the new science of systems ecology.

Athens, Georgia
August, 1972

BERNARD C. PATTEN

Contents of Volume I

PART I

INTRODUCTION TO SYSTEMS SCIENCE

Systems ecology is, or should eventually be, a hybridization of both ecology and systems science, assuming that ecological systems are subsets of general systems. The derived principles of systems ecology should come from general systems principles, and its methodology should be a systems methodology specifically adapted to ecological levels of organization. In the single chapter of this section, the concepts upon which systems science as a discipline is built are outlined for ecologists at an introductory level. The newness of the material ecologically is perhaps appropriate to the newness of the senior author, who was an undergraduate when the chapter was prepared.

Chapter 1 begins with a definition of *system* identifying, as the complement of system, *environment*. The role of the observer and the significance of objectives in defining a system are underscored. The internal structure of a system includes objects (which behave) and couplings (which constrain behavior). An object's behavior is comprehended in terms of *behavioral features* which may be oriented into two types, stimuli and responses. In *free-body* form, an object is coupled to no particular environment and is describable by a set of all possible behavioral time series. Actual behavior is defined by the so-called *behavioral equation* of the object.

The behavioral equation deals with time series rather than with instantaneous values of stimuli and responses. A relation that is a function capable of mapping instantaneous excitations into responses is required, however. The problem to be solved is that of uniqueness. In general, an object can respond in many ways to a given instantaneous stimulus. The concept of *state*, it is developed, is sufficient to remove this indeterminacy and provide a unique mapping of excitations into responses. This leads to a *state-space description* for the object, consisting of (1) a state variable and (2) a stimulus–response–state relation, the latter replacing the behavioral equation. A state-space model of the object is thus arrived at, consisting of a pair of equations: the *state equation* and the *response equation*.

State variables are of central significance because they provide the information required to make both the response and state equations determinate. The question of choosing adequate state variables for objects is taken up in connection with the familiar population models of ecologists. In general it is shown why population size $N(t)$, by itself, cannot serve as a state variable except for a restricted class of problems. Ecologists have already known this, of course, but it is good to be shown the formal reasons and to have the formal conditions for adequate state variables identified.

Objects interact to form systems. This interaction is through the behavioral features, where a feature of one object is required to be equal to a feature of another. Such coupling constraints are represented by a set of *constraint equations* for the system. Two types of elementary constraints are recognized, distributive and conservative. The interconnection pattern between objects is referred to as the "topology" of the system. An aggregation of objects, a system, can also be viewed for a useful class of purposes as a single object. Thus, an object is a system and a system an object, alternately, in a hierarchical sequence of *levels of organization*. Rules governing a lower level are said to be necessary but not sufficient to describe a higher level, leading to consideration of the so-called *emergent properties* of systems. The authors purport emergent properties to be fully explained by systems science, but one suspects that the definitive arguments are not developed here. Those that are lend greater clarity and remove some of the mysticism with which the emergence phenomenon is often viewed.

Energy systems form an important class of systems of ecological interest for which an advanced methodology has been developed in the physical sciences. When this methodology is expanded to deal with social, economic, and biological questions, an underlying structure (well hidden in most physical applications) becomes apparent. Considering the flow of energy or material within a system, two types of variables are identified. Flux appears as an extensive or "through" variable, and its associated potential, cost, or some other useful property as a corresponding intensive or "across" variable. These paired variables are always considered together in a method of analysis that involves an *object graph* and a *system graph*. In the latter, two types of interaction constraints appear, "continuity constraints" for the through variable, and "compatibility constraints" for the across variable. Objects may be of three types, for transformation, transport, and storage. The examples used to illustrate this energy system methodology, including an application to a compartment model (Appendix II), serve to convince that this new set of ideas which the ecologist has not seen before will surely contribute to systems ecology in the future.

The final sections, devoted to general discussion, present several compelling ideas. One is a "body/mind" separation of ecological systems posited in analogy to the classical control system model of systems science. This discussion should be taken under consideration by all ecologists who would distinguish "control" components of ecosystems from "process" components. Projecting beyond modeling to problems of control, design, and synthesis of systems, the authors suggest that these engineering concepts (though not necessarily the associated methods) may eventually be applied to total ecosystem management. Particularly cogent is the concept of an "industrialized ecosystem": "The idea of designing and synthesizing ecosystem structures is repugnant to some ecologists whose first interest is ... to better understand the natural functioning of ecosystems. Unfortunately, there is very little choice since the design and synthesis activities are in full swing within the various components of the industrialized ecosystem."

Systems scientists are not just developing an esoteric theory of systems for its own sake. Many of them are becoming aware of the life-and-death problems of environment and of the potential of systems science for offering superior insights and, possibly, solutions. Ecologists also are beginning to realize the relevance of systems thinking, so that real possibilities for synergistic mixing of the two fields are materializing, as is in the nature of hybridization.

1

An Introduction to Systems Science for Ecologists

HAL CASWELL

DEPARTMENT OF ZOOLOGY

HERMAN E. KOENIG AND JAMES A. RESH

DEPARTMENT OF ELECTRICAL ENGINEERING AND SYSTEMS SCIENCE

AND

QUENTIN E. ROSS*

DEPARTMENT OF ZOOLOGY

MICHIGAN STATE UNIVERSITY, EAST LANSING, MICHIGAN

* Present address: Department of Biology, University of Notre Dame, South Bend, Indiana.

I. Introduction

Ecologists have long known that they deal in their everyday work with complex systems, systems that exhibit rich and varied behavior governed by complex networks of causation. It is not surprising, therefore, that they have come to be interested in systems science, a discipline supposedly devoted to the study of just such systems. But there is so little agreement about what systems science really is that many ecologists entering the field for the first time return with a bewildering array of analytical and computational tools, and very little understanding of a scientific discipline as such.

The objective of this paper is to introduce the concepts upon which systems science as a discipline is built. Referring to Fig. 1, most of the emphasis will be on the process of model-building (upper half of the figure). All other aspects of systems science (analysis, control, design, etc.) presume a validly structured model as a starting point. Such models and the rules for structuring them have not as yet appeared in the systems ecology literature. The literature on physical systems often neglects the process of developing a model. Instead, it is common to begin with, "Consider a system having a model of the form... ." This is often a valid starting point because physical systems are usually designed to have tractable behaviors, models of which can be developed by well-known procedures. Until the study of ecological systems reaches this point, the process of structuring models desperately needs attention.

We will not try to provide a set of computational tools, or attempt to review the engineering literature. Rather, this chapter emphasizes background concepts of systems science, and hopefully will provide a structure that can aid in evaluating the potential utility of some of the tools developed in the physical sciences.

II. Structuring the System: The Role of the Observer

A useful starting point for an introduction to systems science is a definition of the word "system." Such definitions are many and varied; discussion here will be structured around the following:

> A *system* is a collection of *objects*, each *behaving* in such a way as to maintain behavioral *consistency* with its *environment* (which, of course, may include other objects in the system).

In this section we will investigate the concepts of *system*, *environment*,

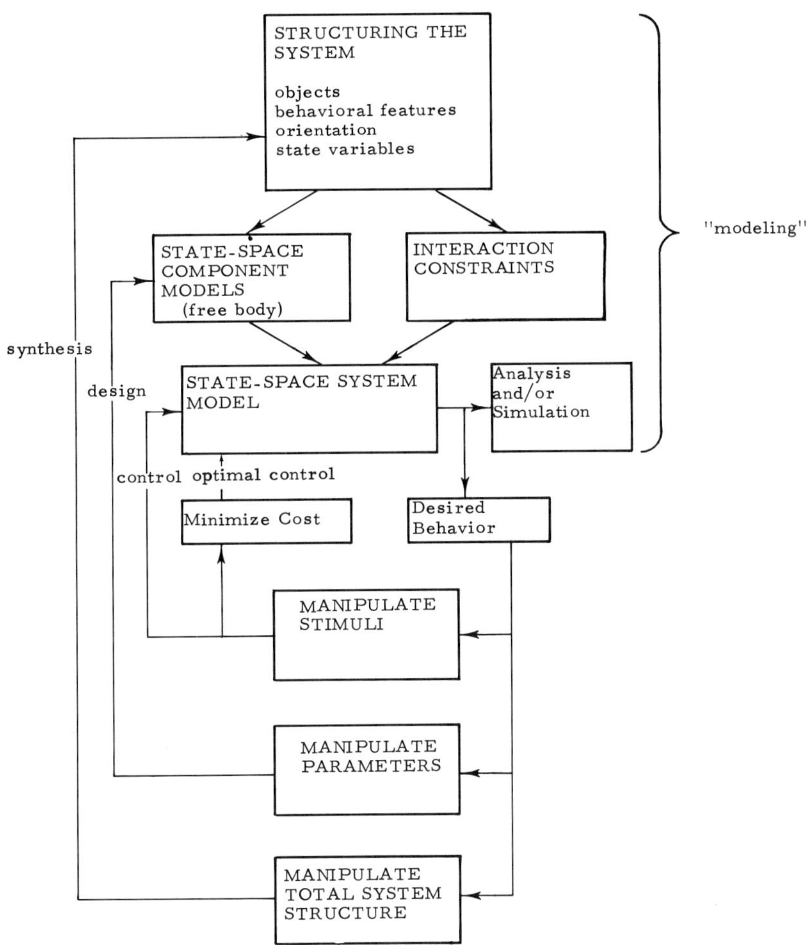

FIG. 1. Outline of systems science.

and *object*. The concept of *behavior* and the generation of behavioral *consistency* is discussed in Section III.

The methodological treatments of systems science (this includes most of the systems engineering and systems ecology literature to date) usually assume a particular system as a starting point, and go on to consider what can be done with it. In fact, arriving at a specification of the system under investigation is the first step in any systems study. Like all first steps it is also one of the most critical, for its influence is manifested at all succeeding stages. One of the basic assumptions around which systems science is built is that the observer has a goal, a question he wants to answer by his study of the system. This is a crucial requirement, for many steps in a systems study involve subjective choices that can be made only in the context of a specific goal. A scientist approaching a system without a specific goal in mind can easily end up wandering without direction through a maze of mathematical techniques. The uncomfortable feeling expressed by some ecologists that systems science does not seem to have answered any interesting questions is partly a result of the fact that specific questions have not often been asked. Because of the central importance of the observer's goals, there are many places in a systems study where the concepts of "right" and "wrong" choices must be replaced by "more useful" and "less useful."

The beginning of a system study is essentially a sequential structuring of the observer's perceptual universe. It is important to remember that this structure is imposed on the universe by the observer; it is not an inherent characteristic. The first step in this process is the partitioning of the perceptual universe into a system and its environment, an explicit expression of what the observer wishes to study, and in what context he wishes to study it.

Our definition of a system imputed to it an internal structure by defining it as a collection of objects. This internal structure must be specified by the observer; it is not determined by the choice of the system and its environment. The observer is at liberty to specify as few or as many objects in his system as he likes, including the option of considering the system as a single object. An example demonstrating the subjective nature of object specification is provided by something as common as a TV set. Because of their differing goals in studying this system, the TV set would be structured very differently by a repairman (objects: capacitors, tubes, etc.), a TV designer (objects: amplifiers, filters, etc.), a furniture mover (objects: wooden casing which may get scratched, a picture tube which may break, etc.), and a child watching cartoons (set as a single object). To state that any of these internal structures is wrong is nonsensical; in order even to comment on their utility one must take into account the objectives of each observer. An analogous biological case is provided by an anthill. To a taxonomist such a system might be composed of individual ants, each of which represents a point in a multidimensional taxonomic space. To a behaviorist it might be useful to consider the colony as made up of groups defined by social functions. An ecologist might be aided by considering the system to be composed of sets of groups which play different roles in energy flow through the population. The fact that the taxonomist, behaviorist, and ecologist might be the same person considering three different questions (perhaps even simultaneously) is a point worth noting. While all three questions seem to deal with an anthill, they lead to studies of three different systems.

After the observer has imposed an internal structure on his system, he is faced with another subjective choice, again depending on the question under consideration. Our definition of "system" indicated that the component objects exhibit behavior. One of the important steps in answering questions by a system study is to arrive at a description of that behavior. Such a description requires a vocabulary, which is provided by specification of a set of *behavioral features* for the object. The construction of a description in terms of this vocabulary forms the subject of the next section. These behavioral features, selected from the infinite array of features which could be measured on the object, are those characteristics of the object in which the observer is interested. We have hinted at this in our discussion of the TV set. The components considered by the furniture mover (casing, picture tube) were described by a behavioral feature representing the type of damage to which they were sensitive. This points out another fact; the behavioral features need not be quantitative. An immediate example of the selection of behavioral features available is provided by the ubiquitous white rat. Among the innumerable features that can be (and are) measured on this object are posture, activity, body temperature, weight, concentration of any number of chemicals in the bloodstream, electrical activity in the nervous system, etc. Each of these features ranges over some set of possibilities; some are quantitative and some qualitative. A scientist studying the rat as an object in some system can legitimately choose a vocabulary composed of any of these features, or others. His choice is dictated not by any innate features of the rat but by his questions and goals.

The behavioral features chosen are used to develop a behavioral description of the object. Thus, as we will see, they are used to express all the object–environment interactions that the observer wishes to consider. The choice of behavioral features is usually made based on the knowledge accumulated by the observer about the system he is studying. As in the choices of system and objects, utility criteria rather than correctness must be used in arriving at a choice of behavioral features.

Because the structure created by the observer in the initial stages of a system study will greatly affect the entire outcome of that study, any discussion of a system should be prefaced by an explicit statement of the interests and goals of the observer and of how the structure he has imposed on the system is related to those goals. Such a statement can provide a common framework within which all concerned can discuss the system, and can eliminate much vacuous argument.

More important even than reduction of the level of polemics is the potential role of such discussion in the development of ecological theory. One of the hallmarks of a well-developed theory is the elucidation of sets of components and behavioral features that are useful in broad and important classes of questions. These provide a common vocabulary for anyone working in the field, and help make the results of one study immediately applicable in another. Moreover, there is a considerable saving in time and effort in not having to go through the selection process for each study. No electrical network theorist today worries about behavioral features. He knows that he can (for a very wide range of questions) use voltage and current. While it is not likely that ecology will ever reach such a stage, it can certainly benefit from the discovery of clusters of important questions and their associated behavioral features. Such discovery will be greatly facilitated by explicit discussions of the choices made in structuring the system (including choices that were tried and rejected as not useful).

We should emphasize at this point the iterative nature of the structuring process. Structures and their resulting behavioral descriptions are evaluated by utility criteria in terms of questions of the observer. There is every reason to expect that the first structure attempted will fail miserably. Thus, in practice, the system is structured in an iterative manner. At any point in the system study, the observer may choose to redefine his system, his components, or their behavioral features. Or at later stages in the study he may modify the behavioral descriptions of his components, or the way they interact. This process may be iterated as long as is necessary to satisfy the observer's questions and goals. There is, of course, no guarantee of convergence to satisfactory results; a fact of life that must be faced by any scientist.

This section has dealt with the imposition of a structure on his perceptual universe by the observer. We turn now to the problem of describing the behavior of the system in terms of this structure.

III. Describing the System

A. Characterization of Objects

The first step in obtaining a behavioral description of a system is to describe the behavior of its component objects. A behavioral description of an object specifies its behavior through time in terms of the behavioral features selected by the observer.

It is one of the major insights of systems science that this description of the object must be in what is designated here as *free-body* form. A free-body description is one which is valid for a wide variety of environments, not just the specific set provided by the other objects in the system of interest. This explicitly prohibits making the behavior of the object a function of the behavioral features of some other object. Just how wide the "wide variety" of environments must be is again a function of the needs of the observer. As we will see, the description must be valid in a class of environments which includes at least all the environments generated by the behavior of the other objects in the system. This class is large for a system with even a fair level of behavioral richness. While the observer can say that it is large, he cannot specify it, hence free-body models are usually formuated for a variety of environments large enough at least to guarantee inclusion of the system at hand. It is useful to build free-body models for a still wider class of environments. This will make the object description valid in systems other than the one under immediate consideration, one of the most important benefits of the free-body approach. It allows information obtained in the study of one system to be utilized in the study of a variety of others. Information can thus be accumulated and used repeatedly in the study of a whole class of systems. This capability is responsible for the success of the design and control aspects of physical sciences.

The shift to free-body descriptions generates a new emphasis for scientific explanation. It moves the focus away from the specific facts of a particular behavior and toward the processes which select that behavior from all other behaviors the object is capable of exhibiting. Later, two other notions basic to this selection process will be introduced: that of stimulus–response orientation and that of interactions among objects. First, however, we will take a closer look at free-body descriptions.

As a framework for discussing free-body descriptions, let us expand the definition of a system given in Section II. The notation used throughout this section is summarized in Table I.

A system is a collection of *objects*... .
An *object* is described by a set of *behaviors*.
A *behavior* is a time series of *acts*.
An *act* is an instantaneous tuplet of *behavioral features*.

We will begin at the bottom of this list and work upward. The behavioral features of the object, chosen by the observer, are represented by $b_1, b_2, \ldots, b_n$. Each of these features ranges over a set of possible values $B_1, B_2, \ldots, B_n$, respectively. So, at any time t, the object can be described (for the purposes of the study in question) by specifying the values of

TABLE I
SUMMARY OF NOTATION

Entity	Notation	Range (if applicable)	Description of Entity
Behavioral feature	$b_1, b_2, \ldots, b_n$	$B_1, B_2, \ldots, B_n$	An element of a set
Stimulus behavioral feature	$e_1, e_2, \ldots, e_j$	$E_1, E_2, \ldots, E_j$	A regrouping of $b_1, b_2, \ldots, b_n$ and $B_1, B_2, \ldots, B_n$
Response behavioral feature	$r_1, r_2, \ldots, r_k$	$R_1, R_2, \ldots, R_k$	
Act	$(b_1, b_2, \ldots, b_n)$	$(B_1 \times B_2 \times \cdots \times B_n)$	Behavioral feature tuplet
Stimulus behavioral feature tuplet	$(e_1, e_2, \ldots, e_j) = \mathbf{e}$	$(E_1 \times E_2 \times \cdots \times E_j)$	
Response behavioral feature tuplet	$(r_1, r_2, \ldots, r_k) = \mathbf{r}$	$(R_1 \times R_2 \times \cdots \times R_k)$	
Behavior	$\beta: (t_0, t_1) \rightarrow (B_1 \times B_2 \times \cdots \times B_n)$		Time series of acts
Stimulus behavior	$\epsilon: (t_0, t_1) \rightarrow (E_1 \times E_2 \times \cdots \times E_j)$		Time series of stimulus behavioral feature tuplets
Response behavior	$\rho: (t_0, t_1) \rightarrow (R_1 \times R_2 \times \cdots \times R_k)$		Time series of response behavioral feature tuplets
Behavioral variable	β or (ϵ, ρ)	$\mathscr{B} = \{\text{all } \beta: (t_0, t_1) \rightarrow B_1 \times B_2 \times \cdots \times B_n\}$	Element of a set of time series

each of the behavioral features. This instantaneous tuplet* of behavioral features is referred as an *act*. Using the laboratory rat for an example again, one might choose behavioral features as follows:

Behavioral features	Range
b_1 = body temperature,	B_1 = all real numbers;
b_2 = activity,	B_2 = {running, walking, sitting, lying down}.

An act at time t is specified by the tuplet $[b_1(t), b_2(t)]$; for example,

$$[b_1(t), b_2(t)] = (37°\text{C, sitting}).$$

Acts range over the set of all possible tuplets of behavioral features. This set is the so-called Cartesian product of the ranges of the features, and is indicated by $(B_1 \times B_2 \times \cdots \times B_n)$. Thus one can think of describing the object in terms of n behavioral features $b_1, b_2, ..., b_n$, ranging over sets $B_1, B_2, ..., B_n$; or in terms of a single multivariate behavioral feature $(b_1, b_2, ..., b_n)$ which ranges over the set $(B_1 \times B_2 \times \cdots \times B_n)$. In either case an act consists of an instantaneous specification of the value of all behavioral features.

A *behavior* of the object is a time series of acts. It could be described in the form of a list,

Time	Act
⋮	⋮
t_0	$[b_1(t_0), b_2(t_0), ..., b_n(t_0)]$
⋮	⋮
t_1	$[b_1(t_1), b_2(t_1), ..., b_n(t)]$,

or a graph, as is shown in Fig. 2 for an object with two behavioral features. A behavior, then, is a time series or time function whose domain is a time interval [here denoted as (t_0, t_1)] and whose range is the set of all acts, shown above to be $(B_1 \times B_2 \times \cdots \times B_n)$. We can denote such a time series, mapping a time interval into the set of all acts, by

$$\beta: (t_0, t_1) \rightarrow (B_1 \times B_2 \times \cdots \times B_n),$$

or β for short. This notation states that the time series β maps a time interval (t_0, t_1) into the set of all acts $(B_1 \times B_2 \times \cdots \times B_n)$.

* A tuplet or n-tuple is an ordered set (vector) of n elements. For example, the position of a point in standard three-dimensional space is specified by a three-tuple (or triplet) of numbers (x, y, z).

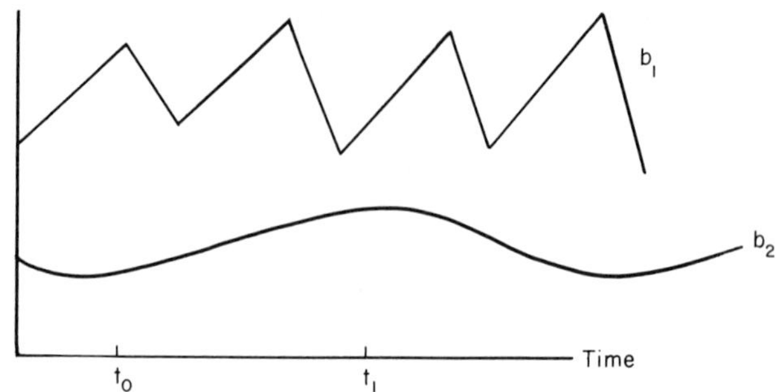

FIG. 2. Hypothetical behavior of an object described by two behavioral features b_1 and b_2.

Any behavior β is only one out of a large set of possible time series. Hence we will define β as the *behavioral variable* of the object. Its range $\mathscr{B}$ is the set of all time series with domain (t_0, t_1) and range $(B_1 \times B_2 \times \cdots \times B_n)$. In the Fig. 2 example, it includes all the curves that can be drawn on the graph. This set clearly includes all the behaviors that can be exhibited by any object described in terms of behavioral features $b_1, b_2, \ldots, b_n$. It is with the behavioral variable β that the behavior of objects will be described.

Since the set $\mathscr{B}$ includes all possible behaviors of the object, a complete behavioral description is obtained by specifying the set of behaviors which actually do occur in some environment. This set of actually occurring behaviors is a subset of $\mathscr{B}$, denoted by $\mathscr{B}'$. Conceptually, this set can be thought of as obtained by placing the object in *all* possible environments and recording the resulting time series of acts. Obtaining the set $\mathscr{B}'$ is the goal of the entire modeling process. In practice, of course, the object is placed in only a small subset of all possible environments. This subset is selected in such a way that mathematical and statistical techniques can be used to extrapolate the results to nonsampled environments.

Description of the behavior of an object by specification of the set of actually occurring behaviors generates the *behavioral equation* of the object,

$$\beta \in \mathscr{B}' \qquad \text{where} \quad \mathscr{B}' \subset \mathscr{B}. \tag{1}$$

Equation (1) merely states that the behavioral variable β is an element of the set $\mathscr{B}'$ of actually occurring behaviors. It is worth noting here that the behavioral variable is a time series, and that the object has been described by specification of a set of these time series. This may appear as an unnecessary complication; it is more natural to most people to think in terms of momentary acts. Such an approach, however, does not allow for the *dynamic* nature of the object's behavior. By making the basic behavioral variable a time series, explicit allowance is made for the temporal organization of sequences of momentary acts.

The behavioral equation (1) above provides a clear picture of the ultimate goal of experimentation on the object. In practice, however, yet another degree of structure is imposed by the observer on the system. This structure—which represents a basic outlook of scientific explanation, and which will affect the entire experimental program for studying the object—is referred to as *stimulus–response orientation* of the object. Like the previous structuring processes this one is largely subjective and determined by goals of the observer.

Orientation consists of subdividing the behavioral features $(b_1, b_2, \ldots, b_n)$ of the object into a *stimulus* (or *excitation*) set and a *response* set, $(e_1, e_2, \ldots, e_j, r_1, r_2, \ldots, r_k)$. This generates a similar subdivision of the behavioral variable β into

$$\beta = (\epsilon, \rho), \tag{2}$$

where ϵ represents a time series of excitation features and ρ a time series of response features (see Table I).

The usual criterion in the choice of ϵ and ρ is that the observer perceives a "cause–effect" relation of some sort between them. The choice is not, however, unique and (like all the other subjective choices made by the observer in structuring the system) the orientation of the object will have significant effects on subsequent interpretation of the object's behavior. As an example, consider an object with two behavioral features, b_1 and b_2, which range over the set of real numbers. Suppose investigation of the object in a wide variety of environments revealed behaviors (time series of acts) of which the following sequence was typical:

Time	Act $= [b_1(t), b_2(t)]$
$\vdots$	$\vdots$
t_0	(12, 72)
t_1	(72, 13)
t_2	(13, 24)
t_3	(24, 6)
t_4	(6, 99)
$\vdots$	$\vdots$

If the observer selects b_2 as the stimulus and b_1 as the response, it appears that the object is a one-time-unit delay mechanism, with the response time series ρ lagging one unit behind the stimulus time series ϵ. If the other (equally valid) orientation is chosen, the object becomes a perfect predictor; ρ anticipates by one time unit the value of ϵ. These are radically different interpretations of the behavior of the same object. This phenomenon crops up repeatedly in the study of a system; certain orientations of an object will almost always be preferable to others, usually for interpretive reasons.

Given an orientation for the object, a logical next step is to rewrite the behavioral equation (1) as

$$(\epsilon, \rho) \in \mathscr{B}' \qquad \text{where} \quad \mathscr{B}' \subset \mathscr{B}, \tag{3}$$

and then to attempt to find a relation R such that (3) can be expressed as

$$\rho \in R(\epsilon). \tag{4}$$

Equation (4) has taken the set $\mathscr{B}'$ of time series of acts and found a relation R between the time series of stimulus features and the time series of response features. In the delay/predictor example above, the two orientations were compared on the basis of the relation R they produced. In the first case this relation was that of a one-time-unit delay; in the second that of a one-time-unit predictor.

It has probably been noted with dismay that the behavioral equation (1) and its oriented form (4) deal with entire time series, not with instantaneous values. If this were as far as systems science could carry the description of objects, it would be a useless exercise in abstraction. The task of operationally describing object behavior in terms of sets of time series is nearly hopeless for even the simplest types of behavior. The obvious avenue of escape from this dilemma is to seek a relation between the *instantaneous* values of stimuli and responses (i.e., a relation between acts rather than behaviors):

$$[r_1(t),..., r_k(t)] = H[e_1(t),..., e_j(t)], \tag{5}$$

or, in simpler notation

$$\mathbf{r}(t) = H[\mathbf{e}(t)]. \tag{6}$$

One would hope that the relation H would be a function (i.e., a many-to-one relation) so that given an instantaneous excitation $\mathbf{e}(t)$, the instantaneous response $\mathbf{r}(t)$ would be uniquely determined. Models of this form find widespread use in certain applications (e.g., linear and nonlinear regression models, economic input–ouput analysis (Leontief, 1970, etc.). However, Eq. (6) is worth examining carefully, for it tells something about the way in which the object's behavior (as a time series) is organized. The response at any time is completely determined by the instantaneous excitation; the object itself imposes no temporal organization on the fluctuations of its environment. Rather its behavior is elicited only by the here-and-now condition of its stimuli, knowing no past or no future.

Unfortunately, in the vast majority of cases (particularly those likely to be of interest to systems ecology) models of the form of Eq. (6) are found to be indeterminate; for any given excitation $\mathbf{e}(t)$ there exist not one, but many responses. Thus the relation H in Eq. (6) is not single valued, and attempts to describe object behavior with Eq. (6) are unsuccessful. All, however, is not lost. It will not be necessary to describe objects in terms of equations like (1) or (4). The concept of *state*, one of the most important in systems science, arrives at this point to provide determinate, dynamic behavior descriptions without resorting to the procedures of Eqs. (1) or (4).

B. The Concept of State

The generation of multiple responses by the same stimulus is often due to the fact that the object's behavior depends not only on the instantaneous values of its stimuli, but on at least some facets of its *history*. The theory of state attempts to express this dependence of behavior on the object's history in a way that will generate a determinate stimulus–response relation and retain the dynamic character of the object's behavior. This is not immediately obvious from the following definitions, but becomes clear from the later discussion of Resh's (1967b) method of contructing state descriptions.

We have made a distinct effort to make this presentation as intuitively clear as possible, at the cost of much of its mathematically rigorous formalism. This is justified since the rigor is not needed in any of our subsequent usage of the concept, although intuitive clarity is. For precise statements of the concepts, in a somewhat different vocabulary than we use here, see Zadeh and Desoer (1963), Zadeh (1969), Resh (1967a), and Resh's forthcoming book. The first definition of state that was really useful for biological applications was presented by Zadeh (1963), and although it has had some technical modifications (Resh 1967a) the basic idea is still intact.

The immediate utility of the state concept is to construct a *state-space description* of an object. Such a description has two basic parts, both of which are constructed by the observer: (1) a *state variable* $\psi(t)$ which ranges over a *state space* Σ, and (2) a *stimulus–response state relation* $A(\psi, \mathbf{e}, \mathbf{r}, t)$, which will replace the stimulus–response relation $H(\mathbf{e}, \mathbf{r}, t)$ in Eq. (6). We will require the state-space model to fulfill the following three conditions:

Condition 1:

A pair $[\mathbf{e}(t), \mathbf{r}(t)]$ are instantaneous values of some time series pair (ϵ, ρ) which is an element of $\mathscr{B}'$ *if and only if* there exists some ψ in the state space Σ such that $(\psi, \mathbf{e}, \mathbf{r}, t)$ satisfies the relation A.

This condition guarantees that the state-space description will agree with the original behavioral equation description. If a behavior sequence $(\epsilon, \rho) \in \mathscr{B}'$ includes the instantaneous value $[\mathbf{e}(t), \mathbf{r}(t)]$, obviously, this fact should be reflected in either description of the object.

Condition 2:

For every ψ in Σ and every excitation $\mathbf{e}(t)$, there exists *at most* one response $\mathbf{r}(t)$ satisfying the relation A.

Condition 2 guarantees that specification of both the state and the stimulus uniquely determines the response of the object. Failure of the static model in Eq. (6) to do this is what led to the necessity of state descriptions. We can thus write

$$\mathbf{r}(t) = G[\psi(t), \mathbf{e}(t), t] \tag{7}$$

secure in the knowledge that G is (at most) a single-valued function. The next condition is concerned with calculation of the value of $\psi(t)$, which is required for evaluation of Eq. (7). Let $\mathbf{e}(t_1) \cdot \mathbf{e}(t_2)$ represent the stimulus $\mathbf{e}(t_1)$ followed immediately by $\mathbf{e}(t_2)$. Then,

Condition 3:

If $G[\psi_1(t_1), \mathbf{e}(t_1) \cdot \mathbf{e}(t_2)] = \mathbf{r}(t_2)$, then there exists a state $\psi_2(t_2)$ such that $G[\psi_2(t_2), \mathbf{e}(t_2)] = \mathbf{r}(t_2)$.

This guarantees that there is a state $\psi(t)$ that can take the place of *any* previous history of excitations. If we consider a time interval (t_a, t_z) from t_a to t_z, we can show what this means. We could place an object in a state ψ_a at time t_a, subject it to the stimulus sequence $\epsilon(t_a, t_z)$, and observe its response sequence $\rho(t_a, t_z)$. Condition 3 states that we can get equivalent results by placing it in a state ψ_b at time t_b and subjecting it to $\epsilon(t_b, t_z)$, or in state ψ_c at time t_c and subjecting it to $\epsilon(t_c, t_z)$, or in state ψ_d,... . Condition 3 guarantees that the states $\psi_a, \psi_b, \psi_c, \psi_d$,... all exist. It also implies that if the state and stimuli are known at one instant, the state at the next instant is determined. This allows us to write

$$\psi(t+1) = F[\psi(t), \mathbf{e}(t), t],$$

or

$$d\psi(t)/dt = F[\psi(t), \mathbf{e}(t), t]. \tag{8}$$

The state-space model for which we are searching is formed from Eqs. (8) and (7):

$$\begin{aligned} d\psi(t)/dt &= F[\psi(t), \mathbf{e}(t), t], \\ \mathbf{r}(t) &= G[\psi(t), \mathbf{e}(t), t]. \end{aligned} \tag{9}$$

The first of these is referred to as the *state equation* of the object, and the second as the *response equation*. Given the state of the object at any time t_0 and the stimulus time series from t_0 to any time t_1 the information in (9) specifies the entire time series of object behavior from t_0 to t_1. Thus, the goal of retaining the dynamic aspects of object behavior without working explicitly with time series has been achieved. However, this has been done by assuming that the observer has constructed a state variable and a stimulus–response state relation that satisfy three conditions. The problem of finding such a state variable still remains. Although it is occasionally presented as such in the ecological literature, it should be obvious by now that the choice of state variables is *not* by any means arbitrary. However, it is not unique either, so that the observer again has a certain subjective freedom. In general, the construction of a satisfactory set of state variables is one of the major and possibly most difficult steps in the experimental study of an object. Resh (1967b) has provided an abstract approach to this construction that also sheds more light on the role of the state variables in object description.

The state variables of an object provide the information necessary to make the response equation (7) determinate. This information concerns the previous stimulus history of the object. Most objects, however, "forget" certain facets of their past, so the entire stimulus history is not needed. (Note that Condition 3 allows the relation between stimulus history and state to be many to one, so that different stimulus sequences may leave the object in the same state.) Resh's construction utilizes this fact in deriving state variables.

Consider two specific stimulus series $\epsilon_1(t_0, t_1)$ and $\epsilon_2(t_0, t_1)$ from time t_0 to time t_1. Subject the object to $\epsilon_1(t_0, t_1) \cdot \epsilon(t_1, t_2)$ and $\epsilon_2(t_0, t_1) \cdot \epsilon(t_1, t_2)$ and record the responses $\mathbf{r}_1(t_2)$ and $\mathbf{r}_2(t_2)$. If $\mathbf{r}_1(t_2) = \mathbf{r}_2(t_2)$ for *any* $\epsilon(t_1, t_2)$, then $\epsilon_1(t_0, t_1)$ and $\epsilon_2(t_0, t_1)$ are defined to be *equivalent stimulus histories*. In other words, an object with history ϵ_1 is indistinguishable from an object with history ϵ_2 by any experiment performed at a time after t_1. If we now go through all the possible stimulus sequences and group together those that are equivalent by this definition, we will generate a set of, say, n equivalence classes of stimulus histories. Resh (1967b) proves that *any* set with n elements in it will then serve as a state space for the object, satisfying Conditions 1–3. This is so because it is not necessary to know the precise stimulus history of the object, but only the equivalence class in which it falls, since the stimulus histories within a class are indistinguishable. Thus, each of n elements of some set can serve as a "tag" (Zadeh, 1969), uniquely specifying an equivalence class of stimulus histories. Specification of this tag leaves no ambiguity about the future behavior of the object. Since any set with as many elements as there are equivalence classes will serve as a state space, the state variables are obviously not unique. Certain choices of state variables can usually be defended on the grounds that they give more appealing explanations of behavior than others, or because the form of the resulting model is more convenient.

A very important point should be made here. It is obvious that a change in behavioral features of the object or a change in orientation will change the sets of equivalence classes of stimulus histories, and thus the state variables. It is common for an object in one orientation to have a completely different set of state variables from the same object in another orientation. The form of the resulting state space may be a good reason for reorienting the object or changing its behavioral features. It is also important to note that the equivalence classes were defined by indistinguishability of behavior under experiments involving $\epsilon_1(t_0, t_1) \cdot \epsilon(t_1, t_2)$ and $\epsilon_2(t_0, t_1) \cdot \epsilon(t_1, t_2)$ for *any* $\epsilon(t_1, t_2)$. In practice this is any $\epsilon(t_1, t_2)$ generated by the class of environments selected by the observer at the beginning of the free-body modeling process (referred to as the set of *admissible stimuli*). A change in the admissible stimulus set will necessitate changes in the state-space of the object. Hence both the orientation and the class of admissible stimuli must be explicitly specified *before* a state space is constructed. A state space that is valid for one admissible stimulus set will, in general, fail to fulfill Conditions 1–3 with another stimulus set. If the state variables are specified *before* the admissible stimulus set, they will generate such a set, but it may not be the one the observer had in mind. In the next section we will investigate the restriction of admissible stimuli by the a priori choice of state variables in the context of classical population modeling.

C. Population Models

In the previous section we discussed the construction of state descriptions of object behavior. One of the conclusions of immediate relevance to any actual study is that the state variables should be found *after* an orientation and an admissible stimulus set have been specified. The expected consequences of reversing this procedure are a restriction of the admissible stimulus set. Stimuli outside this set will elicit indeterminate behavior—the same state–stimulus pair evoking more than one response. One of the best examples of the difficulties generated by this phenomenon is found in classical population modeling. A wide variety of what are in fact state variables have been considered for populations, and they are customarily differentiated on the basis of the time series of stimuli that they adequately partition into equivalence classes. It is safe to say that none of these models was developed within the framework we have presented here. A reinterpretation in these terms can, however, serve to illustrate some important aspects of the theory of state.

The key to what follows is this: If a state variable is to satisfy Conditions 1–3, it must partition stimulus histories into equivalence classes. The stimulus histories "tagged" by a particular value of the state at t_1 must be indistinguishable by any experiments performed after t_1. We will be comparing the range of behaviors that the environment of the population is capable of exhibiting with the range of such behaviors that are properly partitioned into equivalence classes. The larger the second set is in relation to the first, the "better" the state variable is customarily considered to be.

Populations as objects are usually viewed as collections of structured material made up of many similar components (e.g., individuals). The stimulus behavioral features are usually chosen to represent the influence of the environment (in the ecological sense) on the population; often simplified into such macroscopic measures as "carrying capacity," "intrinsic rate of increase," etc. The stimulus and state variables are of primary interest, and the response variables are usually identified with the states.

The simplest starting point is the exponential model of population growth (Lotka, 1925). Here the state description is

$$dN(t)/dt = r(t)\,N(t), \tag{10}$$

with behavioral feature (stimulus) $r(t)$, the intrinsic rate of increase, summarizing all the environmental effects on the object. The state variable is $N(t)$, the number of individuals in the population. The second half of the state description, the response equation, is omitted, since it would merely set the response equal to the state. Thus $F[\psi(t), \mathbf{e}(t), t]$ in Eq. (9) has in this case the form

$$F[\psi(t), \mathbf{e}(t), t] = r(t)\,N(t). \tag{11}$$

In what situations does $N(t)$ fulfill the conditions for a state variable? In the biologically nonexistent case in which all individuals are identical, $N(t)$ is a valid state variable for any time series of stimuli, $r(t)$. Knowledge of the number of individuals and the intrinsic rate of increase generated by the environment specifies the new state and output uniquely. The more interesting case is that in which all individuals are not identical. This is usually considered to be a result of age differences, although the concept can be easily extended to other causes. In this case there are only two very restricted classes of stimuli $r(t)$ for which $N(t)$ successfully partitions stimulus histories into equivalence classes, and hence is a valid state variable. The first and most commonly cited case is that in which $r(t)$ is a constant and the population has and maintains a specific structure (the "stable age distribution"). In the second, less familiar case, the population also has a stable structure. The class of admissible stimuli includes all time series of $r(t)$ that vary only in the death process, and do so in a manner identical for all classes in the population (see Lopez, 1960; Leslie, 1948). This is really not a separate case from the previous two since it allows one component of $r(t)$ to change only if it treats all individuals identically, while requiring that any component which discriminates between individuals remain constant. Situations which fail to fulfill these requirements lead to indeterminacies; the same stimulus resulting in completely different behaviors (e.g., Mertz, 1969).

The logistic equation is the next step in complexity for population models. Its state description, which again lacks a response equation, is

$$\frac{d}{dt}N(t) = r\left(N(t) - \frac{N^2(t)}{K}\right). \tag{12}$$

Here the state variable is again $N(t)$, the number of individuals in the population. The behavioral features (again, stimuli) are r and K. K is the "carrying capacity" of the environment, expressed in units of individuals it will support, and is assumed constant. The feature r is related to $r(t)$ of the previous model. It is the value of $dN(t)/dt$ in the limit as $N(t) \to 0$, and is also assumed to be constant. Essentially this model represents a special case of Eq. (10) in which a time-varying $r(t)$ is generated by a hypothesized relation between r, N, and K. Hence the restrictions on $N(t)$ as a state variable are the same as those given above for varying $r(t)$: stable population structure and $r(t)$ varying only in the death process.

The primary reason that $N(t)$ fails to fulfill the requirements for a state variable is that it ignores the internal structure of the population. Populations with the same number of individuals but possessing different structures react differently to the same stimulus, violating Condition 2 of a state description. A step toward the development of a state variable that would reflect the internal structure of the population was taken by Bernardelli (1941) and Lewis (1942), and thoroughly developed by Leslie (1945, 1948, 1959). This model, which is widely used by demographers

(Keyfitz, 1968) utilizes a difference rather than a differential state equation:

$$\mathbf{N}(t+1) = A(t)\,\mathbf{N}(t). \tag{13}$$

Here $\mathbf{N}(t)$ is an $n \times 1$ column vector whose elements are the numbers of individuals in each of n classes in the population. $A(t)$ is an $n \times n$ matrix which expresses the effect of the environment on the growth of the population. In its original formulation $\mathbf{N}(t)$ divided the population into age classes, and $A(t)$ was a constant matrix. However, $A(t)$ can be allowed to vary with time (Skellam, 1966) or in a feedback manner similar to the logistic (Leslie, 1959) and, as shown below, $\mathbf{N}(t)$ may have to be divided on bases other than age.

Because $\mathbf{N}(t)$ takes into account the age structure of the population, it is a valid state variable over a larger set of admissible stimuli than is $N(t)$, the number of individuals. It will partition into equivalence classes any stimuli which affect all members of an age class identically, although their effect may vary from one age class to another.

Many organisms have the ability to change their growth or development rates in response to changing environmental stimuli (so-called "plastic growth"). The age of such an individual tells relatively little about its response to any given stimulus; so the age structure of such a population is invalid as a state variable.

Several approaches to this problem have been taken in the biological literature. One approach is to modify the age-class model in Eq. (13), grouping the individuals by some criterion, other than age, that is relevant to the contribution of the individual to population growth (e.g., size, instar, etc.). Transitions from one class to another are then expressed in terms of environmental stimuli just as in the age-class model. Examples of such state variables can be found in the work of Lefkovich (1965), Ashford *et al.* (1970), and Frederickson *et al.* (1967), although they use different methods for mathematically expressing the transition of states. It should be noted again that the choice of a state variable is not unique; the approach via equivalence classes of stimuli admits the use of *any* set as a state space, as long as it is of the same cardinality as the number of equivalence classes of stimulus histories. Thus, two structures, e.g., size and instar distributions, might be equally valid as state variables for a given population. They would, of course, result in different behavioral descriptions.

A second, much more restricted approach to the problem of plastic growth is to create a new time scale, "physiological time." This is done by identifying a set of relevant environmental variables and the form of their interaction with the developmental process and expressing the history of the population in these terms. This was done by Hughes (1962) in a study of aphids. He found that the factor that generated the plasticity in growth was temperature, and that its effect was additive over time. Hence he expressed the "physiological age" of an individual in "degree-days." This method of handling growth plasticity is obviously limited to situations in which the relevant environmental factors and the form of their influence on growth can be easily determined.

It may well be that no grouping by a single criterion (age, instar, size, etc.) can be found that will validly partition stimulus histories into equivalence classes. In such situations, some authors have classified individuals by more than one factor. For example, Slobodkin (1953, 1954) found that the physiological condition of an individual *Daphnia* was adequately expressed by specification of both its age *and* size. In this situation one can express the state of the population by its age-size distribution, and express the state equations by a generalization of the method of Eq. (13). This was done by Slobodkin (1953), and recently by Sinko and Streifer (1967, 1969, 1971) who changed the mathematical form to partial differential equations. In similar developments in studies of human populations Rogers (1968), Keyfitz and Murphy (1967) and Feeney (1969) have extended Eq. (13) to distributions based on age and spatial location, and age and sex, utilizing partitioned matrix difference equations. Such multidimensional state variables are valid for larger classes of stimuli than the single-dimensional ones. A single classification, say age, will not function as a state variable for stimuli which affect all members of an age class identically but vary in respect to a second factor, say, spatial location. Adding a spatial dimension to the state variable will allow inclusion of these stimuli.

Studies of plant populations are also confronted with the phenomenon of plastic growth. Rabotnov (1969) has suggested a set of developmental "age states" for plants, ranging from viable seeds through senile to dormant plants. His justification for this turns out to be precisely that which we have used here: Populations with identical age distributions respond differently to the same stimulus, depending on their past history. He feels that populations with identical "age state" distributions will not. An additional problem in the study of plant populations, even using such developmental stage state variables, is that it is often unrealistic to attempt to identify individuals within the population, due to the phenomenon of vegetative reproduction. In light of both of these problems, a state variable such as the "leaf area index" which seems to form a valid state variable, at least for problems concerned with productivity (Harper, 1968), might be developed.

Time-lag models can also be interpreted in terms of the theory of state. These models are often generalizations of the logistic equation, with the form

$$\frac{d}{dt} N(t) = rN(t) \left(1 - \frac{f(N)}{K}\right), \tag{14}$$

where the interpretations on r, $N(t)$, and K are as in Eq. (12). The function $f(N)$ expresses some aspect of the past history of population size. In the continuous-lag models of Volterra (1931), Brelot (1931), Donnan (1936, 1937), and Caswell (1972), $f(N)$ is an integral of a function of the previous population size. Discrete-lag models such as those of Cunningham (1954). Wangersky and Cunningham (1956, 1957*a*, *b*) and Leslie (1959) use the population size at two distinct points, t and $t - \tau$, in a function

$$f(N) = aN(t) + bN(t - \tau). \tag{15}$$

The logistic equation incorporates feedback between the population and its environment, the subject of the next section. Suffice it to say at this point that the term $r[1 - N(t)/K]$ can be considered a stimulus from the environment. By substituting $f(N)$ for $N(t)$, an entire time series of stimuli is considered. Thus in the time-lag models, rather than searching for a partitioning of stimulus histories into equivalence classes, the model explicitly includes the stimulus histories. This limits the set of admissible stimuli to those that can be adequately expressed by r, $f(N)$, and K in Eq. (14), and that in addition satisfy the requirement for the ordinary logistic.

The models discussed here suggest a general form for the state variable for populations. In any set of admissible stimuli to a population there will be some factors whose effects on individuals differ, depending on which of several classes the individual falls into. This suggests that a satisfactory state variable for the population would be a p-dimensional density function, expressing the distribution of individuals among classes in terms of p factors. In the matrix model of Eq. (13), $p = 1$ and the density function expresses the distribution of individuals among age classes. Slobodkin's (1954) state variable is a two-dimensional density function whose axes are age and size. This procedure would be generalizable to any number of factors, although we know of no examples for $p > 2$.

The weaknesses of the models discussed here in terms of state variables and restrictions of admissible stimulus sets are nothing new to ecologists. They do exemplify the kinds of restrictions that can be generated by a priori choice of state variables. It should also serve to highlight the fact that finding valid state variables for ecological objects requires a thorough understanding of their behavior. It is more a biological problem than a mathematical one.

D. Characterization of Interactions

Referring to our original definition of a system in Section II, we see that we have now investigated the ideas of system, object, and behavior. We now need to consider the relation between the behavior of the object and the object's environment, including other objects within the system as well as the environment of the system. The interaction of objects results in a degree of behavioral consistency, in the sense that the behavior of an object is affected by the behavior of the objects with which it interacts. We are still investigating the selection of particular behavior sequences out of the set $\mathscr{B}'$ of behaviors available to the object. Stimulus–response orientation was a first step in this direction; such an orientation represents an implicit belief that the object can be described as if its response behavior ρ was elicited by its stimulus ϵ. One of the results of the consideration of interactions will be an explanation for the selection of ϵ.

Objects interact only through their behavioral features (the ability to describe such interactions was the criterion used in choosing the behavioral features). This interaction of the objects can thus be expressed in terms of relations among the behavioral features of the objects; a feature of one object may be required to be equal to a feature of another, or to a sum of several features of another. Each such relation effectively contrains the behavior of the objects, and the entire set of relations among behavioral features generated by the pattern of interaction is specified by a set of *constraint equations* among the behavioral features. The free-body models of the objects and the constraints generated by the pattern of interconnection (the *topology* of the system) together generate a behavioral description of the entire system. As we will see, the old adage that a system is "more than the sum of its parts" is very true; not only must all parts be specified, but their topology as well.

Consider a set of n objects O_i ($i = 1,..., n$) which make up a system. Each object has a behavioral description,

$$\beta_i \in \mathscr{B}'_i, \qquad i = 1,..., n. \tag{16}$$

In the (trivial) case in which there are no interactions among the objects, each object behaves according to its free-body characterization, and the entire system can be described by

$$(\beta_1, ..., \beta_n) \in \mathscr{B}_1' \times \mathscr{B}_2' \times \cdots \times \mathscr{B}_n', \tag{17}$$

or in more compact notation,

$$\bar{\beta} \in \bar{\mathscr{B}}', \tag{18}$$

where $\bar{\beta}$ is the direct sum of the β_i, and $\bar{\mathscr{B}}'$ is the Cartesian product of the $\mathscr{B}_i'$.

When interactions exist, there will be a set of constraint equations among the behavioral variables. In abstract form we can express these equations as

$$\beta_i = \Phi_i(\beta_1, ..., \beta_n), \qquad i = 1, ..., n, \tag{19}$$

which states that each behavioral variable must bear some relation Φ_i to the behavioral variables of all the other objects in the system.

Putting Eqs. (18) and (19) together, we obtain a description of the entire system, which can be indicated by a *constrained behavioral equation,*

$$\bar{\beta} \in \mathscr{B}^*, \tag{20}$$

where $\mathscr{B}^*$ is the intersection of $\bar{\mathscr{B}}'$ with the solution set of the set of Eqs. (19). $\mathscr{B}^*$ is obviously a much smaller set than $\bar{\mathscr{B}}'$, and thus the interactions have severely restricted the behavior of the interacting system compared with the noninteracting one. There are important classes of physical systems (e.g., electrical networks) in which much of the richness of behavior is generated by different system topologies, utilizing only a small number of simple components (e.g., resistors, capacitors, etc.). The importance of system topology, expressing the interactions of component objects, cannot be overemphasized.

Two basic types of constraint relations seem to cover all cases of object interaction. Consider the interaction of several behavioral features as shown in Fig. 3. If these features are behaving according to what we will

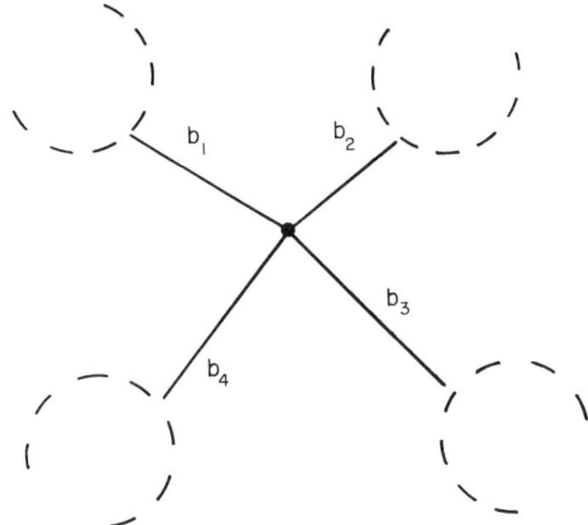

FIG. 3. Interaction of four behavioral features in unoriented form.

call *distributive constraints*, they are required to be equal at all times. If, for example, the four objects shown are individual organisms, we might have a situation as follows. The behavioral feature b_1 describes the mode of activity of individual 1 (e.g., walking, sleeping, eating, etc.). Features b_2, b_3, and b_4 represent the visual inputs to individuals 2, 3, and 4. Each of them perceive, the same activity on the part of individual 1, so the constraint equation is $b_1 = b_2 = b_3 = b_4$. Or, b_1 might represent information about the contents of the gut of a predator. This information is distributed equally to a component regulating body growth, a component regulating willingness to eat and a component regulating activity. Again b_1 through b_4 are all equal. Perhaps the behavioral features represent the energy level of a material flow with respect to some reference potential. At the point of interaction this energy level will by necessity be the same for all the material, since there would otherwise be a potential energy gradient in a homogeneous medium, with no force to maintain it. The key point is that the features are all equal at the point of interaction.

The second class of constraint relations, *conservative constraints*, arises when some function of the behavioral features is conserved at points of interaction. If, for example, the variables represent material flowing along a network, the same mass of material flowing into the junction must flow out. If the behavioral features represent individual organisms moving through space, the number of individuals moving into the junction must also move out of it. Conservative constraints can be expressed computationally by requiring the sum of the features, given a sense according to their direction towards or away from the point of interaction, be zero.

We have not yet discovered any behavioral features whose interaction cannot be expressed in terms of either conservative or distributive constraints. This is not to say that such features do not exist, for there seems to be no deep theoretical reason for the existence of only these two classes of constraints. Nor do we want to imply that the most useful computational expression of the constraint in any specific case will always have the form we have discussed here. In many important instances the constraints are (and should be) operationally expressed in an equivalent but more convenient form. A case in point is that of the potential energy level, mentioned briefly above as an example of a feature undergoing distributive constaints. We will return to this in detail in a later section.

Up to this point we have ignored the orientation of the features being constrained. We will consider this now in terms of the selection of object behavior. The selection of a response sequence ρ has already been resolved; it is selected by the stimulus sequence ϵ. This stimulus sequence is passively accepted by the object, but from where? The answer, of course, is from the environment, in this case another object or objects. Thus we want ϵ to lie in the response portion of the environment's behavior, to be elicited by the stimuli impinging *on* the environment (among which may be the responses of the original object).

The preceding paragraph viewed the orientation process from the point of view of a single object. Recalling that the same considerations apply to all the objects in the system, it is easy to see that there is a whole family of limitations placed on the overall selection of orientation. We can now discuss these *compatibility conditions* as they apply to distributive and conservative constraints.

The primary restriction on an orientation choice has already been mentioned; the object's response must be expressible as a single-valued function of its states and stimuli. Within the bounds of this condition of *object causality* (which applies to free-body as well as constrained objects) there are additional requirements of *interactive compatibility.*

Each point of interaction results in a constraint equation involving response and stimulus features for several objects. From the preceding discussion it is clear that the constraint relations at any interaction should provide specification of all the stimulus features involved as single-valued functions of the responses. If this is not done, some of the objects will have excitation behaviors the selection of which is unaccounted for by the environment.

The requirement of interaction compatibility leads to different requirements for distributive and conservative constraints. Distributive constraints require all the interacting features to be equal. If, in such a case, more than one of the features is a response, the constraints would impose a new selection criterion, independent of the stimuli involved, for two or more objects (Fig. 4a or 4d). If all the features are stimuli (Fig. 4b), then there is no way to specify them as functions of responses. The only way to satisfy the *distributive compatibility* requirement is to have exactly *one* of the interacting features a response, and let that determine uniquely the other (stimulus) features (Fig. 4c).

A similar line of reasoning leads to the conclusion that conservative *compatibility* requires that precisely one feature be a stimulus, which will be uniquely determined by the remaining (response) features (Fig. 4d). In none of the other cases (Fig. 4a–c) is there a unique solution specifying stimuli as functions of responses.

The requirement that the orientation of the system components be compatible with the constraints will place restrictions on the observer's freedom in the orientation process. In systems with few components, or with components that have few or unique orientations (providing the

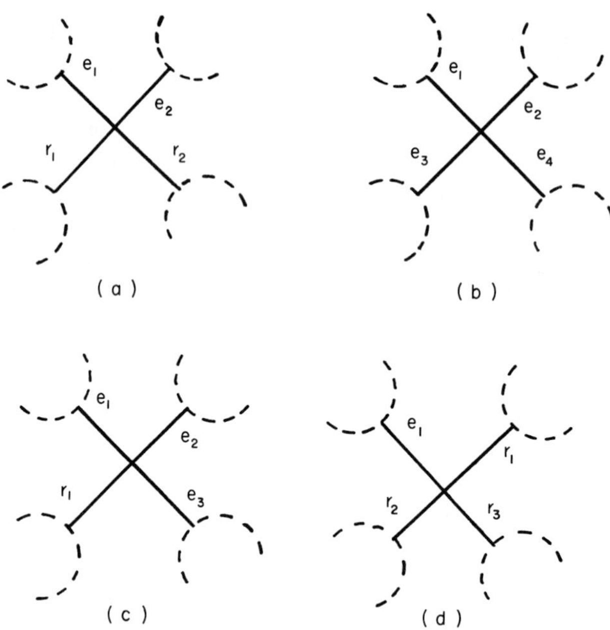

FIG. 4. Constraint relations resulting from different orientation choices for four behavioral features. Stimuli denoted by e_i and responses by r_i. The compatibility requirement demands an orientation that allows specification of stimuli as single-valued functions of responses.

observer is describing an actual existing system), a compatible set of orientations can probably be found by trial and error. For certain classes of more complex systems, algorithms have been worked out to achieve compatible orientations; these will be discussed in Section V.

E. CHARACTERIZATION OF SYSTEMS

Now that the observer has described the system in terms of the free-body models of the objects and the constraints generated by their interactions, he is ready to explore his original concern, the behavior of the system. His next step is to extract information about the overall system behavior from its atomized representation as objects and interactions. Some large portion of the behavioral features of the objects will interact strictly within the system, while some will interact with the environment of the system. The internal behavioral features can be suppressed, thereby producing a description of the overall system in terms of the behavioral features that interact directly with the environment and the states (or some subset of the states) of the objects (see Fig. 5).

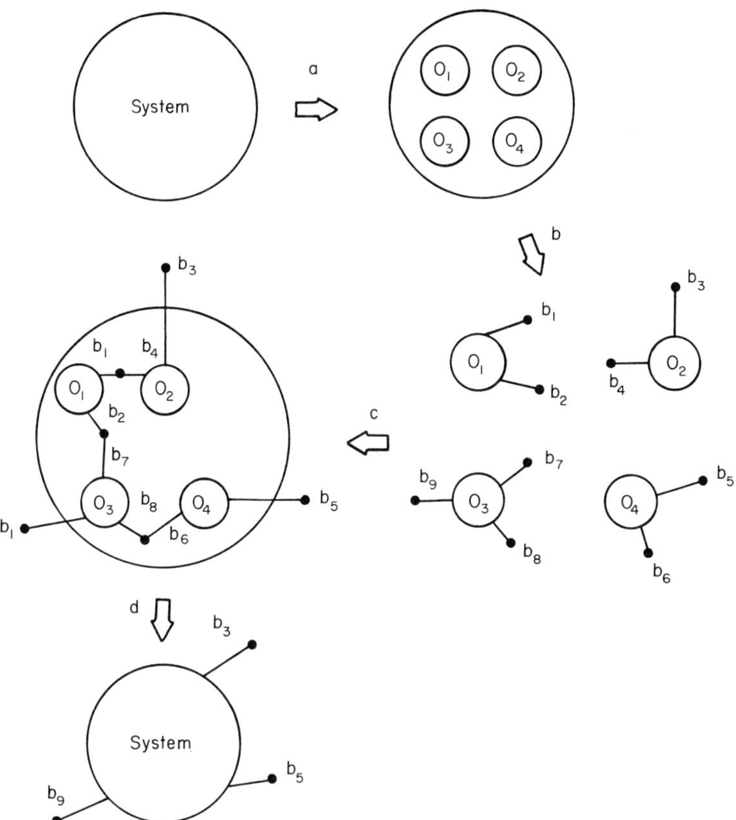

FIG. 5. Steps in the system modeling process: (a) definition of objects; (b) choice of behavioral features and free body modeling; (c) modeling interaction constraints; (d) submerging internal behavioral features to obtain a description of the system as a single object.

Viewed too abstractly it seems a strange exercise indeed. The observer began with an object at one level of resolution (his system), imputed detailed structure to it by resolving it into interacting objects at a lower level, and now turns around and submerges the laboriously obtained detail to produce a description at the original level. Why? Among the several facets of the answer, two are singularly important to ecology.

The first has to do with the experimentation necessary to arrive at a system description. Breaking the system down into component objects for the purpose of study is required by the sheer complexity of large ecological systems. Experimentally studying, say, communities or ecosystems as single objects with a set of behavioral features and states is probably impossible, and certainly inefficient.

More importantly, having modeled a system as a set of interacting objects, the observer can now investigate the effect on system behavior of changing the coupling topology, the behavioral characteristics of an object or objects, or even the effects of adding or deleting objects. This ability is essential to the processes of system design and synthesis. Ecology has always taken as a kind of first principle that the pattern of interaction among the components of the ecosystem is as important as the components themselves. By its strange-seeming approach to system behavior, systems science makes room for explicit and separate consideration of objects, topology, and the resultant behavior of the system.

At this point the information about the system (e.g., Fig. 6) is in the form of a set of free-body object equations,

$$\begin{aligned} d\mathbf{\Psi}_i/dt &= F_i(\mathbf{\Psi}_i, \mathbf{E}_i, t) \\ \mathbf{R}_i &= G_i(\mathbf{\Psi}_i, \mathbf{E}_i, t) \end{aligned} \quad \text{or} \quad \beta_i \in \mathscr{B}'_i, \tag{21}$$

which can be coalesced into a noninteractive system model,

$$\begin{aligned} d\mathbf{\Psi}/dt &= F'(\mathbf{\Psi}, \mathbf{E}, t) \\ \mathbf{R} &= G'(\mathbf{\Psi}, \mathbf{E}, t) \end{aligned} \quad \text{or} \quad \bar{\beta} \in \bar{\mathscr{B}}'. \tag{22}$$

Here $\mathbf{\Psi}$, $\mathbf{R}$, $\mathbf{E}$ are the direct sums of the $\mathbf{\Psi}_i$, $\mathbf{R}_i$, and $\mathbf{E}_i$, respectively, and F' and G' are functions computable from the F_i and G_i.

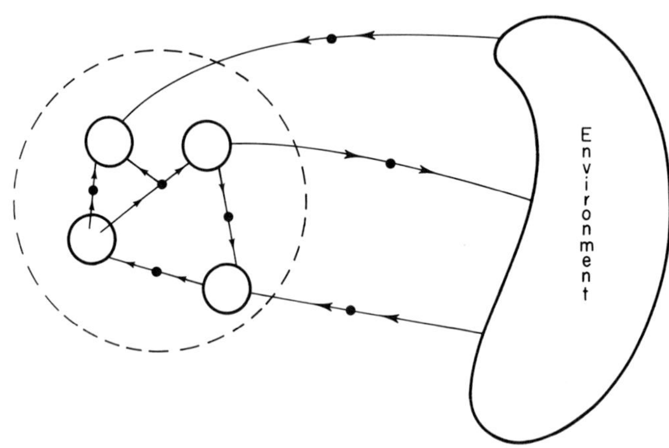

FIG. 6. A system and its environment, showing the features that interact only with other objects and those that interact with the environment.

The constraint equations take the form

$$\mathbf{E} = C\mathbf{R} \quad \text{or} \quad (\beta_1, ..., \beta_n) = \phi(\beta_1, ..., \beta_n). \tag{23}$$

Here C is a rectangular matrix with entries -1, 0, or $+1$. Because the compatibility requirements demand that stimuli be uniquely specified as functions of responses, C is of full rank.

For many purposes it is useful to be able to answer questions of the form: "If the system were placed in an environment that did so-and-so, how would it behave?" In cases like this, the environment can be modeled as if some or all of its behavioral variables are preprogrammed specific time series of interest to the observer. These stimulus-independent responses of the environment will provide specific stimuli to the system, so-called "forcing functions." Some components of the vector $\mathbf{R}$ in Eq. (22) will then be specific time series $\xi(t)$; these will be indicated separately in what follows.

To develop the model of the entire system as a higher level component, we first substitute Eq. (23) into (22), obtaining

$$\begin{aligned} d\mathbf{\Psi}/dt &= F'[\mathbf{\Psi}, C\mathbf{R}, \xi(t), t] \\ \mathbf{R} &= G'[\mathbf{\Psi}, C\mathbf{R}, \xi(t), t]. \end{aligned} \tag{24}$$

At this point the strictly internal stimulus variables have been eliminated. The next step is, if possible, to solve the response equation of Eq. (24) explicitly for $\mathbf{R}$:

$$\mathbf{R} = G[\mathbf{\Psi}, \xi(t), h]. \tag{25}$$

This is by no means a simple task, particularly if Eq. (24) is a nonlinear algebraic equation. This resulting response function is then substituted into the state equation of Eq. (24) to give

$$d\mathbf{\Psi}/dt = F'[\mathbf{\Psi}, CG(\mathbf{\Psi}, \xi(t), t), \xi(t), t], \tag{26}$$

which we redefine as

$$d\mathbf{\Psi}/dt = F[\mathbf{\Psi}, \xi(t), t] \quad \text{or} \quad \bar{\beta} \in \mathscr{B}^*. \tag{27}$$

Equations (27) and (25) form a free-body model of the system as a single object in terms of its states and the stimuli (if any) that the observer wishes to manipulate $[(\xi(t)]$.

As a very simple example of this process, we return to the logistic equation for population growth. We referred to the logistic (12) as a special case of the exponential (10) in which a time varying rate of increase $r(t)$ was generated by a "hypothesized relation" between r, $N(t)$, and K.

We are now in a position to show what that "hypothesized relation" is. We can consider our system to be made up of two objects: the population and its environment, as shown in Fig. 7. We choose as behavioral features for the population its rate of increase $r(t)$ (oriented as a stimulus) and its size $R(t)$ (oriented as a response). We choose $N(t)$ as a state variable, which generates the restrictions on classes of admissible stimuli discussed in connection with Eq. (10). For the environment, we choose two behavioral features: x_1, a measure of utilization of the environment; and x_2, a measure of the environment's ability to support an increase in population. The free-body models for these objects are

$$\text{population} \quad \begin{cases} dN(t)/dt = r(t)\, N(t), & (28) \\ R(t) = N(t), & (29) \end{cases}$$

$$\text{environment} \quad x_2(t) = k_1 \left(1 - \frac{x_1(t)}{k_2}\right). \tag{30}$$

Note that the environment has no state variable, thus assuming that it has no "memory" of its stimulus history. It should also be noted that the

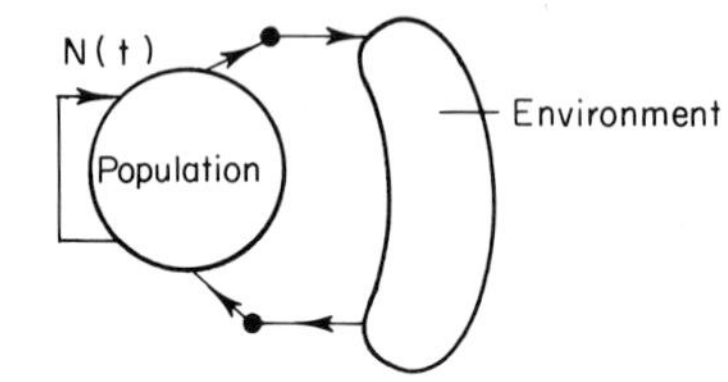

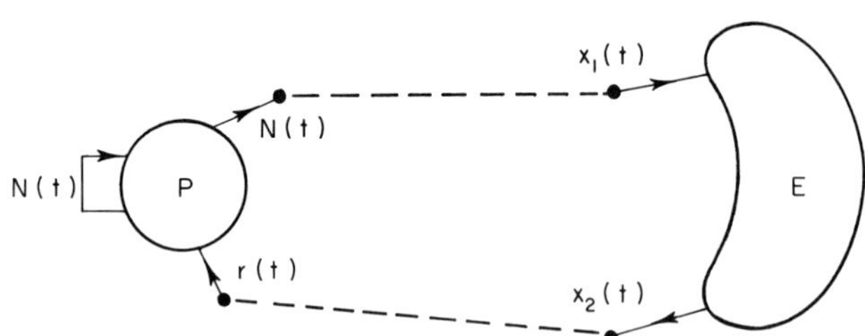

FIG. 7. The logistic equation as a system composed of two objects.

Free-body models:

$$dN(t)\backslash dt = r(t)\, N(t), \qquad x_2(t) = k_1 \left(1 - \frac{x_2(t)}{k_2}\right);$$

Constraints:

$$r(t) = x_2(t), \qquad x_1(t) = N(t).$$

capacity of the environment to support increased population size is assumed to be a specific linear function of the measure of utilization $x_1(t)$. The constraint equations for the system are

$$x_1(t) = R(t), \tag{31}$$

$$r(t) = x_2(t). \tag{32}$$

Equation (31) states that the measure of environmental utilization $x_1(t)$ is equal to the response variable $R(t)$ of the population. Equation (32) expresses the assumption that the rate of increase of the population $r(t)$ is exactly equal to the response $x_2(t)$ of the environment. As required, these equations show stimuli (x_1 and r) as functions of responses (R and x_2).

We now substitute the constraint equations (31) and (32) into the state and response equations (28) and (30), yielding

$$dN(t)/dt = x_2(t)\, N(t), \tag{33}$$

$$R(t) = N(t), \tag{34}$$

$$x_2(t) = k_1\left(1 - \frac{R(t)}{k_2}\right). \tag{35}$$

The next step is to solve the modified response equations (34) and (35) for $R(t)$ and $x_2(t)$ as a function only of the states and time. This is easily done by substitution of (34) into (35), yielding

$$R(t) = N(t), \tag{36}$$

$$x_2(t) = k_1\left(1 - \frac{N(t)}{k_2}\right). \tag{37}$$

The final step is to put these expression back into the state equation:

$$\frac{d}{dt} N(t) = k_1\left(1 - \frac{N(t)}{k_2}\right) N(t). \tag{38}$$

This approach allows many of the standard (e.g., Smith, 1952) criticisms of the logistic equation to be put in perspective. Limitations on the free-body population model (10) have been discussed already. It can also be asked whether the free-body model for the environment, Eq. (30), is reasonable. Does its history really have no impact on its behavior? Is the specified linear relation between utilization (x_1) and ability to support new growth (x_2) reasonable? Are k_1 and k_2 really constant? Independently of this, are the interactions realistically expressed? Is the population size a really reasonable measure of environment utilization? While this is a highly simplified example, it demonstrates nicely the interpretive freedom that systems science provides by keeping track of *both* object models and interaction constraints.

A quick glance through the systems engineering literature reveals at least two approaches to system modeling. Some texts (e.g., Ogata, 1967; Zadeh and Polak, 1969) *begin* with a set of equations like (27) and (25); they assume that the system description has been arrived at, and go on to treat the analysis of the resultant system description. Other texts (e.g., Koenig *et al.*, 1967) begin by assuming the existence of object models, and develop the system description by the methods described here. It is important to remember, when using references in the first class, that the process of arriving at the system model is not always an easy one.

Many simulation computer programs enter the system model in the form of Eqs. (22) and (23). The program carries out the substitutions to find (24), which is then solved numerically, for some (restricted) class of nonlinearities (Brennan and Linebarger, 1964; IBM, 1969). It should be noted that all but a very few such programs require the user to arrive at a compatible set of orientations.

The system state equations (27) may take many forms. For expository purposes, we have been writing ordinary differential equations. This is by no means the only class of interest. Many important systems have state variables which vary in one or more spatial dimensions, as well as time. For these *distributed systems*, the state equations take the form of partial differential equations. In *discrete systems*, difference equations provide an alternative to differential equations. Examples of these have already been given in terms of population models [e.g., Eq. (13)]. As Samuelson (1965) points out, the choice between differential and difference equation form is often a matter of convenience, since numerical solutions of the former are often obtained by rewriting them as the latter. An important situation in many biological problems often leads to another kind of state equation. In writing $d\boldsymbol{\Psi}/dt$ it has been assumed that $\boldsymbol{\Psi}$ can take on an infinite set of values (e.g., all real numbers). This need not be true, and leads to the field of *finite-state systems*. Here the state variable can take on only a finite number of values (e.g., a predator whose "activity state" can take on only the values of digestive pause, search, pursuit, handling, or eating).

Differential systems are often classified according to the form of their state equation. Some important types are

$$d\boldsymbol{\Psi}/dt = P\boldsymbol{\Psi} \qquad \text{Linear, time invariant, homogeneous;} \tag{39}$$

$$= P\boldsymbol{\Psi} + Q\mathbf{E} \qquad \text{Linear, time invariant, nonhomogeneous;} \tag{40}$$

$$= P(t)\boldsymbol{\Psi} + Q(t)\mathbf{E} \qquad \text{Linear, time varying, nonhomogeneous;} \tag{41}$$

$$= F(\boldsymbol{\Psi}) \qquad \text{Nonlinear, time invariant, homogeneous;} \tag{42}$$

$$= F(\boldsymbol{\Psi}, \mathbf{E}, t) \qquad \text{Nonlinear, time varying, nonhomogeneous.} \tag{43}$$

The form in Eq. (43) is the most general, but the analytical techniques for (39) and (40) are the most powerful. Discussions at a variety of levels of the properties and solutions of differential equations are available in the literature.

An important and frequently occurring class of differential systems are those in which all the behavioral features are subject to distributive constraints. Such systems often involve the transmission of a signal of some sort, which at a point of interaction is perceived identically by all the interacting objects. The nature of the features and the mechanism by which a distributive form of interaction is assured may vary, but the resulting constraint relation remains the same.

Such systems are often portrayed by signal flow graphs or block diagrams (DiStefano *et al.*, 1967; Milsum, 1966; Hubbell, 1971) as shown in Fig. 8. In these diagrams stimulus–response orientation is

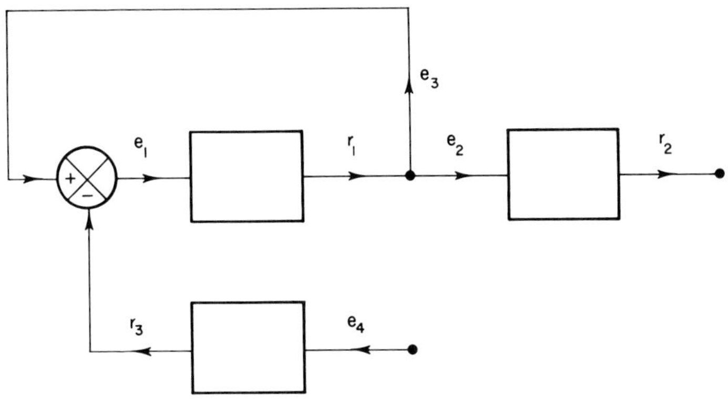

FIG. 8. Block diagram or signal flow diagram, often used to portray differential systems whose behavioral features are subject to distributive constraints.

indicated by an arrow on the lines connecting the objects. *If and only if* the system is linear and time invariant, as in Eq. (39) or (40), mapping the system in block diagram form can considerably simplify the analysis. By the use of the Laplace transform, object behavior can be expressed in the form of a "transfer function." If the system is nonlinear and/or time varying, this approach is next to useless, and the problem must be approached in state-space form.

One important system with purely distributive constraints is the electronic analog computer. In the user's model, the objects making up such a computer are functional units (summers, integrators, etc.) whose behavioral features are distributively constrained energy densities (voltages). In the lower level (designer's) model, however, the component objects are electrical parts whose behavioral features also include conservatively constrained material flows (currents). The designer has, however, taken care to suppress the relevance of these features to the user. If one accepts the manufacturers' approximations and instructions in this regard, the user can consider only the distributively constrained voltages, and work with diagrams like Fig. 8. For an example of how these two views of the analog computer are related, see Patten (1971, pp. 17ff).

Distributive constraint structures of the form shown in Fig. 8 were originally developed in the study of the "signal processing" features of servomechanisms, feedback control devices, and analog and digital computers, and they are the starting point for studies of classical control theory (Ogata, 1967). In general form, these structures are very effective in *conceptualizing* the structure of systems that deal with the flow and exchange of information, such as neurological networks (Kilmer *et al.*, 1969; Kilmer, 1968), or decision-making processes in business management, economic, social, and political systems (Martin, 1968). As previously pointed out, however, their *analytical* utility is limited to a very restricted class of systems. Because texts emphasizing this approach are still very much available to an ecologist looking for "systems science," it is worth quoting a statement by Richard Bellman (1968, p. 17) on the dismal effect that the use of transfer functions had on classical control theory.

> One difficulty, of course, . . . is that a student trained in this school [Laplace transform/transfer function] has a great deal of trouble readjusting his sights to the real world of engineering. If one introduces him to realistic factors of contemporary systems such as constraints and nonlinearity and, what is more, stochastic and adaptive effects, he has a certain tendency to shrug his shoulders fatalistically and say, "I don't understand how to handle these effects in control processes. Therefore I will ignore them." Back to diagrams!

Considering even for a moment how much more prevalent these "realistic factors" are in ecology than in engineering, ecologists should probably think several times before choosing this approach to a problem.

IV. Levels of Organization

The subjectivity of structuring a system description means that an observer is free to choose from a wide range of objects in any study. In biological systems one of the major methods of choosing components is based on the concept of "levels of organization." These levels (molecular, cellular, tissue, organ, organism, population, community, or some modification of this set) are almost universally assumed to exist in biological systems (for extensive reference lists and discussion of related concepts see Whyte *et al.*, 1969). These levels are "strata" (Mesarovic, 1969); different observer-imposed structurings of a system, characterized by physical inclusion of objects at one level within objects at the higher level. Thus, an object at level L can be broken down into components at level $L - 1$, or lumped with other level L objects into a supracomponent at level $L + 1$. This process can be continued in either direction until elementary particles, if such things exist, are reached or until the whole system is treated as a single object. The fact that biological systems are perceived and studied in this manner has important consequences in the selection of objects in the study of a biological system. We will present some of these here in terms of the framework developed so far.

Consider a system composed of objects O_i ($i = 1,..., n$) which can be grouped into aggregations A_j ($j = 1,..., m$; $m < n$) at a higher level of organization. The system can obviously be described in terms of the behavior of the A_j or the O_i. The behavior of each A_j can be described as a system itself, with objects being a subset of the O_i, or as a single object (Fig. 9). The description of A_j as a system will include a set of free-body descriptions for its components and a set of constraint equations.

$$\begin{gathered} \bar{\beta} \in \bar{\mathscr{B}}', \\ \beta_1 = \phi_1(\beta_1, ..., \beta_n) \\ \beta_2 = \phi_2(\beta_1, ..., \beta_n) \\ \vdots \end{gathered} \tag{44}$$

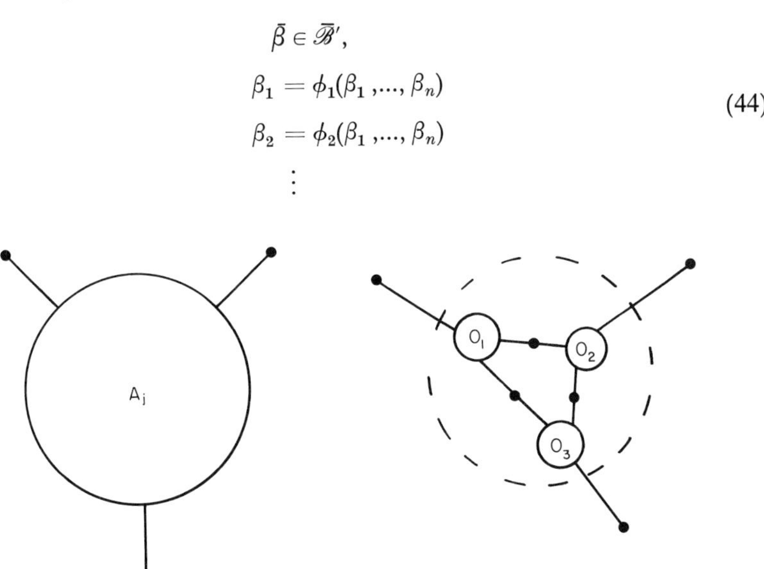

FIG. 9. Two ways of approaching an aggregate A_j : as a single object or as a system of lower level objects O_i.

This information can be condensed into a description of A_j as a single object:

$$\bar{\beta} \in \mathscr{B}^*. \tag{45}$$

Since the set $\mathscr{B}^*$ is a subset of the set $\bar{\mathscr{B}}'$, the behaviors exhibited by an aggregate A_j are a subset of the behaviors open to its component objects. There are *no* behaviors open to A_j that are not open to its constituent objects O_i, but the behavioral descriptions of the O_i are insufficient to characterize A_j because they lack the information on the interaction constraints. The same is true of the behavior of the system in comparison with either the behavior of the O_i or the A_j. The rules governing the lower level are necessary but not sufficient for description of the higher level, a fact about biological systems which has previously been deduced on other grounds (Polanyi, 1968; Botnariuc, 1966; Feibleman, 1959).

This restriction of behavioral possibilities brings us to the question of *emergent properties*. These are differences of some sort generated by the process of moving from a lower to a higher level of organization. Two different phenomena seem to be involved. The first (Fig. 10a) involves the emergence of new behavioral features in the process of switching consideration from a single object to an aggregation of many of those objects. These new features may not be defined for single objects, or they may simply be of interest to the observer at one level and not at the other. For example, each level L object in Fig. 10a might be described by a feature b_1 = size. When switching attention to a level $L + 1$ object a new feature b_2, for example the variance of the values of b_1, might become of interest. Strictly speaking this is not defined for a single object at level L. Or consider an ideal gas molecule in a container. Its relevant behavioral features might be its instantaneous position and momentum. A whole collection of such molecules, in the same container, is likely to be described by new features such as volume, temperature, and pressure. For most purposes the use of these new features is a matter of the observer's interest and convenience.

The second aspect of emergent properties (Fig. 10b) is stated succinctly by Mayr (1965):

> A discussion in this context, of "emergence" is beyond our scope. All I can do here is to state its principle dogmatically: "when two entities are combined at a higher level of integration, not all the properties of the new entity are necessarily a logical or predictable consequence of the properties of the components."

What is emerging in this case is not a new behavioral feature but differences in behavior between two or more level $L + 1$ objects composed of the same set of level L objects.

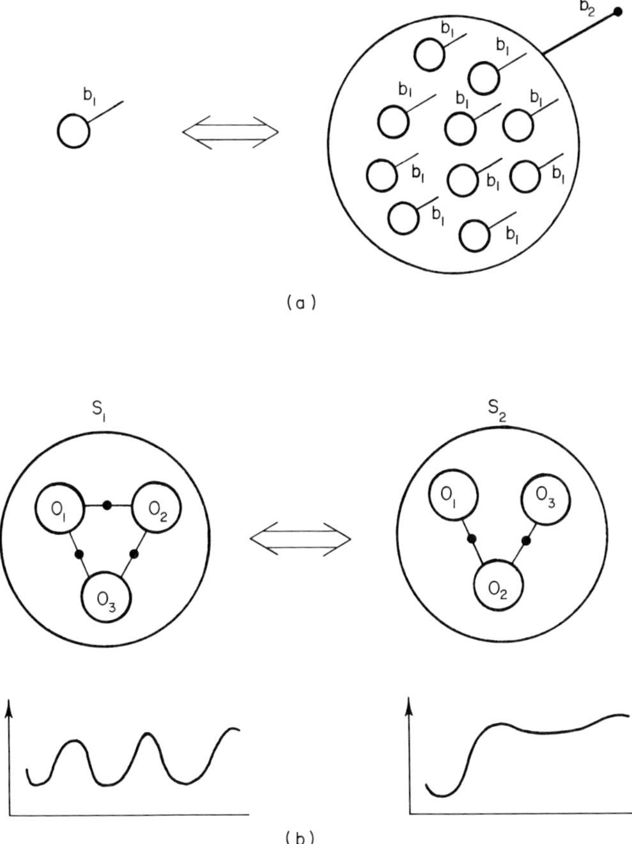

FIG. 10. The two types of "emergent" properties (a) new behavioral features of interest at one level but not at the lower; (b) different behaviors generated by different couplings of the same objects.

It is an unfortunate fact that emergent properties have sometimes been given an almost mystical character in the literature. They *are* real phenomena, which do arise to confront anyone studying a complex system. Thus it is encouraging that systems science removes the mysticism associated with the phenomenon and provides a useful method of handling either aspect of the problem. The first situation, the emergence of new interesting behavioral features, is simply accepted. The behavioral features are up to the observer after all, and if he finds it convenient to describe aggregations in different terms than he describes individual objects, then let him do so. From the viewpoint of systems science the second type of emergence, described above by Mayr, is not surprising at all. Consider system S_1 in Fig. 10b. Its behavioral description in terms of free-body models and constraints is

$$\bar{\beta} \in \bar{\mathscr{B}}', \tag{46}$$

$$(\beta_1, ..., \beta_n) = \phi_{S_1}(\beta_1, ..., \beta_n), \tag{47}$$

while the similar description for S_2 is

$$\bar{\beta} \in \bar{\mathscr{B}}', \tag{48}$$

$$(\beta_1, ..., \beta_n) = \phi_{S_2}(\beta_1, ..., \beta_n). \tag{49}$$

Putting the information in Eqs. (46) and (47) and Eqs. (48) and (49) together, we have for S_1

$$\bar{\beta} \in \mathscr{B}^*_{S_1}, \tag{50}$$

and for S_2

$$\bar{\beta} \in \mathscr{B}^*_{S_2}. \tag{51}$$

If the constraints, ϕ_{S_1} and ϕ_{S_2}, generated by the system topology are different, we *expect* that Eqs. (50) and (51) will describe very different behaviors. The second type of emergent properties, then, are differences in behavior which are unexplainable or surprising *only* if the system topology and its resulting constraints have been ignored. For another clear example, refer back to Eq. (26) which is the system state equation derived from the free-body models and the constraints. Note that the right-hand side is influenced by the free-body state equations (F') and response equations (G) and the constraints (C). If the topology and resultant constraints are ignored, the higher-level behavior is truly "emergent" and unpredictable, but if they are included it becomes completely predictable.

So handily does systems science deal with emergent properties that they actually become the stock in trade of the working systems engineer. The differences in behavior of different kinds of radios, televisions, hi-fi amplifiers, etc. are emergent properties, not predictable from the properties of resistors, tubes, transistors, and so on. But given the methodology of systems science, and taking the effects of the system topology into account, practitioners in the physical systems can actually build systems to meet a set of specified emergent properties.

The fact that ecological systems can be described at many levels of organization, and the relation of the description at one level to that at another leads to some suggestions for the choice of a level in a given study. Refer back to the object-aggregate example of Fig. 9 and Eqs. (44)

and (45). Let us assume that the goal of the observer in this study is to obtain a dynamic behavioral description of the entire system. It would be entirely reasonable in light of this goal to study the system as a single object, but most ecological situations will not allow this because of system complexity. So the observer is faced with a choice of a description in terms of the O_i or the A_j. The first choice would involve development of descriptions for all the objects and constraints in every A_j [the information in Eq. (29)] *and* the constraints among the A_j. The second choice would develop descriptions of the A_j directly [Eq. (30)] and combine this with the constraints between the A_j. Given that the necessary technology exists, it is thus advisable to structure a system at the highest level consistent with the goals of the study.

V. Energy Systems

Many of the most important problems in the physical sciences deal with energy behavior in systems, and a sophisticated methodology exists for dealing with such questions. When this methodology is expanded to deal with social, economic, and biological questions, an underlying structure (well hidden in most textbooks of physical systems science) becomes apparent. In this section we will attempt to convey the conceptual basis of energy systems in the context of a nonrigorous presentation of some of the associated methodology.

Up to now we have placed little emphasis on actual techniques required to utilize the concepts presented. We will depart from that now, for two reasons. While problems involving energy flow have received much attention from systems ecologists, no one has applied either the concepts or the methods of systems science to them. There is thus no background in ecological literature to which we can direct the reader. To make things worse, texts on physical systems, almost without exception, make assumptions that exclude many of the systems of interest to ecologists. Our goal, then, is to bridge the gap between our more general, and hopefully more valid, presentation of basic concepts and the thorough treatments of methodology available in the literature. We do not expect a systems ecologist to function using only the tools we present here, but we hope that he will be able to obtain needed assistance from the engineering literature.

Throughout this section we have utilized extremely simplistic linear object models as examples. We do not want to imply that these can be used as detailed descriptions of biological processes, but they should not be too vigorously scorned. The familiar "compartment model" of the ecological literature uses object models of precisely this level of sophistication. We have worked out such a model in detail (Appendix II) as an example of energy system methodology.

A. Behavioral Features

As a biologically meaningful starting point, consider a flow of material within a system. This flow can be described by a behavioral feature y_i defining the flux rate of material at the point i. Thus y_i will have the units of material/time, and as such it will undergo conservative constraints at points of interaction. Often, particularly in biological systems, material flow is viewed as an energy flux; a carrier of energy in one form or another (e.g., the chemical energy of biomass flowing through a food chain). If equal amounts of material can carry different amounts of energy, we introduce a second behavioral feature x_i, the energy density of the material. The units of x_i are energy/unit material.

The amount of energy in a unit of material can be equated with the energy added to that material in the process of bringing it from some reference point R to the point at which the flows are being considered. This is equivalent to defining x_i as a "cost" to the system of bringing a unit of material to a particular point at a given time. As such it may be either positive or negative; the system may lose or gain by putting the unit of material there. The energy density variable x will be referred to as an *intensive* variable, and the flux variable y as an *extensive* variable.

The intensive (x) and extensive (y) variables are used to describe the dynamics of energy flow in the system. *Energy itself does not appear explicitly as a behavioral feature.* Rather a flow rate of energy is calculated from the flux rate and energy density as

$$e(t) = x(t)\,y(t), \tag{52}$$

with the units energy/time. If the energy density of the material is constant, the energy flow rate $e(t)$ will be proportional to the material flow rate $y(t)$ and it is possible to ignore $x(t)$ and develop the model only in terms of flows. This is not likely to be the case in the majority of ecological systems.

In their original development in electrical field theory, the x and y variables measured the energy required to move a unit charge from infinity to a given point and the current flow across a surface, respectively. They can be generalized to other forms of physical and chemical energy:

Hydraulic and pneumatic	x = pressure,	y = flow rate,
Thermal	x = temperature,	y = flow rate,
Mechanical	x = velocity,	y = force,
Chemical	x = molecular concentration,	y = molecular flow rate.

In economic and ecological modeling, the extensive variable $y(t)$ measures the flow rate of material of a particular form. The intensive variable $x(t)$ measures the cost required to put a unit of material of a given form at a particular point in space and time. It can be measured in a variety of ways: solar energy (e.g., hectare days/unit material), human energy (e.g., manhours/unit material), physical energy (e.g., megawatts/unit material), or in terms of any other nonrenewable resource. Another alternative is to measure the intensive variable in terms of the economic cost in dollars (Koenig *et al.*, 1969; Koenig, 1970). This cost is a scalar function of the various energy costs and other factors, with the different costs weighted by the rules of the economic structure. Such generalized costs will be useful in evaluating the social desirability of alternative industrialized ecosystem structures (Koenig *et al.*, 1971).

Energy concepts and the associated behavioral features are almost sure to prove useful in ecological studies at some level. There are already indications that they will be useful in evaluating the interactions between the man-made and natural components of our life-support system. Their utility in studies of natural ecosystems, communities, or populations is still open to question.

B. Free-Body Models

The model for an object with n points of interaction with its environment is developed in terms of the $2n$ behavioral features x_i and y_i, $i = 1, \ldots, n$. As a working diagram for dealing with these objects, we begin with Fig. 11 in which each point of interaction and a reference point R are indicated by points. These are joined by a set of line segments

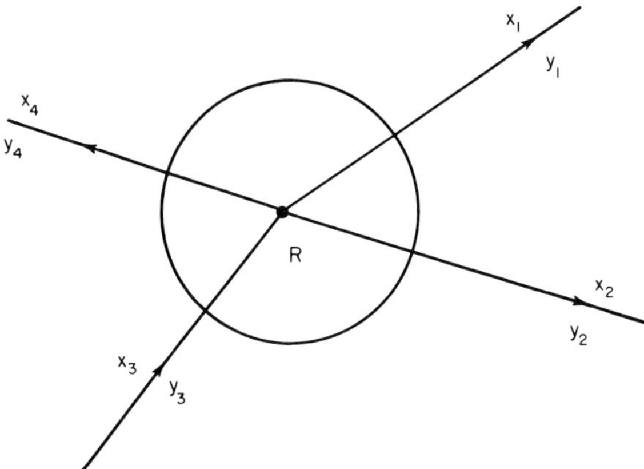

Fig. 11. Object graph for a four-terminal component.

to form the *object graph*. This graph, and the *system graph* formed by connecting a set of object graphs, will be useful in finding constraint equations and compatible stimulus–response orientations. We will see later that in certain cases other object graphs are preferable to that shown in Fig. 11, but since this form is the most general, the initial discussion is structured around it.

If the object interacts with its environment at well-defined spatial areas, as is often the case in physical systems, the n points of interaction are referred to as *ports*. It is assumed that neither the intensive nor extensive variable varies over the area of the port, so that one can validly speak of *the* flow or *the* density at the port. If x and/or y vary over the region of interaction, then the system is referred to as a *distributed* (as opposed to *lumped*) system, and the customary modeling language is partial (as opposed to ordinary) differential equations. In situations where the spatial structure of the objects is less well defined than in many physical systems it becomes impossible to identify actual energy ports (e.g., where is the energy port between a predator population and a prey population?). Even in these cases the idea of an energy port can form a useful abstraction of the actual process.

A direction is associated with the edges in the object graph by addition of an arrow. Unlike the signal-flow or block diagram in Fig. 8, *this arrow has nothing to do with stimulus–response orientation.* Rather it serves to provide a *measurement* orientation for the two behavioral features associated with the edge. The extensive (y) variable associated with the edge represents a flux of some sort. The arrow on the edge provides a sign convention for this flow; a positive value of y indicates a net flow in the direction of the arrow, a negative value a net flow in the opposite direction. The intensive (x) variable must have a reference point associated with it since it represents (in a general sense) the energy cost associated with bringing a unit of material from that reference to the point of interaction. For this reason, the edge is associated with a reference point R. The arrow again defines a sign convention; if the arrow points away from R (as in x_1, Fig. 11), a positive value of x indicates that the system gained energy in moving a unit of material from R to the port (the energy level at R is higher than at the port). If the arrow points toward R the opposite is true. Obviously, the choice of a measurement orientation is arbitrary, and is usually dictated by the observer's desire to consider one flow direction or energetic potential positive and another negative. The extensive variable may be measured at a single point, since it represents a flow through a surface. The intensive variable, on the other hand, is a potential which must be measured across a gap between two points. Hence in many texts dealing with physical systems these variables are defined as "through" and "across" variables, respectively.

The net energy dissipation rate for an object with n ports is

$$E(t) = \sum_{i=1}^{n} x_i(t)\, y_i(t). \tag{53}$$

The energy level of the object is decreasing when $E(t) > 0$ and increasing when $E(t) < 0$. It will be shown later that if the object is a closed system, this term is identically zero, as required by the law of conservation of energy.

Three fundamental classes of objects appear in nearly all energy systems: material transformation, transport, and storage processes.

1. *Material Transformation Processes*

These processes change material from one biological or technical form to another. Biological organisms and industrial manufacturing plants are examples of this type of process. The object represented schematically in Fig. 11, for example, might represent a process which produces substance 4 from ingredients 1, 2, and 3. A very simple model for such a transformation process is

$$\begin{bmatrix} y_1 \\ y_2 \\ y_3 \end{bmatrix} = \begin{bmatrix} k_1 \\ k_2 \\ k_3 \end{bmatrix} y_4 , \tag{54}$$

$$x_4 = -[k_1\; k_2\; k_3] \begin{bmatrix} x_1 \\ x_2 \\ x_3 \end{bmatrix} + f(y_4). \tag{55}$$

Equation (54) states that to produce a unit of y_4 (e.g., plant biomass) requires materials y_1, y_2 and y_3 (e.g., nitrogen, CO_2, and water) in fixed proportions. Equation (55) relates the energy cost of the resulting material (x_4) to the energy cost of the raw materials (x_1, x_2, x_3) and the cost $f(y_4)$ of carrying out the transformation. The energy cost of carrying out the transformation may be positive or negative and depends upon the rate of processing. Economists consider $f(y_4)$ under the heading of "economies of scale." In the special case in which $f(y_4) \equiv 0$, the energy budget for the object is completely balanced since

$$\sum_{i=1}^{4} x_i y_i = 0. \tag{56}$$

In other words, the cost of the product is just the weighted sum of the costs of materials. The reader can verify this important result for himself by substituting Eqs. (54) and (55) into Eq. (56).

Equation (54) represents the "laws of material combination" of the process. They can easily be generalized to an object having any number of ports, as can the energy cost relation in Eq. (55).

2. *Two-Port Material Transport Processes*

When material is transported from one location to another without a change in form, the resulting model is a special case of Eqs. (54) and (55). Using the behavioral features defined by Fig. 12a we have

$$y_1 = y_2\,, \tag{57}$$

$$x_2 = -x_1 + f(y_2). \tag{58}$$

This indicates that the material is transferred unchanged through the object. The unit cost (x_2) of the outflowing material is equal to that of the inflowing material (with a sign change due to the measurement orientation in Fig. 12a) plus the transport cost $f(y_2)$.

New features can be defined in this case (Fig. 12b),

$$y_0 \equiv y_1 = y_2\,, \tag{59}$$

$$x_0 \equiv x_1 + x_2 = f(y_0), \tag{60}$$

where x_0 measures the *change* in energy level between the inflow and outflow. The energy input rate to the transport process is

$$\begin{aligned} E(t) &= x_1 y_1 + x_2 y_2 \\ &= x_1 y_2 + [-x_1 + f(y_2)]\, y_2 \\ &= x_1 y_2 - x_1 y_2 + f(y_2)\, y_2 \\ &= x_0 y_0\,. \end{aligned} \tag{61}$$

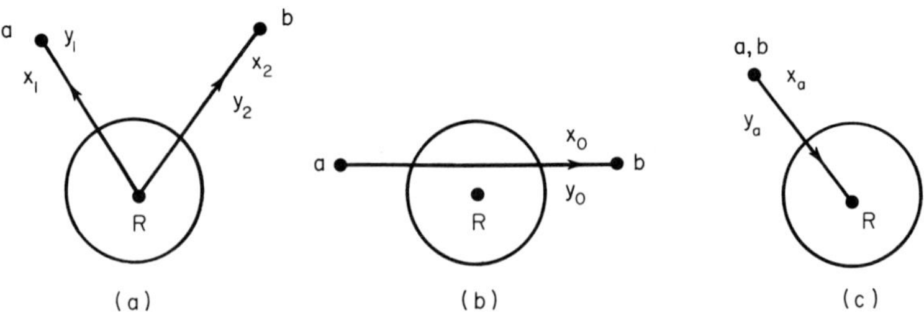

FIG. 12. (a) Object graph for a two-terminal component; (b) modification of (a) for transport processes; (c) modification of (a) for certain special types of storage processes.

Thus the two choices of behavioral features (Fig. 12a and 12b) are equivalent in terms of energy utilization. Either choice can be used, but in many cases the features in Fig. 12b result in improved numerical accuracy, particularly if x_1 and x_2 are large and not very different. If the transport energy costs x_0 are a linear function of the flux rate y_0, then

$$f(y_0) = Ry_0 ;$$

and the constant R is referred to as the resistance of the transport process.

3. *Material Storage Processes*

Material storage objects store material for later redistribution without changing its form. The material in storage is measured by a state variable $\psi(t)$. Use of material storage as a state variable is tantamount to assuming that any object containing ψ units of material is equivalent regardless of the time course over which the material is accumulated.

Using the two-port object graph in Fig. 12a, a simple model is

$$d\psi/dt = y_1 - y_2, \tag{62}$$

$$x_2 = -x_1(\psi, y_1) + f(\psi, y_2). \tag{63}$$

Equation (63) expresses the unit cost of the material being removed as the sum of $-x_1(\psi, y_1)$ and $f(\psi, y_3)$. The former term represents the cost of "putting material into storage," and the latter any additional costs of "removing the material from storage." If $f(\psi, y_2) \equiv 0$, the unit cost of incoming and outgoing materials is the same (except for a sign change), and can be redefined as

$$x_a \equiv -x_2 = x_1(\psi, y_1). \tag{64}$$

The two ports can now be considered as a single port, using the behavioral feature

$$y_a \equiv y_1 - y_2 \tag{65}$$

to measure the *net* flow of material. The energy input rate to the storage component is

$$\begin{aligned} E(t) &= x_1 y_1 + x_2 y_2 \\ &= x_1 y_1 + (-x_1)(y_1 - y_a) \\ &= x_1 y_1 - x_1 y_1 + x_1 y_a \\ &= x_a y_a, \end{aligned} \tag{66}$$

so that the two behavioral descriptions are identical in terms of energy.

If x_1 is independent of y_1, the behavioral description of the object can be developed in terms of x_a and y_a, using the object graph in Fig. 12c. The resulting model is

$$d\psi/dt = y_a, \qquad x_a = x_1(\psi). \tag{67}$$

A description of the electrical capacitor results if

$$x_1(\psi) = k\psi, \qquad k \text{ a constant.} \tag{68}$$

Note that this is the result of two special assumptions; that $f(\psi, y_1) = 0$ and that $x_1(\psi, y_1) = k\psi$. There is in fact no electrical analog of the more general storage component.

C. The System Graph: Orientation and Constraints

The electrical networks to which the energy system methodology was first applied are usually complex connections of very large numbers of objects. The choice of compatible sets of orientations and the specification of constraint equations are formidable projects for such systems. Fortunately, both problems can be solved in a straightforward manner by the use of the *system graph*. The system graph is formed by joining the object graphs at their points of interaction (Fig. 13). Any network theory text devotes considerable attention to the theory of graphs and its application. Treatment at a variety of levels of sophistication are available in the literature; we seek here to provide only enough background for an interested ecologist to begin utilizing that literature. Although these systems are usually formed of two-port transport processes, the results are directly applicable to transformation and storage processes.

The graphs in Fig. 13a, containing transformation, transport, and storage processes, can be redrawn as in Fig. 13b, with the reference point R appearing only once. The two graphs are completely equivalent as far as their uses are concerned.

The extensive variables, representing as they do a flow of material, are subject to conservative constraints at the nodes of the system graph. For a compatible orientation, exactly one of the y variables at each node in the graph must be a stimulus and the others responses. (Note: The arrows on the system graph have nothing to do with stimulus–response orientation.)

The intensive variables represent the energy potential, with respect to R, of a unit of material at a particular point. At a node in the graph then, all the x variables must be equal (to within a sign change if the

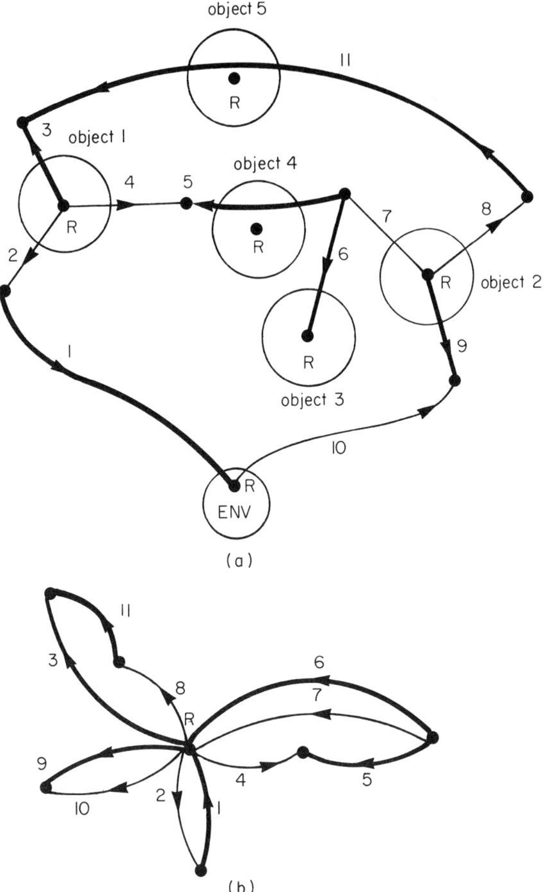

FIG. 13. (a) System graph for a system composed of two transformation and two transport processes and a storage process. (b) Alternative form of the same graph obtained by indicating the reference point R only once.

measurement orientation is different for two of them) and they are thus subject to distributive constraints. To see why they must be equal it may be helpful to remember the assumption regarding the homogeneity of material flux over the area of an energy port. The points of interaction in the system graph represent the fusion of the energy ports of the objects involved. Hence if all the x variables were not equal, there would be an energy gradient being maintained in a homogeneous medium in the absence of any corresponding force to maintain it, in violation of our assumption. Since the intensive variables are subject to distributive constraints, the only compatible orientations are those which result in precisely one response variable at every node.

The compatibility requirements can be satisfied as follows. Select a set of edges in the system graph (Fig. 13a) that connect each node to exactly one reference point R. One such set of edges is indicated by the heavy lines in Fig. 13. If all the reference points are considered the same, as in Fig. 13b, this set of edges forms a *tree*, i.e., it connects all the nodes but forms no circuits. This can be easily verified by attempting to include an additional edge in the tree. In either version of the system graph this forms a circuit.

If all the object models include R explicitly (as in Fig. 12a), there will be exactly one branch of the tree entering each node. In this case the compatibility requirements will obviously be met by orienting the extensive variables in the tree and the intensive variables in the cotree as stimuli, and the remaining variables as responses. When alternative component descriptions, not involving R, are used (as in Fig. 12b) the interacting behavioral features are no longer measured with respect to the same reference point. This means that the distributive constraints on the intensive variables must be expressed in an alternative form which is developed below. However, even in this case the use of a tree and cotree to orient the objects is still valid. While the stimuli at each node may not be determined by responses *at that node*, all of the stimulus variables of the system can be written as unique functions of response variables. This satisfies the compatibility requirements by accounting for the selection of all the behavioral variables. It is not infrequently that one encounters objects whose models are more tractable in one orientation than in another (perhaps because of differences in state spaces in the two orientations) or even objects in which certain orientations are completely inadmissible. These objects can be oriented first by placing the corresponding edges of the system graph in the tree or cotree as required and then filling in the rest of the edges.

We consider now the expressions for the constraint equations on the intensive and extensive variables. An equivalent expression of the conservative constraints is obtained by "cutting" the system graph in two, conceptually breaking any set of edges that separates the graph into exactly two sections. The constraint can be expressed by requiring that the net flow between the two sections of the graph be zero. This must be true for all the possible ways of cutting the graph in two.

The development of an equivalent form for the distributive constraints on the intensive variables is more subtle. Consider the three edges in Fig. 14a. It is immaterial whether the vertices represent points of interaction between objects or if one of the vertices is the reference point R.

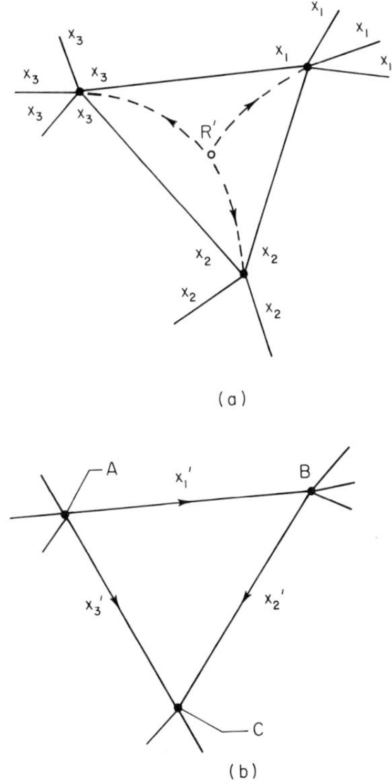

FIG. 14. Generalized graph circuit used to derive an alternative form for the distributive constraint relations on the intensive variables.

The important point is that they form a circuit. At each of the vertices, the intensive variable is measured with respect to a common reference point R'. If $R' = R$, it means only that the value of x at a vertex representing R is zero. The distributive constraints require that all the interacting x variables be equal at each vertex, as shown in Fig. 14a. Now, the reference point R' can be eliminated by defining new variables x_1', x_2', and x_3', as shown in Fig. 14b:

$$x_1' = x_1 - x_3\,, \qquad x_2' = x_2 - x_1\,, \qquad x_3' = x_2 - x_3\,. \tag{69}$$

Each x_i' is now being measured with respect to a different reference point (which may include R if it happens to be one of the vertices). The arrows on the edges indicate the measurement orientation for the new feature. We now begin at any vertex and calculate $\sum \pm x_i'$ around the circuit, using a plus sign if the arrow points in the direction the circuit is being traversed, and the minus sign otherwise. For Fig. 14 this is (starting at vertex A and traveling clockwise)

$$\sum \pm x_i' = x_1' + x_2' - x_3'. \tag{70}$$

Substituting the expressions in (69) we have

$$(x_1 - x_3) + (x_2 - x_1) - (x_2 - x_3) = 0. \tag{71}$$

Thus, for any circuit in the system graph the sum of the intensive variables, taken with a positive or negative sign determined by the measurement orientation arrow, is always zero. This rule, originally developed as Kirchhoff's voltage law in electrical systems, is the constraint relation for intensive variables that is actually used in dealing with the system graph. The modified constraint relations for the y variables are sometimes referred to as *continuity constraints*, those for the x variable as *compatibility constraints* (not to be confused with compatible orientation). It is interesting to note that x_1', x_2', and x_3' represent the difference in energy density between two points. Hence, Eq. (70) expresses the total cost of moving a unit of material around a closed loop, and it is not surprising that it should be zero. It should also be noted that this result is independent of the behavior of the objects in the system, depending only on the distributive property of energy densities.

We now know that, to express the constraints, equations are written summing the intensive variables to zero around loops and setting the net flow from one half of the graph to the other equal to zero, for every loop and for each way of cutting the graph. This is, for a system of any size at all, an enormous number of equations, and is always more information than is needed to specify the constraints since not all of the equations are independent. To present methods of generating a sufficient but non-redundant set of equations we first must introduce some terminology.

A *tree*, as we have already seen, is a set of edges in a graph that connect all vertices but form no circuits. The constituent edges of a tree are *branches*. The complementary set of edges comprise a *cotree*, whose edges are *chords*. A *cut set* is a set of edges whose removal divides the graph into two unconnected parts. A *fundamental cut set* with reference to a specified tree is a cut set exactly one element of which is a branch. A *circuit* is any set of edges that forms a closed path. A *fundamental circuit*, with reference to a specified tree, is a circuit exactly one element of which is a chord.

To arrive at a minimal sufficient set of constraint equations, a tree (any tree) is first specified on the graph. Then the constraints on the extensive variables are written for each fundamental cut set and the constraints on the intensive variables for each fundamental circuit. These equations form a precisely sufficient set of constraints. In Fig. 13a and b, for example, a tree and cotree are defined by

$$\text{tree} = \{1, 3, 5, 6, 9, 11\},$$
$$\text{cotree} = \{2, 4, 7, 8, 10\}.$$

Thus, there will be six fundamental cut set equations (one for each branch in the tree) and five fundamental circuit equations (one for each chord in the cotree).

The six continuity constraints corresponding to the fundamental cuts are

$$\begin{aligned} &\text{Stimulus set} \quad \text{Response set} \\ &\text{(branches)} \quad \text{(chords)} \\ y_1 &= y_2 , \\ y_3 &= -y_8 , \\ y_5 &= -y_4 , \\ y_6 &= y_4 - y_7 , \\ y_9 &= -y_{10} , \\ y_{11} &= y_8 ; \end{aligned} \tag{72}$$

and the five compatibility constraints corresponding to the fundamental circuits are

$$\begin{aligned} &\text{Stimulus set} = \text{Response set} \\ &\text{(chords)} \quad \text{(branches)} \\ x_2 &= -x_1 , \\ x_4 &= x_5 - x_6 , \\ x_7 &= x_6 , \\ x_8 &= -x_{11} + x_3 , \\ x_{10} &= x_9 . \end{aligned} \tag{73}$$

Notice that both Eqs. (72) and (73) express stimuli as functions of responses. It is easier to show some of the important properties of these equations by developing them in matrix form.

To do this we first write the x and y variables as column vectors, arranged to give

$$\mathbf{X} = \begin{bmatrix} \mathbf{X}_b \\ \mathbf{X}_c \end{bmatrix}, \qquad \mathbf{Y} = \begin{bmatrix} \mathbf{Y}_b \\ \mathbf{Y}_c \end{bmatrix},$$

where b indicates the branch variables and c the chord variables. Next fundamental cut set and circuit vectors are defined, and their elements arranged in the same way.

A *fundamental cut set vector* $\boldsymbol{\alpha}$ of a graph with e edges is a $1 \times e$ vector whose ith entry is

$$\alpha_i = \begin{cases} 1 & \text{if edge } i \text{ is in the cut set, and has the same direction as the defining branch,} \\ -1 & \text{if edge } i \text{ is in the cut set, and has the opposite direction of the defining branch,} \\ 0 & \text{otherwise.} \end{cases}$$

The sign convention is arbitrary, and the results would be the same if it were reversed consistently throughout. Similarly, a *fundamental circuit vector* $\boldsymbol{\beta}$ is a $1 \times e$ vector whose ith element is given by

$$\beta_i = \begin{cases} 1 & \text{if edge } i \text{ is traversed in the circuit with the same direction as the defining chord,} \\ -1 & \text{if edge } i \text{ is traversed in the circuit with the opposite direction of the defining chord,} \\ 0 & \text{otherwise.} \end{cases}$$

Again, the sign convention is arbitrary. For example, the vectors defined by branch 1 of the tree and chord 2 of the cotree in Fig. 13 are

$$\begin{array}{lcc} & \text{branches} & \text{chords} \\ & 1\ 3\ 5\ 6\ 9\ 11 & 2\ 4\ 7\ 8\ 10 \\ \boldsymbol{\alpha}_1 = \{ & 1\ 0\ 0\ 0\ 0\ 0 & -1\ 0\ 0\ 0\ 0\}, \\ \boldsymbol{\beta}_2 = \{ & 1\ 0\ 0\ 0\ 0\ 0 & 1\ 0\ 0\ 0\ 0\}. \end{array} \qquad \begin{array}{r} \\ \\ (74) \\ (75) \end{array}$$

The precisely sufficient set of constraints is obtained by setting

$$\boldsymbol{\alpha}_i \mathbf{Y} = \mathbf{0}, \tag{76}$$

$$\boldsymbol{\beta}_i \mathbf{X} = \mathbf{0}, \tag{77}$$

for *each* fundamental circuit and cut set. Note that doing this with the vectors (74) and (76) above leads to the first equations in (72) and (73), respectively.

The entire set of constraints can be written by defining a *cut-set matrix* A and a *circuit matrix* B whose rows are cut-set and circuit vectors. The entire set of constraints is then

$$A\mathbf{Y} = \mathbf{0}, \tag{78}$$

$$B\mathbf{X} = \mathbf{0}. \tag{79}$$

In each cut set vector $\boldsymbol{\alpha}$ the entry corresponding to the defining branch is always 1, and the same is true for the entry corresponding to the defining chord in each circuit vector $\boldsymbol{\beta}$. Thus A and B have the form

$$\begin{aligned} A &= [I \mid A'], \\ B &= [B' \mid I], \end{aligned} \tag{80}$$

where A' and B' are submatrices of A and B, and I is the identity matrix. The constraint relations are thus

$$[I \mid A'] \begin{bmatrix} \mathbf{Y}_b \\ \hline \mathbf{Y}_c \end{bmatrix} = \mathbf{0}, \tag{81}$$

$$[B' \mid I] \begin{bmatrix} \mathbf{X}_b \\ \hline \mathbf{X}_c \end{bmatrix} = \mathbf{0}, \tag{82}$$

or

$$\mathbf{X}_c = -B'\mathbf{X}_b\,, \tag{83}$$

$$\mathbf{Y}_b = -A'\mathbf{Y}_c\,. \tag{84}$$

Notice that both (83) and (84) express stimuli as functions of responses, as required for orientation compatibility.

The constraints for the example in Fig. 13 are

$$\left[\begin{array}{cccccc|ccccc} 1 & 0 & 0 & 0 & 0 & 0 & -1 & 0 & 0 & 0 & 0 \\ 0 & 1 & 0 & 0 & 0 & 0 & 0 & 0 & 0 & +1 & 0 \\ 0 & 0 & 1 & 0 & 0 & 0 & 0 & 1 & 0 & 0 & 0 \\ 0 & 0 & 0 & 1 & 0 & 0 & 0 & -1 & 1 & 0 & 0 \\ 0 & 0 & 0 & 0 & 1 & 0 & 0 & 0 & 0 & 0 & 1 \\ 0 & 0 & 0 & 0 & 0 & 1 & 0 & 0 & 0 & -1 & 0 \end{array}\right] \begin{bmatrix} y_1 \\ y_3 \\ y_5 \\ y_6 \\ y_9 \\ y_{11} \\ \hline y_2 \\ y_4 \\ y_7 \\ y_8 \\ y_{10} \end{bmatrix} = \mathbf{0}, \tag{85}$$

and

$$\left[\begin{array}{cccccc|ccccc} 1 & 0 & 0 & 0 & 0 & 0 & 1 & 0 & 0 & 0 & 0 \\ 0 & 0 & -1 & 1 & 0 & 0 & 0 & 1 & 0 & 0 & 0 \\ 0 & 0 & 0 & -1 & 0 & 0 & 0 & 0 & 1 & 0 & 0 \\ 0 & -1 & 0 & 0 & 0 & +1 & 0 & 0 & 0 & 1 & 0 \\ 0 & 0 & 0 & 0 & -1 & 0 & 0 & 0 & 0 & 0 & 1 \end{array}\right] \begin{bmatrix} x_1 \\ x_3 \\ x_5 \\ x_6 \\ x_9 \\ x_{11} \\ \hdashline x_2 \\ x_4 \\ x_7 \\ x_8 \\ x_{10} \end{bmatrix} = \mathbf{0}. \quad (86)$$

Notice that row 1 of (85) is derived from (75) and row 1 of (86) is derived from (76).

One of the most important theorems dealing with energy systems is a result of the fact that every cut set vector $\boldsymbol{\alpha}$ is orthogonal to every circuit vector $\boldsymbol{\beta}$, so that

$$\boldsymbol{\alpha}^{\mathsf{T}}\boldsymbol{\beta} = \boldsymbol{\beta}^{\mathsf{T}}\boldsymbol{\alpha} = \mathbf{0}. \tag{87}$$

This fact is a property of the system graph and the constraint equations, and has nothing to do with the nature of the components. Since the rows of A and B are cut set and circuit vectors, respectively, it follows from (87) that

$$A^{\mathsf{T}}B = B^{\mathsf{T}}A = \mathbf{0}, \tag{88}$$

or

$$\begin{aligned} [I \mid A']^{\mathsf{T}} [B' \mid I] &= \mathbf{0} \qquad (89) \\ &= \begin{bmatrix} I \\ \hdashline A'^{\mathsf{T}} \end{bmatrix} [B' \mid I] = \mathbf{0} \\ &= B' + A'^{\mathsf{T}} = B'^{\mathsf{T}} + A' = \mathbf{0}. \end{aligned}$$

This implies that

$$B' = -A'^{\mathsf{T}}. \tag{90}$$

That Eq. (90) holds in the example can be verified by comparing B' and A' in Eqs. (85) and (86). This fact can considerably simplify the derivation of the constraint equations, since one set is the negative transpose of the other [see Eqs. (85) and (86)]. Thus only one set really needs to be calculated; the other can be written immediately. More importantly, however, it allows verification that the system obeys the law of conservation of energy. The total net energy input to the system is

$$E(t) = \sum x_i y_i = \mathbf{X}^\mathsf{T}\mathbf{Y}. \tag{91}$$

If the system obeys the law of energy conservation, Eq. (91) must be identically zero. We know that

$$\mathbf{X}^\mathsf{T}\mathbf{Y} = [\mathbf{X}_b{}^\mathsf{T} \mid \mathbf{X}_c{}^\mathsf{T}] \begin{bmatrix} \mathbf{Y}_b \\ \hline \mathbf{Y}_c \end{bmatrix}, \tag{92}$$

and that

$$\begin{aligned} \mathbf{X}_b &= -B'\mathbf{X}_c\,, \\ \mathbf{Y}_c &= -A'\mathbf{Y}_b\,, \end{aligned} \tag{93}$$

so that (92) simplifies to

$$\begin{aligned} \mathbf{X}_b{}^\mathsf{T}[I \mid -B'^\mathsf{T}] \begin{bmatrix} -A' \\ \hline I \end{bmatrix} \mathbf{Y}_c &= \mathbf{X}_b{}^\mathsf{T}(-B'^\mathsf{T} - A')\,\mathbf{Y}_c \\ &= \mathbf{X}_b{}^\mathsf{T}(0)\,\mathbf{Y}_c \\ &= \mathbf{0}. \end{aligned} \tag{94}$$

Thus, it is verified that the system satisfies the law of conservation of energy, and the method of verification is independent of the properties of the component objects.

D. The System Model

So far we have concentrated upon object descriptions and on deriving the constraint equations for the system. To generate a model of the entire system, the series of substitutions described in Section III.E is carried out. An example of this process for a compartment model has been presented in Appendix II. If the reader is interested, he can try his hand on the system shown in Fig. 13. Use the following free-body models:

Objects 1 and 2 (transformation):

$$\begin{bmatrix} y_2 \\ y_4 \end{bmatrix} = \begin{bmatrix} k_{23} \\ k_{43} \end{bmatrix} y_3\,, \qquad x_3 = -[k_{23}\; k_{43}] \begin{bmatrix} x_2 \\ x_4 \end{bmatrix} + f_1(y_3), \tag{95}$$

$$\begin{bmatrix} y_7 \\ y_8 \end{bmatrix} = \begin{bmatrix} k_{79} \\ k_{89} \end{bmatrix} y_9\,, \qquad x_9 = -[k_{79}\; k_{89}] \begin{bmatrix} x_7 \\ x_8 \end{bmatrix} + f_2(y_9). \tag{96}$$

Object 3 (storage):

$$d\psi_3/dt = y_6\,, \tag{97}$$

$$x_6 = f_3(\psi_3). \tag{98}$$

Objects 4 and 5 (transport):

$$x_5 = f_4(y_5), \tag{99}$$

$$x_6 = f_5(y_{11}). \tag{100}$$

Notice that the stimulus–response orientation is defined by the tree in Fig. 13. Notice also that x_1 and y_{10} are *responses* of the environment. Since a free-body model for the environment has not been specified, these will appear as stimulus-independent responses (so-called "forcing functions").

Using the constraint relations of Eqs. (73) and (74) or (85) and (86), the resultant model for the entire system is

$$d\psi_3/dt = -(k_{43}k_{89} + k_{79})\, y_{10}\,, \tag{101}$$

$$y_2 = (k_{23}k_{89})\, y_{10}\,, \tag{102}$$

$$x_9 = (k_{23}k_{89})\, x_1 - f_0(y_{10}) - g_0(\psi_3). \tag{103}$$

The functions $f_0(y_{10})$ and $g_0(\psi_3)$ are

$$\begin{aligned} f_0(y_{10}) = {} & (k_{89}k_{43})\, f_4(-k_{43}k_{89}y_{10}) + k_{89}f_5(-k_{89}y_{10}) \\ & -k_{89}f_1(k_{89}y_{10}) + f_2(-y_{10}), \end{aligned} \tag{104}$$

$$g_0(\psi_3) = (k_{89}k_{43} - k_{79})\, f_3(\psi_3). \tag{105}$$

Equations (101), (102), and (103) express states (ψ_3) and responses (y_2, x_9) of the system as functions of stimuli to the system (y_{10} and x_1). All the internal behavioral features have been submerged.

We should introduce a caveat at this point; the methodology developed for handling electrical networks has some very special properties. These networks are usually composed of a large number of simple components in a complex pattern of interconnection–hence the need for simple ways of arriving at compatible orientations and constraint equations. The situation in ecological studies has not yet reached this point, since we are far from being able to take object models for granted. There are also indications that typical ecological situations will involve a small number of complex components and the constraints may often be obvious without explicit use of the system graph. Ecologists should look to established theory in the physical sciences as a starting point to be modified as required for ecological problems, not as a panacea. One of the additions that will be required by some ecological problems is the subject of the next section.

VI. Some Thoughts on Ecological Systems

The preceding discussions of material transformation, transport, and storage processes have obvious applicability since the exchange of material and the associated energy costs are a very real part of any ecological system. Equally important in many cases is a factor that is not taken into account by these concepts, the flow of information within the system. The processes of feeding and growth of an individual, for example, certainly involve the transformation, transport, and storage of material. As Hubbell (1971) points out, however, the material flow rate is definitely not a function of the material density alone, but involves sensory perception, behavior, choice, strategies, etc. The flow of information in such cases exerts a great deal of control on the flow of material.

It may be that it is neither possible nor desirable to separate completely the processing of materials and information. In such cases it may be useful to subdivide the behavioral features and states of a given object, as in Fig. 15a. The material processing section is described by a set of intensive and extensive variables x, and y, and a state $\boldsymbol{\Psi}_{\mathrm{M}}$. The information processing section is defined by a set of behavioral features b_i and a state $\boldsymbol{\Psi}_{\mathrm{I}}$. Using $\mathbf{E}$ and $\mathbf{R}$ to represent stimulus and response sets (with subscripts to denote the system in question) a model for the object can be written as follows:

$$d\boldsymbol{\Psi}_{\mathrm{M}}/dt = F_{\mathrm{M}}(\boldsymbol{\Psi}_{\mathrm{M}}, \mathbf{E}_{\mathrm{M}}, \mathbf{R}_{\mathrm{I}}, t), \tag{106}$$

$$\mathbf{R}_{\mathrm{M}} = G_{\mathrm{M}}(\boldsymbol{\Psi}_{\mathrm{M}}, \mathbf{E}_{\mathrm{M}}, \mathbf{R}_{\mathrm{I}}, t); \tag{107}$$

$$d\boldsymbol{\Psi}_{\mathrm{I}}/dt = F_{\mathrm{I}}(\boldsymbol{\Psi}_{\mathrm{I}}, \mathbf{E}_{\mathrm{I}}, \mathbf{R}_{\mathrm{M}}, t), \tag{108}$$

$$\mathbf{R}_{\mathrm{I}} = G_{\mathrm{I}}(\boldsymbol{\Psi}_{\mathrm{I}}, \mathbf{E}_{\mathrm{I}}, \mathbf{R}_{\mathrm{M}}, t). \tag{109}$$

The behavior of the energy sector, Eqs. (106) and (107), is influenced by the response of the information sector. This sector, Eqs. (108) and (109), is in turn influenced by the response of the energy sector. The feedback nature of the interaction can be seen in the block diagram symbolism of Fig. 15b. This structure is that of a classical *control system* problem. The energy system is the "plant" or "controlled system," and the information system is the "controller." The laws dictating the impact of the control system on the energy system are referred to as the *strategy* of the control system. In typical engineering applications the controller is designed to realize a preassigned behavior of the energy system, to elicit a certain response or state, or perhaps to bound it away from certain areas of the state space.

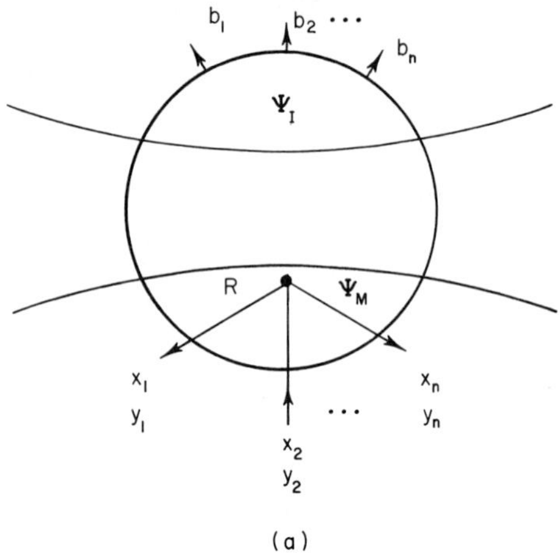

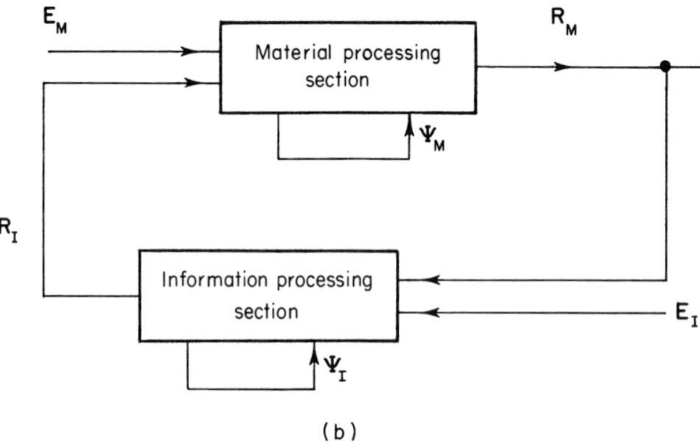

FIG. 15. (a) Ecological component with energy and information processing sections separated. (b) The same object recast in terms of control theory.

This abstraction opens the way for application of control theoretic ideas, both from the systems engineering literature (Appendix I) and from cybernetics (Wiener, 1948; Ashby, 1956), to problems of material and energy flow in ecological systems. The warning of the preceding section, against attempting to force ecological problems into engineering techniques bears repeating here. There is no reason to expect that the methodology developed to handle manmade electrical and mechanical systems will carry over directly to any ecological problem. The basic concepts, however, will certainly provide useful points of departure in developing the tools required by the characteristics of ecological systems.

As an example of the simplification generated by separating, as it were, the "body" and the "mind" of the system, consider the problem of modeling the biomass dynamics of an individual predator. Clearly this is a case in which the concepts of energy systems are applicable; variables of concern are the flow rates and energy costs of the biomass ingested, respired, excreted, and accumulated. The idea of attempting to embed the functional response model of Holling (1965) within the behavioral description of the energy processing system boggles the mind. Yet such a function is needed to generate a decision at each point in time to eat or not to eat. If this problem is attacked within the confines of the energy system framework, the only choice is to include this complex decision process in the state and response equations of the energy system, perhaps as time varying parameters or strange non-linearities. If, on the other hand, the functional response is added to the energy system as a control process, things begin to look more promising. The control system collects information from the material processing system (e.g., material in gut) and from the environment (e.g., density of available food, time of day), processes this information in connection with the state of the control system (e.g., mode of activity) and generates a decision, to eat or not to eat. This decision is transmitted to the energy system and is processed in connection with that system's state and stimuli.

The separation of energy and information processing subsystems has potential application at many levels of organization. Biological communities, for example, may be divided into an energy processing sector (plants, most herbivores, decomposers) and a sector which processes only a small amount of energy (predators, parasites, seed predators, pathogens). This latter group has an impact on the system entirely out of proportion to its role in the processing of energy. Thus, it can be described as a control process acting on the energy process, providing an alternative to the trophic level concept as an expression of community structure.

There is at this point no way to tell how useful these control theoretic concepts will be. Current researches are exploring the problems mentioned above, but firm results have not been obtained. It is clear, however, that these concepts are useful in conceptually structuring systems and planning relevant experimentation. In hopes of speeding the development of these concepts we have included some suggested readings in control theory in Appendix I.

VII. Beyond Modeling

We have presented an introduction to systems science which assumes as a goal the construction of a behavioral description of the system. There are other aspects to systems science that go beyond this construction (see Fig. 1), and have a profound bearing on the future of systems ecology.

The modeling process is essentially a sequential structuring process, which defines a system, objects, behavioral features, orientation, state variables, free-body models, constraints, and, finally, the system once again. This involves continual experimentation with the system and comparison of the qualities of the model with those of the system. This process is very much goal specific, but if the development of the physical system sciences is any clue, there is no need to repeat the entire process for every problem to be investigated. As the field develops, clusters of problems will be found which can be investigated with a particular set of objects and behavioral features. These clusters of important questions and their associated objects and behavioral features begin to provide a common vocabulary for use by anyone in the field. An observer confronted with a system can also be relieved of some of the work of describing it if he can merely parameterize an extant model of some of its objects, rather than deriving these models anew. (Note that this absolutely requires that the models be developed in free-body form.) This represents a significant step toward the development of a theory of the class of systems in question.

This suggests a new activity for systems ecologists, something that might be called descriptive dynamic ecology. If done properly, this will begin to define the clusters of objects useful in particular questions. Essentially it will involve approaching ecological systems (particularly communities and ecosystems, for which the lack of component information is acute) with the very generalized goal of obtaining a structured system model which gives a description of the dynamic behavior of the system. By finding recurrent patterns in these behaviors and the objects in terms of which they can be described, some progress toward valid clusters of objects can be made. This is analagous to the way many parts of current ecological theory were generated by the observation of recurrent patterns in static descriptive ecology. Unless the dynamic descriptions are done in properly structured form, they will be little more than records of the behavior generated by a specific environment and system topology.

It is worth mentioning in passing the controversy between simulation and analysis of ecological systems that is reflected in the title of this volume. It is beginning to be realized that computer simulation studies can be performed that provide little more than a tabulation of a particular behavior sequence of the system. Patten (personal communication) has commented that the traditional data-gathering power of ecology is sometimes wedded to large computers with little or no intervening structuring of a model. The defects of these studies are not a result of the use of simulation as opposed to analysis but rather of the lack of an internal structure to the model. Given a system model developed in the fashion laid out in this paper, a computer simulation is merely one of the methods available for examining behavior sequences. When they are possible, analytical investigations can yield very powerful results. In ecological systems, however, the assumptions necessary to achieve an analytically tractable form may be so gross that the powerful results of the analysis have little relevance to the real world. When the model will not yield to analytical solution, many techniques are available to the investigator—analog, hybrid, and digital simulation; numerical solution; simulation languages; approximate solutions; graphical analysis; etc. The optimum approach may be a combination of these techniques. Starting with a complex nonlinear model, the investigator could perform a linearization around equilibrium points and subject this approximation to analysis. Then several different simplifications of the nonlinear model could be studied by simulation. Comparison of these results would give an idea of the sensitivity of the system to the addition or removal of particular nonlinear or time-varying terms, the effect of including stochastic elements, etc. The important point is that only if the system is developed in a clearly structured form are such investigations able to separate the effects of changes in the free-body models and in topology.

Thus far we have considered the problem of obtaining a systems model, and the resulting study of that model by analysis and/or simulation. Although it has not yet become apparent in the systems ecology literature, there is much more that can be done (Fig. 1), involving *control*, *design*, and *synthesis* of systems.

Control, design, and synthesis activities are all based on the concept of a "desired behavior" of the system (see Fig. 1). This desired behavior may be to realize a particular response series $R(t)$; to attain and hold a "target state" ψ; to follow a particular state trajectory $\psi(t)$; to bound the system away from an undesirable region of the state space, or any other specified response. Control, design, and synthesis represent three different methods of attaining this desired behavior.

Control theory attempts to attain the desired system behavior by manipulation of stimulus time series $\mathbf{E}(t)$. An important branch of control theory, called *optimal control theory*, seeks, out of all the stimulus series that result in the desired behavior, one that is in some sense optimal. This is usually accomplished by defining a scalar *performance index* or *cost function* $J(\mathbf{E}_i)$ which measures the cost associated with the use of stimulus $\mathbf{E}_i$ in achieving the desired behavior. The optimum stimulus is then that which minimizes J. The cost function may be structured, for example, to minimize the time required to move the system from one state to another, or the energy cost of achieving the desired behavior.

In contrast to control theory, design and synthesis deal with the creation of essentially new systems to achieve the desired behavior (see Fig. 1). Although there is no sharp distinction between the two activities, design usually focuses on modification of the components within a specified topology. For example, the problem of determining the size of resistors and capacitors and the type of transistors to use in a given amplifier circuit is referred to as a design problem. The basic topology of the system is fixed, although the design concept may be extended to relatively minor modifications of circuit configuration. Synthesis usually (but not always) refers to the more general and difficult task of building the system from scratch by selecting both the components and topology. There is as yet no unified conceptualization for design and synthesis comparable to that presented for modeling. A few specific techniques are available for problems in some classes of physical systems, but the general technique is repeated analysis coupled with a liberal dose of craftmanship.

The application of design, control, and synthesis has been far more rewarding in the physical systems sciences than the mere description of the behavior of extant systems. The technological advances made by these fields have been due in large measure to the ability to identify and build systems that behave in a specified manner. Again, the need for structuring object models in free-body form is evident; description of objects valid only in a particular system topology are useless to design and synthesis procedures.

The most challenging ecological application of control, design, and synthesis concepts is the management of our total ecosystem. Fig. 16 (from Koenig *et al.*, 1971) presents an overview from a very high level of organization of the "industrialized ecosystem" in which we live.

The industrial sector has long been the subject of control, synthesis and design activities, resulting in highly productive behaviors. It is less widely appreciated that the agricultural sector is similarly the object of design and synthesis activities. Plant and animal breeding for increased yield is clearly a design activity, changing the behavior of the objects

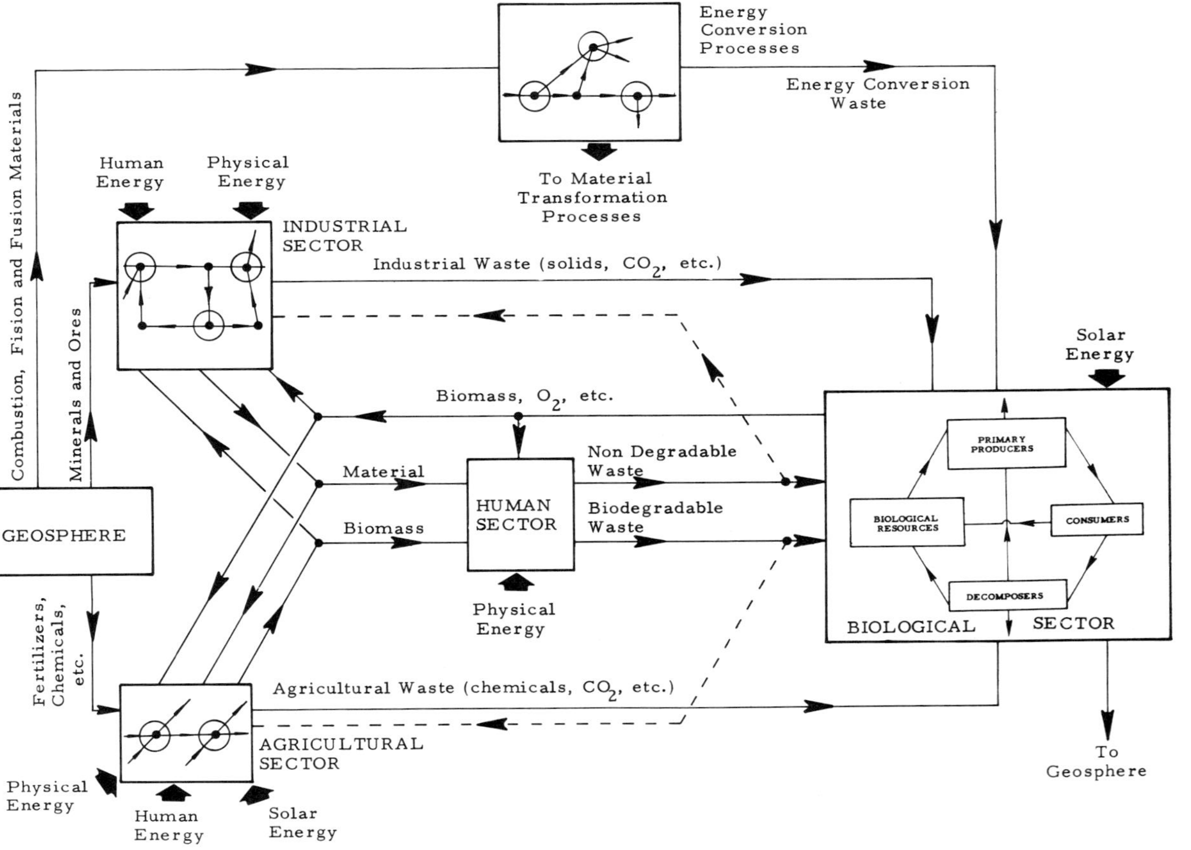

FIG. 16. The material processing sector of an industrialized ecosystem.

already within the system. The introduction of machinery powered by fossil fuels has synthesized a system with new objects and new topologies (e.g., large-scale monoculture) (see, e.g., Odum, 1969; Odum, 1971). To date these design and synthesis activities have been directed at maximizing the input–output efficiency of the objects. Because the interaction topology has been ignored, the emergent property known as pollution is now a serious problem at the next higher level of organization.

Most attempts to deal with this problem have taken the philosophy of control theory, by attempting to manipulate the stimuli impinging on an object (e.g., limiting the flow of biodegradable waste into a lake to avoid eutrophication). Such measures cannot succeed in the long run unless they are coupled with design and synthesis procedures at the ecosystem rather than the component level. The idea of designing and synthesizing ecosystem structures is repugnant to some ecologists whose first interest is (and perhaps rightly so) to better understand the natural functioning of ecosystems. Unfortunately, there is very little choice since the design and synthesis activities are in full swing within the various components of the industrialized ecosystem.

It is unlikely that the level of success in the control, design, and synthesis achieved in the physical sciences will ever be reached with ecological systems; the physical systems scientists have too much of an advantage in their ability to build components with tractable behaviors. But if we discard for a moment the methodology and accept the basic ideas involved in achieving a desired behavior by control, design, or synthesis, two sobering facts emerge. We are far from having even nonquantitative free-body models for the components involved; in many cases we haven't even identified the components. Even more alarming is the fact that there is no agreement, in our society as a whole or within the scientific community, as to what constitutes desirable behavior at this level. Deciding on these desirable behaviors may be the most crucial test faced by our society. An interesting parallel can be drawn with the result of the advances in understanding the biochemical nature of the gene in the early 1960's. There was, at that time, much speculation in the popular press about the possibility of manipulating the genetic makeup of individuals. The question of concern was in what image mankind would structure itself given the opportunity. The question was never resolved, because we still have no means for performing such manipulations. We are now confronted with the same problem at the other end of the spectrum of levels of organization. We have, by sheer weight of numbers and technology, the ability to affect most, if not all, of the components of the biosphere.

This realization is only beginning to reach our society. We are faced with a serious challenge; we have both the power and the need to restructure the ecosystems of which we are a part. If such challenges form turning points in the development of a civilization, as Toynbee claims, the creativity of the response of our society as a whole to this challenge will have much to say about our future.

Acknowledgments

The research leading to this paper has been supported by an NSF grant (GI-20): "Design and Management of Environmental Systems." During the development of these ideas we have all benefited from discussion with and comments from the other participants in the program. Much of the abstract material in Sections II and III is condensed from Resh (1970). A complete version of this material will be submitted shortly for publication in book form. Those who have been stimulated by our treatment of systems science and wish to investigate some of these matters in more detail and with more rigor will undoubtedly find this book interesting.

Appendix I. Suggested Readings

We have made no attempt in this chapter to provide the reader with the tools needed to attack any specific system problem. These readings are suggested entry points into the wealth of such information in the physical systems sciences.

The list is not intended to be exhaustive. We have tried to give several references in each category. Our advice is to shop around before choosing a reference on any particular topic. Pick one that seems to be helpful for the task at hand.

Zadeh, L. A., and Desoer, C. A. (1963). "Linear System Theory." McGraw-Hill, New York.

Zadeh, L. A., and Polak, E., eds. (1969). "System Theory." McGraw-Hill, New York.

> The first of these references contains a complete and rigorous treatment of the theory of state. Zadeh's paper in the second volume gives a different treatment of the same topic. The other papers in the second work cover a variety of topics in system theory. For a treatment of state-space system theory very similar to but more thorough than that in this paper, see the following.

Resh, J. A. "The Systems Approach to Behavior." In preparation.

> The following works all deal with state space system theory from an analytical rather than theoretical point of view. They all treat the relation between state space and transfer function approaches and the analysis of systems of differential equations.

Ward, J. R., and Strum, R. D. (1970). "State Variable Analysis: A Programmed Text." Prentice-Hall, Englewood Cliffs, New Jersey (Has the advantage of being self-teaching, but does not go as far as the others.)

Koenig, H. E., Tokad, Y., and Kesavan, H. K. (1967). "Analysis of Discrete Physical Systems." McGraw-Hill, New York. (Devoted to energy systems, emphasis on graph theoretic material and resulting analysis.)

DeRusso, P. M., Roy, R. J., and Close, C. M. (1966). "State Variables for Engineers." Wiley, New York.

Martens, H. D., and Allen, D. R. (1969). "Introduction to Systems Theory." C. E. Merrill,

Perkins, W. R., and Cruz, J. B. (1969). "Engineering of Dynamic Systems." Wiley, New York. (Also covers topics in simulation and control.)

Timothy, L. K., and Bona, B. E. (1968). "State Space Analysis: An Introduction." McGraw-Hill, New York.

The topic of control theory can be divided into classical control theory (using frequency domain transfer function methods) and modern control theory (using time domain state space methods). An easy introduction to the classical techniques is:

DiStefano, J. J., Stubberud, A. R., and Williams, I. J. (1967). "Feedback and Control Systems." Schaum's Outline Series, New York.

A treatment that covers both topics can be found in:

Ogata, K. (1970). "Modern Control Engineering." Prentice-Hall, Englewood Cliffs, New Jersey.

Good coverage of state space approaches and optimal control problems are:

Ogata, K. (1967). "State Space Analysis of Control Systems." Prentice-Hall, Englewood Cliffs, New Jersey. (Incidentally, this includes a good treatment of state space analysis techniques in general.)

Lee, E. B., and Markus, L. (1967). "Foundations of Optimal Control Theory." Wiley, New York.

Differential equations and matrix algebra are the most often used analytical techniques in system science. There are many treatments of the former; perhaps the best advice we can give is to look for one that seems well-matched with both the systems material you are using and your own level of mathematical sophistication. One work that should be singled out is:

Rosen, R. (1970). "Dynamical System Theory in Biology I: Stability Theory." Wiley (Interscience), New York. (Although its main emphasis is on stability theory rather than on all of differential equations, it is full of biological examples.)

Applied matrix algebra is harder to find in the literature. Fortunately, most of the texts on state space analysis listed above treat the most immediately necessary items. Searle's book (below) is devoted more to algebraic than differential equation systems, but provides an introduction to the basic manipulations of matrices.

Searle, S. R. (1966). "Matrix Algebra for the Biological Sciences." Wiley, New York.

An applied treatment that covers the right topics, but cannot be said to be truly introductory, is in preparation:

Frame, J. S. "Matrix Theory and Linear Algebra with Applications." Submitted for publication.

Appendix II. A Compartment Model

Although compartment models were used in ecological systems at least as long ago as 1935 (Kostitzin, 1935), until recently their main use was in tracer studies of physiological systems (Hart, 1963; Atkins, 1969). They are now perhaps the most widely used type of model in ecology, particularly at the ecosystem level (e.g., Smith, 1970). We will not go into details of the analytical methods that can be applied to compartment models (physiological studies seem more sophisticated than ecological ones here, see Hart, 1963). Instead we will derive the most common and simplest of these models as an example of the energy system methodology in Section V. In the interests of clarity we have been as complete as possible, leaving out a minimum of intermediate steps.

A. Free-Body Models

The system in question is shown in Fig. 17a, consisting of three "compartments" each coupled to all the others and to the environment. An object and its behavioral features are shown in Fig. 17b. The flow y_{ij} represents the flow of material from the environment into object j and y_{0j} the flow of material out of object j to the environment and to the other objects. Flows y_{kj}, $k = 1, 2, 3$, represent flows from compartment k, including that from the object in question itself. The intensive variables x_{kj} represent the density of the material in flow y_{kj}. This might, for example, be directly proportional to the spatial concentration of material of type k; the higher this concentration, the more energy is required to bring a unit of material to its position. A state variable S_j, $j = 1, 2, 3$, will be associated with each object representing the accumulation of material within the component. The environment "pumps" a flow of material y_{aj} into each compartment, and harvests a flow y_{ej}.

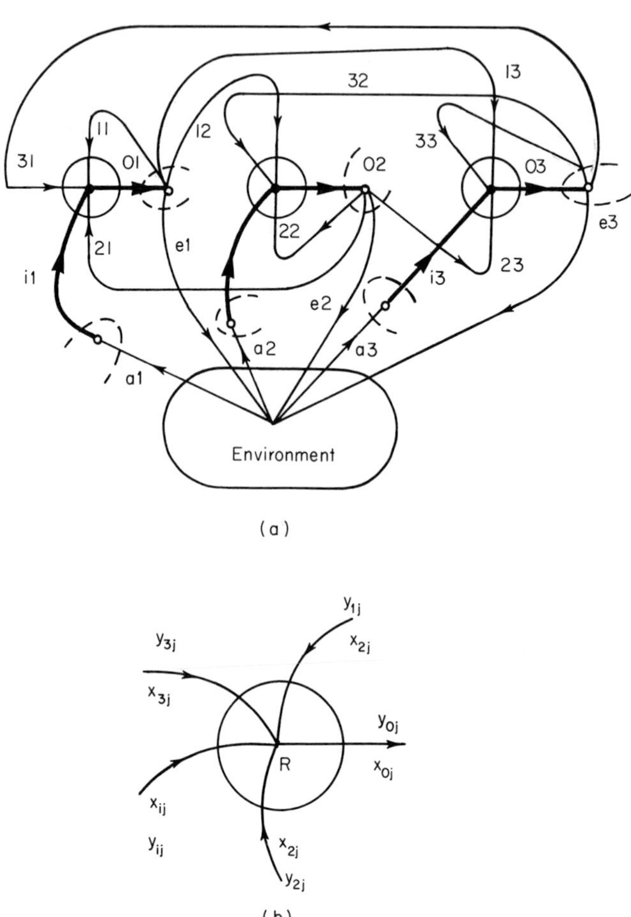

FIG. 17. (a) System of three "compartments." (b) The basic "compartment" object.

The system graph formed from the object graph is shown in Fig. 17a. The double width edges form a tree (try adding any other edge and, remembering that all reference points R are the same, note that this forms a circuit). This tree will be used to assign a stimulus–response orientation to the objects in the system. The extensive variables associated with the tree and the intensive variables associated with the cotree will be taken as stimuli, the complementary set of variables as responses. This orientation is compatible since it results in the presence of one stimulus extensive variable and one response intensive variable at each point of interaction.

For purposes of illustration we will adopt a set of free-body models that result in a "linear donor-controlled" model of the system, a form which appears often in the ecological literature (e.g., Patten and Witkamp, 1967). If we assume that the ingestion of material into any component from the other components is directly proportional to the density of the material in the donor component we have the free-body models (for $j = 1, 2, 3$):

state equation:

$$\frac{d}{dt} S_j = [k_{1j}\ k_{2j}\ k_{3j}] \begin{bmatrix} x_{1j} \\ x_{2j} \\ x_{3j} \end{bmatrix} + y_{ij} - y_{0j}\ ; \tag{110}$$

response equation:

$$\begin{bmatrix} y_{1j} \\ y_{2j} \\ y_{3j} \end{bmatrix} = \begin{bmatrix} k_{1j} & 0 & 0 \\ 0 & k_{2j} & 0 \\ 0 & 0 & k_{3j} \end{bmatrix} \begin{bmatrix} x_{1j} \\ x_{2j} \\ x_{3j} \end{bmatrix}. \tag{111}$$

Equations (110) and (111) are undoubtedly an oversimplification of the behavior of ecological components, assuming as they do that each object passively ingests material forced on it by the concentration of material in the other objects. A point worth noting, however, is that the expression of the model in free-body form helps make these assumptions clear at the beginning of the modeling process.

In the chosen orientation the environment in Fig. 17a has as response variables the flows y_{ej} and y_{aj}, $j = 1, 2, 3$. We will take these to be stimulus independent time functions $f_j(t)$ and $g_j(t)$, respectively. This gives us a model of the environment,

$$\begin{aligned} y_{ej} &= f_j(t), \\ y_{aj} &= g_j(t), \end{aligned} \qquad j = 1, 2, 3, \tag{112}$$

in which the stimuli x_{ej} and x_{aj} do not appear.

B. Constraint Equations

Six fundamental cut sets are indicated by broken lines in Fig. 17a. These generate the equations

$$\begin{aligned} y_{01} &= y_{11} + y_{12} + y_{13} + y_{e1}, \\ y_{02} &= y_{21} + y_{22} + y_{23} + y_{e2}, \\ y_{03} &= y_{31} + y_{32} + y_{33} + y_{e3}, \\ y_{i1} &= y_{a1}, \\ y_{i2} &= y_{a2}, \\ y_{i3} &= y_{a3}. \end{aligned} \tag{113}$$

The nine fundamental circuits in the graph yield the constraint equations for the intensive variables

$$\begin{aligned} x_{11} &= -x_{01}, & x_{12} &= -x_{01}, & x_{13} &= -x_{01}, \\ x_{21} &= -x_{02}, & x_{22} &= -x_{02}, & x_{23} &= -x_{02}, \\ x_{31} &= -x_{03}, & x_{32} &= -x_{03}, & x_{33} &= -x_{03}. \end{aligned} \tag{114}$$

Note that Eqs. (113) and (114) both express stimuli as functions of responses.

C. The System Model

We will now go through a series of substitutions designed to combine the information in Eqs. (110)–(114) into a model of the entire system. First let us write noninteractive state (115) and response (116) equations for the system:

$$\frac{d}{dt}\begin{bmatrix} S_1 \\ S_2 \\ S_3 \end{bmatrix} = \begin{bmatrix} k_{11} & k_{21} & k_{31} & 0 & 0 & 0 & 0 & 0 & 0 \\ 0 & 0 & 0 & k_{12} & k_{22} & k_{32} & 0 & 0 & 0 \\ 0 & 0 & 0 & 0 & 0 & 0 & k_{13} & k_{23} & k_{33} \end{bmatrix} \begin{bmatrix} x_{11} \\ x_{21} \\ x_{31} \\ x_{12} \\ x_{22} \\ x_{32} \\ x_{13} \\ x_{23} \\ x_{33} \end{bmatrix} + \begin{bmatrix} y_{i1} \\ y_{i2} \\ y_{i3} \end{bmatrix} - \begin{bmatrix} y_{01} \\ y_{02} \\ y_{03} \end{bmatrix}; \tag{115}$$

$$\begin{bmatrix} y_{11} \\ y_{21} \\ y_{31} \\ y_{12} \\ y_{22} \\ y_{32} \\ y_{13} \\ y_{23} \\ y_{33} \end{bmatrix} = \begin{bmatrix} k_{11} & & & & & & & & \\ & k_{21} & & & & & & 0 & \\ & & k_{31} & & & & & & \\ & & & k_{12} & & & & & \\ & & & & k_{22} & & & & \\ & & & & & k_{32} & & & \\ & & & & & & k_{13} & & \\ & 0 & & & & & & k_{23} & \\ & & & & & & & & k_{33} \end{bmatrix} \begin{bmatrix} x_{11} \\ x_{21} \\ x_{31} \\ x_{12} \\ x_{22} \\ x_{32} \\ x_{13} \\ x_{23} \\ x_{33} \end{bmatrix}, \tag{116}$$

$$\begin{bmatrix} x_{01} \\ x_{02} \\ x_{03} \end{bmatrix} = \begin{bmatrix} S_1 \\ S_2 \\ S_3 \end{bmatrix}, \qquad \begin{bmatrix} y_{e1} \\ y_{e2} \\ y_{e3} \end{bmatrix} = \begin{bmatrix} f_1(t) \\ f_2(t) \\ f_3(t) \end{bmatrix}, \qquad \begin{bmatrix} y_{a1} \\ y_{a2} \\ y_{a3} \end{bmatrix} = \begin{bmatrix} g_1(t) \\ g_2(t) \\ g_3(t) \end{bmatrix}.$$

We now use information from the constraint equations (113) and (114) to replace the stimulus variables appearing on the right-hand sides of (115) and (116) with functions of response variables. This will involve the following three substitutions:

$$\begin{bmatrix} x_{11} \\ x_{21} \\ x_{31} \\ x_{12} \\ x_{22} \\ x_{32} \\ x_{13} \\ x_{23} \\ x_{33} \end{bmatrix} = \begin{bmatrix} -1 & 0 & 0 \\ 0 & -1 & 0 \\ 0 & 0 & -1 \\ -1 & 0 & 0 \\ 0 & -1 & 0 \\ 0 & 0 & -1 \\ -1 & 0 & 0 \\ 0 & -1 & 0 \\ 0 & 0 & -1 \end{bmatrix} \begin{bmatrix} x_{01} \\ x_{02} \\ x_{03} \end{bmatrix}, \tag{117}$$

$$\begin{bmatrix} y_{01} \\ y_{02} \\ y_{03} \end{bmatrix} = \begin{bmatrix} 1 & 0 & 0 & 1 & 0 & 0 & 1 & 0 & 0 \\ 0 & 1 & 0 & 0 & 1 & 0 & 0 & 1 & 0 \\ 0 & 0 & 1 & 0 & 0 & 1 & 0 & 0 & 1 \end{bmatrix} \begin{bmatrix} y_{11} \\ y_{21} \\ y_{31} \\ y_{12} \\ y_{22} \\ y_{32} \\ y_{13} \\ y_{23} \\ y_{3} \end{bmatrix} + \begin{bmatrix} y_{e1} \\ y_{e2} \\ y_{e3} \end{bmatrix}, \tag{118}$$

and

$$\begin{bmatrix} y_{i1} \\ y_{i2} \\ y_{i3} \end{bmatrix} = \begin{bmatrix} y_{a1} \\ y_{a2} \\ y_{a3} \end{bmatrix}. \tag{119}$$

Substituting the right-hand sides of (117) through (119) wherever their left-hand sides appear in (115) and (116) gives the state and response equations (120) and (121). Note that in both these equations only response variables appear on the right-hand sides.

To complete our development of the state equations for the entire system we must solve Eq. (121) to obtain the response variables as functions only of the states and time. This is simply done by substitution of (121c) into (121a). Substituting the resultant expressions for the response variables into the right-hand side of (120), we arrive at (122), which simplifies immediately to (123). This latter equation is the form in which compartment models are customarily presented in the ecological literature:

$$\frac{d}{dt}\begin{bmatrix} S_1 \\ S_2 \\ S_3 \end{bmatrix} = \begin{bmatrix} k_{11} & k_{21} & k_{31} \\ k_{12} & k_{22} & k_{32} \\ k_{13} & k_{23} & k_{33} \end{bmatrix}\begin{bmatrix} x_{01} \\ x_{02} \\ x_{03} \end{bmatrix} - \begin{bmatrix} 1 & 0 & 0 & 1 & 0 & 0 & 1 & 0 & 0 \\ 0 & 1 & 0 & 0 & 1 & 0 & 0 & 1 & 0 \\ 0 & 0 & 1 & 0 & 0 & 1 & 0 & 0 & 1 \end{bmatrix}\begin{bmatrix} y_{11} \\ y_{21} \\ y_{31} \\ y_{12} \\ y_{22} \\ y_{32} \\ y_{13} \\ y_{23} \\ y_{33} \end{bmatrix}$$

$$+\begin{bmatrix} y_{a1} \\ y_{a2} \\ y_{a3} \end{bmatrix} - \begin{bmatrix} y_{e1} \\ y_{e2} \\ y_{e3} \end{bmatrix}; \tag{120}$$

$$\begin{bmatrix} y_{11} \\ y_{21} \\ y_{31} \\ y_{12} \\ y_{22} \\ y_{32} \\ y_{13} \\ y_{23} \\ y_{33} \end{bmatrix} = \begin{bmatrix} k_{11} & 0 & 0 \\ 0 & k_{21} & 0 \\ 0 & 0 & k_{31} \\ k_{12} & 0 & 0 \\ 0 & k_{22} & 0 \\ 0 & 0 & k_{32} \\ k_{13} & 0 & 0 \\ 0 & k_{23} & 0 \\ 0 & & k_{33} \end{bmatrix}\begin{bmatrix} x_{01} \\ x_{02} \\ x_{03} \end{bmatrix}, \tag{121a}$$

$$\begin{bmatrix} y_{a1} \\ y_{a2} \\ y_{a3} \\ y_{e1} \\ y_{e2} \\ y_{e3} \end{bmatrix} = \begin{bmatrix} g_1(t) \\ g_2(t) \\ g_3(t) \\ f_1(t) \\ f_2(t) \\ f_3(t) \end{bmatrix}, \tag{121b}$$

$$\begin{bmatrix} x_{01} \\ x_{02} \\ x_{03} \end{bmatrix} = \begin{bmatrix} 1 & 0 & 0 \\ 0 & 1 & 0 \\ 0 & 0 & 1 \end{bmatrix} \begin{bmatrix} S_1 \\ S_2 \\ S_3 \end{bmatrix}, \tag{121c}$$

$$\begin{bmatrix} x_{i1} \\ x_{i2} \\ x_{i3} \end{bmatrix} = \begin{bmatrix} 1 & 0 & 0 \\ 0 & 1 & 0 \\ 0 & 0 & 1 \end{bmatrix} \begin{bmatrix} S_1 \\ S_2 \\ S_3 \end{bmatrix}; \tag{121d}$$

$$\begin{aligned} \frac{d}{dt} \begin{bmatrix} S_1 \\ S_2 \\ S_3 \end{bmatrix} = & \begin{bmatrix} k_{11} & k_{21} & k_{31} \\ k_{12} & k_{22} & k_{32} \\ k_{13} & k_{23} & k_{33} \end{bmatrix} \begin{bmatrix} S_1 \\ S_2 \\ S_3 \end{bmatrix} \\ & - \begin{bmatrix} (k_{11} + k_{12} + k_{13}) & 0 & 0 \\ 0 & (k_{21} + k_{22} + k_{23}) & 0 \\ 0 & 0 & (k_{31} + k_{32} + k_{33}) \end{bmatrix} \begin{bmatrix} S_1 \\ S_2 \\ S_3 \end{bmatrix} \\ & + \begin{bmatrix} g_1(t) \\ g_2(t) \\ g_3(t) \end{bmatrix} - \begin{bmatrix} f_1(t) \\ f_2(t) \\ f_3(t) \end{bmatrix}; \end{aligned} \tag{122}$$

and

$$\begin{aligned} \frac{d}{dt} \begin{bmatrix} S_1 \\ S_2 \\ S_3 \end{bmatrix} = & \begin{bmatrix} -(k_{12} + k_{13}) & k_{21} & k_{31} \\ k_{12} & -(k_{21} + k_{23}) & k_{32} \\ k_{13} & k_{23} & -(k_{31} + k_{32}) \end{bmatrix} \begin{bmatrix} S_1 \\ S_2 \\ S_3 \end{bmatrix} \\ & + \begin{bmatrix} g_1(t) \\ g_2(t) \\ g_3(t) \end{bmatrix} - \begin{bmatrix} f_1(t) \\ f_2(t) \\ f_3(t) \end{bmatrix}. \end{aligned} \tag{123}$$

We have essentially reduced the system of Fig. 17 to the single object in Fig. 18. This object has a three-dimensional state variable **S** and stimuli y_{aj} and y_{ej} , $j = 1, 2, 3$. All the other internal dynamics have been suppressed, and the object in Fig. 17 is available for use in the modeling of other systems.

In the linear case considered here there is little to be gained from a simulation of the system; analysis of Eq. (123) reveals everything about the system's behavior.

Rewriting (123) as

$$\frac{d}{dt} \mathbf{S} = \mathbf{PS} + \mathbf{E}, \tag{124}$$

the solution is

$$\mathbf{S}(t) = \mathbf{S}(0) \exp[\mathbf{P}(t)] + \int_0^t \exp[\mathbf{P}(t - \tau)]\, \mathbf{E}(\tau)\, d\tau, \tag{125}$$

or, in the Laplace transform form,

$$\mathbf{S}(s) = (\mathbf{S}I - \mathbf{P})^{-1}\, \mathbf{E}(s). \tag{126}$$

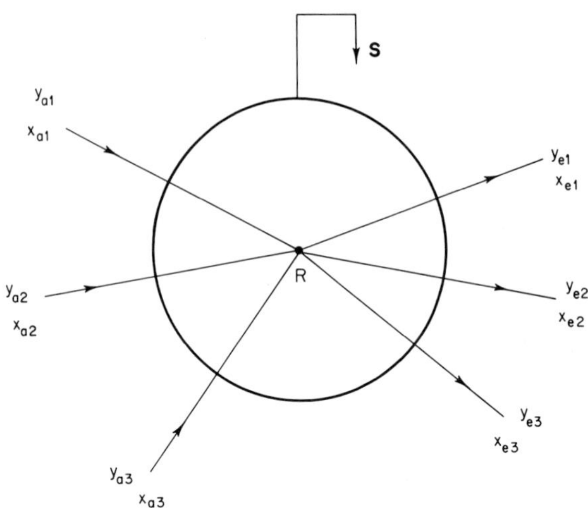

FIG. 18. Three "compartment" system as a single object.

For any specified set of parameters and $\mathbf{E}(t)$, Eq. (125) can be evaluated by standard numerical integration techniques.

REFERENCES

Ashby, W. R. (1956). "An Introduction to Cybernetics." Wiley, New York.

Ashford, J. R., Read, K. L. Q., and Vickers, G. G. (1970), *J. Anim. Ecol.* **39**, 29.

Atkins, G. L. (1969). "Multicompartment Models for Biological Systems." Methuen, London.

Bellman, R. (1968). "Some Vistas of Modern Mathematics." Univ. of Kentucky Press, Lexington, Kentucky.

Bernardelli, H. (1941). *J. Burma Res. Soc.* **31**, 1.

Brennan, R. D., and Linebarger, R. N. (1964). *Simulation* **3**, 22.

Botnariuc, N. (1966). *Yrbk. Soc. Gen. Syst. Res.* **11**, 93.

Brelot, M. (1931). *Ann. Math. Pura Appl. ser.* 4 **9**, 58.

Caswell, H. (1972). *J. Theoret. Biol.* **34**, 419.

Cunningham, W. J. (1954). *Proc. Nat. Acad. Sci. U. S.* **40**, 708.

DiStefano, J. J., Stubberud, A. R., and Williams, I. J. (1967). "Feedback and Control Systems." Schaum, New York.

Donnan, F. G. (1936). *Acta Biotheoret.* **2**, 1.

Donnan, F. G. (1937). *Acta Biotheoret.* **3**, 43.

Feeney, G. M. (1970). *Demography* **7**, 341.

Feiblemen, J. K. (1954). *Brit. J. Philos. of Sci.* **5**, 59.

Fredrickson, A. G., Ramkrishna, D., and Tsuchiya, H. M. (1967). *Math. Biosci.* **1**, 327.

Harper, J. L. (1968). *In* "Population Biology and Evolution" (R. Lewontin, ed.), pp. 139–158. Syracuse Univ. Press, Syracuse, New York.

Hart, H. E., ed. (1963). Multicompartment Analysis of Tracer Experiments, *Ann. N. Y. Acad. Sci.* **108**, 1.
Holling, C. S. (1965). *Mem. Entomol. Soc. Can.* **45**, 1.
Hubbell, S. (1971). *In* "Systems Analysis and Simulation in Ecology" (B. C. Patten, ed.), Vol. I, pp. 269–324. Academic Press, New York.
Hughes, R. D. (1962). *J. Anim. Ecol.* **31**, 389.
IBM (1969). System/360 Continuous System Modeling Program, User's Manual, IBM Appl. Program H20-0367-3. Int. Bus. Machines Corp., White Plains, New York.
Keyfitz, N. (1968). "Introduction to the Mathematics of Population." Addison-Wesley. Reading, Massachusetts.
Keyfitz, N., and Murphy, E. (1967). *Biometrics* **23**, 485.
Kilmer, W. L. (1968). *Ann. N. Y. Acad. Sci.* **161**, 416.
Kilmer, W. L., McCulloch, W. S., and Blum, J. (1969). *Int. J. Man-Machine Stud.* **1**, 274.
Kostitzin, V. A. (1935). "Evolution de l'Atmosphere." Herman, Paris.
Koenig, H. E. (1970). Paper presented at 1970 Joint Automatic Control Conference. Available from author.
Koenig, H. E., Tokad, Y., and Kesavan, H. (1967). "Analysis of Discrete Physical Systems." McGraw-Hill, New York.
Koenig, H. E., Hilmerson, A. M., and Yuan, L. (1969). *Trans. Amer. Soc. Agr. Eng.* **12**, 190.
Koenig, H. E., Cooper, W. E., and Falvey, J. (1971). Industrialized ecosystem design and management. Submitted for publication.
Lefkovich, L. P. (1965). *Biometrics* **21**, 1.
Leontief, W. (1970). *Rev. Econ. Statist.* **52**, 262.
Leslie, P. H. (1945). *Biometrika* **33**, 183.
Leslie, P. H. (1948). *Biometrika* **35**, 213.
Leslie, P. H. (1959). *Physiol. Zool.* **32**, 151.
Lewis, E. G. (1942). *Sankhya* **6**, 93.
Lopez, A. (1960). Problems in Stable Population Theory. Office of Population Res., Princeton Univ., Princeton, New Jersey.
Lotka, A. J. (1925). "Elements of Mathematical Biology." Reprinted 1956 by Dover, New York.
Martin, F. F. (1968). "Computer Modeling and Simulation." Wiley, New York.
Mayr, E. (1965). *In* "Cause and Effect" (D. Lerner, ed.), pp. 33–50. Free Press, New York.
Mertz, D. B. (1969). *Ecol. Monographs* **39**, 1.
Mesarovic, M. D., and Macko, D. (1969). *In* "Hierarchical Structures" (L. L. Whyte, A. G. Wilson, and D. Wilson, eds.), pp. 29–50. American Elsevier, New York.
Milsum, J. H. (1966). "Biological Control Systems Analysis." McGraw-Hill, New York.
Odum, E. P. (1969). *Science* **164**, 262.
Odum, H. T. (1971). "Environment, Power and Society." Wiley (Interscience), New York.
Ogata, K. (1967). "State-Space Analysis of Control Systems." Prentice–Hall, Englewood Cliffs, New Jersey.
Patten, B. C. (1971). *In* "Systems Analysis and Simulation in Ecology" (B. C. Patten, ed.), Vol. I, pp. 1–122. Academic Press, New York.
Patten, B. C., and Witkamp, M. (1967). *Ecology* **48**, 813.
Polanyi, M. (1968). *Science* **160**, 1308.
Rabotnov, T. A. (1969). *Vegetatio* **29**, 87.
Resh, J. A. (1967a). *Proc. 5th Allerton Conf. Circuit System Theory* **112**.
Resh, J. A. (1967b). *Midwest Symp. Circuit Theory, 5th* pp. V-3-1–V-3-11.

Resh, J. A. (1970). Complex Systems. Lecture Notes, Foundations of Systems Science, Michigan State University, mimeo.

Rogers, A. (1968). "Matrix Analysis of Population Growth and Distribution." Univ. of California Press, Berkeley.

Samuelson, P. A. (1965). *In* "Cause and Effect" (D. Lerner, ed.), pp. 99–144. Free Press, New York.

Sinko, J. W., and Streifer, W. (1967). *Ecology* **48**, 910.

Sinko, J. W., and Streifer, W. (1969). *Ecology* **50**, 608.

Sinko, J. W., and Streifer, W. (1971). *Ecology* **52**, 330.

Skellam, J. G. (1966). *Proc. 5th Berkeley Symp. Math. Statist. Probability* **4**, 179.

Slobodkin, L. B. (1953). *Ecology* **34**, 513.

Slobodkin, L. B. (1954). *Ecol. Monographs* **24**, 69.

Smith, F. E. (1952). *Ecology* **33**, 441.

Smith, F. E. (1970). *In* "Analysis of Temperate Forest Ecosystems" (D. E. Reichle, ed.), pp. 7–18. Springer-Verlag, New York.

Volterra, V. (1931). "Leçons sur la Théorie Mathématique de la Lutte pour la Vie." Gauthier-Villars, Paris.

Wangersky, P. J. and Cunningham, W. J. (1956). *Proc. Nat. Acad. Sci. U. S.* **42**, 699.

Wangersky, P. J. and Cunningham, W. J. (1957a). *Ecology* **38**, 136.

Wangersky, P. J., and Cunningham, W. J. (1957b). *Cold Spring Harbor Symp. Quant Biol.* **22**, 239.

Wiener, N. (1948). "Cybernetics." Massachusetts Institute of Technology, Press, Cambridge, Massachusetts.

Whyte, L. L., Wilson, A. G., and Wilson, D., eds. (1969). "Hierarchical Structures." American Elsevier, New York.

Zadeh, L. A. (1963). *In* "Views on General Systems Theory" (M. D. Mesavoric, ed.) Wiley, New York.

Zadeh, L. A. (1969). *In* "Systems Theory" (L. A. Zadeh and E. Polak, eds.), pp. 3–42, McGraw-Hill, New York.

Zadeh, L. A., and Desoer, C. A. (1963). "Linear System Theory: The State Space Approach." McGraw-Hill, New York.

Zadeh, L. A., and Polak, E., eds. (1969). "Systems Theory." McGraw-Hill, New York.

PART II

THE ECOSYSTEM: SYSTEMS ANALYSIS

Systems science includes modeling, simulation, analysis, control, design, and synthesis as distinguishable activities (e.g., Chapter 1, Fig. 1). This section represents half of the simulation and analysis dichotomy formed in the title of the book. The other half appeared as Part III of Volume I. The two papers here concern ecosystem analysis as distinct from simulation. Chapters 3 and 4 of Volume I are also analytical, and could have been grouped with these in another classification.

Chapter 2, by Brylinsky, illustrates the application of *sensitivity analysis* to an energy flow model of a marine ecosystem. Although the model itself is simple, the general approach is applicable to more realistic ecosystem representations. Sensitivity analysis involves determining the amount and kind of change produced in one system parameter or variable by change in another. *Absolute* and *relative* sensitivities are distinguished and defined. Restricting consideration to the steady-state case, these sensitivities are computed by direct differentiation of the equilibrium equations and substitution of appropriate numerical values. Sensitivities of compartmental and total system steady-state energy contents to standing crops, photosynthesis, feeding flows, respiration, mortality, and turnover are presented and discussed. Then, an analysis of benthic fauna is developed in more detail from a resource management point of view. The purpose is to show how sensitivity analysis can provide information for decisions on how to alter an ecosystem to achieve a particular system state. Implicit in the treatment is the fact that the author does not view resource management as an isolated *control* problem, but rather as a problem in total ecosystem manipulation, i.e., a *design* problem. To underscore the point, direct and indirect effects on the benthic fauna are compared, and it is shown that in general indirect rather than direct influences tend to be most important. A systems analyst would appreciate this readily, but not so most resource managers, who traditionally seek to manipulate parameters that are closely associated with target populations. Total ecosystem design is now only a vague concept, but in the future it may be responsible for profound changes in the ways that man relates to his natural resource environment. Sensitivity analysis, in some form, can be expected to play a central role in this evolution.

Chapter 3 explores the application of *frequency analysis* to ecosystems. Frequency response methods of systems analysis are well developed in several fields of engineering, particularly where testing and improving the performance characteristics of physical systems containing feedback are involved. For biological systems frequency methods may be particularly appropriate in view of the preponderance of periodic "signals" in nature, accounting for all sorts of biorhythmic phenomena. The focal system of the chapter is a model of magnesium cycling in a tropical moist forest. Doctors Child and Shugart describe the model, and then the elements of the frequency response method. They develop frequency characteristics of simple and complex systems, and the significance of breaking feedback loops as an analytical procedure. The magnesium cycle of the tropical forest is shown to be heavily overdamped in comparison to most physical systems which the methods were developed to test. The suggestion is that the magnesium dynamics are highly stable, which subsequent, more detailed analysis verifies.

Attention is then turned to the contributions of components to total system frequency response characteristics. The dominant compartments of the ecosystem in magnesium flow are soil, plants, litter, carnivores, herbivores, and detritivores in descending order. The analysis of in- and out-of-system frequency response characteristics of system components is then considered, showing how compartments with small time constants tend to be sensitive to their environment (i.e., coupling), while those with large time constants are little affected.

The closing section considers several aspects of ecosystem self-design in evolution, using frequency analysis as a heuristic point of departure. For example, critical-, under-, and over-damping are identified as the basis for three different evolutionary strategies. Blooming systems, like deserts and plankton, are of the underdamped type and require components with high turnover (hence small time constants and standing crops). Storage systems, maintaining high inventories in relation to inputs, require low turnover components. The composition of turnover types in an ecosystem lends more or less stability. Ecosystem stability is suggested to arise not from the number or diversity of components, but from the nature (kinds and patterns) of couplings. Component diversity is positively distinguished as a concept from coupling diversity (what is now becoming called *connectivity* in systems ecology circles). Finally, a consideration of in- and out-of-system frequency response characteristics leads to the same conclusion as one of the Chapter 1 reasons for emergent properties, namely that the notion that a system is not the "sum" of its parts arises from incomplete knowledge of the couplings. "It is on the subject of couplings," say Child and Shugart, "that ecological systems analysts should focus their attention, for couplings are what separate functional systems from inventories, and no one should expect an inventory to 'sum' to a functional system."

2

Steady-State Sensitivity Analysis of Energy Flow in a Marine Ecosystem

MICHAEL BRYLINSKY*

DEPARTMENT OF ZOOLOGY, UNIVERSITY OF GEORGIA, ATHENS, GEORGIA

I. Introduction

The importance of understanding the flow of energy through ecosystems is well recognized in ecology. Energy flow studies provide a basis for the comparison of different ecosystems as well as a means for assessing the importance of various groups of organisms within an ecosystem. In addition, although energy flow studies emphasize functional aspects of ecosystems, they require a knowledge of ecosystem structure and result in a synthesis of data which often reveals important relationships between ecosystem structure and function. Accordingly, the energetics approach has lead to elucidation of many important concepts and principles of ecology that would not have been as obvious with other approaches aimed at understanding the functioning of ecosystems.

University of Georgia. *Contributions in Systems Ecology*, No. 8.

**Present address*: Department of Biology, Dalhousie University, Halifax, Nova Scotia, Canada.

The energetics approach can be carried one step further by analyzing ecosystem energy flow in terms of its dynamic behavior. This involves developing a mathematical model of the system describing the amounts, rates, and pathways of its energy flows. The mathematical model typically is composed of a set of ordinary differential equations, and the system behavior is studied by simultaneous solution of these equations with the aid of either an analog or digital computer. By altering selected components within the equations numerous experiments can be performed to determine their effects upon the system. In many instances analogous experiments in either the laboratory or field would be very difficult, if not impossible, to perform.

This chapter represents an attempt at combining the energetics and systems analysis approach in a study of a marine ecosystem. Its purpose is to illustrate the use of a systems analysis technique, *sensitivity analysis*, that appears to offer much potential for analyzing the dynamic behavior of complex ecological systems. Sensitivity analysis has been used profitably in various forms by systems engineers but it has not yet been applied very generally in ecological studies. The approach described here was developed by Patten (1969).

II. The Energy Flow Model

An energy flow model of an ecosystem describes the amount and direction of energy transfers occurring within the community itself and between the community and its nonbiotic environment. Its construction requires two basic kinds of information: (1) quantitative data in the form of an energy budget containing values for the standing crops, energy inputs, and energy losses of organisms in the community, and (2) qualitative data on the energy transfer pathways among the organisms themselves and between the organisms and their nonbiotic environment.

The system selected for this study was the marine community of the English Channel. The majority of information needed to construct an energy flow model for this system is contained in a single publication (Harvey, 1950). In a few cases necessary data were lacking and it was therefore unavoidable that either data be substituted from other comparable systems or certain assumptions be made allowing the needed information to be derived from Harvey's data. These instances will be noted in the discussion of the development of the energy flow model which follows.

Harvey divided the organisms of the community into groups with similar energy sources rather than true trophic levels, and his division will be maintained in this study. A summary of the data presented by Harvey is shown in Table I. The necessary energetics data lacking on each group of organisms includes their caloric values, their energy inputs due to ingestion, and their energy losses due to mortality. In addition, there is no estimate of the amount of energy loss by phytoplankton respiration.

TABLE I

SUMMARY OF HARVEY'S DATA[a]

Organism group	Standing crop	Daily production	Daily respiration
Phytoplankton	4	0.4–0.5	—
Zooplankton	1.5	0.15	4
Pelagic fish	1.8	0.0016	1.25
Demersal fish	1.00–1.25	0.001	1.25
Benthic fauna	17	0.03	1.25–2.00
Bacteria	0.14	—	30

[a] Standing crop and production values are in grams dry organic matter per square meter. Respiration is presented as percent of standing crop.

Average caloric values for each kind of organism were obtained from Cummins (1967) and are as follows in kilocalories per gram of dry organic matter: phytoplankton, 5.0; zooplankton, 5.5; fish, 5.0; benthic fauna (assumed to be mainly crustaceans and mollusks), 5.0; bacteria, 5.0.

The amount of energy input to each group of organisms by ingestion is easily derived from Harvey's data by assuming ingestion equal to assimilation and assimilation equal to growth plus respiration. The latter assumption is perfectly justified and the equality between ingestion and assimilation is plausible since egested materials (i.e., ingested but not assimilated) would become available to the detritivores, this portion being represented in either feeding or mortality transfers.

Estimates of energy losses due to mortality were calculated on the assumption that the community as a whole and the individual groups of organisms comprising it were in a steady state for an annual cycle. This means that, for each group of organisms,

$$\sum \text{energy inputs} = \sum \text{energy outputs},$$

and mortality losses were calculated to balance this equation.

Because of the difficulties involved in separating phytoplankton respiration from photosynthesis very few estimates of phytoplankton respiration exist in the literature. Those that do exist suggest a wide range of values from as low as 8% of gross production (Steemann Nielsen and Hansen, 1959) to as high as 50% of gross production (Riley, 1956).

From these studies and others it appears that a value of 25% of gross production is a reasonable average estimate for phytoplankton respiration, and this value is used here.

All values were converted into kilocalories. A summary of the annual community budget is given in Table II.

TABLE II

ANNUAL ENERGY BUDGET FOR THE ENGLISH CHANNEL COMMUNITY[a]

Organism group	Standing crop	Production	Respiration	Ingestion	Mortality
Phytoplankton	20.0	822.0	274.0	—	0.0
Zooplankton	8.3	302.0	120.0	422.0	94.0
Pelagic fish	9.9	3.0	45.0	48.0	3.0
Demersal fish	6.2	2.0	5.0	33.0	2.0
Benthic fauna	85.0	55.0	505.0	560.0	22.0
Bacteria	0.7	—	121.0	121.0	—

[a] Standing crop values are in kilocalories per square meter, all other values are in kilocalories per square meter per year.

With the energy budget completed all that is required to develop fully an energy flow model is information on the pathways of energy transfer occurring between the components of the ecosystem. This is derived partly from the food web of the community and partly from the energy transfers known to occur between each group of organisms and the nonbiotic environment. The latter energy transfers include the sole energy input to the community, photosynthesis, and energy losses due to respiration by each group of organisms.

The energy transfers occurring between the biotic components of the system were obtained from the text of Harvey's paper. In most instances the energy input to each group of organisms is received from only one source. Exceptions to this are the benthic fauna and bacteria. The benthic fauna receive energy inputs from both phytoplankton and zooplankton and the amount of energy transferred to the benthic fauna from each of these sources was calculated on the basis that it was proportional to the productivity of the phytoplankton and zooplankton. The bacteria were assumed to function mainly in the decomposition of dead organisms and therefore received all energy losses due to "mortality."

An energy flow model can be constructed and is presented as a block diagram in Fig. 1. The system is composed of six compartments. Each compartment represents the average annual caloric content, in kilocalories per square meter of the corresponding group of organisms. The

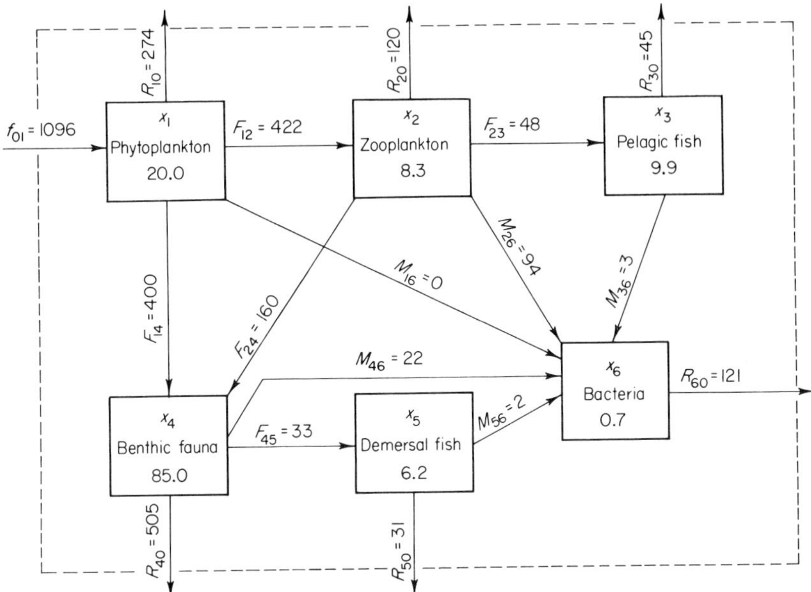

FIG. 1. Block diagram of energy flow for the English Channel. Standing crops are in kilocalories per square meter and energy flows are in kilocalories per square meter per year. See text for further explanation.

energy flows occurring either among individual compartments or between individual compartments and the nonbiotic environment are indicated by the letters f (photosynthesis), F (feeding), R (respiration), and M (mortality). The subscripts of the flows refer to the compartments, the first subscript designating the compartment of origin and the second designating the compartment of destination, where 0 is nonbiotic environment, 1 is phytoplankton, 2 is zooplankton, 3 is pelagic fish, 4 is benthic fauna, 5 is demersal fish, and 6 is bacteria. This model is essentially the same as that presented by Patten (1969) for the same system, with the exception of some minor changes in a few of the standing crop and energy flow values. The present chapter presents a more general analysis of this system, the earlier paper having concentrated mainly on the fish compartments.

III. The Mathematical Model

The formulation of a mathematical model describing the energy flow model thus developed requires that the energy inputs and losses of each compartment be expressed in the form of differential equations. This can be accomplished most conveniently by expressing all energy flows in the form of transfer coefficients. For a linear system, transfer coefficients may be computed for the steady state as the quotient of energy leaving a compartment per unit time by the mechanism involved and the amount of energy contained in the compartment. This is actually a rate constant with units time^{-1}, and represents the number of times a quantity of energy equal to the energy flow in question will flow through the compartment in the unit of time. The following is a listing of all the transfer coefficients for the English Channel model, where the Greek letters ϕ, ρ, and μ indicate feeding, respiration, and mortality transfer coefficients, respectively. Units are year^{-1}.

$$
\begin{array}{lll}
\phi_{12} = 21.10, & \rho_{10} = 13.70, & \mu_{16} = 0.00, \\
\phi_{14} = 20.00, & \rho_{20} = 14.46, & \mu_{26} = 11.33, \\
\phi_{23} = 5.78, & \rho_{30} = 4.45, & \mu_{36} = 0.30, \\
\phi_{24} = 19.27, & \rho_{40} = 5.94, & \mu_{46} = 0.26, \\
\phi_{45} = 0.40, & \rho_{50} = 5.00, & \mu_{56} = 0.32. \\
 & \rho_{60} = 172.86, &
\end{array}
$$

By assuming that the transfer of energy from a "donor" compartment to a "receiving" compartment is directly proportional to the amount of energy contained in the "donor" compartment, and by collecting all the energy inputs and losses of each compartment expressed as transfer coefficients, we can define the system by the following set of equations, each equation corresponding to a single compartment.

$$dx_1/dt = f_{01} - (\phi_{12} + \phi_{14} + \rho_{10} + \mu_{16})x_1\,, \tag{1}$$

$$dx_2/dt = \phi_{12}x_1 - (\phi_{23} + \phi_{24} + \rho_{20} + \mu_{26})x_2\,, \tag{2}$$

$$dx_3/dt = \phi_{23}x_2 - (\rho_{30} + \mu_{36})x_3\,, \tag{3}$$

$$dx_4/dt = \phi_{14}x_1 + \phi_{24}x_2 - (\phi_{45} + \rho_{40} + \mu_{46})x_4\,, \tag{4}$$

$$dx_5/dt = \phi_{45}x_4 - (\rho_{50} + \mu_{56})x_5\,, \tag{5}$$

$$dx_6/dt = \mu_{16}x_1 + \mu_{26}x_2 + \mu_{36}x_3 + \mu_{46}x_4 + \mu_{56}x_5 - \rho_{60}x_6\,. \tag{6}$$

Before proceeding further some comments are in order concerning the validity of the above mathematical model. It is not proposed that this model is a realistic representation of the natural community from which it was derived. It has obvious biological shortcomings. Our knowledge of the food webs of aquatic communities has advanced considerably since the time of Harvey's study, and the representation here is probably not very accurate. The way in which the compartments are mathematically related to each other may also be unrealistic. Numerous other criticisms could also be made. However, it must be realized that the application of systems analysis techniques to ecological systems is relatively new and is quite restricted by not having the kinds of data necessary for a "realistic" simulation of most systems. Indeed, the criteria for determining how "realistic" a model is have yet to be developed. However, the purpose of this chapter is not to present a definitive simulation model, but to describe a technique for analyzing the energy network of ecosystems in the vicinity of the steady state. Linear models are appropriate for this purpose, according to the rationale developed in Volume I, Chap. 2, Section II.A.1.

IV. Sensitivity Analysis

A. Introduction

One of the more interesting ways to analyze the dynamics of a system is by determining how sensitive its various components are to each other. This can be done by a systems analysis technique known as "sensitivity analysis." If the steady-state value of some system parameter, such as a compartment energy content or any energy transfer, is altered, the system will respond by changing the values of its state variables and achieving a new steady state. Sensitivity analysis involves determining the amount and kind of change produced in a given system parameter by a change in another parameter. What makes sensitivity analysis interesting is that it considers the system as a whole and very often a change in one system parameter produces a quite unexpected change in another.

Because all compartments in a system are related to each other and since it is possible to alter any single parameter or group of parameters and determine the effect on any other system parameter, it is readily seen that a great number of sensitivities can be calculated for a system. In the present study only the absolute and relative steady-state sensitivities of compartment energy contents and the total system's energy content, compartment standing crop values, turnover rates, and energy fluxes are considered. *Absolute sensitivity* represents the amount of change produced in one system parameter per unit change in another system parameter. For example, the absolute sensitivity of the zooplankton compartment to the standing crop value of the phytoplankton compartment is $\Delta x_2/\Delta x_1$. A value of 0.4145 for this sensitivity would indicate that the zooplankton compartment increases its steady-state standing crop by 0.4145 kcal m^{-2} for each 1.0000 kcal m^{-2} increase in the phytoplankton compartment. *Relative sensitivity* represents the amount of change produced in one system parameter by a change in another relative to the original steady-state values of both parameters. The relative sensitivity of the zooplankton compartment to the standing crop value of the phytoplankton compartment is expressed as $[(\Delta x_2/\Delta x_1)(x_1(\infty)/x_2(\infty))]$, where $x_1(\infty)$ and $x_2(\infty)$ are steady-state values of compartments 1 and 2, respectively. A value of 1.0000 for this sensitivity would mean that a 100% increase in the standing crop value of the phytoplankton compartment produces a 100% increase in the steady-state energy content of the zooplankton compartment.

There are numerous mathematical approaches that can be used to calculate the sensitivity of one parameter to another. In this study sensitivities were determined by expressing them as partial derivatives, solving for the partial derivatives by direct differentiation of the appropriate steady-state equations, and substituting known values in the resulting equation to obtain the numerical value of the sensitivity. A Fortran program was used to make the necessary calculations. The approach is best illustrated by an example.

Sensitivity of the zooplankton compartment (x_2) to the feeding flow from phytoplankton to zooplankton (F_{12}) can be expressed as a derivative, $\delta x_2/\delta F_{12}$, which is equal to the sum of the partial derivatives representing the effect of F_{12} on x_2. In order to formulate these partial derivatives all the changes in x_2 brought about by a change in F_{12} must be considered. Referring to the block diagram of Fig. 1, it is seen that an increase in F_{12} will produce an increase in x_2 as a result of x_2 receiving a greater amount of energy. However, an increase in F_{12} will also produce a decrease in x_1, and since $F_{12} = \phi_{12}x_1$, this will in turn result in a decrease in x_2. Both the increase and decrease must be considered in formulating the derivative expressing this sensitivity. Using the chain rule for implicit functions from calculus, the absolute sensitivity of x_2 to F_{12} may be expressed (Patten, 1969) as

$$\frac{\delta x_2}{\delta F_{12}} = \frac{\partial x_2}{\partial F_{12}} + \frac{\partial x_2}{\partial x_1}\frac{\partial x_1}{\partial F_{12}}, \tag{7}$$

where the first term represents the *direct* effect of F_{12} on x_2 and the second term represents the *indirect* effect.

Each partial derivative is obtained by direct differentiation of an appropriate steady-state equation relating the variables involved, in this case the steady-state equations for x_1 and x_2. Referring to the original system equations, Eqs. (1) and (2), at steady state $dx_1/dt = 0$ and $dx_2/dt = 0$; therefore,

$$x_1(\infty) = f_{01}/(\rho_{10} + \phi_{14} + \phi_{12} + \mu_{26}) = (f_{01} - F_{12})/(\rho_{10} + \phi_{14} + \mu_{26}),$$

and

$$x_2(\infty) = \phi_{12}x_1/(\phi_{23} + \phi_{24} + \rho_{20} + \mu_{26}) = F_{12}/(\phi_{23} + \phi_{24} + \rho_{20} + \mu_{26}).$$

Direct differentiation of these equations gives

$$\frac{\partial x_2}{\partial F_{12}} = \frac{1}{\phi_{23} + \phi_{24} + \rho_{20} + \mu_{26}}, \tag{8}$$

$$\frac{\partial x_2}{\partial x_1} = \frac{\phi_{12}}{\phi_{23} + \phi_{24} + \rho_{20} + \mu_{26}}, \tag{9}$$

and

$$\frac{\partial x_1}{\partial F_{12}} = -\frac{1}{\phi_{14} + \rho_{10} + \mu_{16}}. \tag{10}$$

Substitution of the numerical values of rate constants in the above equations gives $\partial x_2/\partial F_{12} = 0.0196$, $\partial x_2/\partial x_1 = 0.4150$, and $\partial x_1/\partial F_{12} = -0.0296$. For Eq. (7), then, $\delta x_2/\delta F_{12} = 0.0073$.

Thus, the absolute sensitivity of the zooplankton compartment to the feeding flow from phytoplankton to zooplankton is 0.0073, indicating that for each 1 kcal m^{-2} y^{-1} increase in F_{12} the zooplankton will increase its compartment energy content by 0.0073 kcal m^{-2}. To obtain the relative steady-state sensitivity of x_2 to F_{12}, the absolute steady-state sensitivity is multiplied by the ratio F_{12}/x_2 at steady state: $(0.0073)(422.0/8.3) = 0.3738$. This indicates that a 100% increase in F_{12} will produce a 37.38% increase in x_2 at equilibrium.

As another more complex example we may consider the sensitivity of the benthic fauna compartment (x_4) to the feeding flow from phytoplankton to zooplankton (F_{12}). Referring again to the block diagram of Fig. 1 it is seen that F_{12} can affect x_4 in a number of ways: F_{12} affects x_1 and x_2 which will in turn affect x_4 since it receives its feeding inputs from both x_1 and x_2. In addition, there are two pathways by which x_4 will be affected: x_1 to x_4 and x_1 to x_2 to x_4. All of these interrelationships must be considered in formulating the derivative expressing this sensitivity. Again, using the chain rule for implicit functions,

$$\frac{\delta x_4}{\delta F_{12}} = \frac{\partial x_4}{\partial F_{12}} + \frac{\partial x_4}{\partial x_1}\frac{\partial x_1}{\partial F_{12}} + \frac{\partial x_4}{\partial x_2}\frac{\partial x_2}{\partial F_{12}}, \tag{11}$$

where the first term represents the direct effect of F_{12} on x_4, the second term is the indirect effect as a result of the one-step feeding flow from phytoplankton to benthic fauna, and the third term is the indirect effect of the two-step feeding flow from phytoplankton to zooplankton to benthic fauna.

Following the procedure illustrated in the previous example involving direct differentiation of the appropriate steady-state equations and substitution of known rate constants in the resulting equations, we obtain $\partial x_4/\partial F_{12} = 0$ (since there is no direct effect of F_{12} on x_4), $\partial x_4/\partial x_1 = 4.2170$, $\partial x_1/\partial F_{12} = -0.0297$, $\partial x_4/\partial x_2 = 2.9066$, and $\partial x_2/\partial F_{12} = 0.0073$. Substitution of these values in Eq. (11) gives $\delta x_4/\delta F_{12} = -0.1038$, the absolute steady-state sensitivity of x_4 to F_{12}. The relative sensitivity is $(-0.1038)(422.0/85.0) = -0.5193$.

TABLE III

ABSOLUTE SENSITIVITIES FOR THE ENGLISH CHANNEL MODEL[a]

	x_1	x_2	x_3	x_4	x_5	x_6	Net change	Total change
x_1	—	0.4145	0.4957	4.2170	0.3091	0.3097	5.4760	5.4760
x_2	0.0000	—	1.1959	2.9066	0.2131	0.0732	4.3888	4.3888
x_3	0.0000	0.0000	—	0.0000	0.0000	0.0017	0.0017	0.0017
x_4	0.0000	0.0000	0.0000	—	0.0733	0.0019	0.0752	0.0752
x_5	0.0000	0.0000	0.0000	0.0000	—	0.0018	0.0018	0.0018
x_6	0.0000	0.0000	0.0000	0.0000	0.0000	—	—	—
f_{01}	0.0182	0.0076	0.0090	0.0989	0.0073	0.0015	0.1425	0.1425
F_{12}	−0.0297	0.0073	0.0088	−0.1038	−0.0076	−0.0008	−0.1258	0.1580
F_{14}	−0.0287	−0.0119	−0.0142	0.0052	0.0004	−0.0020	−0.0512	0.0624
F_{23}	0.0000	−0.0222	0.1797	−0.0644	−0.0047	−0.0014	0.0870	0.2724
F_{24}	0.0000	−0.0316	−0.0378	0.0586	0.0043	0.0002	−0.0063	0.1325
F_{45}	0.0000	0.0000	0.0000	−0.1600	0.1762	0.0001	0.0163	0.3363
R_{10}	−0.0243	−0.0101	−0.0121	−0.1319	−0.0097	−0.0020	−0.1901	0.1901
R_{20}	0.0000	−0.0275	−0.0329	−0.0799	−0.0059	−0.0022	−0.1484	0.1484
R_{30}	0.0000	0.0000	−3.3333	0.0000	0.0000	−0.0058	−3.3391	3.3391
R_{40}	0.0000	0.0000	0.0000	−1.4493	−0.1062	−0.0029	−1.5584	1.5584
R_{50}	0.0000	0.0000	0.0000	0.0000	−3.1250	−0.0058	−3.1308	3.1308
R_{60}	0.0000	0.0000	0.0000	0.0000	0.0000	—	—	—
M_{26}	0.0000	−0.2525	−0.0302	−0.0734	−0.0108	0.0037	−0.3632	0.3706
M_{36}	0.0000	0.0000	−0.2198	0.0000	0.0000	0.0054	−0.2144	0.2252
M_{46}	0.0000	0.0000	0.0000	−0.1577	−0.0116	0.0055	−0.1638	0.1748
M_{56}	0.0000	0.0000	0.0000	0.0000	0.2000	0.0054	−0.1946	0.2054
TR_1	−0.3650	−0.1513	−0.1809	−1.9788	−0.0040	0.0000	−2.6800	2.6800
TR_2	0.0000	−0.1629	−0.1948	−0.4734	−0.0347	−0.0132	−0.8790	0.8790
TR_3	0.0000	0.0000	−0.0204	0.0000	0.0000	−0.0035	−0.0239	0.0239
TR_4	0.0000	0.0000	0.0000	−0.0013	−0.9311	−0.0254	−0.9578	0.9578
TR_5	0.0000	0.0000	0.0000	0.0000	−0.1162	−0.0021	−0.1183	0.1183
TR_6	0.0000	0.0000	0.0000	0.0000	0.0000	−0.0041	−0.0041	0.0041

[a] TR_i is the turnover rate of compartment i; see text for further details.

It should be noted that sensitivities can be either positive or negative depending on whether they represent an increase or decrease in the energy content of a compartment. In addition, steady-state sensitivities do not take time into consideration since time is not considered as a variable; i.e., steady-state sensitivities denote the new steady-state values regardless of the time it takes to attain the new steady state.

Table III lists all the absolute sensitivities of compartment energy contents to standing crop values, energy flows, and turnover times that can be computed for the English Channel model. Table IV does the

TABLE IV

RELATIVE SENSITIVITIES FOR THE ENGLISH CHANNEL MODEL[a]

	x_1	x_2	x_3	x_4	x_5	x_6	Net change	Total change
x_1	—	1.0000	1.0000	1.0000	1.0000	1.1074	5.1074	5.1074
x_2	0.0000	—	1.0000	0.2857	0.2857	0.8473	2.4187	2.4187
x_3	0.0000	0.0000	—	0.0000	0.0000	0.0240	0.0240	0.0240
x_4	0.0000	0.0000	0.0000	—	1.0000	0.2201	0.2201	1.2201
x_5	0.0000	0.0000	0.0000	0.0000	—	0.0160	0.0160	0.0160
x_6	0.0000	0.0000	0.0000	0.0000	0.0000	—	—	—
f_{01}	1.0000	1.0000	1.0000	1.2857	1.2857	2.2823	7.8537	7.8537
F_{12}	−0.6261	0.3738	0.3738	−0.5193	0.5193	0.0578	−0.9747	2.4703
F_{14}	−0.5747	−0.5747	−0.5747	0.0246	0.0246	−1.1430	−2.8179	2.9163
F_{23}	0.0000	−0.1286	0.8714	−0.0367	−0.0367	−0.0967	0.5727	1.1701
F_{24}	0.0000	−0.6107	−0.6107	0.1112	0.1112	0.0262	−0.9728	1.4700
F_{45}	0.0000	0.0000	0.0000	−0.0624	0.9376	0.0012	0.8764	1.0012
R_{10}	−0.3333	−0.3333	−0.3333	−0.4286	−0.4286	−0.7608	−2.6179	2.6179
R_{20}	0.0000	−0.3984	−0.3984	−0.1138	−0.1138	−0.3740	−1.3984	1.3984
R_{30}	0.0000	0.0000	−15.1666	0.0000	0.0000	−0.3640	−15.5306	15.5306
R_{40}	0.0000	0.0000	0.0000	−8.6232	−8.6232	−2.0357	−19.2821	19.2821
R_{50}	0.0000	0.0000	0.0000	0.0000	−15.6250	−0.2494	−15.8744	15.8744
R_{60}	0.0000	0.0000	0.0000	0.0000	0.0000	—	—	—
M_{26}	0.0000	−0.0855	−0.2854	−0.0815	−0.1631	0.4867	−0.1288	1.1022
M_{36}	0.0000	0.0000	−0.0659	0.0000	0.0000	0.0224	−0.0435	0.0883
M_{46}	0.0000	0.0000	0.0000	−0.0473	−0.0473	0.1930	0.0984	0.2876
M_{56}	0.0000	0.0000	0.0000	0.0000	−0.0640	0.0150	−0.0490	0.0790
TR_1	−1.0000	−1.0000	−1.0000	−1.2857	−1.2857	0.0000	−5.5714	5.5714
TR_2	0.0000	−1.0000	−1.0000	−0.2857	−0.2857	−0.9388	−3.5102	3.5102
TR_3	0.0000	0.0000	−1.0000	0.0000	0.0000	−0.0240	1.0240	1.0240
TR_4	0.0000	0.0000	0.0000	−1.0000	−1.0000	−0.2361	2.2361	2.2361
TR_5	0.0000	0.0000	0.0000	0.0000	−1.0000	−0.0137	1.0137	1.0137
TR_6	0.0000	0.0000	0.0000	0.0000	0.0000	−1.0000	1.0000	1.0000

[a] TR_i is the turnover rate of compartment i; see text for further details.

same for relative sensitivities. The row headings refer to the parameters that are being altered and the column headings refer to the parameters affected by the alteration. Also shown is the net and total change in the energy content of the *entire* system. Net changes in the system's total energy content are computed as the sum of all changes, both positive and negative, produced in each compartment by a particular system parameter. The total change is the sum of the absolute values of the changes produced in each compartment. These sensitivities can be interpreted in the same manner as individual compartment sensitivities and are considered to represent the sensitivity of the total system to the various system parameters.

The following sections will be concerned with a discussion of the sensitivity structure of the English Channel system in order to illustrate the kinds of information that can be obtained by sensitivity analysis.

B. Sensitivity Structure of the English Channel Model

1. *Sensitivities of Compartments to Standing Crop Values*

Most compartments exhibit comparatively high sensitivities to changes in standing crop values. This is particularly true of compartment sensitivities to phytoplankton and zooplankton standing crop values.

The phytoplankton compartment is independent of energy sources from other compartments in the system and, therefore, exhibits no sensitivity to their standing crop values. However, since all other compartments in the system are dependent on the phytoplankton for their energy supply, either directly or indirectly, an increase in the standing crop value of this compartment results in an increase in all other compartments. In all cases except one, compartments show the greatest absolute sensitivity to phytoplankton energy levels, the exception being the pelagic fish compartment which exhibits a higher absolute sensitivity to zooplankton energy levels. In relative terms all compartments, with the exception of the bacteria, are equally sensitive to the energy content of the phytoplankton compartment, the bacteria being slightly more sensitive than other compartments.

The demersal fish compartment is more sensitive, on an absolute basis, to the phytoplankton standing crop than that of the benthic fauna from which it receives its energy directly. On the other hand, the pelagic fish compartment is almost three times more sensitive to its direct energy source, the zooplankton, than it is to the phytoplankton. In the case of the benthic fauna which receives direct energy inputs from both the phytoplankton and zooplankton, the greatest absolute sensitivity is to the former from which it receives the larger part of its energy.

All compartments receiving energy inputs from only one direct source (zooplankton, pelagic, and demersal fish) have relative sensitivities of 1.0000 to the compartment providing their direct energy input. The benthic fauna are more than three times more sensitive, on a relative basis, to their direct input from phytoplankton than to their direct input from zooplankton.

The bacteria compartment exhibits sensitivities to all other standing crop values in the system. Absolute sensitivities are greater for the zooplankton and lowest for the demersal and pelagic fish compartments. Relative sensitivities are greatest for the phytoplankton and zooplankton compartments.

An increase in a standing crop value represents an increase in the total energy content of the system. As a result both net and total changes in the system's energy content are equal and positive for each increase in standing crop. Because of its early role in the food web of the community, changes in the phytoplankton energy level produce the greatest change, both in absolute and relative terms, in the system's total energy content. This is also true of zooplankton to a lesser extent. When compared to the total system's sensitivities to other components of the system, absolute sensitivities to the standing crop values of the phytoplankton and zooplankton compartments are found to be highest. However, relative sensitivities are only of intermediate magnitude.

2. *Sensitivities of Compartments to Photosynthesis*

All compartments, as expected, exhibit positive sensitivities to the single system forcing, photosynthesis. The benthic fauna compartment is more sensitive to photosynthesis, on an absolute basis, than any other group. In relative terms only the bacteria show a greater sensitivity than the benthic fauna compartment. In general, absolute sensitivities of compartments to photosynthesis are low; however, relative sensitivities are comparatively high for both individual compartments and for the system as a whole.

3. *Sensitivities of Compartments to Feeding Flows*

The relationship of a feeding flow to a particular compartment can take numerous forms. It may be a direct energy input or a direct energy loss from a compartment, or it may represent an indirect energy input or loss being mediated one or more steps from a compartment. In some instances it can be viewed as a competitive feeding flow. Often there may be a combination of relationships involved. For example, the relationship of the feeding flow from phytoplankton to zooplankton (F_{12}) to the benthic fauna compartment can be considered as an indirect energy input since energy travels to the benthic fauna by flows F_{12} and F_{24}. On the other hand, it can be considered to represent a competitive feeding flow since the benthic fauna "compete" with the zooplankton for phytoplankton. Considering it in still another manner it may represent an indirect energy loss since its increase will decrease the energy content of the phytoplankton compartment and consequently the energy transfer from phytoplankton to benthic fauna. Because of the complexity of feeding transfer relationships a sensitivity analysis is particularly helpful in predicting the outcome of changes in a system's feeding flows.

Intuitively it might appear that a change in the magnitude of a feeding flow would result only in a redistribution, among the different compartments, of the total amount of energy contained in the system, and that the net change in the system's total energy content would be zero. That this is not the case is evident from the values of net change for the entire system produced by a change in a feeding flow. If a feeding flow is increased, the compartment from which it originates will decrease as well as any compartment to which it donates energy other than the compartment receiving the energy of the increased flow. On the other hand, the compartment receiving the energy of the increased flow will increase along with any compartment to which it donates energy. Depending on the direction and magnitude of all changes involved, the system will increase or decrease its total energy content. Flows F_{23} and F_{45} produce positive net changes while flows F_{12}, F_{14}, and F_{24} produce negative net changes. This trend is similar for both relative and absolute sensitivities.

It is difficult to generalize on compartment sensitivities to feeding flows but it appears that all compartments exhibit greatest sensitivity to changes in their own feeding transfers. The phytoplankton compartment is independent of energy inputs from other compartments in the system. However, it donates energy directly to both the zooplankton and benthic fauna compartments and is more sensitive, in both absolute and relative terms, to its loss to the latter, although the difference is small.

The zooplankton compartment receives energy from the phytoplankton and donates energy to both the benthic fauna and pelagic fish compartments. It shows the highest absolute sensitivity to its losses (F_{24} and F_{23}) and the lowest to its direct input (F_{12}). It is moderately sensitive to a competitive flow (F_{14}) from phytoplankton to benthic fauna. In relative terms zooplankton are most sensitive to their loss to benthic fauna, moderately sensitive to the competitive flow from phytoplankton to benthic fauna, and least sensitive to their energy loss to the pelagic fish compartment.

The benthic fauna receive feeding transfers from both the phytoplankton and zooplankton compartments and donate energy to the demersal fish compartment. Like the zooplankton, the benthic fauna exhibits its greatest absolute sensitivity to its own energy loss (F_{45}), intermediate sensitivity to competitive feeding flows (F_{12} and F_{23}), and the least absolute sensitivity to its inputs (F_{24} and F_{14}). It differs markedly from the zooplankton, however, in the order of its relative sensitivities in that it is most sensitive to the competitive flow from phytoplankton to zooplankton (F_{12}) followed by its input from zooplankton (F_{24}). In addition its relative sensitivity to its loss to demersal fish (F_{45}) is only intermediate and it is least sensitive to its input from phytoplankton.

The pelagic and demersal fish compartments present a particularly interesting situation for a comparison of their sensitivities to feeding flows. Both have direct feeding transfers from only one other compartment, no energy losses to other compartments by means of feeding flows, and both are one step removed from an indirect feeding transfer originating at the phytoplankton compartment. They differ, however, in that the demersal fish compartment has an additional indirect feeding source two steps removed which is mediated from the phytoplankton through the zooplankton and in turn the benthic fauna. In addition the nature of the competitive feeding flows is somewhat different in terms of how far removed they are from the compartment.

Absolute sensitivities for the pelagic and demersal fish compartments show the same general trends: highest sensitivity to their direct energy inputs, moderate sensitivity to competitive energy transfers, and lowest absolute sensitivity to energy losses. Relative sensitivities show the same trends as absolute sensitivities for the pelagic fish compartment but not for the demersal fish compartment. The latter is most sensitive to its direct energy input followed by its one-step-removed indirect energy input (F_{12}). It should be noted that the latter feeding transfer, although referred to as an indirect energy input, will actually produce a decrease in the compartment's energy content. This is due to the additional competitive nature of the transfer, the zooplankton feeding directly on phytoplankton and thus competing with the benthic fauna, the direct energy source to the demersal fish compartment. As a result, the demersal fish will benefit from an increase in feeding flows originating at the phytoplankton compartment only if it is mediated to the compartment by means of the benthic fauna as opposed to the zooplankton.

The bacteria compartment receives no direct feeding transfer from any other compartment, nor does it provide energy either directly or indirectly to other compartments in the system. It is thus affected only indirectly by changes in feeding transfers. Both relative and absolute sensitivities show the same trends. All are quite low with the exception of F_{14} to which it exhibits a surprisingly high relative sensitivity. Increases in flows F_{14}, F_{23}, and F_{12} decrease the compartment's energy content while increases in F_{24} and F_{45} increase its energy content.

4. *Sensitivities of Compartments to Respiration*

Respiration represents an energy loss from the ecosystem and accordingly an increase in the respiratory energy flux from a compartment will result in a decrease in its energy content as well as a decrease in the system's total energy content.

In both absolute and relative terms, compartments are particularly sensitive to changes in respiratory energy flows. This is especially true in the case of the two fish compartments, which have relative sensitivities to their own respiration that exceed most other sensitivities by an order of magnitude. In general, each compartment is most sensitive to its own respiration and next to the respiration of the compartment from which it derives its energy directly. The benthic fauna compartment is more sensitive to phytoplankton respiration than it is to zooplankton respiration. On a relative basis, bacteria show their highest sensitivity to the respiration of benthic fauna and lowest to the respiration of demersal fish.

The high sensitivity of the system to respiratory energy flows is illustrated by the large values for both relative and absolute sensitivities of the total system. The two fish and benthic fauna compartments' respiratory flows produce large negative changes in the system's energy content.

5. *Sensitivities of Compartments to Mortality*

Absolute and relative sensitivities of compartments to mortality tend to follow the same trends as sensitivities to respiration in that compartments are most sensitive to their own mortality and next to the mortality of the compartment from which they receive their energy. One difference, however, is that the bacteria benefit from increases in mortality flows and consequently have positive sensitivities. They exhibit about equal sensitivities to the mortality of pelagic fish and benthic fauna compartments. Relative sensitivities of bacteria are greatest for the zooplankton and benthic fauna mortality flows.

Considering the system as a whole, it is of intermediate sensitivity to changes in mortality flows. In all cases except one, an increase in a mortality flow leads to a decrease in the system's total energy content as a result of the increase in the bacterial compartment being offset by the decrease in all other compartments. The exception occurs in the case of the system's relative sensitivity to the mortality flow from the benthic fauna to bacteria. In this instance the increase in the bacteria compartment is greater than the total decrease in all other compartments resulting in a positive value for the net relative sensitivity of the system (Table IV, $M_{46} = 0.0984$).

6. *Compartment Sensitivities to Turnover Rates*

The *turnover rate* of a compartment is the ratio of energy flow through it in a unit time to the compartment's average energy content during the same interval. The *turnover time*, or *time constant*, of a compartment is the reciprocal of its turnover rate: the amount of time required for $(1 - 1/e)$ of the compartment's energy content to pass through it. For each compartment of a system at equilibrium, the sum of its loss rates is equal to its turnover rate and the reciprocal of this is its turnover time. Table V lists the turnover rates and turnover times for each compartment of the English Channel model and for the system as a whole.

TABLE V

TURNOVER RATES AND TURNOVER TIMES FOR THE ENGLISH CHANNEL MODEL[a]

	x_1	x_2	x_3	x_4	x_5	x_6	Total system
TR	54.8	50.9	4.9	6.6	5.3	173.0	8.4
TT	6.6	7.3	75.1	55.0	68.6	2.1	43.7

[a] *TR* is turnover rate per year; *TT* is turnover time in days.

Since an increase in the turnover rate of a compartment is the result of increasing its loss rates, compartment sensitivities to turnover rates will be negative. A compartment will exhibit sensitivity to its own turnover rate and to the turnover rates of compartments from which it receives its energy, either directly or indirectly. On an absolute basis, a compartment tends to show its greatest sensitivity to the turnover rates of the compartments from which it receives its energy directly, moderate sensitivity to the turnover rates of its indirect energy sources, and low sensitivity to its own turnover rate. However, there is some variation in this pattern, particularly in the case of the demersal fish compartment which exhibits a moderate sensitivity to its own turnover rate and a low sensitivity to the turnover rate of its indirect energy source.

In relative terms, each compartment shows a sensitivity of -1.0000 to its own turnover rate. With the exception of the benthic fauna and bacteria, all compartments have relative sensitivities of -1.0000 to the turnover rates of their direct energy sources. The benthic fauna are more than four times more sensitive to the phytoplankton turnover rate than to that of zooplankton, both of which are its direct energy sources.

The turnover rates of the phytoplankton, zooplankton, and benthic fauna compartments produce the greatest changes in the system's total energy content.

C. Detailed Analysis of Benthic Fauna Compartment Sensitivities

The preceeding has been a general account of the sensitivity characteristics of the English Channel model. It may, perhaps, be profitable to attempt a more detailed investigation of a selected compartment of the model from a resource management viewpoint. Sensitivity analysis appears to be particularly useful in this respect since it enables one to compare and evaluate the various parameters in a system as to their effect on a particular compartment. Accordingly, sensitivity analysis can provide a framework upon which to base decisions concerning steps to be taken in altering a system to obtain a particular system state.

The benthic fauna compartment is especially well suited for exemplifying this approach since, having a relatively complex relationship to the rest of the system, the effects of the various system parameters on this compartment are not always intuitively obvious. Assuming that our interest is in harvesting and therefore increasing the biomass (compartment energy content) of the benthic fauna compartment, the changes brought about by alterations in the system parameters affecting this compartment will be compared. This requires the consideration of both absolute and relative sensitivities. Absolute sensitivities indicate the *absolute* effect of a parameter on a compartment's biomass, but do not directly indicate the relative degree of change involved in accomplishing this effect. Relative sensitivities, on the other hand, indicate the *relative* amount of change produced in a compartment's biomass for equal amounts of change in a system parameter, but do not directly reveal the magnitude of change that occurs. To illustrate these differences further, an increase of 1 kcal in an energy flow of 5 kcal involves a 20% increase whereas an increase of 1 kcal in an energy flow of 100 kcal involves only a 1% increase. An absolute sensitivity may indicate that the former change has a greater effect than the latter but it gives no indication that a much greater change, in relative terms, is required to accomplish this effect. A relative sensitivity for the same change corrects for this since it represents a relative change rather than the actual magnitude of change in kilocalories. However, although the relative sensitivity may be high, the actual amount of change in kilocalories may be low. It is therefore necessary to consider both absolute and relative sensitivities together in evaluating the effect of a system parameter on the biomass of the benthic fauna compartment since neither alone provides all the information necessary for assessing a parameter's importance. Tables III and IV should be referred to in the following discussion.

1. *Parameters Directly Affecting the Benthic Fauna Compartment*

Those system parameters having a direct effect on the benthic fauna compartment include its direct energy inputs and its direct energy losses. The direct energy inputs are the feeding flows from the phytoplankton and zooplankton compartments to the benthic fauna compartment. Both absolute and relative sensitivities are considerably higher for the flow originating at the zooplankton compartment, indicating that a greater biomass increase would result from an increase in this flow as opposed to the feeding flow originating at the phytoplankton compartment. The direct energy losses from the benthic fauna compartment are its respiration and mortality losses plus a feeding flow to the demersal fish compartment. The energy loss through respiration is exceptionally high in both absolute and relative terms, indicating that this parameter is one of the more important affecting benthic fauna biomass. Any means of significantly decreasing this energy loss would be reflected in a high increase in the benthic fauna biomass. Sensitivities of the energy losses resulting from mortality and the feeding flow to the demersal fish are approximately equal. Absolute sensitivities are fairly high but relative sensitivities are somewhat low, indicating that large decreases in these parameters are necessary to increase significantly the benthic fauna biomass.

Considering all of the system parameters having a direct effect on the benthic fauna compartment, it is seen that the direct energy losses have a greater effect than the direct energy inputs. This suggests that the benthic fauna biomass would best be increased by decreasing its direct energy losses rather than by increasing its direct energy inputs.

2. *Parameters Indirectly Affecting the Benthic Fauna Compartment*

System parameters that affect the benthic fauna compartment indirectly are those which affect the phytoplankton and zooplankton compartments from which the benthic fauna receives its energy inputs. These include photosynthesis, phytoplankton and zooplankton compartment energy levels and respiration, zooplankton mortality, and the feeding flows from phytoplankton to zooplankton and zooplankton to pelagic fish.

The benthic fauna are only moderately sensitive, on an absolute basis, to photosynthesis. However, it exhibits an exceptionally high relative sensitivity to photosynthesis which is exceeded only by the relative sensitivity to its own respiration. This high relative sensitivity indicates that increasing photosynthesis may be a profitable means of increasing the benthic fauna biomass. In absolute terms, the standing crop values of the phytoplankton and zooplankton compartments have a greater effect on the biomass of the benthic fauna compartment than any other system parameter. On a relative basis the effect of the phytoplankton standing crop is exceeded only by the effect of respiration of the benthic compartment and photosynthesis. The effect of the zooplankton standing crop is exceeded by these and also by the respiration of phytoplankton and zooplankton and the feeding flow from phytoplankton to zooplankton. Obviously, better results in increasing the benthic fauna biomass would be obtained by increasing the phytoplankton standing crop rather than the zooplankton standing crop even though, as previously noted, an increase in the feeding flow from zooplankton to benthic fauna has a greater effect than an equal increase from phytoplankton to benthic fauna.

The feeding flow from phytoplankton to zooplankton has a negative effect on the biomass of the benthic fauna. As a result, any decrease in this flow would be beneficial to the benthic fauna. The absolute sensitivity of the benthic fauna to this feeding flow is not great but the relative sensitivity is high, indicating that a small decrease in this flow produces a significant increase in benthic fauna biomass. The feeding flow from zooplankton to pelagic fish is a competitive feeding flow whose increase decreases the standing crop of the zooplankton compartment and, subsequently, the benthic fauna biomass. Both absolute and relative sensitivities of the benthic fauna to this flow are low and thus, it does not appear to be a very important parameter in controlling benthic fauna biomass. This is also true of the effect of zooplankton mortality on the benthic fauna compartment.

The benthic fauna compartment's absolute sensitivities to the respiration of phytoplankton and zooplankton is quite low. However, relative sensitivities are somewhat greater, and the values suggest that these parameters are more important in influencing benthic fauna biomass than most of the parameters directly affecting the compartment. Any means of decreasing these energy losses would be relatively effective in increasing the benthic fauna biomass.

In summary, alterations in those system parameters having an indirect effect on the benthic fauna biomass, as opposed to a direct effect, tend to produce the greatest changes in the biomass of benthic fauna. Of those parameters having a direct effect, energy losses, particularly due to respiration, have a greater influence than energy inputs. This suggests that benthic fauna biomass would be best increased by decreasing the compartment's energy losses rather than by increasing its energy inputs. Of those system parameters having an indirect effect on benthic fauna biomass, those producing the greatest changes in the phytoplankton compartment also produce the greatest changes in the benthic fauna compartment. This is also true, although to a lesser extent, of the zooplankton compartment. Management practices that would increase the phytoplankton biomass, such as an increase in photosynthesis and a decrease in respiration, will be most beneficial to increases in the benthic fauna biomass. Increases in the biomass of zooplankton would also be beneficial provided that this increase does not occur as a result of an increase in the feeding flow from phytoplankton to zooplankton since this has a negative affect on the biomass of the benthic fauna compartment.

V. Discussion

Sensitivity analysis is one of a number of systems analysis techniques that can be used to study the dynamic structure of a system. If offers a means of assessing the significance of the complex interactions typical of natural ecosystems, and therefore holds much potential in ecological systems analysis. Obviously, we must have reasonably accurate models in order to obtain valid information from such a technique. Problems involved in constructing "realistic" models have been discussed elsewhere in this book. Nevertheless, even at the present state of our knowledge, ecological systems analysis techniques appear to offer many advantages.

If the ultimate aim of ecology is to be able to evaluate factors determining the distribution and abundance of organisms with time, a technique of the general sort described here seems inevitable.

REFERENCES

Cummins, K. W. (1967). Unpublished report (2nd ed.). Univ. Pittsburgh, Pittsburgh, Pennsylvania.

Harvey, H. W. (1950). *J. Marine Biol. Assoc. United Kingdom.* **29**, 97.

Patten, B. C. (1969). *Trans. Am. Fish. Soc.* **98**, 570.

Riley, G. A. (1956). *Bull. Bingham Oceanogr. Coll.* **15**, 324.

Steemann Nielsen, E., and Hansen, V. K. (1959). *Deep Sea Res.* **5**, 222.

3

Frequency Response Analysis of Magnesium Cycling in a Tropical Forest Ecosystem

GEORGE I. CHILD* and HERMAN HENRY SHUGART, JR.†

INSTITUTE OF ECOLOGY, UNIVERSITY OF GEORGIA, ATHENS, GEORGIA

I. General Introduction and Introduction to Frequency Response Analysis

Ecology, even though it is one of the youngest scientific disciplines, is in evolution. Change occurs as ecologists develop a more integrated view of nature, causing a major transition from autecological to syn-

University of Georgia, *Contributions in Systems Ecology*, No. 9.

* *Present address:* University of Guam, Agana, Guam.

† *Present address:* Ecological Sciences Division, Oak Ridge National Laboratory, Oak Ridge, Tennessee.

ecological outlooks. Odum (1971) points out that synecological considerations are fruitful but difficult and typically foreign to most traditionally trained biologists. He adds that to understand the complexities of natural systems requires looking through a "macroscope," not just a microscope. Unfortunately the macroscope may never become commercially available, only available as a way of thinking. In the history of ecology a few notables like Lotka, Redfield, Hutchinson, and Lindeman have had the macroscopic view, and to a large degree are responsible for uncovering fundamental ecological principles.

Surely ecosystems have many properties left to be uncovered and it is our contention that some of these properties will be revealed by adopting a holistic view. To many the holistic (macroscopic) view is unacceptable because it requires compartmentalizing, that is lumping large amounts of information into manageable homomorphic models. This compartmentalization should be viewed negatively only if major operational characteristics are omitted or obscured.

Beginning with a well thought out homomorphic conceptualization of a real system, an investigator can explore basic properties of the system with many techniques. If the model expresses time-related processes of the system, it is dynamic. This single factor, time, gives dynamic models a likeness to most, if not all, natural systems—the ability to change.

Beyond considerations in the time domain, the fundamental property of systems is the existence of interrelationships between components. These interrelationships can be represented by exchanges of signals between components. A signal is a measurable observable. For example, any of the following could be a signal: gravity, incident solar radiation, rainfall, material flow, information flow, visual stimuli, sound, etc. Thus, signals are not necessarily rates but anything which is measurable (influencing) and observable (perceived). Signals are the currency of a system, affecting and being affected by system components and other signals. In general, systems analysis deals with studying the ways signals influence and are influenced in systems of all kinds.

A good signal for analytical purposes should be both time and magnitude varying, as are many biospheric signals. Often these signals are oscillatory, ranging from nearly pure sine waves, such as tides, to irregular curve forms that have sinusoidal elements, such as prey–predator fluctuations, rainfall patterns, and temperatures. In fact, many very irregular curve forms can be decomposed into component sinusoids of various frequencies by Fourier analysis.

As well as being a common natural curve form, the sinusoid is a "common" mathematical function, and is also a manageable function.

For the analysis of a dynamic system with sinusoidal signals there is an analysis technique available. This technique is *frequency response analysis*, and owes its development to engineers concerned with testing and improving the performance characteristics of physical systems, particularly those containing feedback. Over the years these engineers have evolved technology and experience permitting direct experimentation and analysis on real and modeled systems. When dealing with real systems, the results of frequency response analysis provide information for model improvement and forecasting in the real system.

Frequency response analysis has potential utility for the ecologist faced with analyzing extremely complex systems, as it deals with dynamic systems operating on time and magnitude varying signals. To illustrate the potential of frequency response analysis we will attempt: (1) to develop a mathematical model depicting magnesium flow in a terrestrial ecosystem, (2) introduce and apply frequency response analyses, and (3) present ecological interpretations of the analyses.

II. Description of the Ecological System

A. State Variables, Fluxes, and Assumptions

The ecological system under consideration was originally modeled by Child *et al.* (1971) as a conceptualization of steady-state elemental transfer in a tropical forest of eastern Panama. The block diagram in Fig. 1 represents a six-compartment version of their ten-compartment model. The original model included four compartments (roots, stems, leaves, and fruits-and-flowers) which are here combined into a single compartment (plants). Lumping these compartments is justified by the fact that Child *et al.* determined that magnesium concentrations in their four plant compartments were statistically homogeneous. One other compartment in the original model (other systems) was used as a sink in the present model, and was not included as a compartment.

Hence, in this study, the tropical forest is conceptualized as a system containing six compartments with steady-state magnesium levels (in milligrams per square meter) as follows:

$$x_1(\text{plants}) = 5.63 \times 10^4, \qquad x_4(\text{litter}) = 4.54 \times 10^2,$$
$$x_2(\text{herbivores}) = 4.30, \qquad x_5(\text{detritivores}) = 2.86 \times 10^1,$$
$$x_3(\text{carnivores}) = 1.10, \qquad x_6(\text{soil}) = 1.75 \times 10^5.$$

Input of magnesium is reasoned to occur via rainfall, a portion of this input (F_{01}, Fig. 1) being intercepted by foliage, and the remainder

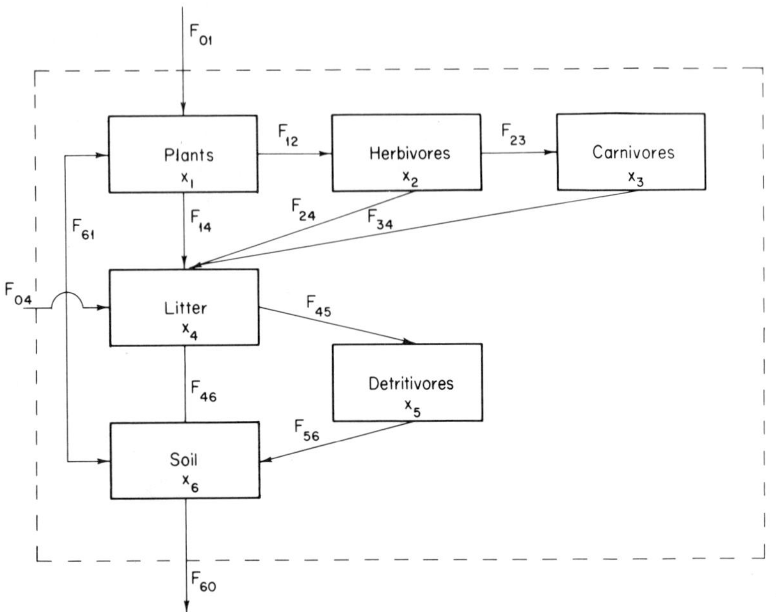

FIG. 1. Block diagram of magnesium flow in a six-compartment representation of a tropical forest ecosystem. Values for flows between compartments may be found in the text. The two forcings (F_{01} and F_{04}) are rainfall inputs of magnesium.

(F_{04}) entering the litter as throughfall. Values of these inputs or forcings in milligrams per square meter per day are

$$F_{01} = 2.25 \times 10^{-1} \qquad \text{and} \qquad F_{04} = 1.10.$$

The intercompartmental transfers (denoted F_{ij}, i the donor and j the recipient compartment) in milligrams per square meter per day, as compiled from Child *et al.* (1971), are

$$\begin{aligned} F_{12} &= 1.379, & F_{34} &= 6.8000 \times 10^{-2}, \\ F_{23} &= 6.8000 \times 10^{-2}, & F_{45} &= 1.275 \times 10, \\ F_{14} &= 1.1514 \times 10, & F_{56} &= 1.275 \times 10, \\ F_{24} &= 1.311, & F_{46} &= 2.714, \\ & & F_{61} &= 1.2668 \times 10. \end{aligned}$$

System output, in the form of runoff, is

$$F_{60} = 1.321 \quad \text{mg m}^{-2}\ \text{day}^{-1}.$$

The underlying assumptions of the original model apply equally well to this model:

(1) The tropical forest is a steady-state system, i.e., magnesium input in rainwater equals output as runoff;
(2) Each compartment of the system is at steady state, with magnesium input–output ratio equal to one;
(3) Intercompartmental transfers of magnesium are linear and controlled by the donor compartments; and
(4) The system has no dead time in any of its compartmental outputs, i.e., the time interval between an input change and an output (response) change equals zero.

B. Formulation of Mathematical Model

A model of magnesium flow in a tropical forest incorporating the actual field measurements, and ecologically founded assumptions, can be expressed as a system of first-order, constant-coefficient, linear differential equations

$$dx_i/dt = F_{0i} + \sum_{\substack{j=1 \\ j\neq i}}^{n} \lambda_{ji}x_j - \sum_{\substack{j=0 \\ j\neq i}}^{n} \lambda_{ij}x_i\,, \qquad (i = 1,\ldots, n;\ \ j = 0,\ldots, n), \tag{1}$$

where dx_i/dt is the instantaneous rate of change in the ith of n compartments in milligrams per square meter per day, λ_{ji} and λ_{ij} are rate constants (per day) and $\lambda_{ji}x_j$ is the flow F_{ji} from the jth to the ith compartment in milligrams per square meter per day.

The rate constants in Eq. (1) were calculated as

$$\lambda_{ij} = F_{ij}/x_i\,, \tag{2}$$

where F_{ij} is the measured flow of magnesium from the ith to the jth compartment in milligrams per square meter per day and x_i is the measured concentration of magnesium in milligrams per square meter in the donor compartment.

In the specific magnesium model for the tropical moist forest, the system of first-order differential equations [Eq. (1)] becomes

$$dx_1/dt = F_{01} + \pi_{61}x_6 - \pi_{12}x_1 - \beta_{14}x_1\,, \tag{3}$$

$$dx_2/dt = \pi_{12}x_1 - \pi_{23}x_2 - \beta_{24}x_2\,, \tag{4}$$

$$dx_3/dt = \pi_{23}x_2 - \beta_{34}x_3\,, \tag{5}$$

$$dx_4/dt = F_{04} + \beta_{14}x_1 + \beta_{24}x_2 + \beta_{34}x_3 - \pi_{45}x_4 - \beta_{46}x_4\,, \tag{6}$$

$$dx_5/dt = \pi_{45}x_4 - \beta_{56}x_5\,, \tag{7}$$

$$dx_6/dt = \beta_{46}x_4 + \beta_{56}x_5 - \beta_{60}x_6 - \pi_{61}x_6\,. \tag{8}$$

The β and π notations [Eqs. (3)–(8)] denote rate constants for biological and physical transfers, respectively. Although all transfers in this model are validly linear and donor-based, the biological transfers might better have been represented as nonlinear or recipient-based transfers. We have, however, chosen to use constant coefficient, first-order linear equations, in part because of the wide spectrum of analysis techniques available, and additionally because with these constraints analyses of a modeled ecosystem yield pertinent information about the system when at or near the steady state (Chap. 1, Vol. II).

To solve Eqs. (3)–(8), rate constants and forcings are necessary. The rate constants, in reciprocal years, from Eq. (2) are

$$\pi_{12} = 2.44 \times 10^{-5}, \qquad \pi_{45} = 2.48074 \times 10^{-2},$$

$$\pi_{23} = 1.58139 \times 10^{-2}, \qquad \beta_{56} = 3.94231 \times 10^{-1},$$

$$\beta_{14} = 2.044 \times 10^{-4}, \qquad \beta_{46} = 5.9713 \times 10^{-3},$$

$$\beta_{24} = 3.04884 \times 10^{-1}, \qquad \pi_{61} = 7.256 \times 10^{-5},$$

$$\beta_{34} = 6.18181 \times 10^{-2}, \qquad \beta_{60} = 7.56 \times 10^{-6}.$$

Once an investigator establishes a mathematical model of some real system, it then becomes possible to analyze the model with many techniques.

III. Concepts in Frequency Response Analysis and Applications to the Modeled Ecosystem

A. Properties of Linear Systems with Sinusoidal Inputs

The behavior of the output from first-order linear compartments or linear systems to sinusoidal input is summarized in Fig. 2. Sinusoidal inputs always produce sinusoidal output of the same frequency. These outputs may differ from the input only in time of peak (phase angle, denoted ϕ) and amplitude (A). For comparative purposes, the differences in input–output amplitudes are often expressed as a ratio (magnitude ratio or gain, output/input, denoted MR). The magnitude ratio and phase angle express characteristics of the compartment through which the sine wave has passed. Later the characteristics that can be determined by input–output relationships of a process block will become clearer.

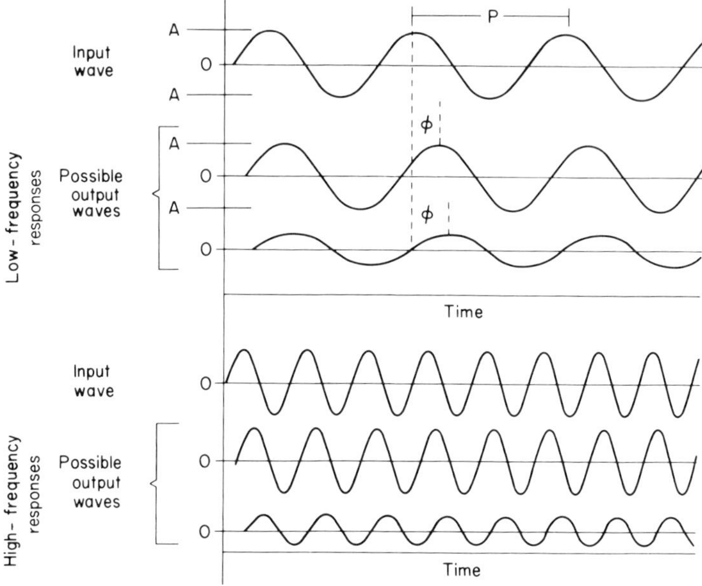

FIG. 2. Typical behavior of sinusoidal inputs in linear systems, where A is amplitude, P is period and ϕ is phase shift.

B. Bode Analysis

When analyzing a model such as the one presented, it is informative to note the changes in magnitude ratio and phase angle in response to a wide range of frequencies. These changes are usually depicted as two curves on a single diagram called a *Bode plot.* A Bode plot for the plant compartment (Fig. 3) illustrates typical changes in MR and ϕ over a range of frequencies. In Fig. 3, the magnitude ratios are presented as decibels ($\text{dB} = 20 \log_{10} \text{MR}$) such that a negative decibel value represents an attenuation of input amplitude, and a positive value amplification.

One property of the process block, attenuation, can be determined in the gain (MR) portion of the Bode plot (Fig. 3). At low frequencies, the compartment output has the same amplitude as the input (zero attenuation). As input frequency increases, the compartment begins to attenuate the signal and the gain curve drops until its slope approaches -1. The gain plot then has two asymptotes: a low-frequency asymptote of 0 dB, and a high-frequency asymptote of slope -1. The point at which the gain curve is 3 dB below the zero line is called the *3-dB down point* or *break point.* The associated frequency (radians per

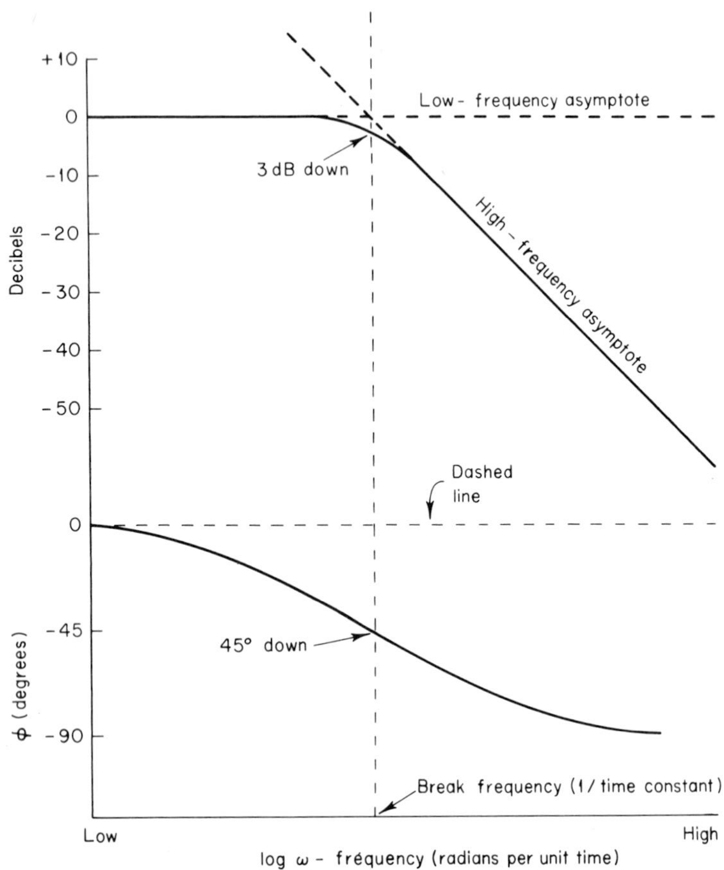

FIG. 3. Bode plot of typical first-order lag behavior. Note the gain portion is plotted in decibels (dB) versus log ω, and the phase portion is plotted as degrees phase shift (ϕ) versus log ω.

day), the *break frequency*, equals the reciprocal of the time constant T. In ecological terms, the time constant is the turnover time.

A second property of the process block, phase shifting, can be determined when the phase angle is plotted against input frequency as shown in the lower portion of Fig. 3. This particular phase curve shows graphically an increasing lag ϕ with increasing angular frequency ω. By lag, it is meant that the output wave peaks at a later time than the input wave (compare input–output waves in Fig. 2). In Fig. 3, the Bode plot depicts a first-order *lag* response. Certain responses are *leads* which will be discussed later. As in the gain plot, the phase plot (Fig. 3) has three distinct regions: a low-frequency asymptote approaching 0°; a high-frequency asymptote which in Fig. 3 approaches −90°; and a transient region between these two asymptotes. The latter has an inflection point at −45° occurring at the same frequency as the break point of the gain plot.

The configuration of the Bode plot reflects the processing attributes of the compartments when treated with a spectrum of frequencies. Figure 3, for example, is a typical first-order lag response characterized by: gain curve less than or equal to 0 with the break point at −3 dB and the high-frequency asymptote slope equal to −1; phase curve less than or equal to 0 with the inflection point at −45°, and −90° marking the high-frequency asymptote. Conversely, a first-order lead response is the negative of each of the first-order lag response properties.

1. *Bode Plots of Multiorder Cascading Systems*

The previous discussion has focused on the typical lag response of a single process block as a means of presenting the frequency response approach. In systems analysis, one is often interested in the behavior of selected blocks of an integrated system or the total system behavior. Such cases are multiorder and easily interpreted by Bode analysis. To illustrate analysis of a multiorder response, a simple, three-compartment cascade (a set of blocks linked in series) has been constructed of the plant–herbivore–carnivore trophic chain from the magnesium model. Figure 4 shows each compartment's response to an input (forcing) to the plants (i.e., $F_{01} = A \sin \omega t$). The response curves for the plant compartment are of the first-order lag type since the signal has passed through only one block. The herbivore response to input to the plant is a second-order lag because the signal has been processed by two first-order blocks in series. Similarly, the output from the carnivores is a third-order lag response. When dealing with n first-order process blocks in series, the high-frequency asymptote for the kth successive phase curve is $k(-90°)$, $k = 1, 2, \ldots, n$. The gain curve (in decibels) and the phase curve for the kth compartment in a cascade is the sum of the individual response curves of the first k compartments. That is: If $P_k'(\omega)$ is the in-system response (MR or ϕ) of the kth compartment in a cascade of n compartments at a given frequency ω; $P_k(\omega)$ is the individual out-of-system response of the kth compartment to an input directly to that compartment at frequency ω; then,

$$P_k'(\omega) = \sum_{k=1}^{K} P_k(\omega), \qquad 1 \leqslant K \leqslant n. \tag{9}$$

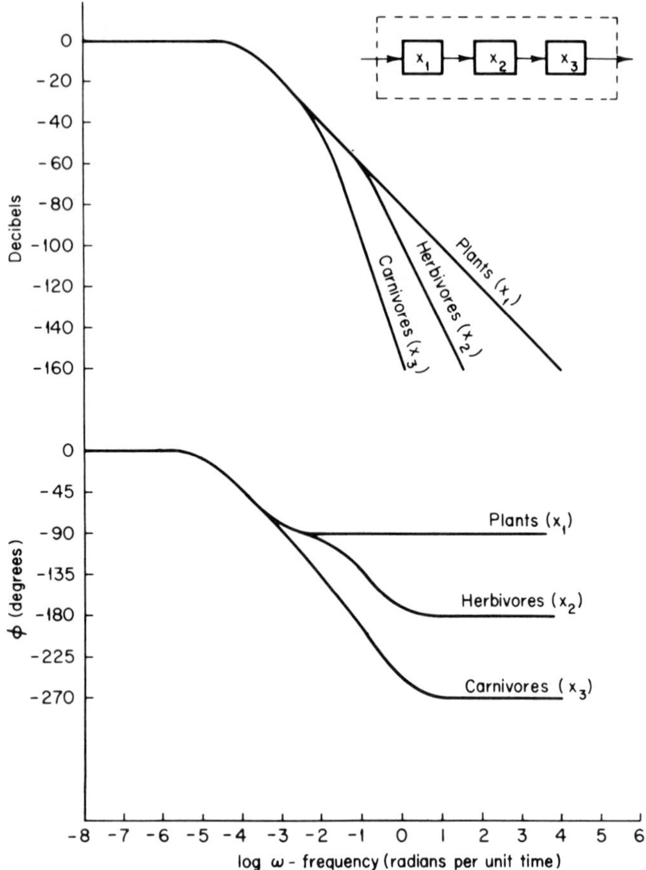

FIG. 4. Bode plot for simple trophic cascade, a subsystem of the tropical moist forest model. Each curve represents a comparison of the input wave to plants to the output from the respective compartments.

Thus, for a cascading system, if one knows the out-of-system responses of each compartment, the in-system phase shift and gain (decibels) of any compartment can be computed using Eq. (9).

An ecologist investigating a complex ecosystem might gain considerable insight to the system's structure and function by reducing the system to a Bode plot. Immediately, the Bode plot shows how the system, or parts thereof, will respond to different periodicities of input. By inspection of a Bode plot, one can observe whether a hypothetical or real oscillatory input to the system will be amplified, not changed, or attentuated as it moves through the system's compartments. This sort of information holds considerable merit for predicting system responses in the vicinity of the steady state to a changed or changing environment.

2. *Ecological Interpretations from Bode Plots*

Unfortunately, very little is known about frequency responses in natural systems. It may be that biological components of an ecosystem coevolve to minimize oscillations thereby preserving internal constancy, or to reproduce or even amplify oscillations (using the oscillations as control signals). Different systems and components have undoubtedly developed different strategies. A given system or component may operate under one strategy for one variable, another strategy for some other variable. One method of approaching questions such as these is through frequency response analysis. For example, if one felt that a particular component of a system possessed first-order lag properties then the determination of the gain for one very high-frequency input wave would allow the investigator to construct the entire gain portion of the Bode plot by drawing a line (slope -1) from the determined gain back to the 0-dB line. This Bode plot can then be used to develop the Laplace transform from the time constant T ($1/T$ is the break frequency), and then the Laplace transform can be used to develop the system equation.

C. Bode Plots and Properties of Complex Systems

The previous discussion has dealt with the construction of Bode plots for simple systems. The ecologist realizes that most biological systems are much more complex than simple cascades. Our intent below is to discuss frequency response analysis of networks more like ecological systems.

1. *Responses of Open and Closed Complex Systems*

Systems can be categorized as either open-loop or closed-loop. In open-loop systems there is no feedback of either material, energy, or information. Conversely, in closed-loop systems a signal once inside can return to its points of origin. Figure 1 depicts a closed-loop system, and the block diagram in Fig. 4 is an open-loop system.

We feel that most ecosystems are closed-loop and thus are of the greatest interest to systems ecologists. It is easy to visualize feedback (closure) in a mineral cycle (Fig. 4), because the mineral elements themselves cycle in the system. In energy flow, however, the unidirectional flow obscures closed-loop characteristics and gives the initial impression that energetic systems are open-loop. In reality, the energy in these systems is probably converted to information and fed back, thus producing a closed-loop system. The existence of homeostasis in natural systems implies regulation which in turn suggests feedback control. This line of reasoning leads us to view most natural systems as closed, where the closure is accomplished by either an information, energy, or material signal.

A closed-loop system may contain compartments linked in series or, in more complex cases, the linkages may branch. The proposed model of magnesium flow in the tropical moist forest (Fig. 1) is a closed-loop system with branching. Analyses of closed-loop systems with internal branching require special considerations not treated in the previous sections.

Again, the objectives of frequency response analysis are to determine properties of individual compartments as well as properties of the entire system. Determining compartmental properties in simple cascades is straightforward, as discussed above, and the system properties are obtained by addition.

2. *Loop Breaking in Closed-Loop Systems*

For a closed-loop system, as exemplified by our model, determination of total system response is accomplished by opening the loop and graphically converting the open-loop response to a closed-loop (total system) response. The open-loop response, or, as it is sometimes called, the "broken-loop" response, is obtained by eliminating the feedback path and treating the resultant open-loop system with a spectrum of sinusoidal inputs. For the tropical moist forest ecosystem model [Fig. 1, Eqs. (3)–(8)], we considered the transfer of magnesium from soil-to-plants to be the feedback path. To open the loop, π_{61} was set to zero and the output from the soil (x_6) and the forcings (F_{01} and F_{04}) were appropriately increased to preserve steady-state values. The resultant open-loop system was treated with a spectrum of sinusoidal inputs using an available software package for frequency response (Kerlin and Lucius, 1966). The values obtained for the broken-loop system actually reflect little of the operating biological system, but are necessary to determine the total closed-loop response. Conversion to the closed-loop response can be accomplished by plotting decibels and ϕ of the open-loop response on a standardized graph, the *Nichols plot*. Figure 5 is a Nichols plot which contains two sets of coordinates. The open-loop responses at selected frequencies are plotted on the rectangular coordinates, and the closed-loop responses for these frequencies are then read from the curvilinear coordinates. The values for the bold curve in

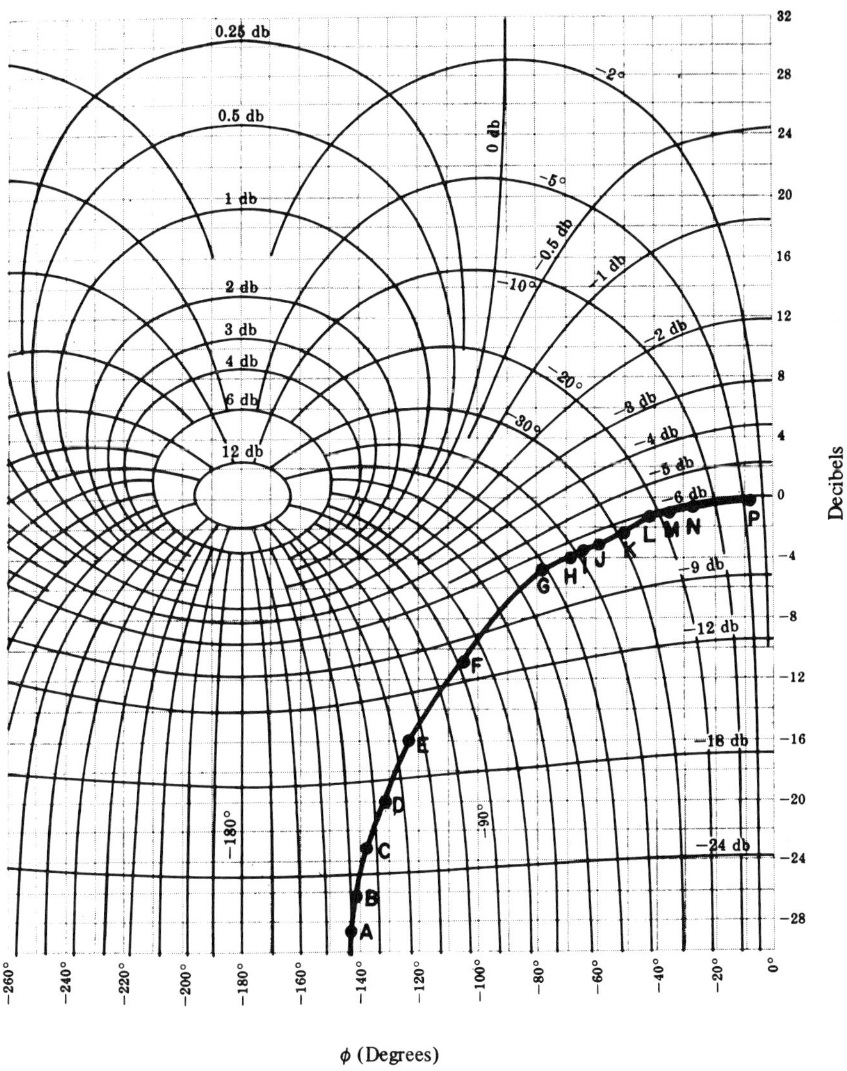

FIG. 5. Nichols chart for determining closed-loop response from open-loop response. Dark line to lower right is the response for the tropical moist forest model where the frequencies (radians per day) for the points marked are: A = 7×10^{-4}, B = 6×10^{-4}, C = 5×10^{-4}, D = 4×10^{-4}, E = 3×10^{-4}, F = 2×10^{-4}, G = 1×10^{-4}, H = 9×10^{-5}, I = 8×10^{-5}, J = 7×10^{-5}, K = 6×10^{-5}, L = 5×10^{-5}, M = 4×10^{-5}, N = 3×10^{-5}, P = 1×10^{-5}. Nichols chart reproduced with permission from DiStefano *et al.* (1967).

Fig. 5 convert the open-loop response to the closed-loop response for the tropical forest model.

For comparative purposes, we have presented both the open- and closed-loop response curves for the tropical moist forest in Fig. 6.

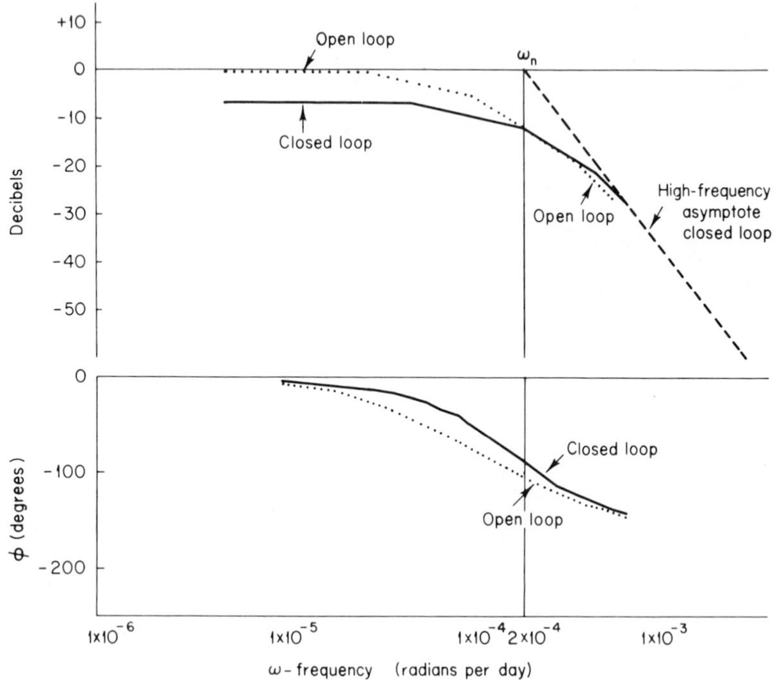

FIG. 6. Total system open- and closed-loop response. Points for Bode plot obtained from Fig. 5.

Clearly, the gain and phase segments differ only slightly when opened and closed. The significance of this similarity will be discussed below. Thus, through the conversion process, one produces a graphical representation of the total system's behavior from which intrinsic system properties can be determined.

D. NATURAL FREQUENCY AND DAMPING FACTOR IN CLOSED–LOOP SYSTEMS

One highly informative property of feedback control systems (with second-order or near second-order responses) is the *natural frequency* ω_n; another is the *damping factor* ζ. Both of these properties are most easily explained by a physical example. Consider a pendulum in a vacuum with friction only at the pivot. The pivotal friction can be adjusted by a control knob. When the friction ζ is set to zero, the pendulum, once set in motion, will remain in motion and oscillate at a constant frequency. This frequency is the natural frequency ω_n. It is characteristically unrestrained, i.e., undamped. If any friction is applied to the pivot, the pendulum will eventually come to rest. If the friction is slight ($\zeta < 1$, *underdamped*), the pendulum will oscillate around the equilibrium point before coming to rest. There is one frictional setting ($\zeta = 1$, *critically damped*) at which the pendulum reaches the equilibrium point without overshoot and in the shortest possible time. If the friction is increased beyond this setting ($\zeta > 1$, *overdamped*), the pendulum will still equilibrate without overshoot, but the time to do so will be longer than if critically damped.

Variables ζ and ω_n should be of particular interest to the biologist; they reveal the nature of a perturbed system's return to equilibrium. Figure 7 shows the effect of ω_n and ζ on the transient response (of a second-order system) to a step forcing.

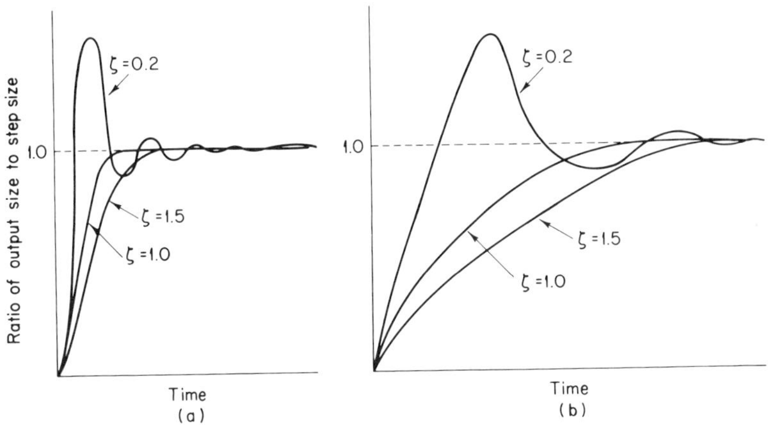

FIG. 7. Response of second-order systems with (a) fast and (b) slow natural frequencies to a step function, where the dashed line is equilibrium.

From a closed-loop Bode plot, as in Fig. 6, the natural frequency for the entire system is read as the frequency at which the high-frequency MR asymptote intersects the 0-dB line. If the system includes no dead time, the phase curve should be $-90°$ at the natural frequency. This check holds only for second-order systems or higher-order systems with typical second-order behavior.* Deviations from $-90°$ at ω_n produce less exact but still useful information. In the case of the tropical moist forest, the phase shift is $-87°$ at ω_n, strengthening an assumption that the flow of magnesium through this system behaves in an essentially second-order fashion.

Given ω_n, ζ can be obtained analytically or graphically. Analytical solution takes the form

$$\zeta = \frac{(\omega_n{}^4 - \omega_n{}^4\,\mathrm{dB}^2 - \omega^4\,\mathrm{dB}^2 + 2\,\mathrm{dB}^2\omega^2\omega_n{}^2)^{1/2}}{(2\,\mathrm{dB}\cdot\omega\cdot\omega_n)}, \tag{10}$$

where $\mathrm{dB} = 20 \log_{10} \mathrm{MR}$, ω is input frequency, ω_n is natural frequency, and ζ is damping factor, and also

$$\zeta = \frac{\omega_n(1 - \omega/\omega_n)^2(\tan\phi)}{2\omega}, \tag{11}$$

where ϕ is phase shift in degrees.

More commonly, where exact solutions are not required, the damping factor (ζ) is estimated by comparing a normalized gain curve with a family of second-order gain curves with known damping factors (Fig. 8). A gain curve is normalized by relating all frequencies to the natural frequency (ω/ω_n). The present system ($\zeta = 2.25$ curve on Fig. 8) is overdamped because its damping factor is greater than 1.

E. Stability Analysis

Probably one of the most topical problems in ecology is how one measures ecosystem stability (e.g., Woodwell and Smith, 1969). Contributing to the problem is the difficulty in defining stability to the satisfaction of a majority of ecologists. Presently, two definitions prevail. One equates stability to persistence, i.e., constancy, the other to resilience, ability to respond. These two approaches to the stability concept have been recently discussed by Margalef (1968) and Lewontin (1969). In control systems analysis, a stable system is defined as one in which the output response is bounded for all bounded inputs (Coughanour and Koppel, 1965). An unbounded response to a bounded input typifies an unstable system. By bounded it is meant that the signal is within fixed limits. For example, yearly rainfall in eastern Panama could be expected

* In practice, control system engineers have found that many high-order feedback systems have essentially second-order behavior. This allows investigators to circumvent the rigors of analytical solutions, and approximate a higher-order system's behavior with second-order analytical and graphical solutions.

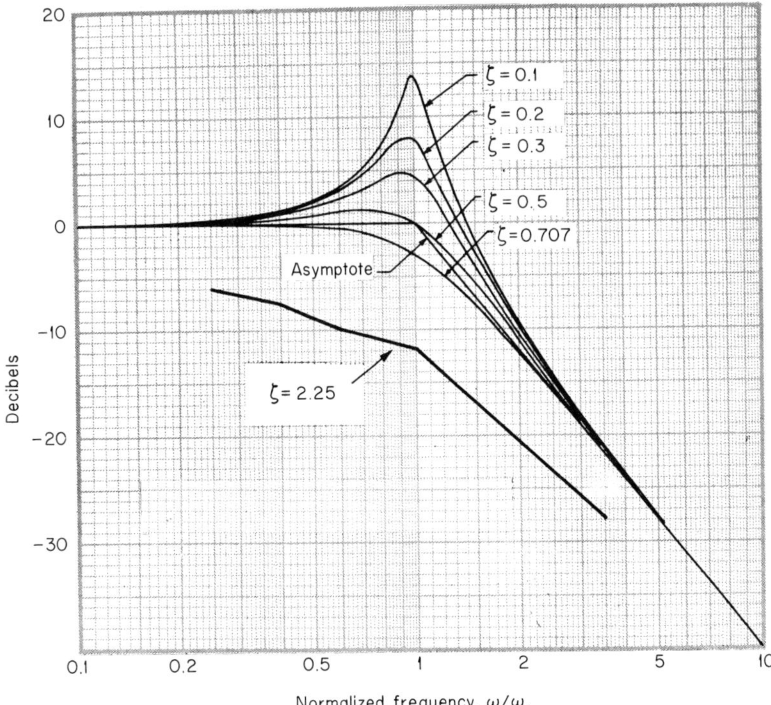

FIG. 8. Family of second-order Bode plot templates. The curve for $\zeta = 2.25$ is for the tropical moist forest model (values taken from Figs. 6 and 7 and scaled by natural frequency to fit the template. Reproduced with permission from DiStephano *et al.* (1967).

to be within the bounds of 0 and 500 cm/year. For all practical purposes, inputs to biological systems are bounded. Although certain bounded input waves may be very irregular, they can be partitioned into sinusoids by Fourier analysis and then responses of the system to these separate waves can be determined by frequency response analysis.

Positive feedback in a closed-loop system produces progressive amplification of the signal resulting in an unbounded output. A progressively amplified signal in a biological system could be expected to destroy some component(s) because biological components usually operate only within fixed limits. A good example of these limits is the zone of thermal tolerance in homeotherms. Temperatures either above or below this zone are beyond the limits of the organism's control and, depending on the magnitude, can produce damage.

The band of frequencies over which a system is unstable can be determined from a Nichols plot (Fig. 5). The point of maximum instability for a closed-loop system on a Nichols chart is the intersection of the open-loop 0-dB and $\pm 180°$ ϕ coordinates. The region around this point is called the *m-circle* (DiStefano *et al.*, 1967). Any frequency which places the system's response in or near the *m*-circle will cause the system to respond in an unstable fashion, i.e., produce unbounded outputs. Our model of the tropical moist forest is by these criteria extremely stable (Fig. 5), with greatest instability at 2.0×10^{-4} rad day^{-1}, i.e., 2.9 cycles per millenium, which is the natural frequency. This is not to say that the tropical moist forest ecosystem is stable under all frequencies tested, and this problem will be discussed later.

Maximum instability in a closed-loop system occurs at the intersection of the open-loop 0-dB and $\pm 180°$ ϕ coordinates because under these conditions the feedback signal would be shifted 180° back into phase and unattenuated. The effect of cycling a wave of 0 dB and 180° phase shift in a closed system is to produce an evergrowing feedback signal, because the feedback wave augments the input wave.

One way of expressing distance from the *m*-circle is by the phase and gain margins, determined from the closed-loop Bode plot. The phase margin is the angular difference between $\pm 180°$ and the close-loop phase shift at the frequency of 0 dB. The gain margin is the MR (in decibels) of the frequency at which the closed-loop phase plot is at $\pm 180°$. We estimated from Fig. 6 the gain margin as +60 dB, and the phase margin as −170°. Both these values are measures of the margin of stability for the tropical moist forest model, i.e., how much phase or gain would have to be added to produce a maximally unstable system.

F. Component Analysis by In- and Out-of-System Comparisons

In the preceding sections we have illustrated how total system frequency response can be determined graphically. The results of this approach satisfy one objective of frequency response analysis, namely, determining total system attributes. It is also important to know how each system component affects the total system response. The total system response cannot be found simply as a sum of the responses of the system components, but a component-by-component breakdown yields a ranking of compartments by relative influence.

Figure 9 shows Bode plots for each compartment of the tropical moist forest as it behaves in the system (solid line), and out of the system (dotted line). An in-system response is obtained by comparing the output from a selected compartment to the system's input. For example, the in-system response of the carnivores (Fig. 9c) is found by comparing the carnivore output to the sum of the two forcings (F_{01} and F_{04}). The out-of-system response is found by comparing the output from a compartment with the input to that same compartment. The out-of-system response gives the intrinsic properties of the process block, while the in-system response gives properties realized as a result of being coupled into the system and, in effect, receiving preprocessed waves.

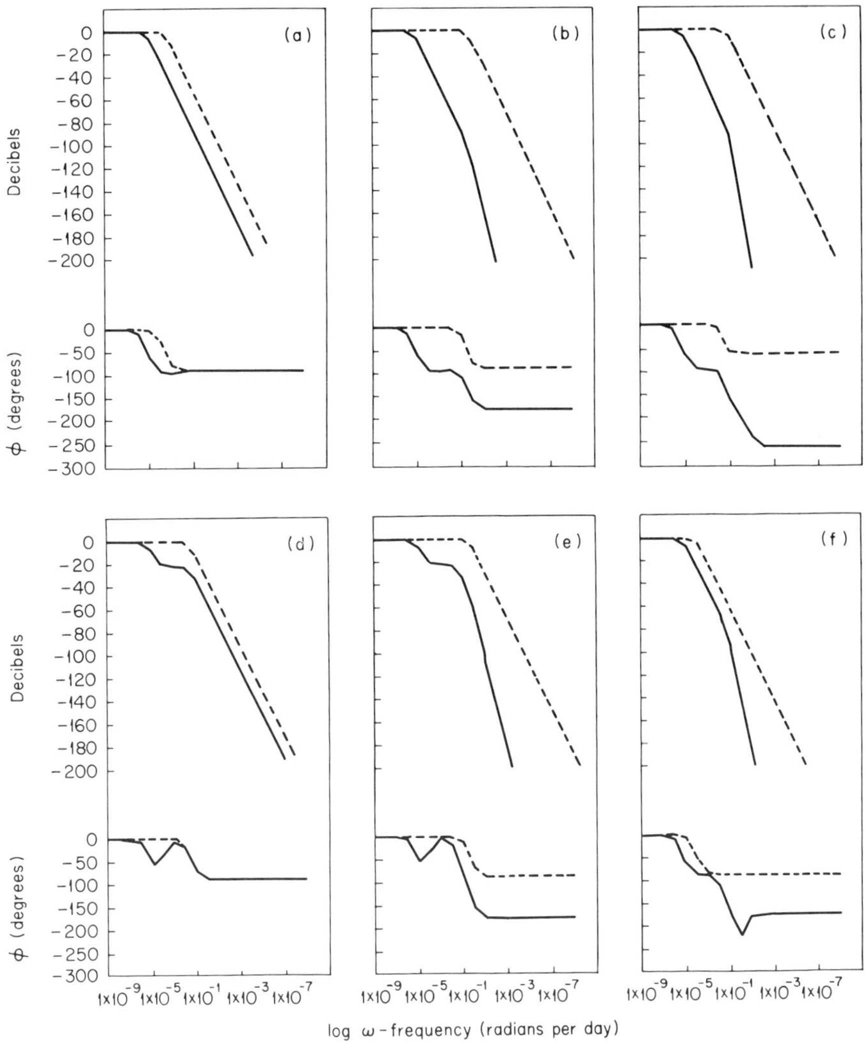

FIG. 9. In- and out-of-system Bode plots for (a) plants, (b) herbivores, (c) carnivores, (d) litter, (e) detritivores, and (f) soil. Solid line is in-system response, dotted line out-of-system response.

The relative influence of each compartment on the entire system's frequency response can be determined by the differences in the in-system and out-of-system Bode plots, particularly in the region of the break frequency. Clearly the in- and out-of-system differences are smaller in compartments x_1, x_4, and x_6 (plants, litter, and soil) than in compartments x_2, x_3, and x_5 (herbivores, carnivores, and detritivores). This can be concluded by comparing the pairs of solid and dotted lines in Fig. 9.

Compartments of the tropical moist forest model which show little differences between in- and out-of-system responses are those with the largest time constants T. Hence, the frequency at which they begin altering the wave (breakpoint is $1/T$) is lower than that of compartments with smaller time constants. To clarify this point, let us compare Figs. 9a, b, and c, which represent the plant–herbivore–carnivore trophic chain. Proceeding down the chain, there is a progressive increase in the difference between the in- and out-of-system responses. This is due to the plants (x_1) attenuating the wave at a very low frequency (3-dB-down point at 2.5×10^{-4} rad day^{-1}). The herbivores (x_2), with a break frequency of 1×10^{-1}, have no influence on the wave at lower frequencies, and only affect the in-system wave at frequencies above 1×10^{-1}. The situation is similar for carnivores (x_3), and the same sort of development can be used to interpret the remaining graphs.

1. *The Determination and Importance of Compartmental Time Constants*

To summarize, the gain portions of the compartment Bode plots are useful in determining the time constants ($1/T$ is break frequency). The time constant can be used as a measure of the effect that a given compartment will have on the gain properties of the system as a whole. In terms of the tropical moist forest model, the dominant compartments (major attenuators) in decreasing order are soil, plants, litter, carnivores, herbivores, and detritivores. The approximate time constants (days) taken from Fig. 9a are

$$T_1 = 4.4 \times 10^3, \qquad T_2 = 3.3, \qquad T_3 = 1.4 \times 10^1,$$

$$T_4 = 3.3 \times 10^1, \qquad T_5 = 2.5, \qquad T_6 = 1.2 \times 10^4.$$

To illustrate the importance of the compartmental time constants, recall that we determined, after accepting the second-order assumption, that the natural frequency (ω_n) of our system was 2.0×10^{-4} rad day^{-1} (see Fig. 6). In any second-order system with no dead time,

$$\omega_n = (T_i T_j)^{-1/2}, \tag{12}$$

where T_i, T_j are time constants of the two compartments and ω_n is natural frequency.

Substituting T_1 and T_6, the time constants for the two most dominant compartments, into Eq. (12) yields a natural frequency of 3.0×10^{-4}. That the actual and estimated natural frequencies are quite close indicate that the plants (x_1) and soil (x_6) influence the system's behavior more than any other compartments. Using any other pair of time constants in Eq. (12), the resultant natural frequency will differ from 2.0×10^{-4} by at least an order of magnitude.

The phase plots in Figs. 9 serve to identify the order of the individual compartment's in-system responses. The order of the response in a lag system is the high-frequency phase asymptote divided by $-90°$. In the plant–herbivore–carnivore trophic chain, the in-system high-frequency asymptotes are $-90°$, $-180°$, and $-270°$, respectively, and typify first-, second-, and third-order lag responses. The order of a response is indicative of the nature of the compartments that came before it. The litter compartment x_4 (Fig. 9d) receives first-order, second-order, and third-order input waves from the trophic chain, plus a pure (zero-order) wave (F_{04}). Since the phase curve of x_4 has a high-frequency asymptote of $-90°$, the input waves from the trophic chain must be of very small amplitude, and the response of x_4 is due mainly to the direct forcing. The deflection of the phase curve (Fig. 9d) at intermediate frequencies can be accounted for by wave input from x_1. As the frequency increases, the waves from x_1 are attentuated and the phase deflection for x_4 is reduced.

In summary, examination of the in- and out-of-system phase curves adds to one's knowledge of the system primarily by indicating the order of the various components.

IV. Discussion

A. Signals in Ecological Systems

Our intent below is to discuss the applicability to ecological systems of ω_n and ζ, stability, and in- and out-of-system differences. We regard these measures as excellent evaluators of signals operating in closed-loop systems. In any dynamic analysis, one is interested in the flux within a system of a signal or signals. In the present case, the signal of interest was magnesium. This element flowed along several pathways, and a portion eventually returned to its point of origin, thus becoming a feedback. Our model deals exclusively with flow of a material, but not all models need be restricted to one type of signal. Basically, there are three types of signals: material signals, as in our model; informational signals; and energy signals. All are common in nature if not omnipresent.

Material and energy flows are usually obvious, and easily envisioned; informational flows are often very subtle, and difficult to quantify but nonetheless important. For example, a bee distributing pollen from flower to flower in the tropical moist forest is, in terms of magnesium, redistributing an extremely small amount of material in the system but is involved in an information transfer process of considerable importance.

B. Significance of Feedback Loops

The significance of a feedback loop is often difficult to ascertain unless the loop is in some way eliminated. In the bee example, if all pollen transferred by the bees were rendered sterile, or if the bees were eliminated from the system, the effect of breaking the flow of viable pollen grains from plant to plant would seriously alter the system. Occasionally an investigator inadvertently breaks a feedback loop in a natural system. A good example is Mason and Odum's (1967) study of the horned passalus beetle (*Popilius disjunctus*). These authors found that passalus beetles when separated from their feces and frasse lost weight and died. Subsequent investigations showed that the beetles reingest the feces and frasse for the nutritional content of the associated fungal growth. Thus, the passalus beetle in nature depends upon a feedback. Of greater importance are feedback controls between predators and prey. Often when these feedbacks are altered or eliminated, there may result unusual oscillations of either population.

In systems with feedback, one is often interested in the extent that feedback operates as a control. In the tropical moist forest it is conceivable that one could interrupt the feedback of nutrients from soil to plants, perhaps by killing the mycorrhiza with a fungicide, and supplement rainfall with additional nutrients to maintain the plants. However, this would certainly be very difficult to do. Alternatively, an idea of the control action of a feedback loop could be obtained through use of frequency response analysis of a system model.

In frequency response analysis, breaking the loop serves to elucidate the closed-loop response, and to demonstrate graphically the influence of feedback control. Loop-breaking is a necessary analytical technique in frequency response analysis, and may or may not concern the practicability of breaking a feedback in a natural system.

1. *Control Action and the Deviation Curve*

There was very little effect on frequency response of breaking the loop in our model. By plotting the ratio of the closed-loop and open-loop responses at different frequencies (a deviation curve, Fig. 10), one can determine the band of frequencies over which the feedback path has possible control attributes. Typically, a deviation curve such as Fig. 10 has three zones:

(1) Zone A, in which feedback reduces signal oscillations,
(2) Zone B, in which feedback increases oscillations, and
(3) Zone C, in which feedback has little effect on a signal.

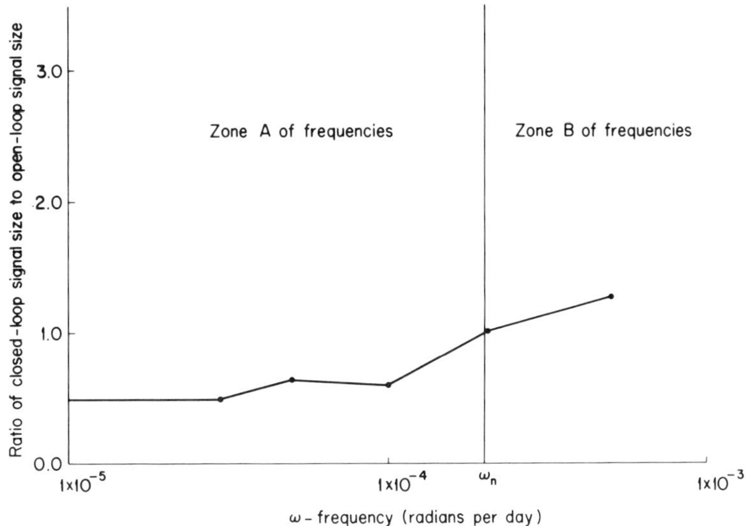

FIG. 10. Deviation curve for tropical moist forest model. Ratio of closed-loop and open-loop response versus frequency.

For the tropical moist forest model, at frequencies above 2×10^{-4} rad day^{-1} (period is 62,830 days per cycle) signal flow along the feedback path is more oscillatory than if the feedback path was absent. If the deviation curve showed an extremely high peak in this, the Zone B range of frequencies, one would predict strong oscillations if the system were exposed to these frequencies of input. In the case of our model, the input frequencies in Zone B are so low that they are not realistic in terms of the tropical moist forest ecosystem. Also, the rise of the deviation curve in the Zone B frequency range is relatively slight (Fig. 10). If the deviation curve had risen sharply in Zone B, or had lower values in Zone A, this would have indicated strong control attributes for the feedback path. This does not imply that this feedback is unimportant for the integrity of the system, in fact, it would be unrealistic to claim that the flow of magnesium from soil to plants has little importance in the magnesium cycling in the tropical moist forest.

In a donor-based model, the output from a particular compartment is "controlled" by the amount of material in that compartment. Instituting control through constant coefficients in effect is a substitution for the action of many undefined informational comparators and relays. Thus donor-based, constant-coefficient formulations might be considered the most elementary type of system control.

As a consequence of this approach, the deviation curve indicates little control action in the present model. If one were seeking to improve the model, control action of a different sort would probably be a logical addition. In other systems where material flow is analyzed by engineers or economists, the control action in the system is often separated from the material flow in the form of informational relays or decision-making processes (Forrester, 1961). This is undoubtedly the way biological systems should be handled as well. Control mechanisms, especially of the informational type, are difficult to measure in nature and difficult to incorporate in a model, but are nonetheless essential in representing the control aspects of a natural system.

2. *Interpretations of the Natural Frequency and Damping Factor*

Regardless of the structural formulation of a model (donor-based control or more complex models with informational feedbacks), dynamic analysis of the frequency response yields insight into the performance of the modeled system. Variables ω_n and ζ summarize the total system response to varying inputs. The tropical moist forest model is highly damped ($\zeta = 2.25$) and has a low natural frequency ($\omega_n = 2.0 \times 10^{-4}$ rad day^{-1}). The model as a representation of magnesium flow in the tropical moist forest predicts the system to be very slow to equilibrate once perturbed, but very difficult to perturb. The modeled system, as a whole, has these properties because its principal components have large time constants (long turnover times) and small turnover rates. Any system, be it an industry or a forest, with large inventories (standing crops) and slow process rates (flows), must have large time constants. The tropical moist forest both intuitively and from the available data is in this category. Exactly how ζ and ω_n calculated for the tropical moist forest compare with other natural systems is presently not known as comparative data are lacking. Nonetheless we view ω_n and ζ as parameters which depict total system characteristics, and thus have great potential for comparative purposes. They provide a summary of entire system behavior regardless of the number or nature of the system's components.

Consider aquatic and terrestrial trophic chains which seem to be inverses with respect to the time constants of at least the first two trophic levels. The primary producers in aquatic systems have relatively small time constants and display blooms and die-offs through time. Time constants become longer and fluctuations less violent in higher aquatic trophic levels. In many terrestrial systems, on the other hand, the standing crop biomass of primary producers is comparatively less fluctuating and usually has a larger time constant than the aquatic counterpart. One is tempted to postulate different total system behavior for these cases based on the attributes of primary producers, but the total system behavior as reflected in ω_n and ζ may be identical for these two types of ecosystems, suggesting some similar overall strategy.

Ecosystem strategies or optimization schemes we hypothesize to fall into three categories with respect to damping factors (ζ). Referring to Fig. 7, if critically damped, the system as a whole equilibrates to a changed input in the shortest possible time. If underdamped, the system maximizes on response time and necessarily sacrifices equilibration ability. An overdamped system requires a long time to reach equilibrium but is the most difficult to perturb. These strategies might be respectively termed *response-time and input-optimization strategy* (critically-damped); *response-time minimization strategy* (underdamped); and *input-maximization strategy* (overdamped).

Systems operating under these strategies should possess certain attributes as a consequence of developing under different environmental conditions. For a response-time minimizing system, the components should respond very rapidly to any change in input, which in turn demands large turnover rates (small turnover times and relatively small standing crops). Such systems might be expected where limiting factors or critical inputs are rapidly changing or unpredictable. The desert, for example, typically responds very rapidly to inputs of rain, and thus might be a time minimizer with respect to water.

Conversely, input-maximizing systems should have relatively long turnover times, and for this reason should show very slow responses to input changes. Systems with these properties would "warehouse" inputs (maintain large inventories relative to inputs). The oceans have been receiving increased inputs of minerals due to pollution since the time of the industrial revolution. Only recently has any comment been made on hypothetical or realized changes in oceanic-system behavior. This may be a possible indicator that the seas are input maximizers.

Response-time and input-optimizing systems must possess components which in the net effect return the entire system to a maintained equilibrium in the shortest possible time after a perturbation. Engineers often manipulate control systems (by changing time constants of appropriate components) so that the systems become response-time input optimizers ($\zeta = 1$). Unfortunately, at this time, there is insufficient specific data available to ascertain what ecosystems might be of this class. One is tempted to suspect that evolution at the ecosystem level may favor response-time input-optimization attributes. Clearly this area needs investigation.

Variables ω_n and ζ provide a measure of a system's resistence and persistance, in that they reflect the difficulty of perturbing a system, as well as the time required to return to equilibrium once perturbed. Therefore, ω_n and ζ can be used to delineate both types of stability generally defined by many ecologists (Margalef, 1968).

C. Stability in Dynamic Systems

It is important to reemphasize that stability as discussed in this chapter refers to the behavior of signals in closed-loop systems. Since each system component as well as the system as a whole has specific limits, any test or measure which reflects changes in signal oscillations within a closed-loop system can serve as a measure of relative system stability. It is highly unlikely that any organism or ecosystem is capable of processing all frequencies of input signals. This constitutes the foundation of frequency stability analysis. As we demonstrated for the tropical moist forest model, there is a frequency which places the open-loop response closest to the *m*-circle (zone of maximum instability). The important consideration here is not the specific range of this frequency, but the fact that an unstable frequency exists for our model and for any closed-loop system.

1. *Factors Contributing to System Stability*

Equally important is an understanding of what factors contribute to a system's stability or lack thereof. Initially, any homeostatic system must have control. Therefore, within a system positive or negative signals must flow, be read, and promote responses. Systems with control also typically have feedback of some sort. Since for the tropical moist forest model, magnesium was the only signal flowing, we will limit our discussion of factors affecting stability to amplification, attentuation, and dead time of a single signal.

Signal amplification from a single process block occurs if the out-of-system gain (decibel) curve has positive values. Amplification of a material signal can occur in the whole system when the feedback signal augments the input. Signal amplification or addition of signals in a closed-loop reduces the gain margin and, depending on the amount, can move the system into relatively unstable conditions. For this reason, the presence of feedback can be considered a necessary but not sufficient condition for stability and control. A gain margin of zero produces maximum instability.

Attenuation of input signals operates as the inverse of amplification. Signal attenuation affects the gain margin and generally moves the open-loop response away from the *m*-circle, especially if the feedback in the system is highly attenuated.

Dead time affects system stability primarily by altering the phase margin. A pure time delay reduces the phase margin algebraically; the phase margin is reduced by subtraction of the total pure time delay in the system. Variable time delays operate in a more complex fashion. They can reduce the phase margin but they can also create amplifications.

Forrester (1961) states that to exhibit oscillation a feedback system must contain amplification. He adds that a variable delay can create such amplification. Additionally Forrester sees one generality emerging from considering the stability of economic systems, specifically that managerial actions are more apt to reduce amplifications in a system than forceful countercyclical policies. In an ecosystem, it is at present difficult to determine whether "managerial policies" or "forceful countercyclical policies" are operating. The latter are appealing in a coevolutionary sense because populations are constantly interacting and adjusting flows.

2. *Suggested Uses and General Discussion of Stability Analysis*

Stability analysis can be used in field studies provided specific data are collected or available. Since a stable system produces bounded outputs for bounded inputs, one need only establish these bounds. Biological systems differ from physical systems in that during periods of stress they change not only their composition but many of their compartment properties as well. If one has determined established limits for a given biological system, he can field test the system by comparing the output responses to varying input frequencies. Outputs exceeding established limits (bounds) would more than likely indicate some internal breakdown or major change. An example might be to measure and vary the phosphate input to a pond and compare to the dissolved organic matter (DOM) at the output weir. An increase or decrease of DOM beyond established bounds would indicate an unstable or changed system. For stability analysis of this sort, baseline information on output bounds are essential. Although only a few stability analyses of biological systems have been attempted, there are good baseline data sets available. For example, records of solar input to lakes and ponds and their biological oxygen demands (BOD), frequency and duration of light and dark regimes administered to migratory birds and measures of their migratory restlessness (*Zugenruhe*) or degree of fattening, and frequencies of fertilizing and crop yields. Comparing inputs and outputs has considerable applicability to pollution research and abatement programs. Stability analysis could help answer questions such as: "At what frequencies (if any) can an industry or a municipality discharge its effluent without altering a river's biological stability?"

Although stability analysis conducted by comparing input and output limits will almost conclusively denote an internal breakdown or change, the exact location or compartment is not elucidated. To determine the compartment, one has two options; construct a mathematical model based on the best information available on the principal components of the system and follow the loop-breaking techniques, or examine the individual response of each system component in nature. Both methods will indicate which components contributed to the instability or are likely suspects.

We feel that stability analysis by frequency response techniques is considerably more meaningful than many other techniques, especially those which hold as their major tenet that stability and diversity are related. Regardless of one's explicit definition for stability there must always be included either implicitly or explicitly an element dealing with dynamics. Almost everyone would agree that a large lead ball lying at rest in the bottom of a deep well is a "stable" object. The ball owes its "stability" to its mass, the earth's mass, the attraction between the two and a whole array of other impinging forces. The point is that the ball is stationary or "stable" because all the forces acting together create this situation. Any attempt to measure whether a system is stable must consider the forces, not just the statics. Inventories, like species and individual counts expressed in diversity indices, or frequencies, or densities, are static measures and exist because of some dynamic forces. Measures such as these are used commonly in ecological circles to express species apportionments, and all too often are related to the community or system's stability. In our opinion, stability arises not necessarily as a function of the number of components, or their richness, or diversity, but rather from the nature of the couplings, i.e., the types of interactions and communications between components. It might be that systems with higher species diversity are more stable than ones of lower diversity simply because there are more couplings. But it seems more reasonable to suspect that what makes one system more stable than another are the internal interactions and abilities to process environmental inputs of specific limits. In terms of species diversity a mangrove forest is considerably less diverse than a temperate deciduous forest. But is the mangrove forest less stable? Species and the diversity of species no doubt affect stability by the way they process and generate signals. As Hairston *et al.* (1968) concluded, much more experimental and observational work is necessary before the range of any functional relationship between diversity and stability can be claimed with confidence.

In the present case the tropical moist forest *model* was shown to be stable over all frequencies tested. This is not, however, to say that the tropical moist forest *ecosystem* would necessarily be stable over these input frequencies. That is, the stability analysis conducted is necessary to determine ecosystem stability but not sufficient. Sufficiency depends on the model being an isomorph or homomorph of the real system. We are confident that, in the broadest sense, the model represents the behavior of the tropical moist forest ecosystem at and near steady-state conditions. The greatest uncertainty with respect to its output centers on the data base.

According to Child *et al.* (1971) the standing crops (compartment contents) for all compartments were taken in the field, but not all the fluxes were. Some were determined algebraically, using steady-state assumptions, and others were estimated from the literature. So the frequency which set the open-loop response nearest the *m*-circle (which happens in this case to be ω_n) should not be regarded as conclusive or absolute for the tropical moist forest ecosystem. The real point of interest is that there are input frequencies which create less stable conditions and some that create more stable conditions.

There is yet another value to stability analysis as alluded to earlier, and that involves its potential as a model tester. We have shown that, in closed-loop systems, stability is dependent on amplification, attenuation, and dead time. Where one has the luxury of many input and output data points for various frequencies but only a conceptualization of the mathematical model, frequency response analysis and stability analysis offer a powerful means of perfecting such a conceptualization. Frequently, systems engineers use this approach when constructing models. Tucker and Wills (1962) lucidly present the procedures for making adjustments in a model to fit values observed from the system.

D. In- and Out-of-System Behavior

The difference between in- and out-of-system behavior for a selected component reflects the effects of coupled interactions. One might rightfully expect almost any entity to perform differently when tested uncoupled than when tested in some system context; this is the essence of systems thinking. Wheel balancing of an automobile tire is an example. Should the balancing be done off the car, on some test rack, the effects of drum rotational properties, the difference in lug weights, the weight characteristics of the hubcap, influences due to air flows, weight loading on the tire as influenced by shock absorbers or springs, etc. are not considered. The influencing variables are many, each operating to add, subtract, or magnify the wheel's rotational performance. The point is, to obtain the best wheel balancing, an in-system test is preferable. But to determine the influence of the car on the wheel an out-of-system comparison is necessary.

Not only are many automobile parts frequently tested and "tuned" out-of-system and then expected to operate in-system, but so are many ecosystem components. Thus, when an entire set of components is investigated out-of-system, and then coupled together, it is often found that the coupling (sum) of the parts does not equal the "whole" system. In these cases, the values for the parts are incorrect. We believe that the notion that a system is not the "sum" of its parts arises from incomplete knowledge of the couplings, i.e., operational variables. It is on the subject of couplings that ecological systems analysts should focus their attention, for couplings are what separate functional systems from inventories, and no one should expect an inventory to "sum" to a functional system.

Interestingly, in-system/out-of-system discrepencies are not new considerations in biology. Physiologists, embryologists, pathologists, and many others have, for years, exercised considerable caution when interpreting *in vivo* phenomena from *in vitro* data. The ecologist faces the same problem in ecosystem analysis, and should operate as cautiously. Systems analyses, where an integrated view is essential, can elucidate the effects of an entire system upon the properties of a single compartment and vice versa.

Two main points emerge from a consideration of in- and out-of-system comparisons. First, dominant components are quantitatively ranked, and second, the effects of interactions are measured.

1. *Dominant Components*

Dominance shows up, as in the case of the tropical moist forest model, as the compartments which have the greatest effect on the circulating signal. Dominance as used herein equates more or less with control over signal flow, but since there are no specific set points or control mechanisms in our system, dominance denotes a component's effect upon the signal due exclusively to its process properties. The plants and soil through such properties control the amplitude and phase of the circulating signal to a greater extent than the other compartments. Because the tropical moist forest model is purely linear and also contains no dead time, behavior of the circulating signal must be dominated by the component(s) with the largest time constant(s). This would be so for any linear system, especially open-loop systems. Where nonlinearities or delays are contained in the model or system, dominance over the signal is not necessarily associated with the components having large time constants. Nonlinear couplings and delays can amplify and significantly shift the phase of a signal. Therefore, a component which amplifies or shifts a signal can also be dominant in the system.

As pointed out earlier, material signals like magnesium, while in transit through a single linear process block cannot be amplified but can be delayed. A material signal can, however, be amplified in a single block if nonlinearities are present or in an entire closed-loop system containing feedback. The individual compartments of the tropical moist forest model contain neither amplification nor delay time, yet the plants and soil were found to dominate the flow of magnesium, a point which is intuitively satisfying. The findings, furthermore, supplement our conviction that the model presented is, in the broadest sense, adequate to describe steady-state or near steady-state behavior.

2. *Measured Effects*

A second point of interest arising from in- and out-of-system comparisons deals with the influence of the couplings or interactions. As shown for the tropical moist forest model the deviation between in-system and out-of-system Bode plots was a direct reflection of the system's effect on each component and vice versa. Empirical alterations of process block properties or linkages to other compartments can be interpreted through in- and out-of-system comparisons.

The in- and out-of-system Bode plots (Fig. 9) were determined entirely from computer simulations. That is, no tests were conducted in the field. In- and out-of-system analysis should be considered as a powerful technique for determining dominance and effects of interactions. Studies like that of Patten and Whitkamp (1967) and Ragsdale *et al.* (1968) are good examples of the kinds of information obtained by analyzing a system in varying degrees of complexity. Data such as these are suited for in- and out-of-system comparisons. In general, measures of dominance and the significance of interactions establish foundations on which one can make predictions about the system's performance when stressed (subjected to differing input frequencies). These measures are not entirely unlike sensitivities (see Vol. II, Chap. 2).

V. Summary

The magnesium cycling in the tropical moist forest, as represented by our model, is an extremely stable closed-loop system. The system behavior is dominated by compartments with large time constants (T), notably soil and plants. The natural frequency (ω_n) and damping factor (ζ) associated with the system indicate that the magnesium cycle should be difficult to perturb and slow to equilibrate once perturbed.

The model is felt to characterize the dynamic properties of the rainforest at steady state or near steady-state conditions. The model's behavior shows that the feedback loop (magnesium transfer from soil to plants) has little control action. It is postulated that regulatory control is due to other signals more informational in nature.

REFERENCES

Child, G. I., Duever, M. J., and McGinnis, J. T. (1971). *In* "Mineral Cycling in Tropical Forest Ecosystems" (F. B. Golley, ed.), pp. 112–121. Univ. of Georgia Press, Athens, Georgia.

Coughanour, R., and Koppel, L. B. (1965). "Process Systems Analysis and Control." McGraw-Hill, New York.

DiStefano, J. J., Stubberud, A. R., and Williams, I. J. (1967). "Feedback and Control Systems." Schaum, New York.

Forrester, J. W. (1961). "Industrial Dynamics." MIT Press, Cambridge, Massachusetts.

Hairston, G. N., Allan, J. D., Colwell, R. K., Futuyma, D. J., Howell, J., Lubin, M. D., Mathias, J., and Vandermeer, J. H. (1968). *Ecology* **49**, 1091.

Kerlin, T. W., and Lucius, J. L. (1966). The SFR-3 code. A Fortran program for calculating the frequency response of a multivariable system and its sensitivity to parameter changes. ORNL-TM-1575. Oak Ridge, Tennessee.

Lewontin, R. C. (1969). Brookhaven Natl. Lab. Symp. 22 [BNL 50175 (c-56)], pp. 13–24.

Margalef, R. (1968). "Perspectives in Ecological Theory." Univ. of Chicago Press, Chicago, Illinois.

Mason, W., and Odum, E. P. (1967). Proc. Second Natl. Symp. Radioecology. Conf. 670503, pp. 721.

Odum, H. T. (1971). "Environment, Power and Society." Wiley, New York.

Patten, B. C., and Witkamp, M. (1967). *Ecology* **48**, 813.

Ragsdale, H. L., Witherspoon, J. P., and Nelson, D. J. (1968). The effects of biotic complexity and fast neutron radiation on cesium-137 and cobalt-60 kinetics in aquatic microcosms. ORNL 4318, VC-48 Biol. & Med. 174 pp. Oak Ridge, Tennessee.

Tucker, G. K., and Wills, D. M. (1962). "A Simplified Technique of Control System Engineering." Honeywell, Fort Washington, Pennsylvania.

Woodwell, G. M., and Smith, H. H. (eds.). (1969). *Brookhaven Natl. Lab. Symp.* **22** [BNL 50175 (c-56)], 264.

PART III

THEORY

Systems ecology is at present a hypothesis, namely that in the general theory of systems there is a way of considering phenomena that differs sufficiently from established modes of ecological analysis to deserve a special recognition. Underlying this hypothesis is an abiding belief in "Gestalten" properties of higher levels of system organization, elucidation of which reveals the general in the particular, a goal of scientific understanding. This section presents a group of papers labeled "Theory" not because they represent some consistent set of formal characterizations of some central themes of systems ecology, but rather because they are "theoretical" in tone and intent. It could be argued that so are the majority of chapters making up these two volumes, to which it would be replied that what the chapters here have in common is mainly the fact that they appear together, here.

Chapter 4 is by an author who is without peer in energetically pursuing "gestalten" properties of ecological systems. One of the boldest innovators in ecology, Dr. H. T. Odum has persistently focused on *what* systems do rather than *how* they do it. In recent years this focus has brought him independently to an energy circuit language of his own derivation [e.g., Odum, H. T. (1971). "Environment, Power, and Society." Wiley (Interscience), New York] that bears a remarkable resemblance to the elements of systems science outlined in Chapter 1. The present chapter describes the theoretical basis for each of the dozen or so modules of this language. These modules are unmistakably particular versions of "objects" from Chapter 1 whose "behavioral features" come from the worlds of thermodynamics, physics, and chemistry. Odum presents them as a mutually exclusive and exhaustive set sufficient for the functional representation of macroscale systems. Just as the Chapter 1 objects are coupled by "conservative" and "distributive" constraints, the energy circuit modules are joined by additive (corresponding to conservative) and multiplicative (not corresponding) interactions. This chapter may be found difficult to read because it demands a breadth and depth of background that few readers can call forth with facility. Where its material will come to rest in the future sifting and sorting of systems ecology is hard to say, but several things are clear: It may not assimilate easily because too many frames of reference are involved and need to be resolved, and it is too important to just go away. If the specifics of the energy circuit language are not retained, the principles that motivated it will be well expressed in the definitive methodology for macrosystems just as they are in the methods for physical systems.

Chapter 5 is an investigation into the feasibility of several nonlinear formulations for use in compartment modeling of ecosystems. The criterion of feasibility is the existence of at least one mathemtatically and ecologically satisfactory steady state in which all compartments and flows have real, positive, and finite values. Simple systems with unbranched and branched topologies are compared under conditions where they receive "forced" vs. "self-generating" inputs, and where intercompartmental flows are "uncontrolled" vs. "controlled." It is shown in a rather lucid, if not completely general, way for ecologists that not every nonlinear system which might be designed on paper can serve as a model for a natural system.

Chapter 6 takes up the important topic of *connectivity* in systems. It does so in an ecologically interesting way by focusing attention on food webs, abstracting them as mathematical *graphs*. A graph is a collection of points (termed *nodes* or *vertices*) that may be connected to each other in some pattern by a set of lines (*links, edges*). The points may correspond to Chapter 1 "objects" and the edges to "behavioral features," in which case the graph representation has the effect of suppressing object characteristics and emphasizing the coupling topology. "Oriented" behavioral features are represented by arrows instead of lines. In the present application, lines and arrows in a graph denote *binary relations*, lines when the relations are symmetric and arrows when they are not. The two relations of food-web interest are the *trophic relation* (unoriented) and the *eating relation* (oriented). Their corresponding graphs are referred to as *undirected* and *directed*, respectively.

After the framework and terminology of graph theory are sufficiently developed, Dr. Gallopin proceeds to the description of food web properties. Graph concepts like *chains* and *circuits* lead naturally to more familiar food chains and feedback paths. Food webs may be partitioned into *trophic partitions*, which reduce the total number of possible interactions, and thence into *trophic levels*, which account for a further reduction. *Subwebs* can be defined in respect to any arbitrary node, and correspond to the kinds of functional units identified by, e.g., radioactive tracer studies. The formal definitions definitely serve to clarify often hazy concepts. Among food web properties considered are complexity, the maximum number of links in a web, the maximum length and number of food chains, etc. "Predator and prey similarities" of any pair of nodes are computable and used to define their *trophic distance* from each other in the web. Connectivity properties are expressed as *incidency* and *adjacency* matrices, and a measure of *status* in food webs viewed as organization charts is defined. A richly original contribution, one nevertheless senses that only obvious and in some cases ecologically uninteresting structural properties have so far been revealed, and that the real power of graph theory applications lies ahead now that some first connections have been made.

Chapter 7 is concerned with quantification of the ecological niche, one of the most heuristically useful but operationally elusive concepts in ecology. A multivariate treatment of habitat selection by winter sparrows in north Georgia provides a data set as well as identifying four different strategies of habitat selection. Three niche measures are developed, one depicting Hutchinson's (1957. *Cold Spring Harbor Symp. Quant. Biol.* **22**, 415) "realized niche," and the others representing two aspects of his "fundamental niche," niche width and relative abundance in optimal habitat. These measures are used to define coordinate axes in a three-space in which can be plotted a point for each species. The distribution of points comprises a *niche pattern* for the collection of taxa. Most of the bird species were observed to cluster in a region of the niche pattern space corresponding to wide realized niches, good regional adaptation, and typically low numbers at optimal sites. The ability of the niche pattern concept to characterize the niche structure of a regional avifauna suggests wider applications. Hypothetical niche patterns are constructed to compare a variety of communities, including tundra vs. rainforest consumers, island biota, and desert plants. Time-related processes such as succession and evolution are indicated to be characterizable as trajectories in niche pattern space. The utility of niche patterns in identifying good or bad management practices and schemes, in assessing the potential success or epidemic threat of introductions, and in evaluating the possibility of extinctions are briefly discussed. If the niche quantifiers developed here for the first time are not definitive, the niche pattern idea which emerges from them will undoubtedly persist and be modified into a variety of useful forms for future studies of biotic designs in communities, both theoretical and applied.

4

An Energy Circuit Language for Ecological and Social Systems: Its Physical Basis

HOWARD T. ODUM*
ENVIRONMENTAL ENGINEERING SCIENCES,
UNIVERSITY OF FLORIDA, GAINESVILLE

Based on studies supported by the U.S. Atomic Energy Commission, Division of Biology and Medicine, Contract At-(40-1)-3666. Manuscript submitted in 1969.

I. Introduction

Phenomena of the macroscopic world of ecology and sociology have been represented for many years by energy diagrams showing caloric flows. These diagrams portray the fate of energy in converging and branching networks, and suggest to varying degrees the kinetics, energy laws, and compartments of these systems (e.g., Zimmerman, 1933; MacFayden, 1963; Odum, 1957; Odum and Odum, 1959; Kormondy, 1969; and Philipson, 1966). While compartments and flows of matter, dollars, minerals, populations, etc. are frequently considered, only energy is a sufficiently common denominator to include all the forces, factors, and units of complex systems of man and nature.

In several previous papers describing complex systems (Odum 1967a, b; 1968), a symbolic language was introduced which combined energy laws, principles of kinetics, and some philosophical tenets of electrical systems. The language consists of a dozen basic modules, each having a mathematical definition. The symbols representing these modules are shown in Fig. 1. This chapter describes this energy systems language and its use. Many basic principles and equations of physics and chemistry will be cited to show their incorporation into the language

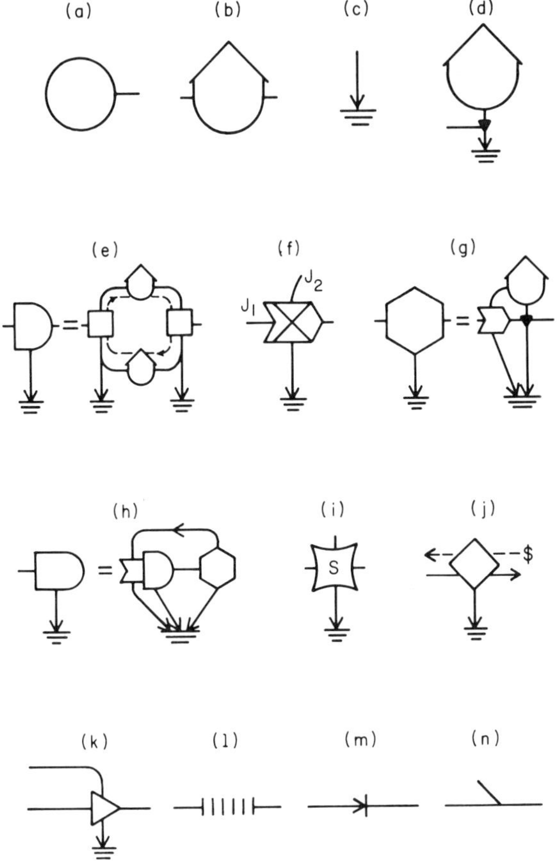

FIG. 1. Symbols of the energy circuit language. Each symbol is discussed in a separate section of the text. (a) source, (b) passive storage, (c) heat sink, (d) potential generating work, (e) cycling receptor, (f) work gate, (g) self-maintenance, (h) green plant, (i) switch, (j) transaction, (k) constant gain amplifier, (l) active impedence, (m) one-way valve, and (n) adding junction.

and its basis in them. The additive nature of energies of all types is the basis for a single network representing all processes. For a book account of energy language written for the general reader, see Odum (1970). The language is offered as an interface language for simulation.

Energy Diagram

Illustrated in Fig. 2 is an energy diagram for a house. Potential energy is derived from outside sources that have characteristic programs of delivery. It flows through the system along indicated flow lines and ultimately passes out as dispersed heat (heat sink symbols). Along the way energy may be temporarily stored as potential energy, may do various forms of work, may loop back upstream for action from a downstream position, or may interact with other energy flows. When energy transformers are connected by pathways, simple systems are

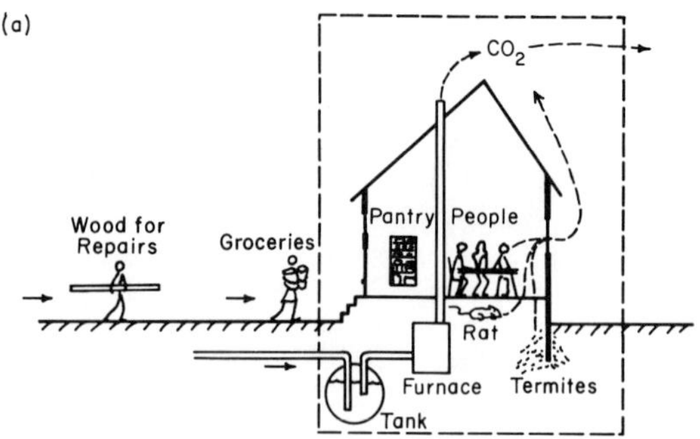

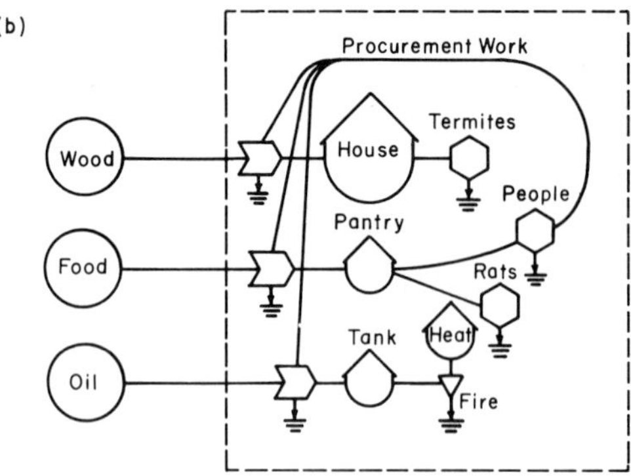

FIG. 2. Diagram of a human house system. (a) Schematic diagram, (b) energy diagram.

formed, some of which occur frequently and are basic building blocks for more complex networks (e.g., Figs. 1e, g, and h). In a qualitative and superficial way, the energy diagram helps to keep track of all possible flows and parts of complex systems, and aids in computing budgets. Beyond this, however, a degree of rigor derives from the fact that implicit in the energy circuit language are many rules and constraints derived from long-established principles of physics, chemistry, and biology.

The basic types of circuits can be found in many fields of science where the relationships have often been rediscovered independently. Rashevsky (1960) considers chains of physiological processes involving reactions and diffusion. Levenspiel (1962) summarizes chains of processes in chemical reaction engineering. Goodwin (1963) develops parallel, feedback, and more complex circuits to account for gene control of enzyme processes in biochemistry. There are many texts on electrical network elements. The same network units are considered by different writers under different names such as bionics, cybernetics, network theory, and reaction kinetics. By placing some of these concepts in the language of energy circuits and generalizing somewhat, basic combinations also may be applied to environmental systems and society.

II. Energy Source Module

Energy sources from outside the defined boundaries of a system of interest are indicated by a circle (Fig. 1a). To specify a source completely, the nature of the energy flow (as light, heat diffusion, flow of organic matter, etc.) must be indicated as well as its mode of delivery. In many systems languages "forcing functions" are described as outside programs affecting temporal patterns inside. Independent outside patterns of force delivery and energy flows are forcing functions and require independent energy storages.

One class of energy sources includes those which exert a constant driving tendency even when large flows are being drained. Examples are pressure from very large reservoirs, voltages from large batteries, and chemical reactions under conditions where concentrations of reactants are maintained. Another class consists of sources which have a constant flow of energy regardless of drains to using systems. Water flowing past a waterwheel, the flow of fuel from coal or oil fields under some conditions, and light flux are examples.

Whether a source is of constant force, constant flux output, or follows some other program, its representation is incomplete until defined, e.g., by some graphical function or explanatory equation.

Energy sources may deliver energy in pure forms such as light, sound, and water waves, or they may deliver energy with matter flow as with a flow of fuel or water. Sometimes one energy source pumps in a second source by doing work on its flow. The energy source symbol has one or more flow lines representing the pathway of potential energy delivery and also the pathway of action of driving forces which may be part of the energy delivery.

Another kind of energy source has a repeating pattern of energy delivery. Examples include sine waves, square waves, impulse functions, etc. A source may be stochastic, resembling noise and describable by statistical parameters.

III. Force and the Energy Pathway

Rooted deeply in human cultural origins are concepts of causal force and energy reflecting early qualitative recognition of fundamental laws. The formalization of qualitative concepts into quantitative definitions of force and power of mechanics in nineteenth-century physics did not eliminate the variety of causal force and energy concepts in everyday life, but merely added rigor to part of the area of previous application. For example, the numerical product of force and velocity is power in Newtonian mechanics, but political causes and cultural energies are still discussed in vague terms.

With the development of concrete formulas for driving tendency in various areas of evolving science such as electricity and thermodynamics, a generalized quantitative concept of force emerged. In considering various phenomena of the environmental systems, we can extend further the concepts of generalized force as a driving impetus and power as a measure of its delivery rate. Denbigh (1952) and Prigogine (1955) summarized the generalized concept of force as applied in physical science (see examples in Table I). Jammer (1957) gives the earlier history of the force concept and its use in such areas as theology and literature where its application is still qualitative.

In this chapter two kinds of force laws are identified in energy transformations of the macroscopic environment (Fig. 3). Both are incorporated into circuit notations. Although physical forces are at the root of all cause, the phenomena of interest are often the expressions of populations of forces combined in various ways whose average effects serve as a causal function in steady state. Causal force concept is generalized for groups of processes under the name "population force." Ultimately, through understanding of the distribution of input forces, population force, and loads, one hopes to understand all flows of energy through ecological and social networks.

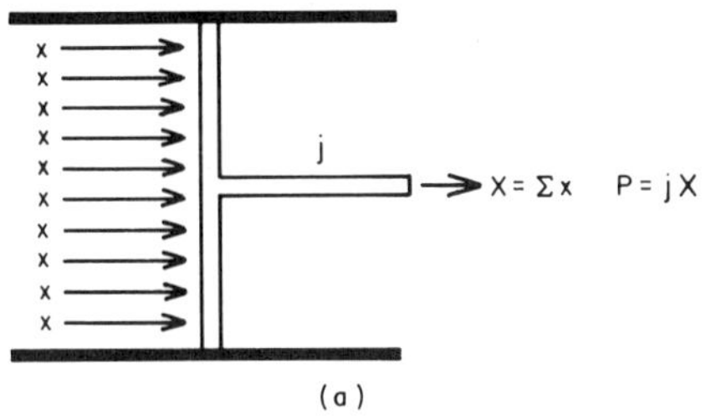

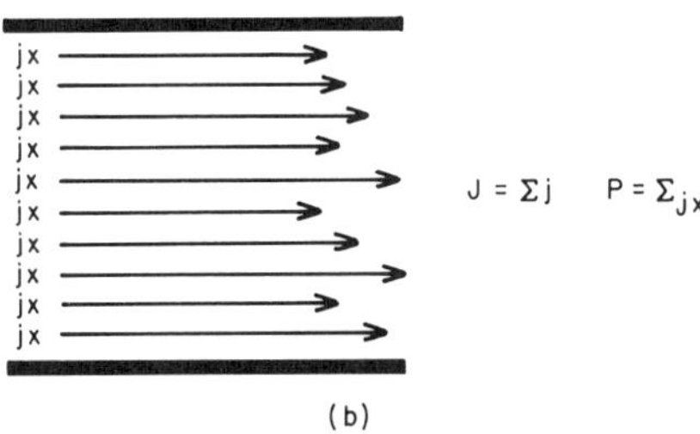

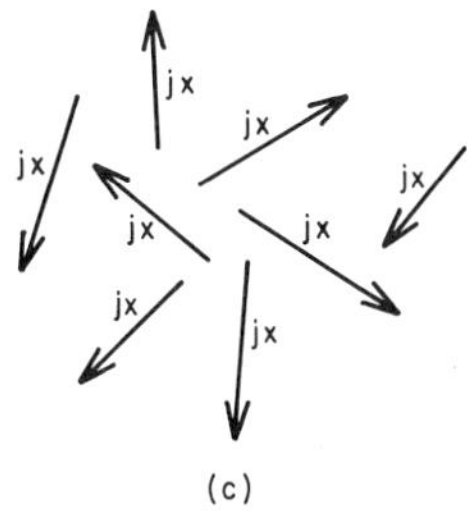

FIG. 3. Three types of organization of component power transmissions. (a) Component forces add to form a single force that determines flux; (b) component fluxes add in parallel delivery of power because of channelling; flux depends on population force; (c) component processes without organization disperse heat.

A. THE ENERGY PATHWAY

In the energy circuit symbols, solid lines represent flows of potential energy from a source. The source may be an outside program of energy input, indicated by the source module, or it may be derived from a potential energy storage unit within the system (indicated by the tank symbol). Potential energy sources deliver their energy flows by directing a force. The line is the pathway of the force from the energy source.

An opposing force is directed backwards along the line in opposition to the driving force. The opposing force may be from a downstream storage, from friction of the pathway, or from inertial backforce if acceleration is involved. The pathway satisfies the equal and opposite force requirement of classical physics. The energy pathway therefore is actually an equation in which the forces are set equal at every point. It is also a pathway of flow of an energy quantity that does not increase or decrease, but only changes in form. The constraint of the first law of thermodynamics is recognized by providing that all inflows be balanced by storages and outflows.

In electrical networks, the ordinary wires are instantaneously in steady state between the force of the voltage and the backforce of the frictional resistance to the passage of electrons, but the modules that the wires connect may not be in steady state. This property is adopted for the energy network language. The isolated pathway line is always in steady state, although the system of pathways and modules may not be. The energy language by this device, as in electrical networks, has transient phenomena such as growth and decay of storages while the flow lines remain in a steady state balance of driving and frictional forces. The pathways are defined as the balance between force X from the energy source and the frictional backforces RJ, where backforce is proportional to the flux (of matter or energy, J),

$$X = RJ, \tag{1}$$

and R is the constant of proportionality (called resistance in electric circuits). Equation (1) rearranged shows flux proportional to net driving force, which is Ohm's law in the electrical case and more generally the basic force–flux law of irreversible thermodynamics,

$$J = \frac{1}{R} X, \tag{2}$$

$$J = LX, \tag{3}$$

where L is the conductivity describing frictional backforce tendencies of the pathway. The absence of a line indicates absence of a pathway. All pathways involve some kind of structure, temporary or permanent, for which potential energy must be spent for maintenance. Each pathway requires a work expenditure to maintain the upstream, downstream, or some lateral module. Each pathway shows the path of potential energy crossing energy barriers passing against frictional dissipation, or doing work against opposing forces of downstream storages. The requirement that potential energy be dispersed as unusable heat is indicated by the heat ground diversion (heat sink symbol, Fig. 1c) of some potential energy by whichever modules are responsible for the particular pathway.

In practice, the heat sink energy diversions are drawn at one end or the other of many circuits. For example, a fish swims to obtain food and establishes by his movement a pathway of food to him, but the energy is shown in the diagram as a heat sink on his main module since energy costs are mainly derived and distributed there.

The Newtonian convention that every force has an equal and opposite force is represented by the energy pathway as a balance of forces. If acceleration occurs, it is considered within one of the modules as an energy storage rather than within the energy flow line. The energy dissipated in a pathway includes loss of potential energy in overcoming friction plus loss in specific work processes designed to maintain organized pathways as pipes, wires, spending channels, organisms, food chains, etc.

In Table I are given some of the single forces that may be driving energy flows along the pathways. For these kinds of processes Eqs. (1)–(3) are descriptive.

B. Coupled Forces

When there are more than two forces entering from energy sources or storages along or intersecting the same pathway, then there is a force component of one affecting the other and vice versa. Forces are coupled. This situation is described in the notation of irreversible thermodynamics by Eqs. (4) and (5).

$$J_1 = L_{11}X_1 + L_{21}X_2 \tag{4}$$

$$J_2 = L_{12}X_1 + L_{22}X_2 \,. \tag{5}$$

It is the Onsager reciprocity principle that the component L_{12} of force X_1 on the flow of the second J_2 is equal to the component L_{21} of the second force X_2 affecting the first flow J_1,

$$L_{12} = L_{21} \,. \tag{6}$$

This principle may be an extension of the Newtonian principle of equality of opposing forces. Thus X_1 and X_2 may be directed in the same direction along an energy pathway, in opposing directions along the pathway, or one may cross the other, doing work on it as described later for the work gate module.

TABLE I

FORCES AND FLUXES FROM POTENTIAL ENERGY SOURCES

Type of energy	Force[a] (calories per unit displacement)	Flux (displacement per time)
Symbol	X	J
Electrical	emf (V)	Current (A); Charge per time
Mechanical	Newtonian force	Velocity (d/t)
Rotary motion	Torque	Angular velocity
Heat gradient	$\Delta T/T$	Calories per time
Hydrostatic	Pressure	Compression rate (v/t)
Surface energy	Surface tension	Area per time
Light	Radiation pressure (cal/volume)	Volume per time
Molecular diffusion	Gradient of chemical potential	Molecules per time
Individual molecular chemical reaction in two parts[b]	X	J
(1) Energy activation crossing force barrier	Deceleration of momentum	Molecular velocity
(2) Chemical combination	Intermolecular and interatomic forces	Molecular velocity
Steady state population of chemical reaction processes. Two conventions, neither have both $JX = P$ and $J = LX$		
A. Free energy convention[c]	Chemical potential u	Rate of reaction dn/dt
B. Population of forces acting separately[d]	N, proportional to number of component reaction pathways	Number of reactions per time

[a] Potential is defined as the ratio of the power delivered to the accompanying flux of matter or pure energy. Many of the forces in this column are potentials by this definition.

[b] Single molecular process involving accelerations–decelerations and thus not a steady state until considered as a population.

[c] JX is power but J is not proportional to X. This has been used in chemical thermodynamics erroneously on the justification of an approximation.

[d] J is proportional to X, but JX is not power.

Fluxes J on the energy diagram change along the routes, being material in one segment, but pure energy such as heat and light in others. The inclusion of more than one kind of energy makes energy networks differ from material network diagrams, such as those for dollars or the nitrogen cycle, where by definition there is only one kind of item flowing.

Whereas forces represented by vectors in Euclidean space have a magnitude and a direction, force vectors implied or drawn along the energy flow lines are in topological space and have only a positive or a negative direction, a force from the left and from outside sources being taken as positive. Inputs and outputs are the pertinent features in most networks.

C. Dissipative Forces

In Table II are given some forces which serve only against driving forces. The first three are frictional resisting forces. The last three are populations of resistive forces associated with some energy flows of more complex systems. They act in reverse to energy sources although they may consist of more than one process and resistive force. Some of these forces are necessary and may be useful only because they permit a low cost steady state. Others may be directly useful. A grindstone is an example of dissipative work which is useful to its system.

TABLE II

Forces and Fluxes in Dissipative Work Functions[a]

Type of work	Force (calories per unit displacement)	Flux (displacement per time)
Frictional processes		
Electrical resistance	Back emf	Current (A)
Mechanical resistance	Friction	Velocity
Rotary resistance	Braking torque	Angular momentum
Structural controlling processes		
Pattern changing	Start–stopping force of spatial patterning per unit mass	Mass arranged per time
Organizational connections of units	Calories per circuit connection	Connections between units established per time
Maintenance	Calories per gram replacement	Rates of replacements and reorganization; (g/time)

[a] For these forces the flux J is always against the force X and does not exist without an outside potential energy inflow.

As listed in Table II there are several kinds of dissipative work concerned with change of structure in which potential energy is passed into heat with changes in structure resulting. As separated in Table II these flows of energy can be thought of as displacements against structural forces, which like friction oppose the process. The forces exert more back resistance as the speed of the process is increased.

With *pattern-changing work* objects are rearranged. If rearranging automobiles in a parking lot, it takes the coupling of free energy to start them rolling and then when they are stopped, this energy of momentum goes out as heat in the stopping friction of the brake. If one rearranges the automobiles rapidly, one uses more energy flux for the job than if one does it slowly. As the time used approaches infinity, the energy costs approach zero, neglecting friction of the machinery. Thus, the amount of potential free energy that must be expended in pattern-changing work for speed tax is a function of the rate of rearranging. The final product involves a new state of arrangement, but no appreciable storage of the energy that went through the system to accomplish the work. Pattern-changing work as here defined, like friction, is a function of the path, not the initial and final states alone. However, there may be a path that is optimal for maximum accomplishment.

Another kind of structural work is *organizational work*. This work involves making connections between parts to form a system, thus eliminating uncertainty in relationships and decreasing the possible operation alternatives of the whole system.

Maintenance work (Fig. 1g) is a form of structural work in which some parts are replaced, rearranged, and organized so as to balance exactly the dissipative tendencies for loss of necessary system structure.

D. Energy Flow in Structure Arranging Processes with Acceleration

An important process in macroscopic systems necessary to their survival involves spatial arrangements of parts. These involve acceleration–deceleration actions and friction. The energy flows can be visualized for a simple limiting case, which is found to be a type of dissipative energy transformation.

Consider the successive arranging of identical weights in space without ordinary friction. A force is exerted on each weight, accelerating it as it moves toward its final position where it is stopped abruptly, the kinetic energy going into heat. As soon as one weight stops the force accelerates a second so that the process is steady, but pulsing. Such a system is a straightforward problem in mechanics. The input force is balanced on average by the resistive force of the stopping-heat dispersals, so that the process can be regarded in the long range as a steady-state process. This is something like friction, except that the input force first does inertial work and then at a later time the kinetic energy goes 100% into heat. For a practical example visualize the arranging of trucks in a parking lot, arranging troops in a field, or collecting materials for construction. These examples also have ordinary friction involved although the correction is not necessarily large. The energy per parcel is proportional to the square of the average velocity. The total power is the product of energy per parcel times flux of parcels. The overall power dissipation is thus proportional to the cube of flux.

In Fig. 4c is plotted the relation of power and velocity for the arranging process showing the much increased energy flux with increased speed when the input force is increased successively. Transformation is to inertial work, which then goes into heat. The entire flow can be useful and at the same time is speed tax.

A comparison of curves in Fig. 4c shows the advantage of a steady frictional load in an arranging process as compared to an accelerative and stop type. For the same flux achieved, the power requirement as the cube is vastly greater with accelerative arranging. Friction is a very useful property to complex systems, as the astronauts found.

E. Generalized Force–Flux Concepts of Steady-State Energetics

The energy circuit language represents steady state thermodynamics, which concerns inflows and fates of various kinds of potential energy in spontaneous processes. For each kind of process whether chemical, mechanical, thermal, magnetic, electrical, etc., each potential energy flow is expressed as the product JX of the two quantities which have the dimensions of power P: J, the rate of flow (flux) of the material concerned; and X, one of the energetic forces. The J's and X's are chosen so that the product JX is power and so that each J is proportional to an X affecting it with L, the conductivity, a constant of proportionality. For example, the power entering from an energy source given in Table I is

$$JX = P \tag{7}$$

[Eqs. (1)–(6)]. (See Table I for forces and fluxes which fit this definition.) Some of the flows of potential energy are positive, representing power added to the system by input forces from potential energy sources. Others represent power delivered in overcoming backforces in the process of storing or exporting potential energy. As required by the second energy principle (degradation law) some of the inflowing power becomes a flow of irreversible heat, often termed "speed tax." A heat dispersal symbol (Fig. 1c) is defined for this flow. As required by the first energy principle (conservation law), the sum of the positive and negative flows of potential energy JX is the speed tax (P_t in Eq. (8) and Fig. 4b),

$$P_t = J_1X_1 + J_2X_2. \tag{8}$$

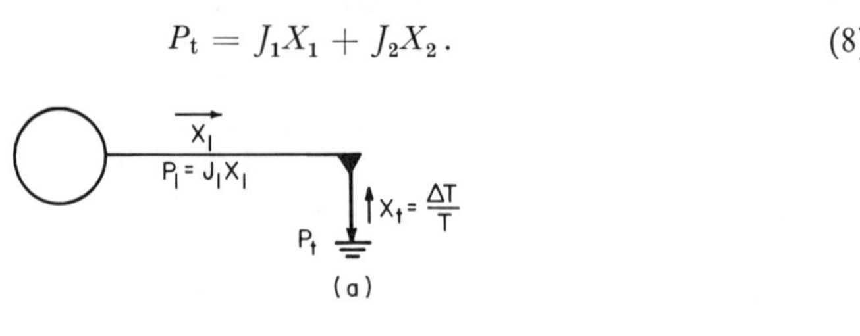

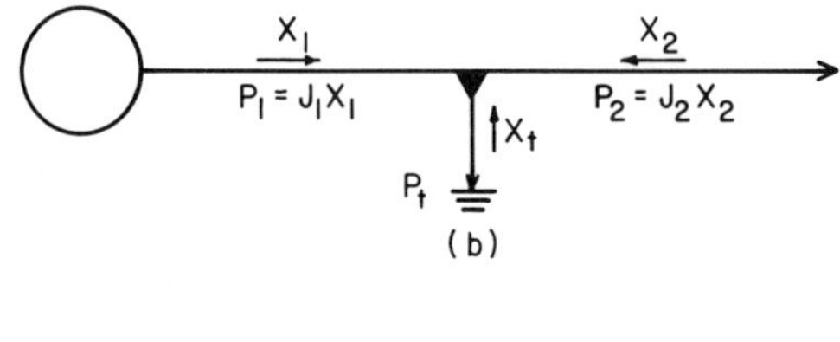

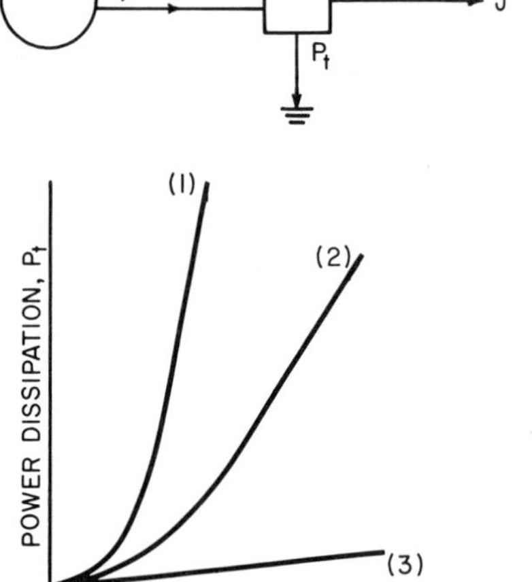

FIG. 4. Identification of forces, flux, and power flows in dissipational energy process. (a) Input power equals entropy-generating speed tax; (b) input power exceeds speed tax; (c) power dissipation of three classes of transport pathways: (1) dissipation in proportion to cube of flux, as in mechanical acceleration followed by abrupt deceleration ($P = kJ^3$); (2) dissipation in proportion to the square of flux, as in electrical flow in wires or swimming fish ($P = RJ^2$); (3) dissipation in proportion to flux, as in sliding friction ($P = kJ$).

The potential energy lost is equal to the heat dispersed in the environment by the first energy principle. This heat dispersed by diffusion processes is in proportion to the heat gradient ($\Delta T/T$), one of the thermodynamic forces (Table I). Defining calorie flow as the flux J_t of the speed tax, the power dispersed is the product of the force and flux, $J_t \, \Delta T/T$. By definition, the quotient of the heat dispersed, ΔH, divided by the absolute temperature T is the entropy, ΔS, generated in the environment, $\Delta H/T$. This entropy is a measure of the randomness produced from more ordered molecular patterns by statistical motions of molecular energy. The second energy principle requires that there be heat dispersal and entropy increase in the environment for any process to go spontaneously. Stating the principle in another way: the process must be probable. The speed tax power dispersal may be expressed in terms of entropy generated as in Eq. (9),

$$P_t = J_t \, \Delta T/T = T \, dS/(dt). \tag{9}$$

F. Force and Flux with Energy in Packets

The discrete inflows of energy in such unitized forms as photons, sound waves, water waves, cannonballs, electrical pulses, nuclear particles, and molecules of organic matter are a familiar part of many systems. The number of energy packets transformed depends on the number caught by the energy receptor, which depends on the density of the packets in that vicinity. The input of power from a process of catching such energy flows is the product of the energy packet density in the receptor vicinity and the volume of that vicinity swept per time. One may identify a force and flux for such flows by choosing the density of energy packets as the force and the volume of the receptor vicinity swept as the flux. Then the flux is proportional to the force and the product of the force and flux has the dimensions of power. For example this convention has been used for light in Table I. The density of light is sometimes called "radiation pressure."

If a system is catching all of the packets incoming, then the power input is identical to the passing flow of packets. In other instances the passing energy is a reservoir of energy whose density encourages the capture of some but not all of the packets as input flux J. As long as there is incident flow in the system vicinity, it is self-renewing and acts as much as a reservoir of constant force as the infinite water table or electric battery. Energy passing without capture is unused potential.

G. Force and Flux in Diffusion of Molecules

A common area of application of the force–flux concept is diffusion. Fick's law states that the rate of diffusion flux of chemical molecules, J, is proportional to the gradient of concentration ([ ]).

$$J = k\, d[\;]/d\delta, \tag{10}$$

where δ is distance. Since pressure is a function of the concentration of molecules, the chemical potential energy (free energy per mole, μ) can be rewritten as Eq. (11) (μ refers to one molecular species; ΔF (see below) refers to all substances present),

$$\mu = \mathbf{R}T \ln [\;] \tag{11}$$

[See Eqs. (14) and (31)–(33)]. Taking the derivative of (11) with respect to distance δ, one finds

$$\frac{d\mu}{d\delta} = \frac{\mathbf{R}T}{[\;]}\frac{d[\;]}{d\delta}. \tag{12}$$

Combining (10) and (12), one writes the expression for diffusion in terms of the gradient of chemical potential,

$$J = \frac{k[\;]}{\mathbf{R}T}\frac{d\mu}{d\delta}. \tag{13}$$

If the flux of molecules is taken as J and the gradient in chemical potential as the force X, JX is power and J is proportional to X, consistent with other forces and fluxes in Table I. For an example of application of these concepts, see Best and Hearon (1960). While chemical potential is a force in diffusion, it may not be in chemical reactions.

H. First Convention for Choosing Force and Flux in Chemical Reaction Systems

There are special problems in identifying force and flux in chemical systems so that the flux J is proportional to the force X [Eq. (3)] and the product of force and flux JX is power flow [Eq. (7)]. In chemical thermodynamics one convention, as given in Table IA, is to choose the chemical potential energy (ΔF or μ, free energy per mole) or its negative, the affinity, as the force, and the reaction rate as the flux. When this is done, Eq. (7) holds, but flux is proportional to potential only in special cases. Chemical reactions go at speeds proportional to concentrations, not in proportion to free energy available. As derived in Eqs. (31)–(33), the potential energy (constant pressure) per mole is proportional to the *logarithm* of the concentrations of reactants and products of the process concerned:

$$\Delta F = \Delta F_0 + \mathbf{R}T \ln \frac{\text{products}}{\text{reactants}} \tag{14}$$

(ΔF_0 is for the standard state).

If the chemical potential ΔF is the logarithm of the concentration, the flux, rate of reaction J, is proportional to the concentration; then the relation between flux and chemical potential is tested by combining Eqs. (14) and (3), letting X be concentration,

$$J = \frac{1}{R} e^{\Delta F/RT}, \tag{15}$$

where R is the chemical resistance. The flux is not proportional to ΔF and hence ΔF is not correct as a choice of force.

The reasons that there is difficulty in choosing force and flux in chemical systems is that one is not dealing with a single process but with a statistical sum of separate physical processes. As considered subsequently there may be a more accurate causal concept than chemical potential for statistical assemblages.

I. Semantic Problem of Linearity

Many processes of the macroscopic world when expressed in the form of Eq. (1) are not linear, i.e., R is not a constant. However, some of these processes are actually compound linear circuits by which energies are involved in feedbacks and other arrangements. It is a basic aspect of the energy language that there is a linearity in the simple single pathway, although the compound modules which these simple pathways form may be nonlinear. It is customary in many fields such as electrical engineering to define the linear process as one in which the flux is proportional to upstream storage, referring to other processes as nonlinear, even though these other processes are made up of components that are all linear. For those in other fields this custom is sometimes a source of semantic misunderstanding. A radio with many pathways each of which is following linear processes has many modules, whose overall equations are nonlinear. For steady-state thermodynamics and its macroscopic application here using a concept of a linear population force, all complex nonlinear processes may be dissected into linear component pathways. In this sense all processes are linear and all processes are in proportion to a force, either a single force or a population force. Compound modules are nonlinear, but their nonlinearity is a result of the configurations of more than one linear pathway.

For example, in micrometeorology and plant physiology the diffusion of carbon dioxide in a forest is often represented with Eq. (1); *R* is described as a parameter that varies with place, height, and time of day. It is not really a constant but a variable because the process is not a simple pathway; *R* is actually being misused. In energy language there is a multiplicative intersection of energy flow from an energy source in turbulent eddy flux. The component processes are linear, but the combination is not. Whereas the use of variables as parameters has been an objective process, the energy circuit language has a more basic ideal of breaking compound processes into their linear component circuits or if compound modules are used, they should have clear definitions in relation to elementary units and with the corresponding differential equations.

IV. Population Force

The energy circuit language recognizes two levels of force, only one of which has been discussed in the previous paragraphs. The classical physical concept of a single force, such as a force of gravity, an electromotive force, a magnetic field force, etc., is represented by the energy pathways as the line of action of these forces, and the movement of energy along the pathway is in response to these forces. As frictional backforce adjusts to equal the driving force [Eqs. (1)–(3)] a balance is achieved by which the flux is proportional to the force. Work is done. For these regularly recognized physical forces, the energy circuit language is suitable for showing the network of actions, pathways, and energy sources. These forces may be made up of smaller component forces, as when water is additive in the water tank or electrical cells are additive in voltage in the series battery, but the output pathway has one resultant force, where the component forces are additive.

The energy circuit language also recognizes another level of force which is pertinent to the population sciences, defining population sciences as those in which the phenomena of interest result from parallel actions of a population of small forces. Chemical reactions, biological processes, sociological phenomena for example, are mainly concerned with the sum of a population of separate actions. Whereas people can pull together to develop one force on a battering ram, their usual actions are to work in parallel each contributing to a group output.

For these types of energy flows, each separate action has a force and flux of a regular physical type (previous paragraph), but the *flux of output of the group is proportional to the number* of simultaneously acting units. Flux is proportional to the number of forces (population force), which is designated N. Whereas the energy per component action is as previously discussed, the total energy is proportional to the number of actions. The variable is N, not X.

The energy circuit language where populations are involved expresses the action of N, the population force, in driving the flow of energy bearing flux J,

$$J = LN, \tag{16}$$

which is as fundamental a proposition as the force–flux relationship of the single force [Eq. (3)]. Both types of force operate in the energy circuits, one compounded of the other. Both are linear propositions and hence the energy circuit language has as its basic feature a linearity of flow in response to a force from an energy source or storage. In some cases the line represents a single force X (where its component forces are adding) and in other instances the line represents a sum of parallel drives, the component forces being according to the population force expression N.

A. Components of Population Force

Many processes in nature combine the action of populations of separate events that individually are one-way actions not in steady state. The overall changes in chemical reactions are the sum of actions of populations of individual molecules; food chain flows are the sum of individual food flows in populations of organisms; the flow of fuel through the transportation industry is a sum of flows through populations of motorized units. For each component event we can identify a true force x and flux j as in Table I for single chemical events. We can then add the component forces and component fluxes to obtain a group number for each quantity, but if we do, the product of the sum of fluxes and sum of forces is not the sum of the power that one obtains by summing the total of the individually separate jx products. Arithmetically, the sum of products is different from the product of the sum. It is simply not mathematically possible to represent the population of chemical reaction processes by a flux proportional to a force whose product is power although one can do this for each individual reaction. Erroneous early misapplication of steady-state thermodynamics to chemical reactions resulted from this error (see correction by Denbigh,

1952.) The erroneous choice of force and flux is given in Table IA. In order to distinguish the components from the population quantities we designate small letters for the component force x, component flux j, and component power jx. We use X' as the sum of the component forces and J' as the sum of the component fluxes. The following derivation [Eqs. (17)–(22)] is made for populations of equal components, a special case of a population of unequal components.

For the separate component processes we rewrite Eqs. (3) and (7) with the component symbols (small letters) including l as conductivity of the component process,

$$j = lx, \tag{17}$$

$$P = jx = lx^2. \tag{18}$$

When the component processes are acting separately, there are numerous parallel flows whose total flux is given in Eq. (19) by multiplying the flux per unit by the number of components n and by the probability p of their operation at any one time:

$$J' = pnj. \tag{19}$$

Combining Eqs. (19) and (17) the flux resulting from the components is given in terms of the component properties:

$$J' = pnlx. \tag{20}$$

For a population process in which p, l, and x are constant, the population flux J' is proportional to the number of units n, which may thus be regarded as a linear causal factor. In the chemical case x is chemical potential, Δu (Table I).

In general both the mass q and the number n as measures of quantity of component fuel units acting in parallel are in proportion so that the flow may be written as a linear function of either if appropriate constants are used. Because the equations follow closely some analogous ones for electrical analogs, it is convenient to introduce the quantity N as population force, defining it in terms of number and mass as in Eq. (21). Population force is proportional to the quantity of matter driving the population flow, but differs numerically from n or q according to the constants of the system involved:

$$N = Q/C = nq/C. \tag{21}$$

The ratio of mass Q to population force N is the constant C analogous to electrical capacitance, and q the weight per individual unit enumerated by n. See also Eq. (34).

Combining Eqs. (20) and (21) and using the constant L' to represent the group of constants $plCx/q$, one may write the causal equation of population flux as a function of the population force in a form analogous to that for the force–flux equation:

$$J' = (plCx/q)\,N = L'N. \tag{22}$$

Thus we define a causal factor N for a population phenomenon using a different convention (B in Table I) from that often used in chemical systems. Many chemical, biochemical, biological, ecological, and industrial systems can be studied with this expression in which flux is proportional to the cause N. However, the product of population force N and flux J' is not power. The quantity N was previously discussed as "ecoforce" (Odum 1960, 1967a), but since it applies to chemistry and cellular biology as well as ecosystems and sociology and is not a single force, it now seems better to use a different name, "population force."

In the special case in which the potential energy W is in proportion to the number of component processes (as when grain is stored in an elevator), the flux in Eq. (22) is also proportional to potential energy.

B. Proportionality of Potential Energy, Potential, Force, and Power

Sites of potential energy storage (symbols in Fig. 1b and d) have potential and exert force. Potential is the work done per unit moved from equilibrium. Force as defined broadly here is that function to which the flux is proportional and may be a single force or population force. In some systems such as electrical ones, the flux of amperage is proportional to the force of voltage, which turns out also to be a potential so that flux is proportional to potential. With chemical reactions, flux is proportional to the population force, a concentration gradient, but not to potential. Chemical potential is a logarithmic function of concentration ratios. However, other systems in which flow is proportional to population force in the macroscopic dimensions, transmit packages of energy.

The energy storage is made up of these packages of energy set so that the flow is proportional to the population of packages and can at the same time be proportional to the potential energy storage and the potential. Thus, many ecological and social systems operating with these relationships are simpler than the electrical and molecular ones. In energy circuit language, the pathways and storages represent energy storages and flows. The flows are proportional to the type of force prevailing (single force or population force). Potentials, storage functions, and relationship of flux and power to these vary with the type of energy storage and flow and must be specified if this becomes of interest.

In the energy circuit language the pathways are always flows of power. Flux, however, is variously defined in different segments of the network. The flux may be defined as identical with energy flow in heat and other pure energy flows. In other systems it may be the flow of a material such as electrons or water (Table I). In some systems, those that fit the group of thermodynamic forces and fluxes in Table I, the product of single force and flux is power. In other population flows the power in the pathway is not the product of population force and flux. It is the product of the energy per component μ and rate of flow of components,

$$P = \mu J = \mu L'N. \tag{23}$$

C. Power and Forces Delivered by Populations in Series and Parallel

The power and force delivered by a population of separate duplicate processes depends on the manner of combining the components (Fig. 3). If the flows are parallel, the fluxes add; and if the components are joined the forces add. A familiar example is a population of electrical dry cells. In either case the total power for the population of processes is the sum of the powers of the components jx:

$$P = \sum jx = pnjx. \tag{24}$$

Note in Eq. (24) that the power is proportional to the population size n as well as to the magnitudes of component force and flux.

In processes such as radioactive decay and metabolism of populations of people, the flows are separate, the forces act separately, and Eqs. (19) and (24) can be combined as (25):

$$P = J'x. \tag{25}$$

The flux J' is the sum of the components, $\sum j$, and the force is that of each component.

In other processes, such as living electrochemical cells uniting electrical outputs in series in an electric eel or people exerting pull on the same rope to combine forces with one flux and one group force, Eq. (24) can be expressed in terms of a single group force X' substituted for pnx:

$$P = jX'. \tag{26}$$

There is one flux j that goes through higher resistance and the force X is the sum of components $\sum x$. The regular definitions of force and flux in Eqs. (3) and (7) are appropriate.

When power flow in chemical systems that follow Eq. (25) (for separate action) is divided by the flux J, according to the procedure for calculating potentials, the component force x results. This quotient, power per unit flux, is μ, chemical potential. The erroneous convention A in Table I for selecting force gives the component force rather than the variable quantity that determines rate of flow, because in this class of chemical reactions the potential function [Eq. (27)] is not linear with concentration. Convention B is more appropriate for selecting cause where populations are in parallel action.

Populations of chemical molecules can act either in series or with forces united depending on the process. When there is a chemical reaction where molecules are reacting by collisions in many separate actions the forces are separate and convention B is appropriate, since the rate is proportional to the population force N [Eq. (22)]. However, when a surface area is provided against which the molecular moments may sum to produce a pressure or through which they may thrust an osmotic pressure, the forces add, and the rate is proportional to the group force X' [Eq. (26)].

When an electrode is provided and equilibrium achieved (potential energies equal), a third type of force expression results because electrons derived from the population of molecular reactions at the surface act together to develop an electrical potential E that is equal to the chemical potential of the solution ΔF as given in Eq. (27). In this expression, f is Faraday's constant and n the number of moles. Due to the self-repulsing charged nature of electrons their potential is the same as the electromotive force. The force thus expressed when the electrode and solution are at equilibrium is equal to the potential and hence to the logarithm of the number of components. When the electrode is connected so as to flow electrons externally there is no equilibrium and the flow may then follow Eq. (26). Thus, the type of force provided by a population of component forces depends on the circuit connections of their action.

Figure 3 diagrams the parallel and combined force arrangements for two processes with the same overall power output. Both can be represented by the same energy circuit:

$$nfE = \Delta F = \Delta F_0 + \mathbf{R}T \ln \frac{\text{products}}{\text{reactants}}. \tag{27}$$

Where the forces combine there is one force and the flux passes along one route. Where the forces are acting separately, we combine them in our energy study by placing them in a compartment because they are indistinguishable and for the larger system are of little separate interest. Hence we can represent the population as one group flux J' and identify the force involved as the component force x, while using the population force N as the linear causal factor of importance affecting the flow. In Fig. 3c the forces are not channeled but dispersed in stochastic manner, interfering with the opportunity of the parts to accomplish group action. Such a system is normally nonadaptive and is excluded by selection unless such disordering serves some role, as it sometimes does. Examples of dispersive actions which are adaptive are: urban renewal, dispersal of seeds, and distribution of rain.

Summarizing, the pathway of the energy language always represents the flow of energy from a potential energy source, a balance of forces, and a steady state in that segment when its inputs and outputs are steady. The nature of the causal force is not automatically defined by the pathway line. One must add vectors or other auxiliary notation (i.e., Fig. 3) to specify whether a given segment is a single Newtonian force proportional to potential of the driving source or the causal linear drive of a population force proportional to number of component driving process elements. When diagramming and analyzing simple systems such as electrical networks and groundwater, the Newtonian drives are involved [Eq. (2)]; when diagramming and analyzing ecological and social systems, population force drives predominate [Eq. (16)]. Both can be used in the same energy network system. One may quantitatively diagram flows and storages without necessarily knowing which applies if power data are available.

In analyzing systems of energy pathways and their modules differential equations can be written following either of two procedures: (1) The rate of change of a storage in terms of input and output energy flows can be written substituting expressions for each flow in terms of other source and storage forces. (2) Forces that converge on a point can be set equal, and expressions substituted for each force in terms of sources and flows. These procedures are employed in commentary that follows on equations and behavior of each module in terms of the internal networks they represent.

V. Heat Sink Module

The second energy principle requires that all spontaneous processes include dispersal of potential energy as heat P_t distributed into the environment, unavailable further as a driving impetus for processes.

Wandering of molecules makes the process go and by definition disperses its molecular motions collectively called "heat." The heat sink symbol (Fig. 1c) represents this energy dispersal, which must occur spontaneously from energy operating module of an energy system. When some simple system such as a water flow or electrical resistance is being represented, the heat sink receives flows directly from the pathway (Fig. 4). In complex modules that represent groups of processes, the heat sink is a miscellaneous conduit of heat dispersal of the many processes, grouped together for convenience. The respiration of an organism is the sum of the many processes of work and heat dispersal. Respiration is easily measureable by the oxygen consumption or carbon dioxide production which accompanies these processes, and measurement of total flow to the heat sink is a good starting point in evaluating the energy flows of complex systems. Whenever of interest, particular component processes may be isolated, labeled, measured, and represented by separate flow lines and heat sinks, leaving by subtraction, a lesser flux of heat dispersal represented by the main miscellaneous heat flow of the module (Fig. 1c).

As shown in Fig. 4, heat dispersal into the heat sink can be visualized as a flow from a source of potential energy which exerts a driving force in the form of a thermal gradient and is opposed by a frictional force inherent in the property of entropy increase (X_t in Fig. 4). See Eq. (9).

The heat sink symbol represents the last stages of potential energy dispersion when molecular and thermal diffusion is dispersing any potential energy remaining in molecular component populations. When heat is dispersed by eddies and other self-generating thermal engines, an outpumping subsystem exists that requires identification, diagramming, and evaluation. The symbol for this is more than the simple heat drain. Additional modules with energy storage and pumping work are required, as described for Fig. 1g.

VI. Passive Energy Storage Module

When potential energy is stored within the defined system it is represented by the tank symbols (Fig. 1b and d). Energy storages can exert true forces either together as a single Newtonian force or in parallel as a group population force. A potential energy storage is not completely defined until the quantity of potential energy in calories is expressed and its storage form defined by an appropriate equation. Water stored in a tank against gravity has energy and force both related to its hydrostatic head; energy in compressed gas is related to pressure by a logarithmic function. Energy and population force delivered from a silo of stored grain is mainly proportional to the number of grains. Potential energy is stored by doing work against the resisting force of the potential storage just as one stores potential energy when one presses a coiled spring against its resisting force. The more work that is done against the backforce of the storage unit, the more energy there is stored and the more force that unit can deliver when a pathway of outflow is provided.

In Fig. 1b passive storage is indicated by which the potential energy developed by work elsewhere is moved into a storage location without creating new potential energy. Such a flow requires an energy diversion or a second energy flow to move the potential energy into position, doing work against friction and energy barriers, but not increasing the quantity of potential energy by moving it into situation. Pumping gasoline, stacking potatoes in a store, or placing dynamite in an explosive shed are examples.

When an energy storage reservoir consists of a collection of component identical energy-containing packages which are not exerting forces on each other, the potential energy of that reservoir is the product of the mass of storage units Q and the energy per unit mass μ. This is the population potential energy W:

$$W = \mu Q. \tag{28}$$

For example, a bin of potatoes constitutes a group of units each with chemical potential energy, the total potential being the product in Eq. (28). If one were computing the potential energy within each potato in terms of the molecular composition one would use a different formulation for the potential in terms of storage forces overcome during synthesis. For macroscopic processes, however, the potential energies of storage reservoirs are given in proportion to the number of storage units present. The energy of storage may be delivered from the site as packages, or it may be provided with a pathway that expresses its component forces so as to discharge the energies stored within each package.

Storage Functions

Following are some of the storage functions by which energy may be stored against backforces of the storing process. A storage module is not adequately described unless its storage function is indicated. In the passive module, the storage work is done before the potential energy is brought into the module in packages.

Where the addition of successive stored units does not involve changing forces x or displacements h against which the work is done, the potential energy is proportional to their number n:

$$W = nxh. \tag{29}$$

Elevation of successive equal weights to a platform against gravity involves a constant gravitational force and the same vertical distance of displacement. Storing of compounds in a chemical industry is another example, as is the reproduction of offspring of equal size which may serve as potential energy to other food chains.

Where the back force of the potential is increasing as a function of the energy units as they are stored, summation requires integration. In pumping water into a vertical cylinder, the force (gravity g) against which work is done increases as the weight of the accumulated water increases with height. Water density is s, and area is a:

$$W = a \int sgh\, dh = \tfrac{1}{2}asgh^2. \tag{30}$$

Where the force of the potential is exerted due to the combined actions of accumulating molecules behaving as ideal gases in a gas phase or a dilute solution, the sum of the work done is the sum of the product of compressed volumes and pressures overcome with pressure and volume responding inversely according to the gas law,

$$Pv = n\mathbf{R}T; \tag{31}$$

$$W = \int P\, dV = \int \frac{nRT}{V}\, dV = n\mathbf{R}T \ln \frac{V_1}{V_2}, \tag{32}$$

$$W = \int V\, dP = \int \frac{nRT}{P}\, dP = n\mathbf{R}T \ln \frac{P_1}{P_2}. \tag{33}$$

Because of the reciprocal relation of pressure and volume or any molecular concentration and its group thrust across a plane surface, increase of molecular concentration adds less and less potential energy in the logarithmic relation of Eq. (33). This applies to pumping gas molecules into a tank, to the thrusting of molecules through a membrane, or to the packing of electrons from an electrode into a solution reaction. Thus the group force of molecules on a plane surface is in proportion to their number, but the accumulated potential energy stored there is in proportion to the logarithm of these numbers. The water vapor potentials existing between various components of the soil such as roots, clay, and organic matter is not in proportion to the vapor pressure, but logarithmically related as in Eq. (33).

The change in potential energy (excluding that required to change volumes while adapting to constant pressure) was also called "free energy, ΔF." Equation (33) is the expression for free energy change in terms of molecular concentrations or pressures. Free energy is appropriately used where processes, as in the biosphere, are under a constant pressure to which any gases appearing adjust volumes until they have equal pressure. Work of such volume adjustments is omitted when free energy is calculated, but has to be included in a complete energy budget.

In storage of electrical charges (Q) on condensers with electron flow in metallic conductors, the charges have electrostatic self-repelling field forces (voltage $E = Q/C$) so that the force overcome in adding an electron to storage is proportional to charges already accumulated:

$$W = \frac{1}{C} \int_0^Q Q\, dQ = \frac{Q^2}{2C} . \tag{34}$$

VII. Potential-Generating Storage Module

When the work of generating potential energy is done within the system it is represented with the modular symbol of Fig. 1d, which shows heat dispersal into the heat sink accompanying the pumping of potential energy into storage against backforces of that storage. Depending on the kind of work being done, the nature of the storage function may differ and a full description of the system requires an expression of this. The potential generating process involves at least three forces: the driving input force, the backforce from the potential energy storage being increased, and the frictional backforces of heat dispersal (Fig. 5). When the input and potential generating forces are equal, the system does nothing and is defined as a thermodynamically reversible one that can go either way with slight addition of energy to either of the two balanced forces. The second energy law requires that for spontaneous movement there be some heat dispersal and thus for the input force to exceed the potential generating backforce (Fig. 5). As described previously (Odum and Pinkerton, 1955; Tribus, 1961), selection for maximum rate of power storage adjusts the backforce to be half that of the input force. The maximum power property is implied by the modular symbol whenever power selection has been attained as theory predicts for real self maintaining systems. Where useful power expenditure is storage, maximum power selection of a one-step process leads to load ratios and storage efficiencies of the middle point in Fig. 5. The non-steady-state Atwood's machine has maximum power efficiency higher than 50 %.

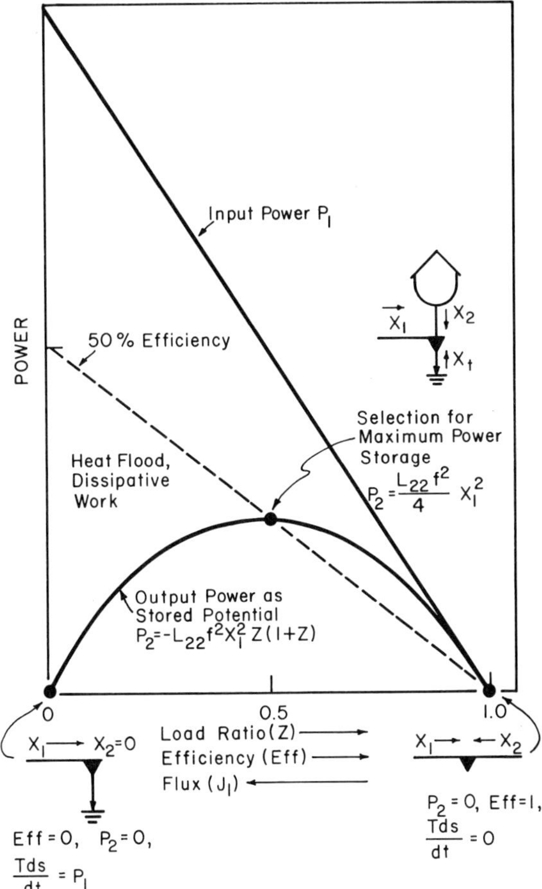

FIG. 5. Power input and useful power output as a function of load ratio, efficiency, and input flux, when input force X_1 is constant and output load X_2 is varied.

When provision for storage of materials and energy is inserted in a flow so that there is an opportunity for generation of potential energy W characteristic of the storage, then the potential energy of storage exerts a force depending on the potential force against which the work is done.

When two different kinds of storages are connected so that one can transfer energy to the other, unequal storage site forces will shift energies until equilibrium is attained. The potentials are then equal. For example, metallic electrical systems can be connected with chemical processes in solutions and gravity potentials can be connected with gas pressure systems in hydraulic systems.

A. One–Way Valves

Adopting the diode symbol (Fig. 1m) for a valve that permits flow in only one direction, the property of undirectional flow may be indicated in the energy circuit diagrams. Many of the flows have this property because outside work is required to traverse pathways but energy for backflow is unavailable. The diode symbol need not be drawn if the module on the end of the pathway has the one-way property as part of its interior structure. The self-maintaining module, for example, has work gate feedbacks controlling inflow and thus has little backflow. The value does allow backforce to affect flow.

B. Linear Decays of Storage

In Eq. (35) P represents the power delivered by potential energy source W which declines in potential energy dW/dt by the power delivered:

$$P = dW/dt = u\,dQ/dt. \tag{35}$$

In simple potential energy storages, decline in potential energy W may be accompanied by decline in output force of that storage X, but in compound population storages, it is the number of energy storage units that declines and the force that declines is the causal group action of the population N. The decay of energy follows the flux in this way only when the energy storage function is also in proportion to the quantity of units stored Q. An electrical capacitor is an example of one that is not. The mass of charge on an electrical condenser discharges with pattern of Eqs. (35)–(37), but not its energy, which is proportional to the square of the charge stored [Eq. (34)].

One of the well-established experimental confirmations of the linearity of many complex population storages is the linear decay when inputs to storage are eliminated. The decay rates of radioactive populations, of sewage solutions, of soil litter, or starving animals, and many others have as first approximation flow dependent on storage, the outflow being equal to the rate of decrease:

$$dQ/dt = J = L'N = (L/C)Q = (1/RC)Q. \tag{36}$$

Here, L/C is the fractional turnover rate and C/L the time constant. The integrated form of Eq. (36) is

$$\frac{Q}{Q_0} = \exp - \frac{L}{C}\,t. \tag{37}$$

Energy storages W follow Q in those population examples with linear storage functions only [Eq. (28)].

The constant L/C may be evaluated from a semilog graph of mass with time.

C. Storage Kinetics

The energy storage has characteristic Von Bertalanffy kinetics as it achieves a steady state when an inflow J is supplied. Equations describing the storage with time are similar for three cases: (1) the energy storage may be passive but filling faster than its outflow which depends on amount stored; (2) in filling a storage in a potential-generating situation, increasing potential is developed against increasing static back force; (3) the driving force is accelerating matter or accelerating electrical flow so that potential energy is developed in the inertia of mass or the electrical-magnetic field which opposes any acceleration until it ceases, after which the energy of the inertia or of field that has been stored drives the current. The latter property is used in electrical circuits with wire coils as inductances to add time lags that are out of phase with those achieved by filling a tank against static potential generating forces. This impeding property may be generalized in complex living systems by behavior programs which have the same property of resisting input while storing energy in proportion to the original impetus, later diverting the energy storages into actions that do later what the original force might have done had it not been blocked. (A symbol is given in Fig. 11.)

Whatever the process, storage rate is a balance of inflows and outflows. Sometimes referred to as the Von Bertalanffy equation, the storage tank is appropriately called a "Von Bertalanffy module" with saturation equations (38) and (39). Energy is proportional to mass Q only in those systems with linear storage functions as in many population examples (see Fig. 6):

$$dQ/dt = J - L'N = J - (L'Q/C). \tag{38}$$

If initial storage is zero, the integrated equation takes the form

$$Q = \frac{CJ}{L}\left(1 - \exp - \frac{L}{C}\,t\right). \tag{39}$$

This is the charge equation of electrical circuits where C/L is the time constant.

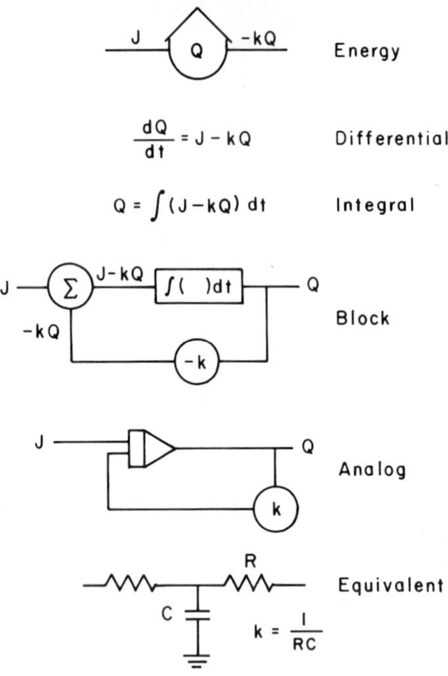

FIG. 6. Von Bertalanffy storage module stated in six languages.

D. Efficiency in Potential-Generating Circuits

The loading of backforces from an output process which stores potential energy (Fig. 4b) is in a different class from passive dissipative loads. A potential load has an independent value that is not dependent on the flux but is determined by its own energy storage level or by an external potential in the case of an export circuit. The amount of loading can vary from zero to a stopping load which converts the process into a reversible stationary situation. The load may be greater than the input force so that the process reverses with energy flow and flux in the opposite direction.

Unlike the dissipative flows, power output against a potential load is not a conversion of potential energy into dispersed heat and thus by the second energy principle will not take place unless there is at the same time arrangement for some diversion of heat flow for speed tax. Thus the speed with which power output is directed into a potential circuit depends on the amount of speed tax, other things being equal. The efficiency of power transmission into the new potential reservoir from the old is given in Eq. (40) by the ratio of expressions for power given in Eq. (8):

$$\text{Eff} = \frac{P_2}{P_1} = \frac{J_2X_2}{J_1X_1} = \frac{J_1X_1 - P_t}{J_1X_1}. \tag{40}$$

Classes of power output in potential generating circuits are those given in Eqs. (28)–(34). An output of heat which develops a reservoir of heat at a temperature higher than the surroundings is potential energy, but an output of heat which disperses without changing temperature gradients is dissipative and serves as speed tax.

In order to measure the potential force loading, let us define a ratio Z as the quotient of the potential generating force X_2 and the input force X_1, inserting f to place both forces in the same dimensions so that Z is dimensionless:

$$Z = X_2/fX_1. \tag{41}$$

When the output potential force X_2 balances the input force X_1 the process is stalled as on the right in Fig. 5. Then the force ratio Z is one, the flux J_2 is zero, and Eq. (4) may be combined with (41) and rearranged to provide:

$$f = \frac{X_2}{X_1} = \frac{L_{12}}{L_{22}} = \frac{L_{11}}{L_{21}}. \tag{42}$$

The ratio of conductivities f can then be evaluated. Energy flows vary with load ratios Z (Fig. 5). Any leakage is regarded as a separate circuit.

E. Energy Flow in Lever Transformations

As in elementary science books, levers such as the inclined plane, screws, pulleys, gear wheels, lever arms, etc., can transmit power with little loss while transforming the relative magnitudes of flux and force reciprocally. Thus, in Eq. (7) the force X may be varied with a compensatory change in the flux J so that power expressed is unchanged except for losses to heat dispersal. An example of a lever transformation is change from situation (a) to (b) in Fig. 3.

The concept of torque for the force on a rotary system is usually taught as the product of linear force and the radius, similar products exerting similar rotary accelerations, even though force and radius vary. The torque concept is derivable from the concept of power conservation during a lever transformation. Since the velocity of a point on a wheel is proportional to the radius, the compensatory change of J and X is equivalent to a compensatory change of J and radius.

F. The Possibilities of Loading

The backforce against which a process does work is the loading. These load forces determine the partition of power budget among several flows and whether there is optimal loading of each flow (Fig. 5).

If the work is one of storing energy, it is potential generating, and according to the second energy principle if the potential-generating output load equals the input load, no power flows. Otherwise there could be use and reuse of energy. The size of the potential-generating load affects the speed and power transformed through that flow. It affects the energy restored as compared to that dispersed as speed tax,

$$J_1X_1 + J_2X_2 = P_t = P_1 + P_2 , \tag{43}$$

where the fluxes are related to the forces as in Eqs. (4) and (5). Substituting (4) and (5) into (43) and also the definition of force ratio Z from Eq. (41), we obtain

$$P_2 = -L_{11}X_1^2Z(1 + Z). \tag{44}$$

With forces to the right considered positive X_2 is taken as negative, and Z is negative. This convention is different than that used previously (Odum and Pinkerton, 1955) in order to simplify application to the energy circuit diagrams. Equation (44) expresses the useful power output in terms of the loading ratios of the output to input forces Z, where the input force X_1 is constant and L_{11} is the input conductivity.

If one sets the derivative equal to zero so as to find the value with maximum output, one obtains $Z = -\frac{1}{2}$ and the equation for maximum power is

$$\underset{\substack{(\max^2 X_1\\ \text{const})}}{P_2} = (L_{11}/4)\, X_1^2. \tag{45}$$

Substituting (4), (42), and (44) into Eq. (40) for efficiency of restoration, one obtains Z for the efficiency and 0.5 for the efficiency at maximum power output:

$$\underset{(\text{power max})}{\text{Eff}} = P_2/P_1 = -Z = 0.5. \tag{46}$$

Evaluation of constants in a particular problem can be made with data in particular cases. With output force set to balance input force so that the process stops, the ratio f is evaluated from Eq. (42). If while stalled there is a flux J_1, it occurs through some other energy flow circuit which has not been stalled by backforce. One may either include such flows in the equation as a leakage term, as done previously (Odum and Pinkerton, 1955), or exclude them from consideration as a separate circuit, as done here. Later the overall input and output efficiencies can be computed for a network if there are several diverging flows. Hence, in describing the power-efficiency relations of a single circuit, J_1 is zero when Z is one. The conductivities L_{22} and L_{11} as previously discussed may be regarded as the flux-proportional coefficients of resistive force with X_2 or X_1 as input forces, and the conductivities may be evaluated with flux measurements in situations with one force input against resistive drag.

Thus L_{22} is equal to the coefficient of a resistive force when a force X_2 is imposed on the system while X_1 is zero and flow is in steady state. The other conductivities are related to L_{22} and are evaluated as follows: When the output force has stalled the input force as in Eq. (41), f is computed relating dimensions of L_{12} to L_{21}. With X_2 zero and considering all leaks as part of other circuits, L_{11} is the coefficient of resistive force after a flow from X_1 is in steady state through the system. When the input flow is stalled with backforce X_2 so that J_1 is zero, then the conductivities L_{11} and L_{21} are related as their dimensions f in Eq. (41) and summarized in Eq. (42). Some other conversions among conductivities may be written as

$$L_{11} = L_{22} f^2 = L_{21} f. \tag{47}$$

Equations (44) and (45) are plotted in Fig. 5. The point of maximum power may be chosen by natural selection under the conditions discussed in which the power is flexibly usable as specific competitive criteria require. Selection of the maximum power point also sets the force ratio and hence the efficiency at 50% [Eq. (46)]. Although one might theoretically load a system so that the force ratio (and efficiency) was at some other value, it would not be a competitive arrangement where power storage was useful. The loading would either be too slow or waste too much energy as speed tax. One may state this as a fourth energy principle that natural selection adjusts the speed tax and efficiencies of potential output to 50% for a single transformation. Whereas power going into heat was found to be linearly proportional to flux when the input force was constant in single dissipative flows, the speed tax to heat on a potential conversion goes up as the square of the flux. It is the area between curves in Fig. 5.

G. Power Output When Potential Loading is Constant but the Input Force is Varied

The relationship of output power with varying load and constant input force was summarized as a parabolic relation in Fig. 5. Equally important is the pattern of power output that results when the input force is varied while the output potential load force is held constant (Fig. 7). Increasing the input voltage on a battery charging process is an example.

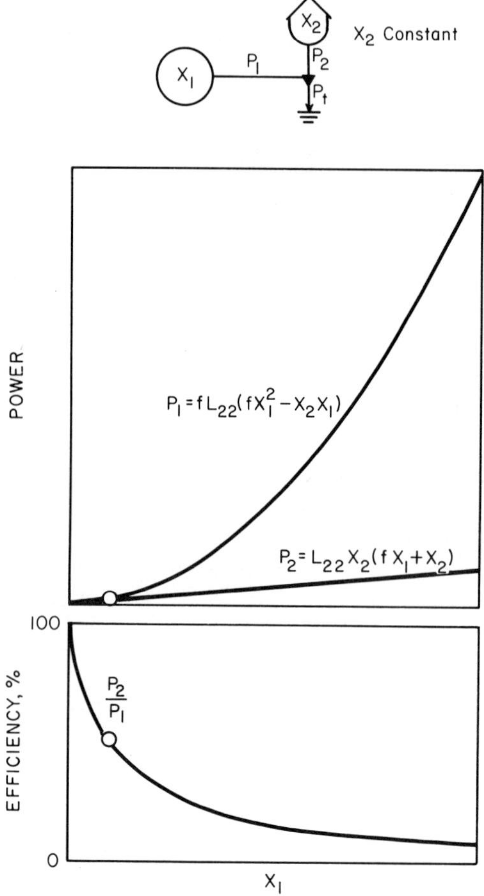

FIG. 7. Power storage P_2, efficiency of power storage, and input power P_1 when output loading X_2 is constant and input force varied. Loading of 50% is encircled.

In a procedure similar to that used for derivation of Eq. (45), an expression for power output against the fixed potential load X_2 is developed as a function of the variable input force X_1. The result is a linear equation,

$$P_2 = L_{22}X_2(fX_1 + X_2), \tag{48}$$

which is graphed in Fig. 7. The power input P_1 varies as the square of the input force,

$$P_1 = fL_{22}(fX_1^2 - X_1X_2), \tag{49}$$

and is also graphed in Fig. 7. The ratio of power flows is the efficiency, in Eq. (50) as a function of the load ratio Z, defined in Eq. (41):

$$\text{Eff} = \frac{P_2}{P_1} = Z. \tag{50}$$

This efficiency is graphed also in Fig. 7. When f is one, the efficiency is the force ratio. From the steeply declining hyperbolic pattern of the efficiency plotted as a function of the increasing input force X_1, it can be observed that under increasing input force the system is using greater and greater quantities of input power, and greater and greater amounts are going for speed tax. Beyond the point of optimum efficiency marked with a circle in Fig. 7, the system is very uncompetitive.

If a system of varying inputs were set up so simply, it would rarely function effectively. It will be shown that biological systems with varying inputs have various devices for maintaining a more optimum loading ratio by varying the output loads and by other means. Like automobiles they have their gears changed as the input force is increased. Photosynthesis is one example.

Because some properties of speed, starting characteristics, and other aspects are selected by engineers in the design of such engines as electric motors, there results the same kind of selection for power as exists in the natural selective process. In both kinds of systems there are various devices for regulation, starting, and control so that other loadings are not allowed. It is thus not always possible to vary load experimentally in order to observe the changes in percentages of speed tax, efficiency, etc. in the basic speed regulator curves (Fig. 5).

H. Resistance Matching of Dissipative Processes in Flux Series

When there are two dissipative processes whose fluxes are in series, such as two resistances in an electrical circuit, the one with the greater resistance draws the greater proportion of the power dissipation of the two, but it is also the rate-limiting process. If the larger one diminishes its resistance, it then allows the flux through both units to increase and hence its own power increases, but its proportion of the total power then diminishes. When the two resistances are equal, the power is maximal relative to other combinations with the same total resistance. If one resistance is lowered below the other it increases total power drain, but the proportion shifts to the other so that the smaller one has less power dissipation than when it was equal.

A familiar example of dissipative processes in flux series is the battery with external resistive circuit. The battery acts as if it has a certain internal resistance. Maximum power is drained to the outside when the resistance outside is made identical. A higher resistance slows down the flow too much. A lower one dumps energy within the battery.

The relation of power and load on the second of a pair of dissipative transformations in series is similar to that of power and potential conversion load following the hump shape of Fig. 5. However, selective processes need not necessarily favor maximization of a resistive dissipation. If the second of the two processes is a useful one, selection processes may act to regulate its speed and thus its loading. If the second process is not by itself a useful one it may serve a useful function to the first process if that one is useful, regulating its power flow. If selection causes a system's loading to be set so that the second process is at maximum power output, this makes the first process subject indirectly to speed control by maximum power selection. A single dissipative process has no maximum in its relation of load to power, but the double series system with input force held constant does have a power-loading curve with a maximum.

I. Power and Efficiency of Populations of Processes Each with Constant Force Loading

Equations (39)–(46) related power and efficiency where forces were varying. Somewhat analogous equations may be written for systems where the individual component forces are acting separately, are not varying, and have each backforce opposite an input force. The total flux, power, and efficiency are then the effect of population force and population backforce, the numbers of actions being the variable. Input forces n_1 are more numerous than opposition forces n_2. The ratio is $z = n_2/n_1$.

An input system with n_1 separate input actions may have some of its population of input forces x_1 opposed by n_2 negative backforces x_2, whereas others may be opposed by no backforce other than resistive forces implied in the component conductivity l_{11}. The number of unopposed inflows is $n_1 - n_2$. The opposed and unopposed flows are given in Eqs. (51) and (52):

$$\underset{(\text{unopp})}{J_1'} = l_{11}x_1(n_1 - n_2), \tag{51}$$

$$\underset{(\text{opp by } x_2)}{J_1'} = (l_{11}x_1 + l_{21}x_2)\, n_2 . \tag{52}$$

By adding opposed and unopposed populations the total input flux J_1' results:

$$J_1' = l_{11}x_1n_1 + l_{21}x_2n_2\,. \tag{53}$$

Since n_1 is greater than n_2, there are no unopposed component forces x_2 and we may write for the output flux

$$J_2' = (l_{12}x_1 + l_{22}x_2)\,n_2\,. \tag{54}$$

For the simple case of all component loads ratios represented by one Z value, expressions for power delivery follow from Eq. (18) with substitution of expressions (41), (42), and (47) for the group flux J':

$$P_2 = J_2'x_2 = -l_{11}x_1^2n_1zZ(1 + Z), \tag{55}$$

$$P_1 = J_1'x_1 = l_{11}x_1^2n_1(1 + zZ). \tag{56}$$

Where the component forces, fluxes, and loading ratios are constant, the power delivered is proportional to the ratio of populations z opposed as well as the population of input forces n_1. The input flow is proportional to the size of the population of actions n_1 diminished by the ratio of opposing thrusts, where z and Z are both negative. The ratio of the power flows is the efficiency:

$$\text{Eff} = \frac{P_2}{P_1} = \frac{-zZ(1 + Z)}{1 + zZ}\,. \tag{57}$$

Increases in the ratio z of opposed forces of a population of useful outputs result in increases in flux, power, and efficiency without maxima up to a value of $z = 1$. In Eq. (57), for example, efficiency is maximum when the force ratio of the component processes Z is 0.5 and when the ratio of number of processes z loaded is 1 (100% loaded). The overall efficiency is the 50% for maximum power storage possible.

VIII. Intersection and Feedback

The intersection of two flows involves phenomena that depend on the kind of reaction. If the two flows are of similar materials and energy types, they may add forces and flows as in the examples of two converging water pipes or copper electrical conductors. Such *additive junctions* are shown in our notation (Fig. 1n) as simple intersections that may exist without any heat losses. If there are no one-way gates, the flows may subtract forces and reverse the flows that existed before junction.

If, however, an intersection involves different kinds of flows that react in definite proportions, the output may depend on the product of the two driving forces and may be called a *multiplicative junction* (Fig. 1f). The intersection may have a constant effect, drawing power as needed (Fig. 1k) as a constant gain *amplifier*. Figure 1i is an intersection operating switching action of one flow on another. In the notations for environmental networks the kind of functions involved at each junction must be indicated by an appropriate symbol.

If a flow of energy or materials has a circuit connecting a down-circuit section to an up-circuit section so as to form a circular loop, it is called "a feedback circuit." For example, there is a feedback of electrical power from the generator to the water pump in Fig. 8f. The connection of the feedback with the main flow forms a converging junction, but the closed loop causes the loop system to exhibit characteristic properties that are different from intersecting junctions without feedback.

There are many kinds of feedback. Some circuits return materials without much force or energy; others feed back both force and energy. The feedback may serve as transport and energy barrier boost without otherwise affecting the force, energy, or materials flowing. Amplifier flows may be involved in some part of the loop so that the feedback has more force and energy than the original up-circuit inflow. In other systems the feedback has a diminished force and serves as a gate on the upstream flow.

IX. Constant Gain Amplifier Module

If one energy flow undergoes a transformation so as to increase the force and energy exerted through a second-flow by a constant factor, the combination constitutes a constant gain amplifier. The interaction of the two energy flows is drawn with the symbols shown in Fig. 1k. The combination follows the second energy principle by dispersing more potential energy into heat in the amplifier flow than is stored in the flow amplified.

If either the increased force X_2 or the population force N_2 after amplification is proportional to the input force, one may represent the change with Eq. (58) where g is the amplification factor often referred to as gain. The triangular amplifier symbol of Fig. 1k implies that power is drawn as needed to make the Eqs. (58) and (59) hold:

$$X_{\text{out}} = gX_{\text{in}}, \tag{58}$$

$$N_{\text{out}} = gN_{\text{in}}. \tag{59}$$

There are other amplification functions also. The symbol g may be constant or variable. If the gain is high, the down-circuit process may be guided by the up-circuit process without drawing much power from the input.

A. Feedbacks That Add or Subtract

Shown in Fig. 8 are some feedbacks that pass the same type of force and energy flow back to an upstream site so that they add or subtract. Examples are electrical circuits that feed back electrons and voltage to upstream junctions, or fluid pipes that return downstream fluids to upstream sites in chemical reaction industries. The loops closed by these circuits may have outside energy sources and amplifier actions as shown in Fig. 9d.

Feedbacks may be classified according to the nature of the feedback intersection as additive or multiplicative and positive or negative. The feedback may be a pathway of flow of force and energy as shown in energy circuit language in Fig. 8. Also there are feedbacks of effect without a special pathway. Negative feedback is implied in the Von Bertalanffy storage module in which the outflow is proportional to the storage. In Fig. 6 the energy storage symbol with a steady inflow J and density dependent outflow kN is shown. The differential equation for the storage is a balance of the two flows. Converting to integral equations and diagramming the terms in block language, a classical negative feedback, stabilized system results. In other words any storage unit is a negative feedback stabilized integrator of energy flows even though there are no actual pathways of force or material feedback. The property in ecology of density-dependent population regulation turns out therefore to be classical negative feedback. Also shown in Fig. 6 is the electrical analog equivalent of one storage compartment. Much ecological and tracer simulation so far has involved chains of these units.

B. Feedback Loops Which Add Forces

One may characterize the general pattern of additive (or subtracting) feedback pathways with equations relating forces and transfer functions. Using symbols in Fig. 9d, the force X_3 or population force N_3 exerted by a feedback loop is given by

$$N_3 = BN_4 \qquad \text{or} \qquad X_3 = BX_4\,, \tag{60}$$

where B is the feedback function, which may be a simple constant or more complex function and may be positive or negative. If gain is

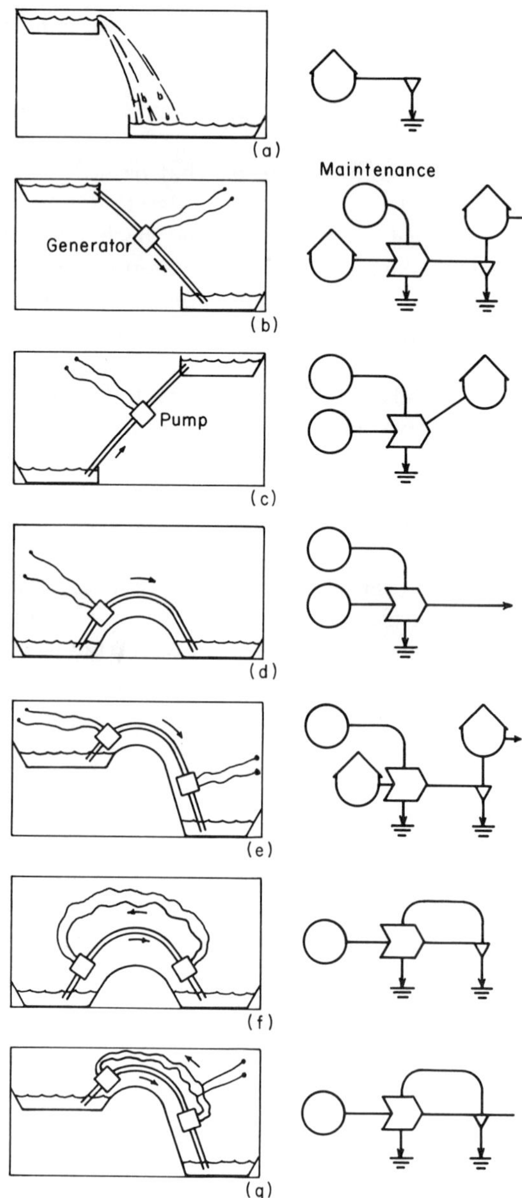

FIG. 8. Some hydrologic examples which illustrate energy transport pathways and their expression in energy circuit language.

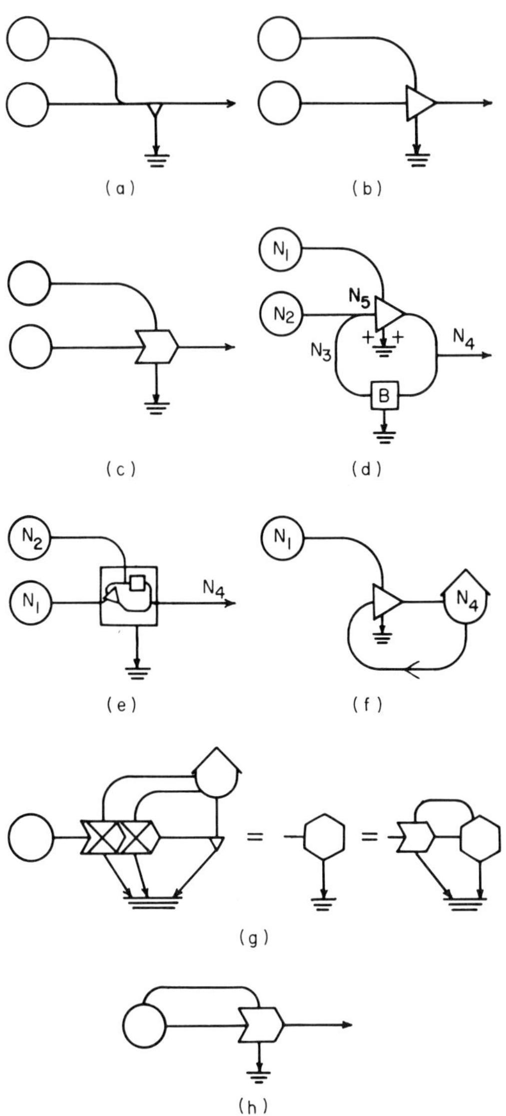

FIG. 9. Types of feedback, intersections, and multiplicative junctions. (a) Additive junction, (b) constant gain amplifier, (c) work gate, (d) and (e) additive feedback, (f) Malthusian growth, (g) logistic loops, (h) self-pumping.

involved, an outside power source may be operating (Fig. 9b). In Fig. 9d B is a box symbol (a square box is reserved as a general purpose symbol where a function is unspecified or one for which a special symbol has not been assigned). In Fig. 9e an abbreviation has been drawn from Fig. 9d emphasizing the role of force N_2 modulating a power flow from N_1.

In the circuit notation of this chapter, a constant gain amplifier stage is indicated, as in Figs. 1k and 9b, with the power supply from the top, the controlling force from the left and the feedback flow extending from the down-circuit to the up-circuit junction. The energy circuit usage differs slightly from the well-established usages in electronics. The symbol does not mean a switch in sign (plus–minus) and it must carry a heat sink and power supply, features that are not necessarily shown in diagrams of electron flows.

Stable feedback loops apply a force at the feedback junction with opposite sign so that the net force is the algebraic sum of the forces at that point, as stated in Eq. (61) using symbols shown in (Fig. 9d):

$$N_5 = N_2 + BN_4 . \tag{61}$$

Combining (61) with Eq. (57) for the transfer function of a linear amplifier, gX_5, one obtains the following basic equation for servomechanisms:

$$X_4 = \frac{g}{1 + gB} X_2 , \qquad N_4 = \frac{g}{1 + gB} N_2 . \tag{62}$$

The force X_4 beyond the down-circuit loop is related to the input force X_1 according to the fraction that contains the gain factor g and the feedback function B. With values of g and their algebraic sign, the steady state output of the loop may be calculated. If the loop X_2 is negative to X_1, there is a stabilized steady state. If, however, the feedback is positive, Eq. (62) provides only an instantaneous transient value in a rapid growth sequence.

When the feedback circuit acts to diminish the force, flux, or energy of the main flow, it is a negative feedback. The more one increases forces at the junction, the more feedback force there is in the role of a cancelling force, regardless of variations in the energy inputs to loop amplifiers. Whereas positive additional feedback is unstable, negative feedback provides a mechanism for stability of flows in spite of noisy variations in participating flows.

With positive feedback (Figs. 9d and e) an increase of force at the feedback junction produces an output increase at N_4 which then reinforces the drive at ($N_3 = BN_4$), which then increases the force at N_4, etc. Such an arrangement draws increasing amounts of power from storage or inputs within the loop, N_1, and accelerates. It is unstable, since no steady state is reached until ultimately, the acceleration ceases due to limitations in the power sources.

The Malthusian circle is the special case of positive feedback drawn as Fig. 9f, in which the output of an amplifier is fed back for amplification again with no outside flows other than the supply of power and materials entering at the amplifier as needed to provide gain. Such a system accelerates growth of materials, forces, storages, and energies.

X. Work Gate Module

Shown in Fig. 1f is the work gate module that represents the work of one flow of energy controlling and facilitating conductivity of a second. If the effect of the second flow on the first is linear, the module in effect has an output proportional to the product of the driving forces of the two pathways, and in that case has a multiplication sign × on it. By custom the main power flow is shown as the horizontal one and the limiting and controlling lesser energy flow is shown passing in from above. The heat sink represents the usual spontaneous heat increases associated with any process including those from the coupling of the two forces. The smaller flow is the signal in an amplifier effect in controlling a larger low, but the gain is a variable, proportional to the forces. The smaller control flow may come from another part of the network or from outside energy sources:

$$J = kN_1N_2\,, \tag{63}$$

$$J = kN_1X_1\,, \tag{64}$$

$$J = kX_1X_2\,. \tag{65}$$

If one energy flow facilitates the transport of a second flow without increasing or decreasing the latter's force or energy, the first flow is accomplishing work of transport. Transport junctions are amplifiers with gain equal to the ratio of the control flow to power flow. In several natural systems it seems to have a value of the order of 100.

The transport may involve temporary amplification followed by a loss of the temporary gain. The transport may involve structural maintenance work, work of accelerations and braking, and other means for supplying auxiliary energy to pay for the energy transport of the second circuit. For convenience of recognition the transport junction is indicated by the circuit work function in Fig. 1f if no constant amplification is involved. Amplifier and transporter junctions are given in various combinations for some hydroelectric systems in Fig. 8.

A. Crossing Energy Barriers

The existence of circuits in environmental systems often depends on auxiliary sources of potential energy that allow transport flows to cross energy barriers. For example, the maintenance flow of a repair man (Fig. 8b), permits a generator flow on a permanent basis. Without barrier-crossing transport junctions, complicated environmental networks could not exist.

According to the second energy principle a pathway proposed so as to involve an elevation of potential will not flow unless more than equal expenditures of other potential energies are provided. Thus, water does not flow uphill without a pump, and electrons will not cross a vacuum tube without an energy source to heat the filament. A circuit with a route through a state of higher potential thus has an energy barrier that requires an auxiliary energy source as a pump, although the ultimate source of auxiliary booster circuit may be the same.

In Fig. 8 are schematic sketches and energy diagrams for water pumping situations, illustrating energy barriers and energy pumps for barrier circuits. In (b) is the coupling of gravitational energy generating electrical potential energy with accompanying speed tax. In (c) is the reverse with electric potential energy generating gravitational potential in raising water with accompanying speed tax. In (d) the flow crosses an energy barrier with the combination of pumping of sketch (c) and a frictional dissipation of energy after crossing the barrier. In (e) the energy on the falling side regenerates some electrical potential with some energy for speed tax and friction. In (f) there is a feedback of the electric potential generated to the pumping flow with levels and loads so adjusted that there is no net potential. Finally (g) shows the flows of (f) but with some net generation of electrical potential.

For situations (c)–(f) as shown, there are three ways of drawing the energy circuit depending upon one's point of view. If the water flow is of principal interest, it is shown with the horizontal input line and the electrical input is shown as a work circuit flow with down-directed arrow across a circuit box.

Where there is an energy barrier not involving thereafter a net gain or loss of potential energy of the flow, the auxiliary energy source supplying energy to pass the barrier is represented as a wholly dissipative work function, as in Fig. 8d. All of the potential energy gained in crossing the hill in the example is dissipated by friction in coming back down.

In (f) there is a feedback to the barrier pump of the energy generated by the fall of water downstream. Like the somewhat simpler siphon which has the same energy diagram, the process, if interrupted, requires outside energy for restarting.

Using the terminology and diagrams of Fig. 8, one may characterize energy diagrams by the work necessary to maintain flow. Much of the energy of complex systems is dissipated in pumping other flows over energy barriers and gaps. For complete representation the diagrams should also have maintenance work flows for any structures involved.

Since the work flux of maintaining flow across energy barriers involves dissipation of potential energy into heat either from the outside or from a feedback, work flows are not reversible. Cross-barrier systems are ordinarily unidirectional. Transporting and amplifying junctions are usually one-way gates. Since the transport circuits involve the interaction of the flow transported and the work flow of the pump, they are usually double-flux reaction systems with kinetics as discussed subsequently below.

B. Diagramming Pathway Detail with Work Gates

As first presented with Figs. 3 and 4 energy flow pathways were diagrammed as simple lines plus some heat sink dispersion either along the pathway or localized in a module at one or both ends of the segment, these indicating the main association of the entropy increasing process by which the segment was spontaneously driven. The box in Fig. 4c left the functions as an unspecified operation of the input energy. This procedure can represent simple or compound processes vis-à-vis their contribution to the overall system. The single line compartmentalizes much detail of the pathway. This procedure is useful where pathway details are not of interest.

If, however, the pathway details are of interest, they may be diagrammed further by adding those modules that show the energy flows more explicitly. For example, Figs. 8f and g show details of water flow between two lakes in which the energy resources available to that pathway drive the flow through various work diversions on the pathway. These could be represented by the simple lines of Fig. 4b if the detail were not of interest.

The magnitude of an energy barrier is measured by the temporary potential energy which a unit of flux will develop in passing over the barrier. In chemical systems this potential energy quantity is the free energy of activation. In chemical systems the stochastic distribution of energies among component molecules produces some molecules at higher energies with enough impetus to cross energy barriers that would block other molecules. In chemical reactions repulsive forces of molecular charge often constitute energy barriers. In effect a subpopulation of the molecules is being pumped over a barrier by power transformations within the others. The energy circuit is Fig. 9h. The Arrhenius relationship of reaction rate and temperature reflects the nonlinear distribution of excessive molecular energies with temperature.

Whereas classic teaching of thermodynamics relates rates of reaction to the energy barriers and not to the potential energy of initial and final states, most real systems of the natural world develop networks in which there may be feedback of some of the potential energy of the overall reaction into maintaining the pathway or aiding the crossing of the energy barriers (Fig. 8f). The amount of energy which may be derived from the two states in this feedback arrangement is a function of the nature of the pathway. Classical thermodynamics emphasizes the equilibrium state at which there is no pathway and only the potential energy independent of path is appraised.

C. Multiplicative Junctions with Limiting Flows

If two flows intersect at a junction where they react in definite ratios and the output is the product of the two driving influences [Eqs. (63)–(65)] the processes are said to be second order. Many biochemical reactions are double flux reactions as are processes on a larger scale, such as the flows of parts to be combined in an industrial assembly line. The amplifier and energy barrier transporter transformations are special cases.

Multiplicative junctions are diagrammed in Fig. 10. Figure 10a is the energy diagram showing two converging energy flows intersecting in the work module (symbol marked $\times$) that represents the multiplicative effect of the separate influences at the reaction site. Also shown (Fig. 10a) are two auxiliary flows under driving influences X_5 and X_6, which serve as transport pumps in supplying the two energy flows to the reaction site developing a force due to the potential energy in the states of the reactants assembled. The rates of pumping by the auxiliary flows control the rates of supply of the reactants. At the steady state the rates of energy transformation conform to the rates of supply rather than vice versa. In Fig. 10a, if the inflows are fast the concentrations in the site storage increase and the thrust increases the reaction and/or the bypass flow. If the inflows are small, the concentrations at the site are diminished. The energetic cost of the whole process may be measured by the energetic cost of the auxiliary pump of the limiting process.

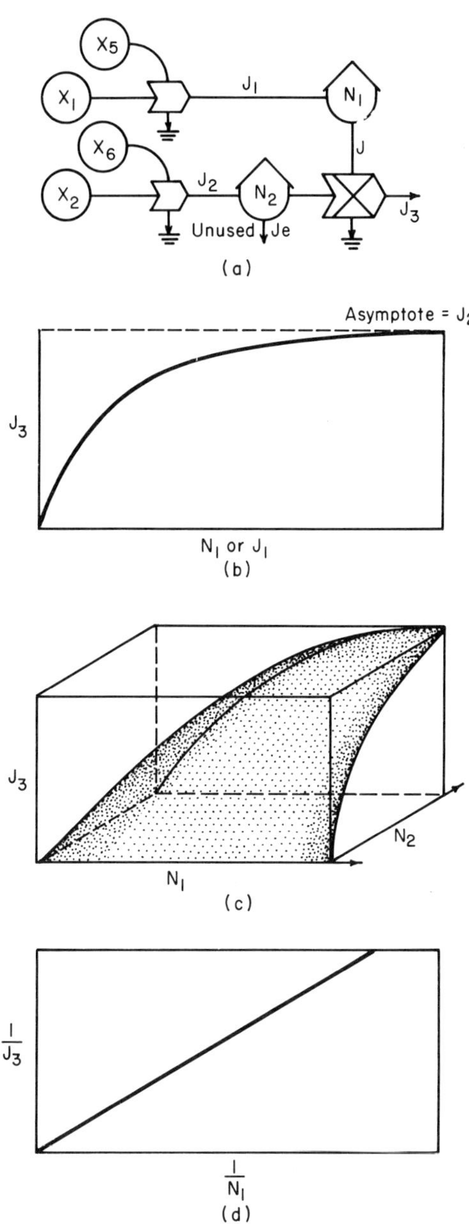

FIG. 10. Limiting factor intersection and responses in energy circuit language.

These auxiliary costs may be derived from outside as shown in Fig. 10 or from a feedback.

In Fig. 10a is a material diagram for those examples that show the convergence of two material flows J_1 and J_2 to form one flow J_3 plus some bypass of excess as J_e. There are, however, many double-flux processes where only one of the flows contributes material to the outflow, the other inflow contributing only energy. Examples are the transport transformations that were already discussed. Figure 8c shows the simple flow of material where the second branch of the inflowing reaction involves no material contribution.

Equations (63) and (65) are familiar relations from mass action chemical kinetics for the reaction flux when the combination is in proportion to the opportunity for combinations according to the concentrations at the reaction site. A double flux process multiplies the forces of N or X.

For some macroscopic flows, N_1 and N_2 are the populations of reacting units each exerting component forces and energy transformations. For the system of flows drawn in Fig. 10 the concentrations N_1 and N_2 increase or decrease until the flux J_3 is consistent with the inflows J_1 and J_2 as driven by remote forces X_1 and X_2 and, if present, their outside-supported pump drives X_5 and X_6. Thus J_3 cannot exceed the stoichiometric requirement of the smaller of the two inflows. Changing the junction process by changing k in Eq. (63) only modifies the concentrations N_1 or N_2 necessary at the junction for steady state. Equations (66) and (67) describe material flows in the reservoirs at the reaction site in Fig. 10a:

$$dN_1/dt = J_1 - J = J_1 - sJ_3, \qquad (N_1 \text{ limiting}), \tag{66}$$

where s is the stoichiometric ratio J/J_3 of necessary use, and

$$dN_2/dt = J_2 - SJ_3 - J_e. \tag{67}$$

At steady state the rates of change in Eqs. (56) and (57) become zero with Eqs. (68) and (69) resulting:

$$J_1 = sJ_3, \tag{68}$$

$$J_2 = SJ_3 + J_e. \tag{69}$$

Except for the special simple case of J_1 and J_2 in the ratio of their reaction formulas, one flow is less than the other relative to the reaction outflow and hence is limiting. If J_1 is chosen for the limiting flow, it has no excess beyond that necessary for the reaction. Outflow J_e exists only for the reactant inflowing in excess, J_2. Equations (68) and (69) thus define the concept of limiting factors.

In double-flux flows the energy may be associated more with one flow or the other or with the relation of the two. The viewpoint as to which flow is the energy is sometimes relative. With a flow of fuel and oxygen reacting in a flame, one might regard the energy as in the fuel out of mental habit, since the fuel is ordinarily the rare quantity and the atmospheric oxygen the more commonplace. Yet with a fire within an oil mass deep in the earth one may prefer to regard the oxygen as the energy source. Actually both contribute.

D. Limiting Concentrations to Double Flux Reactions

Involving extensive experimental data in several sciences are studies of double flux reactions in which the quantity of limiting reactant at the reaction site is controlled as an independent variable. Thus N_1 in Fig. 10a is varied while the influx J_2 is in constant excess. For this special case the export of the excess J_e is proportional to the quantity N_2:

$$J_e = LN_2 . \tag{70}$$

Combining (63) with (68), (69), and (70), one obtains a rectangular hyperbola, Eq. (71), relating output flux to limiting concentration:

$$J_3 = kJ_2N_1/(L + SkN_1). \tag{71}$$

The relationship is graphed in Fig. 10b. When N_1 is small, J_3 is almost a linear function of the limiting concentration at the junction site. As N_2 becomes larger, the reaction flux approaches an asymptote determined in part by J_2, the other input flux. Authors have sometimes described the curve artificially in three zones: linear, limiting-nonlinear, and nonlimiting. In this derivation N_1 was limiting. If N_2 is limiting, a similar equation results with different constants and N_2 is the variable. However, it is simpler to place the usual limiting source in the position as controller from above in Fig. 10a, swapping designations.

Expression (71) derived above from the circuits is the equation of Monod (1942) for the effect of limiting growth requirements on bacteria (Novick and Szilard, 1950). It is the relation to which many plant physiological data and agricultural experiments on plant requirements have been related. The flows of nonlimiting constituents such as J_2 have not always been controlled or kept constant so that many experiments do not precisely follow the relation.

Shinozaki and Kira (1961) and Ikusima (1962) expressed, from experimental data, the limiting effects of water, carbon dioxide, nutrients, and light on plant growth as a reciprocal equation,

$$\frac{1}{J_3} = \frac{k_1}{N_1} + k_2 . \tag{72}$$

Here J_3 is the rate of gain in weight of plants, k_1 and k_2 are constants, and N_1 is the limiting factor. Rearranging this expression, the hyperbolic form is again found,

$$J_3 = N_1/(k_1 + k_2N_2). \tag{73}$$

An advantage to writing the rectangular hyperbolas such as Eq. (71) in reciprocal form, as Eq. (74) is the straight line relation which simplifies the fitting of observed data:

$$\frac{1}{J_3} = \frac{L}{kJ_2N_1} + \frac{S}{J_2} . \tag{74}$$

As in Fig. 10d, $1/J_3$ is plotted versus $1/N_1$. Ikusima (1962), for example, found the equation for duckweed population growth as a function of various limiting factors using this procedure.

Verduin (1964) discusses the Baule–Mitscherlich equation that was offered earlier to explain the asymptotic graphs from limiting factor studies. The equation is a form of the saturation equation (39) with limiting concentration substituted for time. Although the exponential graph empirically resembles the rectangular hyperbola, the theoretical pertinence has apparently not been shown.

E. Multiplicative Flows Limited by Diffusion

Consider next the special cases in which diffusion is the source pumping in the limiting reactant (no X_5 in Fig. 10a). Equation (16) may be rewritten as (75). Then N_1 exerts backforce, and X_1 in Fig. 10a becomes N_0:

$$J_1 = L_1(N_0 - N_1), \tag{75}$$

where N_0 is the concentration of the limiting material outside and N_1 the concentration at the reaction junction. If the concentration of the nonlimiting reactant N_2 is maintained constant at the junction, one may write Eq. (76) for the steady state by combining equations (63), (68), and (75):

$$\underset{(N_2\ \text{Const})}{J_3} = \frac{L_1kN_2}{L_1 + ks_1N_2} N_0 . \tag{76}$$

Changes in the outside concentration of the limiting flow by this relation produce linear increases in the process. However, in most situations N_2 is not and cannot be maintained constant since it is used at greater rates as reaction speed increases.

If the nonlimiting reactant N_2 is being supplied at a constant rate by some mode of transport J_2, there may be an excess J_e that flows out according to Eq. (69) when J_3 is small. According to Eq. (70) this excess is removed in proportion to its concentration at the junction N_2. Combining equations we obtain (77) for the multiplicative outflow in terms of the external concentration of limiting reactant N_1 and the nonlimiting flow J_2 where $r_1 = 1/L_1$:

$$J_3 = \frac{N_0 J_2}{r_1 s J_2 + SN_0 + (Le/k) - r_1 S J_3}. \tag{77}$$

Rearranging (77) as (78), one may recognize the form of Rashevsky's (1960) equation for respiration J_3 as a function of external concentration N_1 of a limiting requirement being supplied by diffusion:

$$N_0 = r_1 s J_3 + \frac{Le J_3}{k(J_2 - sJ_3)}. \tag{78}$$

Equation (79) is copied from Rashevsky (1960) for comparison without changing his symbols,

$$X = Y + \frac{Y}{1 - Y}, \tag{79}$$

where X is concentration and Y is the fraction of maximum metabolism to which the curve is asymptotic. This function like the simple hyperbola also rises to an asymptote as the outside concentration of limiting reactant N_0 is varied. At low concentrations the effect is almost linear, but at higher concentrations the asymptote is reached where J_2 is then limiting. Thus, varying the concentration of limiting reactant at the junction as in Eq. (71) produces somewhat similar results to varying the concentration at some distance so that it affects inflow as in Eq. (77).

Many special cases of double flux reactions in physiology were studied by Rashevsky, Landahl, and associates (Rashevsky, 1960). In one case flux J_2 was the flow of a fuel like glucose and J_1 was the oxygen. In another case the oxygen diffused in, but the fuel was supplied from internal storage as in some yolk-rich eggs during early embryonic development. Rashevsky's equations become more complex than the basic form in Eq. (77) above because of spherical curvatures introduced for cells, for permeability considerations, and other special complexities applicable to cell metabolism. The theoretical expressions were fitted with considerable success to various experimental data from many authors on respiration of sea urchin eggs, luminescent bacteria, and *Chlorella* cells.

For purposes of recognizing general systems functions, we indicate again that the asymptotic form of all of these graphs is due not to peculiarities of cell metabolism, but will result from any double flux situation. The form of an asymptotic limiting factor graph is given in Fig. 10c, and the effect of varying both flows is shown on the quadratic surface. As one increases the limiting factor, the reaction output approaches an asymptote at which the other flow is then limiting.

The equations for double flow interactions with limiting aspects resemble those of the Michaelis–Menten kinetics of the cycling receptor module but there is a fundamental difference since no recycling materials are required for the work gate performance as a limiting hyperbola.

F. Work Gates As Valves

Before they meet one of the two multiplicative flows clearly carries energy, whereas the second flow, also essential, seems to involve relatively small flows of mass or energy. Regardless of its weight or energy contribution, the limiting and controlling flow is the one inflowing without excess, N_1. Examples are trace elements required for photosynthesis and relay circuits in power stations. In such systems the small flow serves as a gate controlling large energies. Complex environmental systems have their power circuits modulated by control systems that operate gates, some of which are multiplicative double flux junctions. Double flux junctions are network valves.

There are two energy values which may characterize the limiting flow: one its energy contribution to the module before reaction and the other the amplified value of its energy role.

Sugita (1961, 1963) and Sugita and Fukuda (1963), discussing enzyme–chemical circuits, term the regulating flow at a double flux junction a "throttling factor." An enzyme participating in a reaction and being regenerated does follow the Michaelis–Menten relation which is a special case of a double flux junction involving feedback as considered below with Fig. 11. Sugita discusses negative and positive catalyzers. Auxiliary flows are compared to logic circuits where go or no-go throttles are operated by one circuit upon another. In the macroscopic world both on–off regulators and varying-rate regulators occur. A limiting and controlling flux can serve as a rate-controlling valve; but if action thresholds are involved, off–on valve actions may result.

XI. The Switch Module

Logic circuits are a special case where work done by one or more energy circuits controls another with only on and off positions. The energy symbol for on–off valves is in Fig. 1i. A valve is a double flux multiplicative junction with one of the two flows regarded as the control on the conductivity on the other.

Figure 1i is the general energy switch symbol used for all digital logic functions. The program of output response to input energy flows and forces must be specified in detail. These switches may serve as *and*, *nand*, *or*, or *nor* gates, etc.

On and off switching may represent birth and death of pathways of organisms. The switch symbol, for example, may show the convergence of work energy of one male and one female, the output being proportional to effective coincidence of the two input flows (forces), the output being reproduction. Another example is the requirement which controls converging voting work actions of people in elections. A logic function for a system of two individuals becomes a finite population transformer as in population genetics.

The switch requires maintenance energy by at least one of the inflowing processes and like other complex modules the switch function has a heat drain associated with its pathway maintenance.

XII. Self-Maintaining Module

Figure 1g shows the self-maintaining module. In its simplest form there is potential energy input being in part restored against potential generating forces, and then routed to increase flow from the upstream source by at least one kind of work on the upstream flow. Cells, organisms, populations, cities, and other units with respiration may be represented by modules of this class although most involve many kinds of storage and self-affecting work. To completely specify the characteristics of even the simplest module requires the storages to be characterized by its storage-to-force ratio [(C in Eq. (21)], by its work gate module [k in Eqs. (63)–(65)], and by the ratio of potential generating work to dissipational flows.

The growth of the storage with this arrangement is logistic. Component equations are given in Fig. 11, which also shows variations such as incorporation of several positive feedback loops.

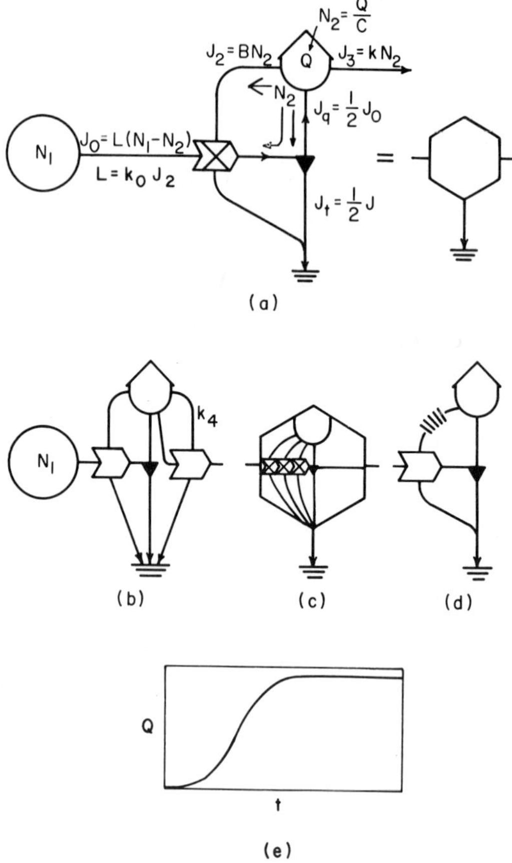

FIG. 11. Component processes of the self-maintaining module with logistic growth form.

FEEDBACK LOOPS WHICH MULTIPLY

When a feedback intersection involves a multiplicative reaction with the input shown in Fig. 11, the multiplicative loop is sometimes said to be autocatalytic by chemists and logistic by population biologists. To obtain an equation for the loop transfer function, one writes expressions for each arm of the circuit shown (Fig. 11) and combines them. The input force N_1 minus the backforce from N_2 is given by

$$J_0 = L(N_1 - N_2). \tag{80}$$

The effect on the feedback flow is

$$J_2 = BN_2, \qquad L = k_0 J_2. \tag{81}$$

At maximum power adjustment J_q is half J_0:

$$J_q = J_0/2. \tag{82}$$

The down-circuit drive is given by

$$J_3 = kN_2. \tag{83}$$

The quantity being stored Q delivers force to its pathways according to Eq. (21). The rate of change of the storage Q is the balance between the inflow to storage J_q and the outflows from the storage to feedback J_2 and downcircuit J_3:

$$dQ/dt = J_q - J_2 - J_3. \tag{84}$$

Substituting in (84) from (81), (82), and (83) one obtains a differential equation, (85), which can be recognized as logistic by comparing with Eq. (87) when N_1 is constant:

$$\frac{dQ}{dt} = \frac{1}{C}\left[\frac{k_0 BN_1}{2} - B - k\right] Q - \frac{k_0 B}{2C^2} Q^2. \tag{85}$$

Since the export flux J_3 is in proportion to the storage force, the transfer function for the multiplicative loop is also logistic. Setting Eq. (85) equal to zero and solving for Q, one obtains the steady-state carrying capacity as an energy-dependent expression, $CN_1 - [2C(B + k)]/k_0 B$.

The logistic equation has been much used for study of the macroscopic world and its usual form in ecology [Eq. (87)] may be compared with Eq. (85). If the magnitude of the constant percent change rate is linearly and negatively diminished as the quantity q approaches a limiting value ($q = K$), then the differential equation, (86), results, the logistic equation:

$$dq/q\,dt = k(K - q)/K. \tag{86}$$

Sometimes it is useful to rewrite Eq. (86) in the form

$$dq/dt = kq - (kq^2/K). \tag{87}$$

For Eq. (87) a verbal statement can be made that the rate of change of the quantity q is proportional to the quantity q except decreased in proportion to the chance of interactions q^2. If there is no backforce, the module may still operate by logistic form if the outflows are being pumped out by one interior circuit pumping or gating another as shown in Fig. 11b. Energy draining interactions may be proportional to mass action and hence to the square of the number of population units. If there are neither backforce or forward force interactions that produce negative square terms and the energy source is of constant force type, then the module may grow exponentially (Malthusian). If the energy source is of constant flow type, the growth reaches an asymptote determined by the level of input energy in a Von Bertalanffy charge pattern.

Integrating both sides of (86) with respect to time, one obtains the exponential form, Eq. (88) where $a = \ln(K - q_0/q_0)$ and q_0 is the starting storage, when $t = 0$:

$$q = K/(1 + \exp a - kt). \tag{88}$$

An s-shaped curve results, shown in Fig. 11, in which the value of the quantity q levels off at K. This equation has often been used to represent growth of a population of animals under the simplifying conditions stated above. One set of experiments whose behavior fits these simplifications is given by Ikusima (1962), working with duckweed.

The form of the logistic equations in Eqs. (86) and (87) has been related to biological premises such as an intrinsic rate of natural increase, implying that k can be constant for a population. The derivations in Eqs. (80)–(85) show the same type of equation emerging from different component assumptions about circuits, forces, and flux which provides a more general interpretation. The percent growth when storage is small is a function of the energy source and input force and is not really intrinsic. The ecological forms of the equation [(86) and (87)] are special cases of the more general circuit (Figs. 9g and 11).

These derivations illustrate the many response possibilities of the simplest self-maintaining circuits, which are relatively easily represented in circuit language but become cumbersome as differential equations. Most real self-maintaining modules (organisms, populations, human groups, etc.) have much more complexity in internal programs of time lag, multiple loopbacks, multiple storages, and packaged subroutines. It remains to be seen how much error there is in visualizing such complex modules as though they had the performance of the simplest case. Ultimately any features of the real module which vary should be diagrammed as part of the energy network representation and analysis.

XIII. Cycling Receptor Module

Shown in Fig. 1e is the module which includes a feedback of downstream energy interacting multiplicatively and positively with the upstream flow but with the additional constraint of being dependant on the recycling of a material which is constant over short periods. The receipt of light energy by an electrical photocell, by photosynthesis, and other processes is included as shown in Fig. 12. The kinetics of this module are also given in Fig. 12, which has an equation for output as a hyperbolic function of the input, the output leveling off as the recycling process becomes limiting.

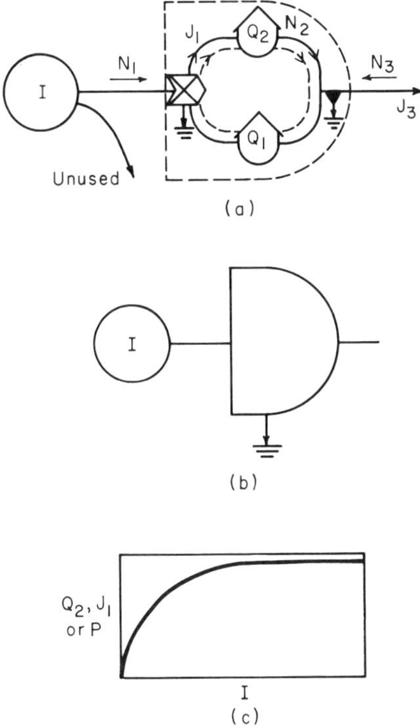

FIG. 12. Cycling receptor module and component processes.

Receptor circles are an important class of circular systems including closed material cycles that receive inflows of pure energy, transforming and storing them often as the first stage of other complex networks. The recycling material reacts with the energy flow to form an activated energy-rich state, that is stored temporarily as a potential. Then as the receptor is returned to its receptive state, the potential energy may drive other processes down-circuit. The sketches in Figs. 12 and 14 illustrate operation of receptor circles. Balls falling into a bowl or water waves on a rocky shore splash water from the low energy state into an elevated reservoir from which it may flow down again doing work. The receptor material is water.

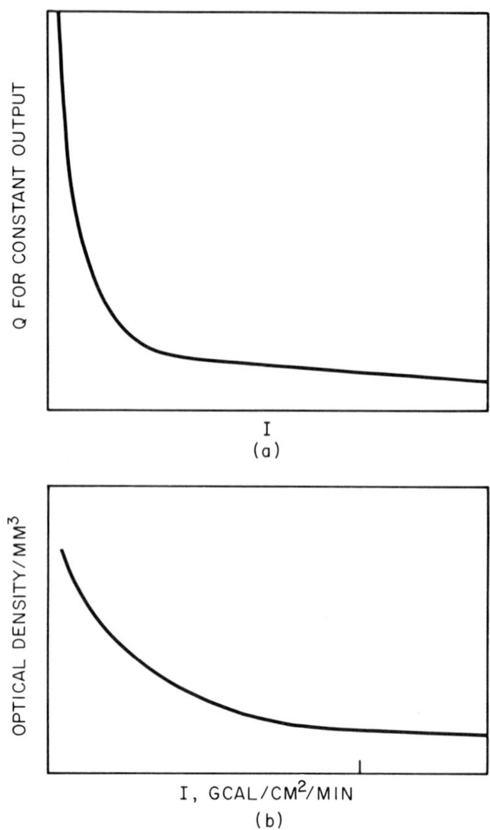

FIG. 13. Quantity of cycling receptor for delivery of constant output load with force varying. (a) Theoretical prediction of cycling receptor module, $Q = (K + kI)I$; (b) chlorophyll as a function of light intensity in stacked dishes of *Chlorella* cultures (Phillips and Myers, 1954).

The energy and material circuits (dashes) for receptor circles are included as (a) in Fig. 12. Since the feedback junction involves reaction of the energy flow and the receptor material, the flow is multiplicative as indicated by the × symbol. The loop arrangement tends to smooth out variation. When energy inflows are excessive, the receptor is kept in the upper reservoir, unavailable for further input reaction; but when inflows of energy are small, most of the receptor is back in receptive state increasing the probability of reaction.

Although quite different examples are given in Fig. 14, the energy circuits are similar. In (a) ocean waves throw water on an elevated reef from which it flows back to the sea level. In that example the receptor water is unlimited. In (b)–(d) electrons are the recycling material, receiving sound and light energies; (d) has pulsed electrical flow in an input circuit storing energy in a second circuit by induction. In (e) mercury vapor is the receptor, and in (f) the revolving cups of the wheel are the receptor. The Michaelis–Menten formulation for an enzyme-substrate reaction is shown in (g). Most interesting of all are the photosynthetic receptor systems involving chlorophyll. For simplicity all the loads drawn in Fig. 14 are dissipative drains although other kinds may be substituted.

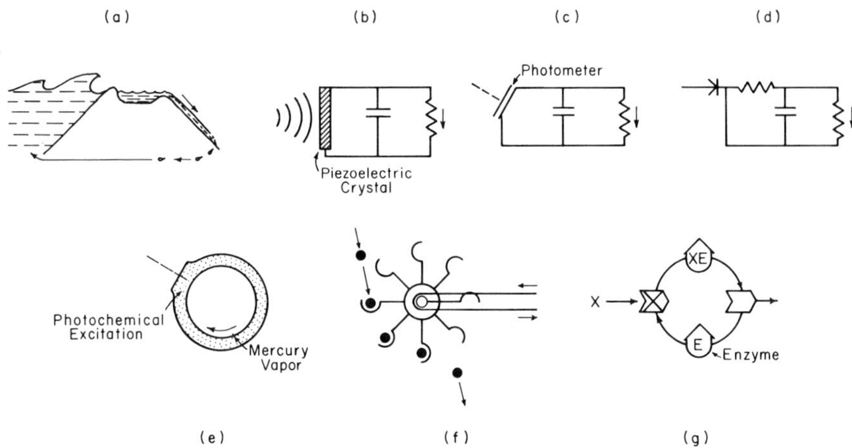

FIG. 14. Examples of cycling receptor systems receiving trains of pulsed energy. (a) water waves, (b) sound, (c) light, (d) pulsed voltage, (e) light, (f) cannon balls, (g) substrate packages.

Cycling receptor systems convert pure energy flows like sound, light, and other wave trains into energy storages associated with matter. Pulsed energies with matter also follow this scheme. As the kinetic derivation shows, the circular aspect improves the efficiency as an energy receiver.

A. Kinetics of Receptor Circles

The kinetic behavior of a cycling receptor circle is given next in a general way applicable to all of the examples cited. Derivations for particular examples were independently given long ago. For example, Michaelis and Menten (1913) and Hearon (1949) provided expressions for enzyme reactions, Mitchell and Zemansky (1934) gave equations for light reception by mercury vapor, Rose (1961) expressed the photochemical activation of anthracene, and Lumry and Reiske (1959) provided formulations for chloroplasts. Consider the notations written for the energy diagram in Fig. 12.

Energy is inflowing past the receptor zone with intensity I so that N_1, the concentration of energy packets in the range of reception, is in proportion:

$$N_1 = k_0 I. \tag{89}$$

The total quantity of receptor material is the sum of that in the receptive low energy state Q_1 and that in high energy storage reservoir Q_2:

$$Q = Q_1 + Q_2 . \tag{90}$$

The multiplicative reaction of the energy and the receptor material produces a production reaction flux J_1 carrying activated receptor and energy into the storage reservoir marked Q_2:

$$J_1 = k_1 N_1 Q_1 . \tag{91}$$

If the force exerted by the storage is in proportion to the amount of receptor activated there, then

$$N_2 = Q_2/C. \tag{92}$$

Flows out of the reservoir go in proportion to the number and force of the activated receptor units there. Hence the recycling flow of receptor J_2 is in proportion to N_2. It is a dissipative work:

$$J_2 = LN_2 = (L/C)Q_2 . \tag{93}$$

For a particular inflow condition a steady state flux J may be established when the flow into the reservoir J_1 equals the recycling flow J_2. Setting (91) equal to (93) and substituting (89) and (90) results in

$$Q_2 = Ck_0 k_1 QI/(L + Ck_0 k_1 I). \tag{94}$$

The population force N_2 from the upper potential energy storage is obtained by combining Eqs. (92) and (94):

$$N_2 = k_0 k_1 QI/(L + Ck_0 k_1 I). \tag{95}$$

The steady-state flux J in the cycling material is obtained from Eqs. (93) and (95):

$$J = Lk_0 k_1 QI/(L + Ck_0 k_1 I). \tag{96}$$

For systems which have energy storage in linear proportion to flux, as in many biological storage processes, power delivery P into productive storage compartment is proportional to potential energy value per unit flux μ:

$$P = \mu Lk_0 k_1 QI/(L + Ck_0 k_1 I). \tag{97}$$

The form of this expression is shown in Fig. 12, the familiar graph for photosynthesis with light or enzyme reaction as a function of input fuel substrate.

If there is an output circuit with loading backforce N_3 driven by energy in Q_2, the output flux is the sum of force N_2 from the reservoir and load, the latter being negative (see Fig. 12a):

$$J_3 = L_{23} N_2 + L_{33} N_3 . \tag{98}$$

Combining (95) and (98) one obtains the output in terms of the energy inflow to the receptor circle:

$$J_3 = \frac{L_{23} k_0 k_1 QI}{L + Ck_0 k_1 I} + L_{33} N_3 . \tag{99}$$

Rose (1961) provided a procedure for analyzing data for such hyperbolic relations [Eqs. (94)–(98)] plotting reciprocals to obtain a straight line. Thus, the storage potential N_2 and the pertinent fluxes J and J_3 of a receptor circle vary with the input energy flow I according to rectangular hyperbolas like that plotted in Fig. 12. This relation is for steady state.

Passive or operational electrical analogs can be arranged. The photometer in Fig. 14 is a special case and may also serve as a passive analog. All these relationships define the symbol in Fig. 12b.

B. Power and Efficiency of Receptor Circles

Figure 7 shows the efficiencies and power properties of a simple transformation system in which the output load X_2 is held constant and the input load X_1 varied. Except for one particular zone of loading indicated by a circle, the transformation was either very inefficient or not drawing much power. In a regime of varying input force, the system would rarely be set at the optimum loading for maximum power transmission of potential.

In comparison, consider the graph of the hyperbolic transfer function (Fig. 12c) for the same variables and coordinates. The presence of a cycling receptor Q which is limited in quantity causes higher efficiencies by allowing excess energy to pass without reception for use by other units. The cycling receptor serves as a feedback governor on the loading. Energy bypassed is available for use in other units. Both P_1 and P_2 follow the shape of the graph in Fig. 12.

C. Loading and Power Transmission Optima in Photosynthesis

The input–output loading arrangements in the cycling receptor systems of Figs. 11 and 12 at steady state permit the same kind of rate variations with loads as discussed for potential generation work modules. For maximum conversion of energy into storage potential there is a particular loading ratio. If the input light is held constant so that the density of photons available to the receptor sweep is constant, the input force is also constant [Eq. 89]. By varying the output load from zero to a stopping load one obtains the hump-shaped output curves already given for other systems in Fig. 5. Photosynthetic examples were represented (Odum, 1968) where data from Lumry and Spikes (1957) and Clendenning and Ehrmantraut (1951) for the Hill reaction with varying concentrations of reactants were graphed in a hump shape. A blue–green algal mat under constant light provides a hump-shaped power delivery with varying external loading (Armstrong and Odum, 1964). The receptor system follows the optimum efficiency-maximum power principle.

However, in most photosynthetic situations the input varies diurnally and seasonally with the march of the sun. For optimal loading to be maintained in spite of varying inputs, some flexible ways of changing the loading with input are required. Gears have to be changed as in the accelerating car. Whereas a simple potential-generating transformation has a very poor response under such conditions with a poor setting most of the time, as shown in Fig. 7, the cycling receptor system has a better regulated response (Fig. 12).

The single cycling receptor system draws less power and has a higher average efficiency under the varying input regime. Systems of several consecutive loops have even more stable patterns, as discussed by Heinmets and Herschman (1962). Plants have additional special mechanisms for maintaining favorable loading.

D. Changes in Pigment Cycling Systems for Load Adaptation

A number of mechanisms are known which help to adapt the plant receptor system to its loading. Some are rapid responses characteristic of the loop; some are adaptive physiological responses involving additional special circuits; and some are adaptations accomplished by substitution of species in the system. The coupling of flows from one pigment to another as drawn by French and Fork (1961) provides means for stabilizing output where wavelengths are varying.

The adaptive properties of the hyperbolic loop function can be recognized by study of Eq. (99) for downcircuit flow. When light input is small, Ck_0k_1I is small so that the relation approaches that in Eq. (100):

$$\underset{(I\ \text{small})}{J_3} = k_0k_1QI + L_{33}N_3\,. \tag{100}$$

When light intensities are high, Eq. (100) approaches Eq. (101), the asymptote of the hyperbola:

$$\underset{(I\ \text{large})}{J_3} = (L_{23}Q/C) + L_{33}N_3\,. \tag{101}$$

When Eq. (99) is solved for Q, the quantity of total receptor cycling, Eq. (102) results:

$$Q = (k_0k_1CI + L)(J_3 - L_{33}N_3)/L_{23}k_0k_1I. \tag{102}$$

At any light intensity the output may be increased by augmenting the amount of chlorophyll and other pigments that can be brought into action. Limits to adding pigment are the physical limits of space for more pigment, the energetic costs of synthesis and maintenance, and the need to match input to the output loadings.

From Eq. (102) Q, the pigment required to maintain a constant output of photosynthesis J_3 at constant load N_3, is in inverse relation to the light intensity. As graphed in Fig. 13, the drive applied to downstream loading can be adapted to increases in light intensity by diminishing the pigment. The effect is large only at small light intensities. Shown also in Fig. 14 is a graph of experimental data for *Chlorella* from Phillips and Myers (1954). Wassink *et al.* (1956) report such adaptations in maple leaves, Gessner (1937) in land and water plants, and Steeman-Nielsen (1962) in plankton, to name a few.

XIV. Production and Regeneration Module (P–R)

Given in Fig. 1h is the module formed by combining a cycling receptor module, a self-maintaining module which it feeds, and a work feedback loop which controls the inflow process by multiplicative and limiting actions. An example is the green plant, which has a respiratory system. More complex examples have more than one respiratory system (Fig. 15). An example is the plant and animal symbiosis of corals and zoochlorellae-containing hydra. On a large scale the module represents plants and consumers of ecosystems or agriculture and cities. It was already shown that the cycling receptor module had a hyperbolic transfer function due to the limits that develop by necessary internal cycles. The P–R module adds more cycles. Generally, the form of the output of a group of cycling processes coupled into one system behaves with an asymptotic pattern not unlike that of the simpler single recycling system. For example, analog simulation of models in Figs. 15–17 show this stability.

XV. Economic Transactor Module

For human systems with circulating money currency, Eq. (103) and the module in Fig. 1j pertains:

$$J_{\text{energy}} = -kJ_{\text{currency}} \cdot \tag{103}$$

Flows of currency move opposite in direction to the flow of potential energy and work but are at each transactor point regulated to go in proportion to the price established for the goods or services. The money flow must balance and this happens if there are loopbacks of valuable work services that have upstream amplifying effects that are worth as much money as those with more power. The money system ensures some distribution of closed loops of service organizing each compartment of the system in exchange for return payments and energy flows. Shown in Fig. 16 are some usages of the economic transactor system.

The relationship of energy flow and money flow shows that in a macroeconomic sense, the energy and money flows are coupled and neither can be understood without the other. Since there is a balance of work services of comparable value among components and a balance of money payments it is possible to determine the ratio of one to the other in a given economy at a given time by dividing the total energy budget by the total money budget. This ratio allows one to convert the dollar value to the energy value at its loop amplified maximum.

Control circuits thus have two energy values, their own and that which they obtain in their work gate amplification. The relationships of a simple economic cycle are given in Fig. 16, which expresses the relationships in differential equation form as implied by the modules.

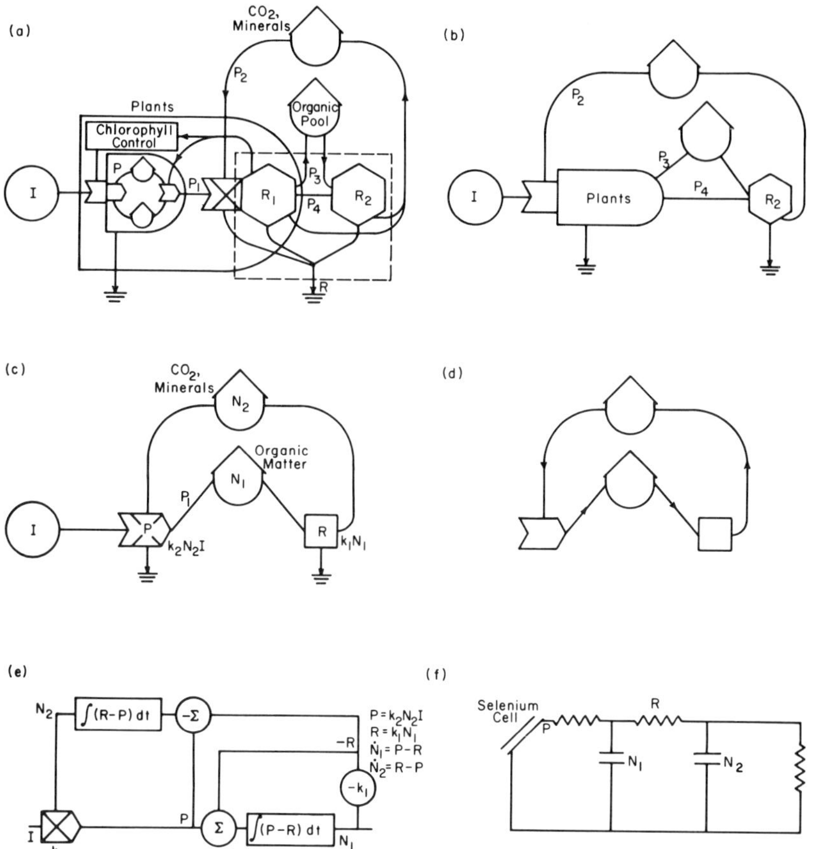

FIG. 15. System of production and consumption in energy circuit language illustrating cycles of work feedback loops within cycles. The process of modeling, simplification, and simulation is illustrated in the sequence from (a)–(e) or (f). (a) Energy #1. An ecological system showing plants and consumers doing respiration and maintenance R. Within plants is a cycling receptor module, a workgate receiving control loop from mineral-CO_2 mix pool, and the plant's own respiratory system R_1. Total respiration in the dashed box includes plant and animal–microbe respiration units which feed by-products into nutrient-CO_2 pool. (b) Energy #2. Diagram (a) is further abbreviated by eliminating detail within the plant (P–R) module. (c) Energy #3. Diagram (a) simplified further into two von Bertalanffy storage modules, one work gate control of mineral feedback isolating the essence of a productive and respiratory process of the ecosystem as measured when studies are made of photosynthesis and respiration of microcosms. (d) Mineral cycle in simplified model. (e) Block diagram of differential equations showing the double loop of negative feedback characteristics providing self regulating homeostasis. This diagram is also the one used to program an analog computer. (f) Passive analog circuit (equivalent circuit) for simple simulation.

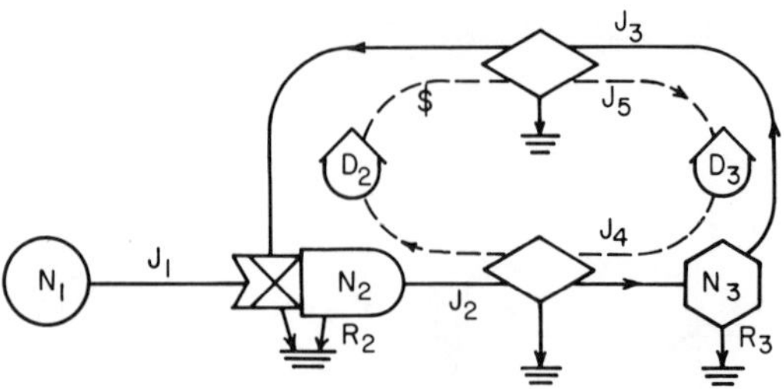

FIG. 16. A system of recycling currency showing component processes and modules, where N_2 is the agriculture sector and D_2 its capital; N_3 is the urban sector and D_3 its capital.

The ratio of energy flow to currency flow is maximum near the source of energy decreasing downstream, but the energy value as amplifier control increases downstream in any surviving system which has its downstream units reward-looped to the upstream inflow modules. The energy definition of value of a flow is the sum of the two energy components (actual flow and amplified loop effect). If the downstream units are people, their salaries are highest in proportion to their amplified loop effects, since their actual power consumption is small and little varying. Macroeconomics is a system of currency and energy, the full analysis of which is not possible using one of the component networks alone. The diagram for a simple two-module simplification of an economy with agricultural and city sectors in Fig. 16 implies the following equations:

Work gate on agriculture:

$$J_1 = k_0 N_1 J_3 ; \tag{104}$$

Rate of population force accumulation in agricultural sector ($N_2 = Q_2/C_2$):

$$\frac{dQ_2}{dt} = \frac{Lk_1 Q J_1}{L + k_1 C J_1} - J_2 - R_2 ; \tag{105}$$

Price control modules:

$$J_2 = k_2 J_4 , \qquad J_3 = k_3 J_5 ; \tag{106}$$

Rate of population force accumulation in city sector:

$$dQ_3/dt = J_2 - J_3 - R_3 ; \tag{107}$$

Currency flows from capital storages D:

$$J_4 = k_4 D_3\,, \qquad J_5 = k_5 D_5\,; \tag{108}$$

Rates of change of capital:

$$dD_2/dt = J_4 - J_5\,, \qquad dD_3/dt = J_5 - J_4\,; \tag{109}$$

Constant total currency in system:

$$D_2 + D_3 = K; \tag{110}$$

Energy costs of maintenance:

$$R_2 = k_6 N_2\,, \qquad R_3 = k_7 N_3\,. \tag{111}$$

On analog, the model's behavior depends on functions used to control prices k_2 and k_3.

XVI. System Examples and Their Simulation

Some energy circuits with various degrees of complexity are given in Figs. 15–17. Included are some for very simple physical processes and others for systems of man and nature. The reader will recognize various degrees of combining and compartmentalizing of detail, a practice which hides detail and creates a model to help the human mind visualize broad features of the system and its possible responses.

For simulation, modules of the energy language carry with them that characteristic performance as often written with differential equations. The energy diagrams are themselves mathematical expressions of the network or our model view of it. It is almost an automatic process to substitute difference equations or differential equations for each module. Then analog or digital means are used to keep a running computation of the stocks, flows, and rates of the many parts of the system under testing with various forcing functions and other experimental manipulations that answer questions about the system's performance. When the temporal responses of the model have similarity with that of the measured system, one is encouraged that there is truth in the model and further details are added and tests are made to determine if the similarities are not fortuitous. An example of this process is the balanced ecosystem studies in small chambers containing rain forest soil, litter, herbs, small animals, and microorganisms. The model is given in Fig. 15c–f. When simulated by Odum *et al.* (1970), graph properties were obtained which were similar to the graphs of the diurnal variation of carbon-dioxide in the small atmosphere of the microcosms.

The coefficients were evaluated by substituting known values of flux when storages were known. The steady-state annual flows and fluxes may be used for this purpose as a first approximation. It is often convenient to write steady-state values of stock and flux on the diagram as in the example, Fig. 17. The outflow coefficients are the ratios of flux to stock, the fraction passed per unit time (see k in Fig. 6). The reciprocal is the time constant, the steady state turnover time.

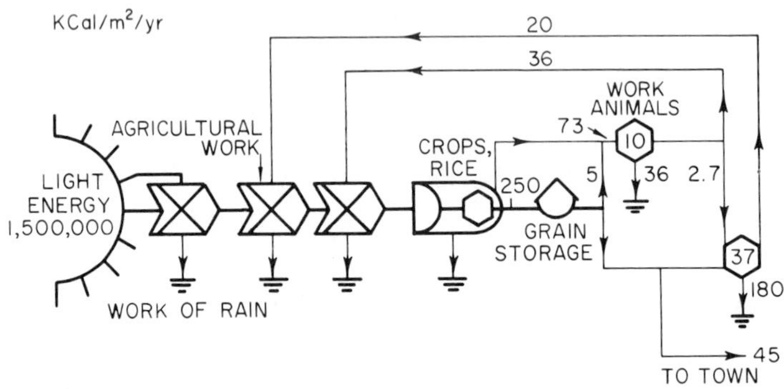

FIG. 17. Example of a system of man and nature in energy circuit language with stocks and steady-state flows indicated (Odum, 1968). Monsoon agriculture in India is portrayed showing the role of sacred cows.

If passive analog simulation is used, the energy circuit language has its equivalent circuits (see example in Fig. 15f). If operational analog simulation procedure is to be done, the block diagramming of the integral equations are drawn as in Fig. 15e which also shows where there are properties of feedback stabilization. If digital simulation is to be done, the difference equations for each module are written so as to add the changes for an increment of time followed by a print and loop statement. For the example in Fig. 15c, the digital program for varying light I in English becomes:

1. INPUT N_1, N_2, N_3, k_1, k_2, I, P, R, X, Y, I as memory locations,
2. INPUT initial conditions for N_1, N_2, I, k_1, k_2, T,
3. PRINT N_1, N_2, I, k_1, k_2, T,
4. Let $P = k_2 N_2 I$,
5. Let $R = k_1 N_1$,
6. Let $X = N_1$,
7. Let $N_1 = N_1 + P - R$,
8. Let $\dot{N} = N_1 - X$,

9. Let $Y = N_2$,
10. Let $N_2 = N_2 + R - P$,
11. Let $\dot{N} = N_2 - Y$,
12. Let $T = T + 1$,
13. PRINT OUT N_1, N_2, $\dot{N}$, P, R,
14. If T has reached desired stopping place, stop; if not, continue.
15. Return to Statement 4 and repeat with a new input value for I.

Thus the simulation procedure for the energy circuit follows in simple automatic manner from the energy circuit diagram; the thinking on the behavior and structure of the system is done in the diagramming. Since groups of simulation and system characteristics go with each module, the modules are the words of the language which carry group laws and temporal transient characteristics that are helpful in thinking and expression in the same manner as in other languages. Many kinds of systems from the physical and molecular to the ecological and social can be expressed by the same language and the diagramming helps us to recognize many as special cases of a relatively few general system types. The examples used here to illustrate the modules of the energy circuit were single processes that defined the behavior of modules in their simplest form in terms of well-established kinetics and energy laws. In application to more complex systems one recognizes the type of module in the system before the exact kinetic aspects are known. For example, populations of people, cities, land divisions, etc. are self-maintaining modules of the class of the logistic module although complexity in these units makes it unlikely that the module would have an overall behavior exactly that of the simple, single, linear, feedback multiplication. The details would have to be verified by performance studies and diagramming of the within-module circuits. However, recognizing the class of the module does carry some semiquantitative feeling about the nature of the growth, stabilization, and power processing. In time the behavior of more and more configurations will become known, and the meaning to the language reader will also increase with his knowledge.

Offered as a general systems language, the energy language allows all kinds of factors to be included in the same network. One may keep track of the budgets of flowing quantities while at the same time dealing with dynamic aspects such as forces, differential equations, and transients. It attempts to connect the biological and economic worlds to that of electronics and systems science. We have used it in teaching for several years in ecological subject areas. At least it solves many semantic arguments and is efficient in making ideas clear; at best it leads readily from the real world data to realistic computer simulation.

The language has the interesting property of showing many entirely different kinds of systems as similar in type. For example, food chains resemble eddy diffusion chains; hurricanes resemble organisms; war functions resemble carnivore actions on two competing food populations.

Hopefully the language synthesizes many disciplines. For example, the work gate brings in physiological ecology; the cycling receptor module, biochemistry; the configurations of one or two self-maintaining modules, population ecology; the reward loops, mineral cycles; the numerical flows, ecological energetics; the transaction module, economics; the analysis of the diagrams produced, matrix ecology; and the more elaborate control modules, the switch; and combinations, behavioral sciences.

Energy diagrams for rain forest and radiation processes are given in our rain forest book (Odum and Pigeon, 1970), for marine systems in our coastal ecosystems report (Odum *et al.*, 1969), and for photoregeneration, in Odum *et al.* 1970b. For previous applications, see Odum (1967a, b, 1968), and for a general account, Odum (1970). J am grateful for comment from David Cowan, Gettysburg College.

REFERENCES

Armstrong, N. E., and Odum, H. T. (1964). *Science* **143**, 256.

Best, J. B., and Hearon, J. Z. (1960). *In* "Mineral Metabolism" (C. L. Comar and F. Brenner, eds.), Chapter 2. Academic Press, New York.

Caperon, J. (1967). *Ecology* **48**, 709.

Clendenning, K. A., and Ehrmantraut, H. C. (1951). *In* "Photosynthesis" (E. T. Rabonowitch, ed.), Vol. II, Pt. 2, pp. 1617–1623. Wiley (Interscience), New York.

Denbigh, K. G. (1951). "The Thermodynamics of the Steady State." Wiley, New York.

Denbigh, K. G. (1952). *Trans. Faraday Soc.* **48**, 389.

Gessner, F. (1937). *Jahrbuch wiss. Bot.* **85**, 267.

French, C. S., and Fork, D. C. (1961). *Biophys. J.* **1**, 669.

Goodwin, B. S. (1963). "Temporal Organization in Cells." Academic Press, New York.

Hearon, J. Z. (1949). *Bull. Math. Biophys.* **11**, 29.

Heinmets, F., and Herschman, A. (1962). *In* "Biological Systems and Synthetic Systems" (E. E. Bernard and M. R. Kare, eds.) pp. 61–70. Plenum Press, New York.

Ikusima, I. (1962). *Physiol. Ecol.* **10**, 130.

Jammer, M. (1957). "Concepts of Force." Harper, New York.

Kormondy, E. J. (1969). "Concepts of Ecology." Prentice-Hall, Englewood Cliffs, New Jersey.

Levenspiel, O. (1962). "Chemical Reaction Engineering." Wiley, New York.

Lumry, R., and Rieske, J. S. (1959). *Plant Physiol.* **34**, 301.

Lumry, R., and Spikes, J. D. (1957). *In* "Research in Photosynthesis" (H. Gaffron, ed.), pp. 373–391. Wiley (Interscience), New York.

Macfadyen, A. (1963). "Animal Ecology, Aims and Methods." Putnam, London.

Michaelis, L., and Menten, M. L. (1913). *Biochem. Z.* **49**, 333.
Milsum, J. H. (1966). "Biological Control Systems Analysis." McGraw-Hill, New York.
Mitchell, A. C. G., and Zemansky, M. W. (1934). "Resonance Radiation and Excited Atoms." Cambridge Univ. Press, London and New York. Reprinted 1961.
Monod, J. (1942). "Recherches sur la Croissance des Cultures Bacteriennes." Hermann et Cie, Paris.
Novick, A., and Szilard, L. (1950). *Proc. Nat. Acad. Sci.* **36**, 708.
Odum, H. T. (1957). *Ecol. Monogr.* **27**, 55.
Odum, H. T. (1960). *Amer. Sci.* **48**, 1.
Odum, H. T. (1967a). *In* "Pollution and Marine Ecology" (T. A. Olson and F. J. Burgess, eds.), pp. 99–157. Wiley (Interscience), New York.
Odum, H. T. (1967b). *In* "The World Food Problem," Vol. 3, Chapter 3, pp. 55–94. (Report of President's Science Advisory Committee Panel on World Food Supply). Whitehouse, Washington, D.C.
Odum, H. T. (1968). *In* "Primary Productivity and Mineral Cycling in Natural Ecosystems" (H. E. Young, ed.), pp. 81–138. Univ. of Maine Press, Orono, Maine.
Odum, H. T. (1971). "Environment, Power, and Society." Wiley, New York.
Odum, H. T., and Odum, E. P. (1959). *In* "Fundamentals of Ecology" (E. P. Odum), pp. 43–87. Saunders, Philadelphia, Pennsylvania.
Odum, H. T., and Pigeon, R. F. (eds.) (1970). "A Tropical Rainforest." Division of Tech. Information and Education, U.S. Atomic Energy Commission, Oak Ridge, Tennessee.
Odum, H. T., and Pinkerton, R. C. (1955). *Amer. Sci.* **43**, 331.
Odum, H. T., Beyers, R. J , and Armstrong, N. E. (1963). *J. Mar. Res.* **21**, 191.
Odum, H. T., Copeland, B. J., and McMahan, E. A. (1969). "Coastal Ecosystems of the United States." Report to Federal Water Pollution Control Administration, Institute of Marine Science, Univ. of North Carolina, Morehead City, North Carolina.
Odum, H. T., Lugo, A., and Burns, L. (1970a). *In* "A Tropical Rain Forest" (H. T. Odum and R. F. Pigeon, eds.), Chapter I-3. Division of Technical Information and Education, U.S. Atomic Energy Commission, Oak Ridge, Tennessee.
Odum, H. T., Nixon, S., and DiSalvo, L. (1970b). *Amer. Microscopical Soc. Trans.* (in press).
Phillips, J. N., and Myers, J. (1954). Plant Physiol. **29**, 148.
Phillipson, J. (1966). "Ecological Energetics." St. Martin's Press, New York.
Prigogine, I. (1955). "Introduction to Thermodynamics of Irreversible Processes." Wiley (Interscience), New York.
Rashevsky, N. (1960). "Mathematical Biophysics." Dover, New York.
Rose, J. (1961). "Dynamic Physical Chemistry." Wiley, New York.
Shinozaki, K., and Kira, T. (1961). *J. Biol. Osaka City Univ.* **12**, 69.
Steeman-Nielsen, E. (1962). *Physiol. Plant.* **15**, 161.
Sugita, M. (1961). *J. Theoret. Biol.* **1**, 415.
Sugita, M. (1963). *J. Theoret. Biol.* **4**, 179.
Sugita, M., and Fukuda, N. (1963). *J. Theoret. Biol.* **5**, 412.
Tribus, M. (1961). "Thermostatics and Thermodynamics." Van Nostrand, Princeton, New Jersey.
Verduin, J. (1964). *In* "Algae and Man" (D. Jackson, ed.), pp. 221–238. Plenum Press, New York.
Wassink, H. C., Richardson, S. C., and Preters, G. A. (1956). *Acta Bot. Neerlandica* **5**, 247.
Zimmerman, E. W. (1933). "World Resources and Industries." Harper, New York.

5

Steady-State Equilibriums in Simple Nonlinear Food Webs

RICHARD B. WILLIAMS
BUREAU OF COMMERCIAL FISHERIES
CENTER FOR ESTUARINE AND MENHADEN RESEARCH, BEAUFORT, NORTH CAROLINA

I. Introduction

The representation of an ecosystem as discrete compartments joined by flows of energy or materials abstracts the complexity of nature into something both simple enough to be readily grasped by the mind and suitable for mathematical analysis. This abstraction of nature can take one of two general forms, the linear model in which each flow is a linear function of one of the compartments which it joins, or the nonlinear model in which some flows are not linear functions of the compartments joined. Linear and nonlinear models have different advantages and disadvantages. The mathematics of linear systems is well understood; the construction and analysis of linear models is relatively easy and is described in detail in numerous publications. There is, however, no reason to believe that ecosystems are in fact linear systems. Rather, the living components of an ecosystem appear to interact in a complex and nonlinear manner. Thus, although linear models provide a tool for examining how an ecosystem operates at or near steady state—i.e., how much is transferred per unit time between adjoining compartments—it is unlikely that linear models can be used to simulate the responses of an ecosystem to radically changing conditions such as the seasonal cycle in temperate latitudes. Nonlinear models, on the other hand, have the potential for simulating the responses of an ecosystem to changing environmental conditions. This potential has been little utilized because the mathematics of nonlinear systems is poorly understood, and without knowledge of their mathematics, the construction of satisfactory nonlinear models has proved a difficult undertaking.

The study which forms the body of this chapter was undertaken to determine enough of the mathematical properties of nonlinear systems to permit developing nonlinear models of ecosystems. The key question was, "Is a given system mathematically feasible and thus capable of serving as a model for an ecosystem?" The approach taken was to analyze a variety of simple systems and from these analyses generalize where possible as to the behavior of classes of nonlinear systems. The criterion for feasibility was the presence of at least one mathematically and ecologically satisfactory steady-state equilibrium in which all the compartments and flows could have positive finite values. If a system had no steady state it obviously could not persist. Steady states involving infinite, zero, negative, or imaginary values for the compartments and flows, though mathematically sound, are ecological nonsense. Zero values represent the disappearance of a part of the system under study and thus its conversion to another and simpler system. Negative, imaginary, and infinite values are impossible.

The examination of the systems was thus centered on the steady state. Each system was tested for the presence of a steady-state equilibrium by direct mathematical analysis of its defining equations, and for its ability to return to this equilibrium if displaced from it by simulation of an example of the system on a digital computer. Many of the systems were also analyzed for their response when at steady state to small changes in the values of their coefficients.

II. Systems with Simple Nonlinear Transfers

A. The Forced Two-Compartment System

A simple and mathematically feasible two-compartment system is diagrammed in Fig. 1. Its compartments are joined by a nonlinear flow which is equal to a transfer rate ϕ multiplied by both its donor compartment X_1 and recipient compartment X_2. This flow could,

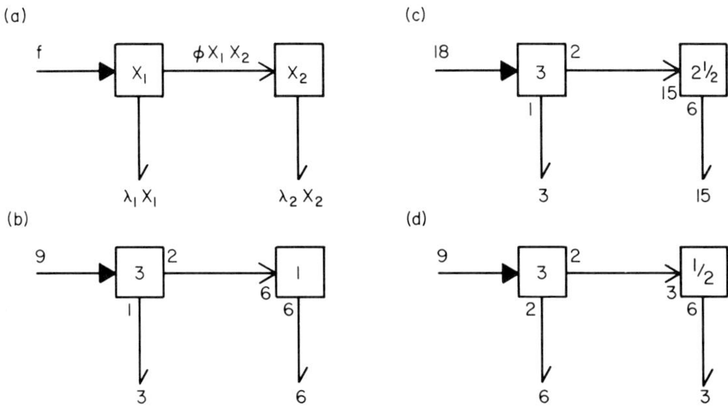

Fig. 1. Forced two-compartment system. (a) Functions defining flows in the system. (b) Example of the system at steady state. The value of each compartment is placed within the block representing it. Values for the constants (forcing function and flow rates) are placed at the bases of the arrows, and values for the flows they generate are placed at the tips of the arrows. (c) Steady state for a system similar to that in Fig. 1b, except that the forcing function is increased from 9 to 18. (d) Steady state for a system similar to that in Fig. 1b, except that the linear loss from compartment 1 is increased from 1 to 2. A key to the arrows used in this and subsequent figures is as follows: forced input ⟶; flow linear with donor compartment controlling ⟶; flow linear with recipient compartment controlling ⟶; simple, uncontrolled, nonlinear flow ⟶; controlled nonlinear flow ⟶; controlled self-generating input ⟶.

for example, represent the transfer of energy or material from prey to predator. The system is fed by a forcing function (or constant input) f which enters X_1. Loss takes place from all compartments through linear flows which are fractions, λ, of the compartment's content. These linear losses could represent some physiological activity such as respiration or excretion.

The steady-state equilibrium for the system is evaluated from the system equations, differential equations representing change in the size (or value) of the compartments:

$$\begin{aligned} \dot{X}_1 &= f - X_1(\lambda_1 + \phi X_2), \\ \dot{X}_2 &= X_2(\phi X_1 - \lambda_2). \end{aligned} \tag{1}$$

Change in a compartment equals its inflow minus its loss. To obtain steady-state values for the compartments, Eq. (1) is set equal to zero—because at steady state inflow equals loss and change is zero—and solved for X_1 and X_2:

$$\begin{aligned} X_1 &= \lambda_2/\phi, \\ X_2 &= (f\phi - \lambda_1\lambda_2)/\phi\lambda_2 . \end{aligned} \tag{2}$$

The solution for X_1 indicates that its steady-state value is a function solely of ϕ and λ_2 and that the steady-state value will always be positive and finite as long as ϕ and λ_2 are positive and finite. The solution for X_2 contains all three transfer rates and the forcing function, and indicates that even though the coefficients are all positive and finite, the mathematically defined steady state for X_2 could be zero or negative—if the product of $\lambda_1\lambda_2$ were equal to or greater than the product of ϕf. Since zero and negative values have no ecological meaning, this system could only serve as a model of a natural system when $f\phi > \lambda_1\lambda_2$. Of interest, and not at all obvious from the block diagram (Fig. 1a) is that the steady state of compartment x_1 is not explicitly dependent on either its own input (f) or loss rate (λ_1).

Figure 1b is an example of this forced two-compartment system with numerical values provided for the compartments, forcing function, and transfer rates. The system is at steady state; input equals loss. The system was simulated on a digital computer by preparing a brief program in Fortran similar to one below.

```
PROGRAM NONLIN 1
T = 0.
N = 0
XA = 3.
XB = 1.
WRITE (3,100) T, XA , XB
DO 2 I = 1, 1000
R = I
T = R * .01
DXA = .09 - XA * (.01 + .02 * XB)
DXB = XB * (.02 * XA - .06)
```

```
      XA = XA + DXA
      XB = XB + DXB
      N = N + 1
      IF (N - 10), 2, 1, 1
    1 WRITE (3, 100) T, XA, XB
      N = 0
    2 CONTINUE
  100 FORMAT (F10.2, 2F10.3)
```

The program operates for ten units of time; computations are made at intervals of 0.01 unit of time and the values for time and the two compartments printed every tenth computation. When set in operation with its compartments not at steady state (for example, XA = 5., XB = 2.), the system quickly returns to steady state.

The partial derivatives of the steady-state equations [Eqs. (2)] are summarized in Table IA. These partial derivatives, or sensitivities, yield the changes in steady-state value for the compartments and flows in response to small changes in coefficients of the system. Information in Table IA is summarized further in Table IB which indicates merely

TABLE I

SENSITIVITIES FOR THE FORCED TWO-COMPARTMENT SYSTEM[a]

	Independent variable	Dependent variable			
		∂X_1	$\partial(\lambda_1 X_1)$	∂X_2	$\partial(\lambda_2 X_2)$
IA	∂f	0	0	$1/\lambda_2$	1
	$\partial \phi$	$-\lambda_2/\phi^2$	$-\lambda_1\lambda_2/\phi^2$	λ_1/ϕ^2	$\lambda_1\lambda_2/\phi^2$
	$\partial \lambda_1$	0	λ_2/ϕ	$-1/\phi$	$-\lambda_2/\phi$
	$\partial \lambda_2$	$1/\phi$	λ_1/ϕ	$-f/\lambda_2^2$	$-\lambda_1/\phi$

	Independent variable	Dependent variable			
		∂X_1	$\partial(\lambda_1 X_1)$	∂X_2	$\partial(\lambda_2 X_2)$
IB	∂f	0	0	+	+
	$\partial \phi$	−	−	+	+
	$\partial \lambda_1$	0	+	−	−
	$\partial \lambda_2$	+	+	−	−

[a] System given in Fig. 1a and Eqs. (1) and (2). Table IA lists the sensitivity equations in matrix form and is read, for example: The partial derivative (i.e., the sensitivity) of X_1 with respect to f equals zero. Table IB lists the effect of an increase in each of the independent variables (the forcing function and the flow rates) on the dependent variables (steady-state values for compartments and flows). Effects are evaluated as no change (0), increase (+), or decrease (−).

the effect of an increase in each of the coefficients on steady-state values of the compartments and flows. Effect is here evaluated as either positive (an increase in the steady-state value), negative (a decrease in the steady-state value), or zero (no change in the steady-state value). An increase in the forcing function f has, for example, no effect on the steady-state value of X_1 or the linear loss from X_1, but increases the steady-state value of X_2 and the flow through X_2. One important generalization may be drawn from Table IB: The direction of the response of this simple system to changes in its coefficients, i.e., whether the new steady-state values are greater, less, or unchanged, is determined solely by the structure of the system and is independent of the numerical value of the coefficients.

It may be hard to accept the idea that a change in the input into a compartment or in the linear loss rate from a compartment might have no effect on the steady-state value of that compartment. The new steady-state equilibriums are diagrammed for the system in Fig. 1b with its forcing function f increased from 9 to 18 (Fig. 1c) and with the loss rate from compartment X_1, λ_1, increased from 1 to 2 (Fig. 1d). In both cases X_1 remains 3, i.e., unaffected by the changes.

B. The Self-Generating Two-Compartment System

A slight modification of the preceding forced system is diagrammed in Fig. 2. The forcing function is replaced with a "self-generating"

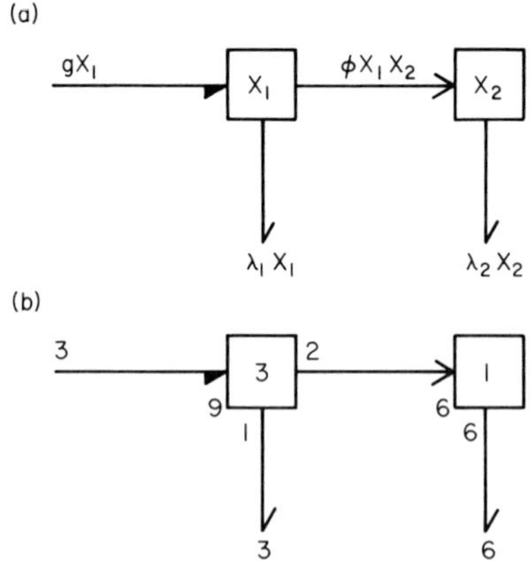

Fig. 2. Self-generating two-compartment system. (a) Functions defining flows in the system. (b) Example of the system at steady state.

function, i.e., a function whose input into X_1 is equal to a constant g multiplied by the value of X_1:

$$\begin{aligned} \dot{X}_1 &= X_1(g - \lambda_1 - \phi X_2), \\ \dot{X}_2 &= X_2(\phi X_1 - \lambda_2). \end{aligned} \tag{3}$$

This system appears a closer approximation of nature than the preceding forced system because inflow into the first compartment (a population of organisms) is now a function of the biomass of that population. The system seems feasible, the steady state equations [Eqs. (4)] obtained from the system equations [Eqs. (3)] indicate that for any set of coefficients there exists a unique steady-state value for each compartment:

$$\begin{aligned} X_1 &= \lambda_2/\phi, \\ X_2 &= (g - \lambda_1)/\phi. \end{aligned} \tag{4}$$

If the coefficients of the system are positive, the steady-state value for X_1 must always be positive, whereas the steady-state value for X_2 will be positive only if g exceeds λ_1. The steady-state sensitivity expressions for the compartments and flows are summarized in Table IIA, and the

TABLE II

SENSITIVITIES FOR THE SELF-GENERATING TWO-COMPARTMENT SYSTEM[a]

	Independent variable	Dependent variable				
		$\partial(gX_1)$	∂X_1	$\partial(\lambda_1 X_1)$	∂X_2	$\partial(\lambda_2 X_2)$
IIA	∂g	λ_2/ϕ	0	0	$1/\phi$	λ_2/ϕ
	$\partial\phi$	$-g\lambda_2/\phi^2$	$-\lambda_2/\phi^2$	$-\lambda_1\lambda_2/\phi^2$	$-(g - \lambda_1)/\phi^2$	$-(g - \lambda_1)\lambda_2/\phi^2$
	$\partial\lambda_1$	0	0	λ_2/ϕ	$-1/\phi$	$-\lambda_2/\phi$
	$\partial\lambda_2$	g/ϕ	$1/\phi$	λ_1/ϕ	0	$(g - \lambda_1)/\phi$

	Independent variable	Dependent variable				
		$\partial(gX_1)$	∂X_1	$\partial(\lambda_1 X_1)$	∂X_2	$\partial(\lambda_2 X_2)$
IIB	∂g	+	0	0	+	+
	$\partial\phi$	−	−	−	−	−
	$\partial\lambda_1$	0	0	+	−	−
	$\partial\lambda_2$	+	+	+	0	+

[a] System given in Fig. 2a and Eqs. (3) and (4). Table IIA lists the sensitivity equations, and Table IIB the effects of an increase in each of the independent variables on the dependent variables (the steady-state values for compartments and flows).

effects of an increase in each of the coefficients on the steady-state value of the compartments and flows in Table IIB. Like the preceding forced system, the direction of change in the compartments and flows in response to a change in a coefficient is determined entirely by the structure of the system and is independent of coefficient values.

There is one flaw in the steady-state behavior of this system which neither the steady-state equations nor the sensitivities hint at. This was found only by simulating an example of the system on a digital computer. When the system is displaced from steady state it begins to oscillate, and the oscillations increase steadily in amplitude. The extreme values for the compartments approach zero as a lower limit and have no upper limit. In the example tested, the period of oscillation was constant. This ability of the self-generating system to enter into self-amplifying oscillations may limit its usefulness in the modeling of ecosystems.

C. Forced Three-Compartment Systems

Increasing the number of compartments in a nonlinear system substantially increases the possibilities for complexity. A forced three-compartment system arranged as an unbranched sequence is diagrammed in Fig. 3. Its system equations [Eqs. (5)] yield complicated steady-state equations [Eqs. (6)]:

$$\begin{aligned} \dot{X}_1 &= f - X_1(\lambda_1 + \phi_{12}X_2), \\ \dot{X}_2 &= X_2(\phi_{12}X_1 - \lambda_2 - \phi_{23}X_3), \\ \dot{X}_3 &= X_3(\phi_{23}X_2 - \lambda_3); \end{aligned} \tag{5}$$

$$\begin{aligned} X_1 &= \frac{f\phi_{23}}{\phi_{12}\lambda_3 + \phi_{23}\lambda_1}, \\ X_2 &= \frac{\lambda_3}{\phi_{23}}, \\ X_3 &= \frac{f\phi_{12}\phi_{23} - \phi_{12}\lambda_2\lambda_3 - \phi_{23}\lambda_1\lambda_2}{\phi_{23}(\phi_{12}\lambda_3 + \phi_{23}\lambda_1)}. \end{aligned} \tag{6}$$

For any set of coefficients each compartment still possesses, however, one and only one steady-state value. The effect of an increase in each coefficient on the steady-state value of the compartments and flows is summarized in Table III. The direction of the change in steady-state values is no longer defined in every case by the structure of the system, but instead depends in three cases partially on the numerical values of the coefficients. These indeterminate cases, indicated with question

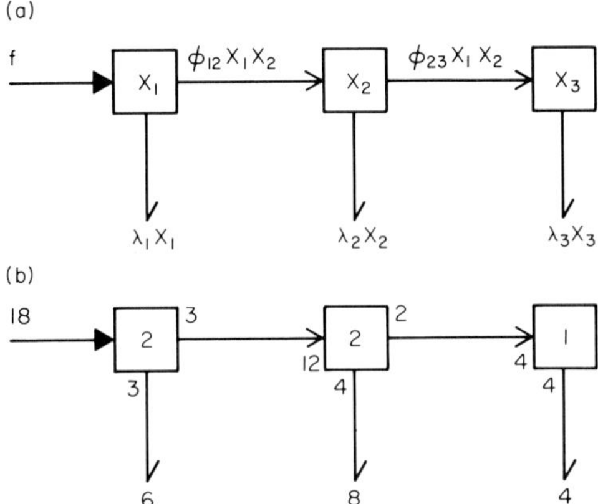

FIG. 3. Forced three-compartment system as an unbranched sequence. (a) Functions defining flows in the system. (b) Example of the system at steady state.

TABLE III

SENSITIVITIES FOR THE FORCED THREE-COMPARTMENT SYSTEM AS AN UNBRANCHED SEQUENCE[a]

Independent variable	Dependent variable						
	∂X_1	$\partial(\lambda_1 X_1)$	$\partial(\phi_{12} X_1 X_2)$	∂X_2	$\partial(\lambda_2 X_2)$	∂X_3	$\partial(\lambda_3 X_3)$
∂f	+	+	+	0	0	+	+
$\partial \phi_{12}$	−	−	+	0	0	+	+
$\partial \phi_{23}$	+	+	−	−	−	?	?
$\partial \lambda_1$	−	+	−	0	0	−	−
$\partial \lambda_2$	0	0	0	0	+	−	−
$\partial \lambda_3$	−	−	+	+	+	−	?

[a] The system is given in Fig. 3a and Eqs. (5) and (6) and is expressed as the effect of an increase in each of the independent variables on the dependent variables (the steady-state values for compartments and flows). Cases in which the effect may be positive, negative, or zero, depending on the numerical values of the independent variables, are indicated with (?).

marks in Table III, result from partial derivatives which contain both a positive and a negative term:

$$\frac{\partial X_3}{\partial \phi_{23}} = \frac{\lambda_2}{\phi_{23}^2} - \frac{f\phi_{12}\lambda_1}{(\phi_{12}\lambda_3 + \phi_{23}\lambda_1)^2}; \tag{7}$$

$$\frac{\partial(\lambda_3 X_3)}{\partial \lambda_3} = -\frac{\lambda_1}{\phi_{23}} + \frac{f\phi_{12}\phi_{23}\lambda_1}{(\phi_{12}\lambda_3 + \phi_{23}\lambda_1)^2}. \tag{8}$$

This three-compartment system may be further complicated by adding a flow from X_1 to X_3, converting the system to a closed network (Fig. 4).

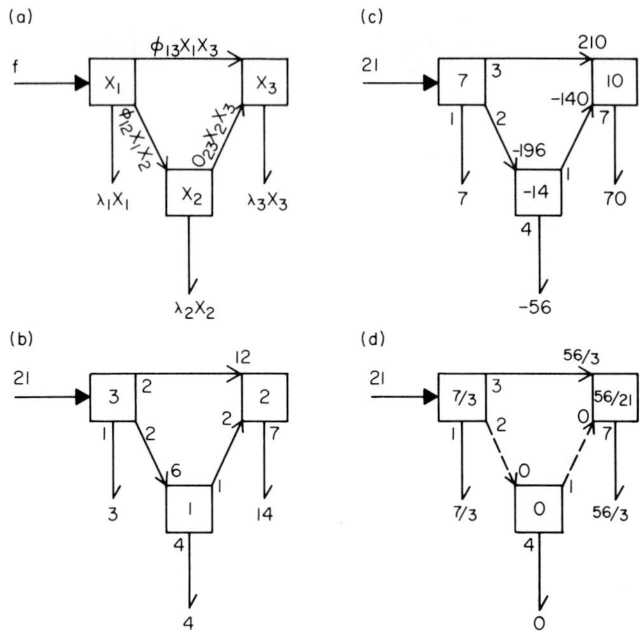

FIG. 4. Forced three-compartment system as a closed network. (a) Functions defining flows in the system. (b) Example of the system at steady state. (c) Mathematically defined state for system in Fig. 1c following an increase in transfer rate ϕ_{13} from 2 to 3. (d) Ecologically reasonable steady state for system in Fig. 1c following an increase in transfer rate ϕ_{13} from 2 to 3.

A closed network such as this might serve as a model of an ecosystem containing a herbivore, a primary carnivore, and a secondary carnivore preying on both the herbivore and primary carnivore. The system equations for the closed network [Eqs. (9)] yield steady-state equations [Eqs. (10)] more complicated than those of the unbranched three-compartment system:

$$\begin{aligned} \dot{X}_1 &= f - X_1(\lambda_1 + \phi_{12}X_2 + \phi_{13}X_3), \\ \dot{X}_2 &= X_2(\phi_{12}X_1 - \lambda_2 - \phi_{23}X_3), \\ \dot{X}_3 &= X_3(\phi_{13}X_1 + \phi_{23}X_2 - \lambda_3); \end{aligned} \tag{9}$$

$$\begin{aligned} X_1 &= \frac{f\phi_{23}}{\phi_{12}\lambda_3 - \phi_{13}\lambda_2 + \phi_{23}\lambda_1}, \\ X_2 &= \frac{\lambda_3 - \phi_{13}X_1}{\phi_{23}}, \\ X_3 &= \frac{-\lambda_2 + \phi_{12}X_1}{\phi_{23}}. \end{aligned} \tag{10}$$

Like the unbranched sequence, the closed network has one and only one steady state for any set of coefficients. With sets of coefficients such that the denominator of the fraction in the first of Eqs. (10) equals zero, however, the mathematically defined steady state requires a value of infinity in all compartments.

The sensitivity expressions for this network system were evaluated with the assumption that the steady-state values of all the compartments were positive and finite. The direction of change in the steady-state values of the compartments in response to increases in the coefficients is indeterminate in many cases (Table IV). Sensitivity equations were not prepared for the flows, but the direction of change in the steady-state values for the flows in response to increases in the coefficients must be indeterminate in five cases and may be indeterminate in still others. As in the case of the unbranched sequence, sensitivity equations for the indeterminate cases contain both a positive and a negative term.

TABLE IV

SENSITIVITIES FOR THE FORCED THREE-COMPARTMENT SYSTEM AS A CLOSED NETWORK[a]

Independent variable	Dependent variable		
	∂X_1	∂X_2	∂X_3
∂f	+	−	+
$\partial \phi_{12}$	−	+	?
$\partial \phi_{13}$	+	−	+
$\partial \phi_{23}$	?	?	?
$\partial \lambda_1$	−	+	−
$\partial \lambda_2$	+	−	?
$\partial \lambda_3$	−	+	−

[a] The system is given by Fig. 4a and Eqs. (9) and (10) and is expressed as the effect of an increase in each of the independent variables on the dependent variables (the steady-state values for the compartments).

It may often be necessary when using nonlinear systems as models of ecosystems to distinguish between a mathematically defined steady state and an ecologically feasible steady state. A mathematically sound and ecologically feasible example of a three-compartment closed network is diagrammed in Fig. 4b. If its transfer rate ϕ_{13} is increased from two to three, the new mathematically defined steady state is ecologically impossible (Fig. 4c). Compartment X_2 is negative, generating negative flows and thus a cyclic movement within the system. The corresponding ecologically feasible steady state is diagrammed in Fig. 4d. Compartment X_2 is driven to zero (i.e., its population is exterminated), the flows through X_2 are eliminated, and the system is shifted to a two-compartment sequence.

D. Self-Generating Three-Compartment Systems

Self-generating systems similar to the forced three-compartment systems considered in the preceeding section are diagrammed in Figs. 5 and 6. Solution of the system equations [Eqs. (11)] for the unbranched sequence (Fig. 5) yields steady-state equations [Eqs. (12)] which define

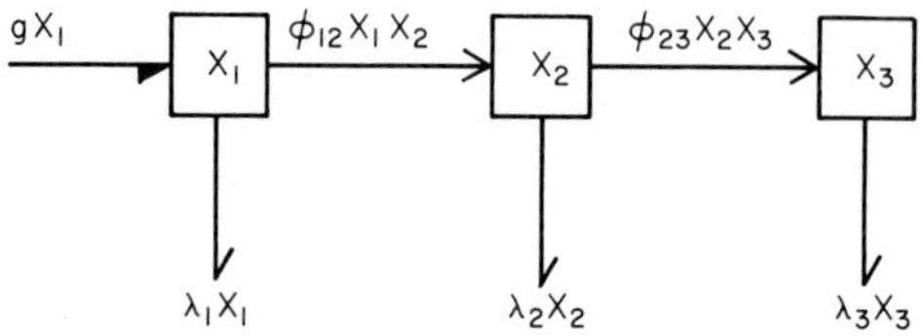

Fig. 5. Self-generating three-compartment system as an unbranched sequence; functions defining flows in the system.

compartment X_2 twice, i.e., with two independent functions, and define compartments X_1 and X_3 only in terms of each other:

$$\begin{aligned} \dot{X}_1 &= X_1(g - \lambda_1 - \phi_{12}X_2), \\ \dot{X}_2 &= X_2(\phi_{12}X_1 - \lambda_2 - \phi_{23}X_3), \\ \dot{X}_3 &= X_3(\phi_{23}X_2 - \lambda_3); \end{aligned} \tag{11}$$

$$\begin{aligned} X_1 &= \frac{\lambda_2 + \phi_{23}X_3}{\phi_{12}}, \\ X_2 &= \frac{\lambda_3}{\phi_{23}} = \frac{g - \lambda_1}{\phi_{12}}, \\ X_3 &= \frac{\phi_{12}X_1 - \lambda_2}{\phi_{23}}. \end{aligned} \tag{12}$$

A steady state can thus be established if and only if λ_3/ϕ_{23} equals $(g - \lambda_1)/\phi_{12}$, thus requiring that the values for one of these rates be dependent on the others. Steady state could not be established if the rates varied independently of one another. There is no reason to think that natural systems possess rates which are either utterly unchanging or dependent on one another. The presence of a compartment whose value is independent of the remainder of the system also seems unreasonable.

The self-generating, three-compartment, unbranched sequence does not have a steady state for every set of rates and a unique steady state for any set of rates. It is thus ecologically infeasible and cannot serve as a model for natural systems.

Converting the three-compartment unbranched sequence into a closed network by connecting compartment X_1 directly to compartment X_3 (Fig. 6) does not produce an ecologically feasible system which might

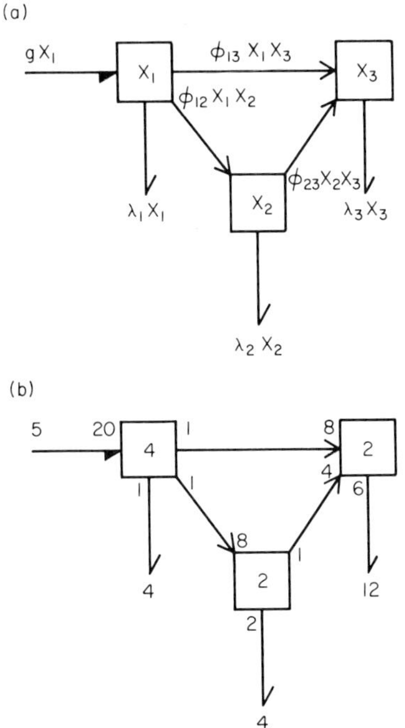

FIG. 6. Self-generating three-compartment system as a closed network. (a) Functions defining flows in the system. (b) Example of the system at steady state.

have some counterpart in nature. The system equations [Eqs. (13)] yield steady-state equations [Eqs. (14)] in which the compartment values still are defined only in terms of each other, and $\phi_{12}\lambda_3$ must equal the sum of $\phi_{13}\lambda_2$ and $\phi_{23}(g - \lambda_1)$ to permit the existence of a steady state.

$$\begin{aligned}\dot{X}_1 &= X_1(g - \lambda_1 - \phi_{12}X_2 - \phi_{13}X_3),\\ \dot{X}_2 &= X_2(\phi_{12}X_1 - \lambda_2 - \phi_{23}X_3),\\ \dot{X}_3 &= X_3(\phi_{13}X_1 + \phi_{23}X_2 - \lambda_3);\end{aligned} \tag{13}$$

$$X_2 = \frac{\lambda_3 - \phi_{13}X_1}{\phi_{23}},$$
$$X_3 = \frac{\phi_{12}X_1 - \lambda_2}{\phi_{23}}. \tag{14}$$

This system, like the two-compartment self-generating system, will engage in persistent oscillations if any of its compartments are displaced from steady state.

E. Feasibility of Unbranched Sequences

1. *Forced Systems*

Based on the above study of simple systems, it is now possible to consider in general terms which kinds of unbranched structures yield mathematically feasible systems and which do not. Consider first a forced unbranched sequence with an even number of compartments, e.g., a four-compartment system (Fig. 7a). Obtaining the steady-state equa-

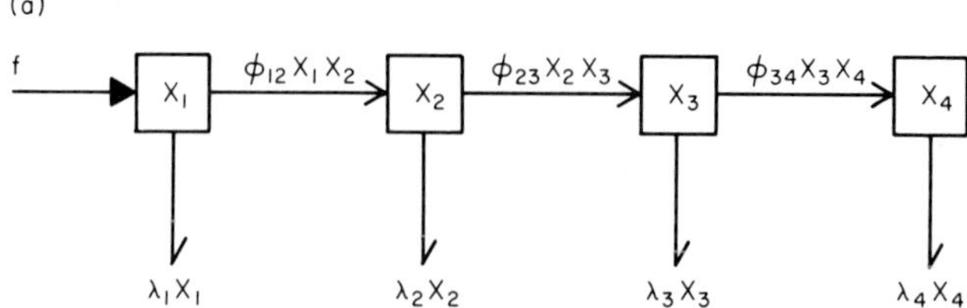

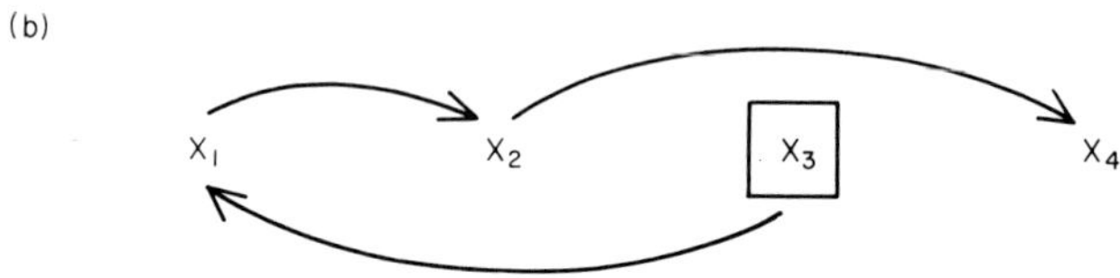

FIG. 7. Forced four-compartment system as an unbranched sequence. (a) Functions defining flows in the system. (b) Sequence in determining steady-state equations for the compartments.

tions [Eqs. (16)] from the system equations [Eqs. (15)] follows the path depicted in Fig. 7b:

$$\begin{aligned} \dot{X}_1 &= f - X_1(\lambda_1 + \phi_{12}X_2), \\ \dot{X}_2 &= X_2(\phi_{12}X_1 - \lambda_2 - \phi_{23}X_3), \\ \dot{X}_3 &= X_3(\phi_{23}X_2 - \lambda_3 - \phi_{34}X_4), \\ \dot{X}_4 &= X_4(\phi_{34}X_3 - \lambda_4); \end{aligned} \tag{15}$$

$$
\begin{aligned}
X_1 &= \frac{\phi_{23}\lambda_4 + \phi_{34}\lambda_2}{\phi_{12}\phi_{34}}, \\
X_2 &= \frac{f\phi_{12}\phi_{34} - \phi_{23}\lambda_1\lambda_4 - \phi_{34}\lambda_1\lambda_2}{\phi_{12}\phi_{23}\lambda_4 + \phi_{12}\phi_{34}\lambda_2}, \\
X_3 &= \frac{\lambda_4}{\phi_{34}}, \\
X_4 &= \frac{f\phi_{12}\phi_{23}\phi_{34} - \phi_{12}\phi_{23}\lambda_3\lambda_4 - \phi_{12}\phi_{34}\lambda_2\lambda_3 - \phi_{23}^2\lambda_1\lambda_4 - \phi_{23}\phi_{34}\lambda_1\lambda_2}{\phi_{12}\phi_{23}\phi_{34}\lambda_4 + \phi_{12}\phi_{34}^2\lambda_2}.
\end{aligned}
\tag{16}
$$

That is, setting the differential equation for compartment 4 equal to zero yields the steady-state equation for compartment 3. This solution for the next to last compartment can be substituted into the differential equation for compartment 2 to obtain a steady-state equation for compartment 1. This solving for the steady-state equations of the odd-numbered compartments can be continued until the first compartment of the system is reached. Substituting the steady-state equation for the first compartment into the differential equation for the first compartment yields the steady-state equation for the second compartment. This can be substituted into the differential equation for the third compartment to obtain the steady-state equation for the fourth compartment. Solving for steady-state equations of the even-numbered compartments is continued to the last compartment of the sequence. The complexity of the steady-state equations increases with each step in the solution process; the first steady-state equation obtained (that for X_3 in this example) has first power terms in both numerator and denominator, the second steady-state equation (that for X_1) has second power terms, and so on.

The above approach also can be applied to a forced unbranched sequence with an odd number of compartments, e.g., a five-compartment system [Fig. 8a and Eqs. (17)]:

$$
\begin{aligned}
\dot{X}_1 &= f - X_1(\lambda_1 + \phi_{12}X_2), \\
\dot{X}_2 &= X_2(\phi_{12}X_1 - \lambda_2 - \phi_{23}X_3), \\
\dot{X}_3 &= X_3(\phi_{23}X_2 - \lambda_3 - \phi_{34}X_4), \\
\dot{X}_4 &= X_4(\phi_{34}X_3 - \lambda_4 - \phi_{45}X_5), \\
\dot{X}_5 &= X_5(\phi_{45}X_4 - \lambda_5).
\end{aligned}
\tag{17}
$$

Again the differential equation for the last compartment yields the steady-state equation for the next to last compartment. This yields the steady-state equation for the compartment located two compartments forward, and so on until the second compartment of the system is reached (Fig. 8b). Substituting the steady-state equation for the second compartment into the differential equation for the first compartment yields the steady-state equation for the first compartment (Fig. 8b). This in turn yields the steady-state equation for the third compartment, the third for the fifth, and so on until the last compartment of the system is reached (Fig. 8b).

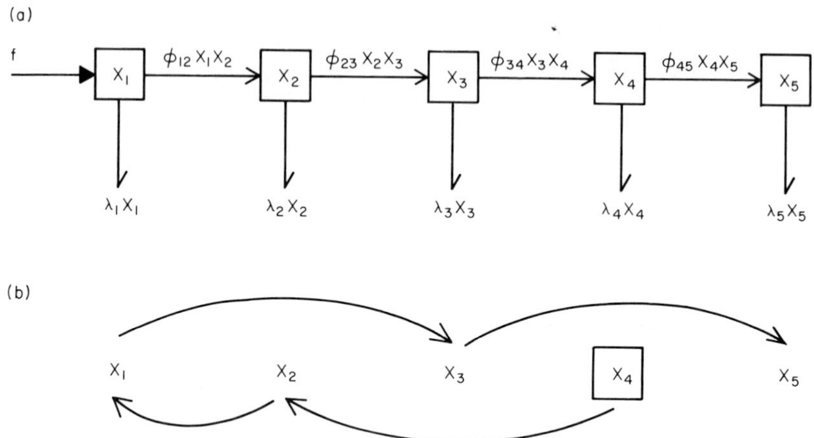

FIG. 8. Forced five-compartment system as an unbranched sequence. (a) Functions defining flows in the system. (b) Sequence in determining steady-state equations for the compartments.

Thus, regardless of the total number of compartments or whether that number is odd or even, the steady-state value of every compartment in a forced unbranched sequence is uniquely defined by the rates of the system. Therefore every forced unbranched sequence is a mathematically feasible system. Whether or not a mathematically feasible system is also ecologically feasible is determined, of course, by its coefficients.

2. *Self-Generating Systems*

Examination of a self-generating unbranched sequence with an even number of compartments—e.g., four compartments [Fig. 9a and Eqs. (18)]—reveals that the steady-state equation for compartment 2 can be obtained from the differential equation for the first compartment, and the steady-state equation for the next to last compartment from the differential equation for the last compartment:

$$\begin{aligned} \dot{X}_1 &= X_1(g - \lambda_1 - \phi_{12}X_2), \\ \dot{X}_2 &= X_2(\phi_{12}X_1 - \lambda_2 - \phi_{23}X_3), \\ \dot{X}_3 &= X_3(\phi_{23}X_2 - \lambda_3 - \phi_{34}X_4), \\ \dot{X}_4 &= X_4(\phi_{34}X_3 - \lambda_4); \end{aligned} \qquad (18)$$

$$\begin{aligned} X_1 &= \frac{\phi_{23}\lambda_4 + \phi_{34}\lambda_2}{\phi_{12}\phi_{34}}, \\ X_2 &= \frac{g - \lambda_1}{\phi_{12}}, \\ X_3 &= \frac{\lambda_4}{\phi_{34}}, \\ X_4 &= \frac{g\phi_{23} - \phi_{12}\lambda_3 - \phi_{23}\lambda_1}{\phi_{12}\phi_{34}}. \end{aligned} \tag{19}$$

The steady-state equation for the second compartment can be substituted into the appropriate differential equation to obtain a steady-state equation for the fourth compartment, and so on to the end of the sequence (Fig. 9b). Similarly, the steady-state equation for the next to last compartment yields the steady-state equation for the third to last compartment, and so on to the first compartment of the sequence. As in the case of the unbranched forced systems, the steady-state equations obtained first (those for X_2 and X_3 in this case) have first power terms in both numerator and denominator, the steady-state equations obtained second (those for X_1 and X_4) second power terms, and so on.

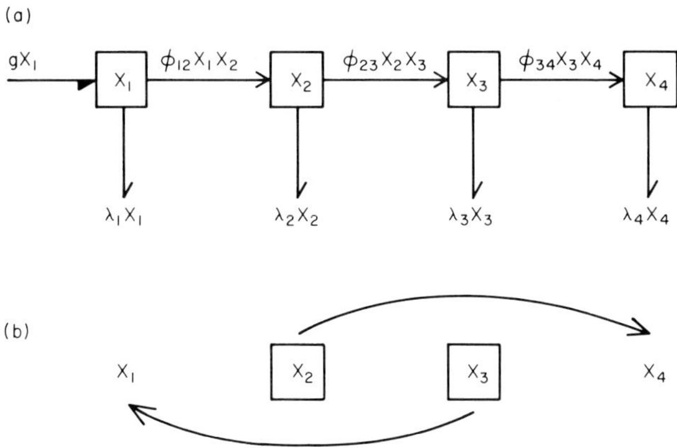

FIG. 9. Self-generating four-compartment system as an unbranched sequence. (a) Functions defining flows in the system. (b) Sequence in determining steady-state equations for the compartments.

If on the other hand, the self-generating unbranched sequence has an odd number of compartments—e.g., five compartments [Fig. 10 and Eqs. (20)]—the steady-state values of the even numbered compartments of the sequence are each defined by two independent functions [e.g.,

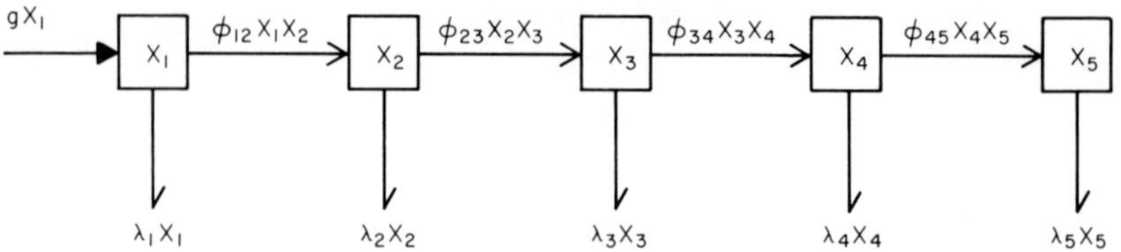

FIG. 10. Self-generating five-compartment system as an unbranched sequence; functions defining flows in the system.

the solutions for X_2 and X_4 in Eqs. (21)], indicating that the presence of any steady state depends on having a particular combination of coefficients:

$$\begin{aligned}
\dot{X}_1 &= X_1(g - \lambda_1 - \phi_{12}X_2),\\
\dot{X}_2 &= X_2(\phi_{12}X_1 - \lambda_2 - \phi_{23}X_3),\\
\dot{X}_3 &= X_3(\phi_{23}X_2 - \lambda_3 - \phi_{34}X_4),\\
\dot{X}_4 &= X_4(\phi_{34}X_3 - \lambda_4 - \phi_{45}X_5),\\
\dot{X}_5 &= X_5(\phi_{45}X_4 - \lambda_5);
\end{aligned} \tag{20}$$

$$\begin{aligned}
X_1 &= \frac{\phi_{23}X_3 + \lambda_2}{\phi_{12}},\\
X_2 &= \frac{g - \lambda_1}{\phi_{12}} = \frac{\phi_{34}\lambda_5 + \phi_{45}\lambda_3}{\phi_{23}\phi_{45}},\\
X_4 &= \frac{\lambda_5}{\phi_{45}} = \frac{-g\phi_{23} + \phi_{12}\lambda_3 + \phi_{23}\lambda_1}{\phi_{12}\phi_{34}},\\
X_5 &= \frac{\phi_{34}X_3 - \lambda_4}{\phi_{45}}.
\end{aligned} \tag{21}$$

Furthermore, the steady-state values for the odd numbered compartments in the sequence are defined in part by each other, as well as by the rates [solutions for X_1 and X_5 in Eqs. (21)]. Like the three-compartment unbranched sequence there is no satisfactory steady state. It can be generalized, therefore, that a self-generating unbranched sequence is mathematically feasible if, and only if, the sequence has an even number of compartments.

F. Feasibility of Complex Systems

1. *Systems with a Single Input*

A variety of systems was examined for the presence of a mathematically feasible steady state. The results are summarized in Fig. 11. Only one

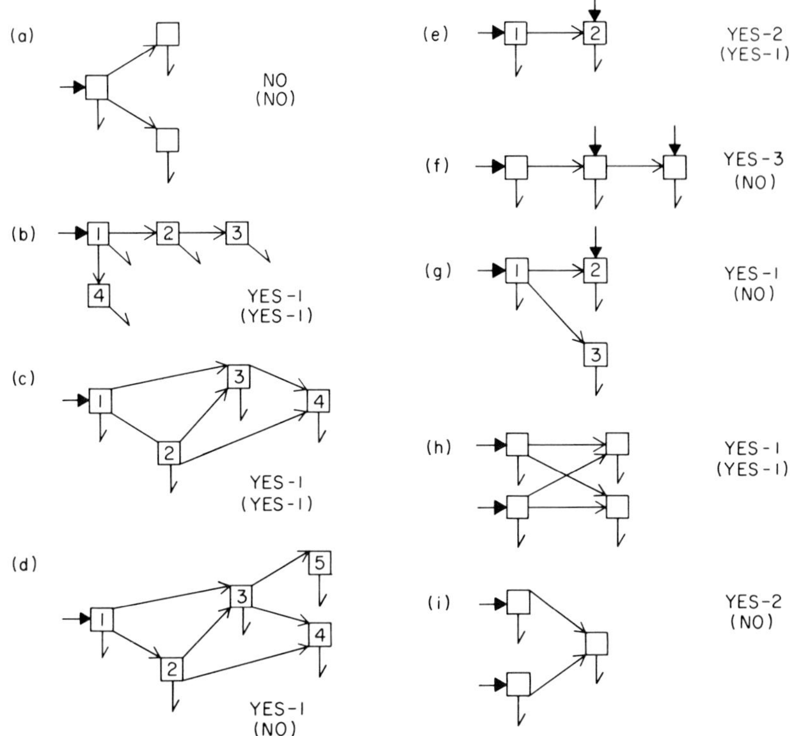

FIG. 11. A variety of nonlinear systems evaluated for feasibility. The words, "yes" and "no," indicate whether or not the system is mathematically feasible and the number following the word is the number of steady states. The upper term refers to the forced system in the figure and the lower term in parentheses to a similar system with the forcing functions replaced with self-generating functions.

type of branching appears infeasible; this is the presence of two (or more) single compartment branches originating from the same compartment (Fig. 11a) because each such branch uniquely defines the steady-state value of its source compartment. All other types of branching appear mathematically feasible (e.g., Fig. 11b) regardless of whether the system is forced [Eqs. (22) and (23)] or self-generating [Eqs. (24) and (25)].

Forced

$$
\begin{aligned}
\dot{X}_1 &= f - X_1(\phi_{12}X_2 + \phi_{14}X_4 + \lambda_1), \\
\dot{X}_2 &= X_2(\phi_{12}X_1 - \lambda_2 - \phi_{23}X_3), \\
\dot{X}_3 &= X_3(\phi_{23}X_2 - \lambda_3), \\
\dot{X}_4 &= X_4(\phi_{14}X_1 - \lambda_4);
\end{aligned}
\tag{22}
$$

$$
\begin{aligned}
X_1 &= \frac{\lambda_4}{\phi_{14}}, \\
X_2 &= \frac{\lambda_3}{\phi_{23}}, \\
X_3 &= \frac{\phi_{12}\lambda_4 - \phi_{14}\lambda_2}{\phi_{23}\phi_{14}}, \\
X_4 &= \frac{f\phi_{23}\phi_{14} - \phi_{12}\lambda_3\lambda_4 - \phi_{23}\lambda_1\lambda_4}{\phi_{23}\phi_{14}\lambda_4}.
\end{aligned} \tag{23}
$$

Self-generating

$$
\begin{aligned}
\dot{X}_1 &= X_1(g - \lambda_1 - \phi_{12}X_2 - \phi_{14}X_4), \\
\dot{X}_2 &= X_2(\phi_{12}X_1 - \lambda_2 - \phi_{23}X_3), \\
\dot{X}_3 &= X_3(\phi_{23}X_2 - \lambda_3), \\
\dot{X}_4 &= X_4(\phi_{14}X_1 - \lambda_4);
\end{aligned} \tag{24}
$$

$$
\begin{aligned}
X_1 &= \frac{\lambda_4}{\phi_{14}}, \\
X_2 &= \frac{\lambda_3}{\phi_{13}}, \\
X_3 &= \frac{\phi_{12}\lambda_4 - \phi_{14}\lambda_2}{\phi_{14}\phi_{23}}, \\
X_4 &= \frac{(g - \lambda_1)\,\phi_{23} - \phi_{12}\lambda_3}{\phi_{14}\phi_{23}}.
\end{aligned} \tag{25}
$$

A four-compartment closed network (Fig. 11c) is mathematically feasible whether forced [Eqs. (26) and (27)] or self-generating [Eqs. (28) and (29)].

Forced

$$
\begin{aligned}
\dot{X}_1 &= f - X_1(\lambda_1 + \phi_{12}X_2 + \phi_{13}X_3), \\
\dot{X}_2 &= X_2(\phi_{12}X_1 - \lambda_2 - \phi_{23}X_3 - \phi_{24}X_4), \\
\dot{X}_3 &= X_3(\phi_{13}X_1 + \phi_{23}X_2 - \lambda_3 - \phi_{34}X_4), \\
\dot{X}_4 &= X_4(\phi_{24}X_2 + \phi_{34}X_3 - \lambda_4);
\end{aligned} \tag{26}
$$

$$
\begin{aligned}
X_1 &= \frac{\phi_{23}\lambda_4 - \phi_{24}\lambda_3 + \phi_{34}\lambda_2}{\phi_{12}\phi_{34} - \phi_{13}\phi_{24}}, \\
X_2 &= \frac{f\phi_{34}}{\phi_{23}\lambda_4 - \phi_{24}\lambda_3 + \phi_{34}\lambda_2} - \frac{\phi_{13}\lambda_4 + \phi_{34}\lambda_1}{\phi_{12}\phi_{34} - \phi_{13}\phi_{24}}, \\
X_3 &= \frac{f\phi_{24}}{\phi_{23}\lambda_4 - \phi_{24}\lambda_3 + \phi_{34}\lambda_2} + \frac{\phi_{12}\lambda_4 + \phi_{24}\lambda_1}{\phi_{12}\phi_{34} - \phi_{13}\phi_{24}}, \\
X_4 &= \frac{f\phi_{23}}{\phi_{23}\lambda_4 - \phi_{24}\lambda_3 + \phi_{34}\lambda_2} + \frac{\phi_{13}\lambda_2 - \phi_{12}\lambda_3 - \phi_{23}\lambda_1}{\phi_{12}\phi_{34} - \phi_{13}\phi_{24}}.
\end{aligned} \tag{27}
$$

Self-generating

$$\begin{aligned}
\dot{X}_1 &= X_1(g - \lambda_1 - \phi_{12}X_2 - \phi_{13}X_3),\\
\dot{X}_2 &= X_2(\phi_{12}X_1 - \lambda_2 - \phi_{23}X_3 - \phi_{24}X_4),\\
\dot{X}_3 &= X_3(\phi_{13}X_1 + \phi_{23}X_2 - \lambda_3 - \phi_{34}X_4),\\
\dot{X}_4 &= X_4(\phi_{24}X_2 + \phi_{34}X_3 - \lambda_4);
\end{aligned} \tag{28}$$

$$\begin{aligned}
X_1 &= \frac{\phi_{23}\lambda_4 - \phi_{24}\lambda_3 + \phi_{34}\lambda_2}{\phi_{12}\phi_{34} - \phi_{13}\phi_{24}},\\
X_2 &= \frac{(g - \lambda_1)\phi_{34} - \phi_{13}\lambda_4}{\phi_{12}\phi_{34} - \phi_{13}\phi_{24}},\\
X_3 &= \frac{(\lambda_1 - g)\phi_{24} + \phi_{12}\lambda_4}{\phi_{12}\phi_{34} - \phi_{13}\phi_{24}},\\
X_4 &= \frac{(g - \lambda_1)\phi_{23} - \phi_{12}\lambda_3 + \phi_{13}\lambda_2}{\phi_{12}\phi_{34} - \phi_{13}\phi_{24}}.
\end{aligned} \tag{29}$$

The removal of the flow between compartments X_2 and X_3 or the addition of a flow between compartments X_1 and X_4 does not affect the feasibility of the system. Examination of several more complex closed networks suggests that all forced but not all self-generating closed networks are mathematically feasible. No general rule was found for predicting the feasibility of self-generating networks.

The presence of side branches is mathematically feasible for forced networks [for example, Fig. 11d and Eqs. (30)], except of course, that two or more single compartment side branches may not originate from the same compartment:

$$\begin{aligned}
\dot{X}_1 &= f - X_1(\lambda_1 + \phi_{12}X_2 + \phi_{13}X_3),\\
\dot{X}_2 &= X_2(\phi_{12}X_1 - \lambda_2 - \phi_{23}X_3 - \phi_{24}X_4),\\
\dot{X}_3 &= X_3(\phi_{13}X_1 + \phi_{23}X_2 - \lambda_3 - \phi_{34}X_4 - \phi_{35}X_5),\\
\dot{X}_4 &= X_4(\phi_{24}X_2 + \phi_{34}X_3 - \lambda_4),\\
\dot{X}_5 &= X_5(\phi_{35}X_3 - \lambda_5).
\end{aligned} \tag{30}$$

Predicting the feasibility of self-generating branched networks, like that of self-generating closed networks, was not reduced to a set of rules. The self-generating five-compartment system corresponding to the forced system in Fig. 11d is infeasible, but lengthening the branch to two compartments produces a mathematically feasible self-generating system.

2. *Systems with Multiple Inputs*

Multiple inputs affect forced and self-generating systems quite differently. A self-generating function acts on a compartment merely as a negative loss rate, and thus has no effect on the mathematical properties of the system. Any mathematically feasible self-generating system with one input will remain feasible with multiple inputs, and still have one and only one steady state (Fig. 11e). On the other hand, any self-generating system with an infeasible structure will remain infeasible with additional inputs (Figs. 11f and g).

Multiple inputs greatly increase the mathematical complexity of forced systems. A forced unbranched sequence will have as many steady states as there are forcing functions (Figs. 11e and f). The equations [Eqs. (31)] representing the system with two forcing functions (Fig. 11e) are not unduly complex. However, they yield complicated steady-state equations [Eqs. (32)] which represent solutions to quadratic equations:

$$\begin{aligned} \dot{X}_1 &= f_1 - X_1(\phi X_2 + \lambda_1), \\ \dot{X}_2 &= f_2 + X_2(\phi X_1 - \lambda_2); \end{aligned} \tag{31}$$

$$\begin{aligned} X_1 &= \frac{f_1\phi + f_2\phi + \lambda_1\lambda_2 \pm [(f_1\phi + f_2\phi + \lambda_1\lambda_2)^2 - 4f_1\phi\lambda_1\lambda_2]^{1/2}}{2\phi\lambda_1}, \\ X_2 &= \frac{f_1\phi + f_2\phi - \lambda_1\lambda_2 \pm [(f_1\phi + f_2\phi - \lambda_1\lambda_2)^2 - 4f_2\phi\lambda_1\lambda_2]^{1/2}}{-2\phi\lambda_2}. \end{aligned} \tag{32}$$

Forced branching and network systems with multiple inputs may or may not have multiple steady states (Figs. 11g, h, and i); the number of steady states will never exceed the number of forcing functions. No general rule, however, has been derived for predicting the number of steady states possessed by such systems. Additional forcing functions cannot render a mathematically feasible system infeasible, but they may make a mathematically infeasible system feasible [Figs. 11a and g and Eqs. (33)–(34)]:

$$\begin{aligned} \dot{X}_1 &= f_1 - X_1(\phi_{12}X_2 + \phi_{13}X_3 + \lambda_1), \\ \dot{X}_2 &= f_2 + X_2(\phi_{12}X_1 - \lambda_2), \\ \dot{X}_3 &= X_3(\phi_{13}X_1 - \lambda_3); \end{aligned} \tag{33}$$

$$\begin{aligned} X_1 &= \frac{\lambda_3}{\phi_{13}}, \\ X_2 &= \frac{f_2\phi_{13}}{\phi_{13}\lambda_2 - \phi_{12}\lambda_3}, \\ X_3 &= \frac{(f_1\phi_{13} - \lambda_1\lambda_3)(\phi_{13}\lambda_2 - \phi_{12}\lambda_3) - f_2\phi_{12}\phi_{13}\lambda_3}{\phi_{13}\lambda_3(\phi_{13}\lambda_2 - \phi_{12}\lambda_3)}. \end{aligned} \tag{34}$$

The presence of multiple steady states is not in general a hindrance in modeling, because commonly all but one will involve negative or imaginary values for some of the compartments and thus be of no ecological significance.

III. Systems with Controlled Nonlinear Transfers

A. The Forced Two-Compartment System

It is unreasonable to assume that the flow between two compartments representing a population of organisms and their food must always be proportionate to the product of the compartments. This assumption predicts, for example, that the flow from food to organisms can be kept constant in the face of a diminished food supply by increasing the population of feeding organisms. In nature, however, at some point these feeding organisms begin to compete with one another for the diminished food supply and thus reduce the rate of feeding of each organism. This is the Lotka–Volterra concept of self-inhibition—that as a population increases in size, the rate per individual for the transfer of energy or material into the population is reduced (Gause and Witt, 1935). To introduce this concept into a nonlinear system, the function used previously to compute the flow between successive compartments, $\phi X_1 X_2$, is multiplied by a new factor, $(1 - aX_2)$. The new coefficient, a, is a measure of self-inhibition. If this transfer is the only input into compartment X_2, X_2 can never exceed $1/a$ because input drops to zero at that value of X_2. The coefficient, a, thus defines the maximum size of the recipient compartment and so controls the flow through the system.

A forced two-compartment system is diagrammed in Fig. 12a. The system is potentially feasible and very stable, i.e., it rapidly returns to steady state if displaced from it. The system has two steady states (Fig. 12b); its system equations [Eqs. (35)] yield steady-state equations [Eqs. (36)] which are solutions to quadratic equations:

$$\begin{aligned} \dot{X}_1 &= f - X_1[\lambda_1 + \phi X_2(1 - aX_2)], \\ \dot{X}_2 &= X_2[\phi X_1(1 - aX_2) - \lambda_2]; \end{aligned} \tag{35}$$

$$\begin{aligned} X_1 &= \frac{\phi af - \phi\lambda_2 \pm [(\phi\lambda_2 + \phi af)^2 + 4\phi a\lambda_1\lambda_2{}^2]^{1/2}}{2\phi a\lambda_1}, \\ X_2 &= \frac{\phi af - \phi\lambda_2 \pm [(\phi\lambda_2 + \phi af)^2 + 4\phi a\lambda_2(\lambda_1\lambda_2 - \phi f)]^{1/2}}{2\phi a\lambda_2}. \end{aligned} \tag{36}$$

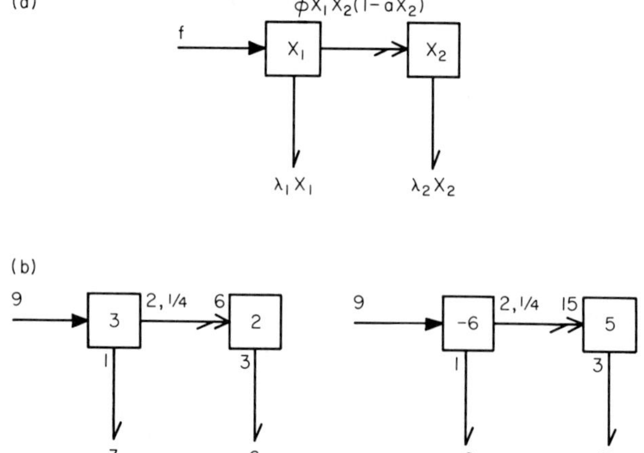

FIG. 12. Forced two-compartment system with a controlled nonlinear flow between compartments. (a) Functions defining flows in the system. (b) Example of the system at both steady states. The first number at the base of the arrow representing the nonlinear flow indicates the value for ϕ and the second number the value for a.

In both steady-state equations all the coefficients in the system appear at least twice, and the square root term in each of the steady-state equations contains all the coefficients. The sensitivity expressions for the system are thus very complex. In no case is the direction of change of the steady-state values in response to changes in the coefficients determined by the structure of the system. All compartments and all flows are sensitive to change in any of the coefficients.

B. THE SELF-GENERATING TWO-COMPARTMENT SYSTEM

If the forcing function in Fig. 12a is replaced with a self-generating function, the new system [Fig. 13a and Eqs. (37)] still has two steady states but is now not entirely stable:

$$\begin{aligned} \dot{X}_1 &= X_1[g - \lambda_1 - \phi X_2(1 - aX_2)], \\ \dot{X}_2 &= X_2[\phi X_1(1 - aX_2) - \lambda_2]; \end{aligned} \tag{37}$$

$$\begin{aligned} X_1 &= \frac{2\lambda_2}{\phi \pm [\phi^2 - 4a\phi(g - \lambda_1)]^{1/2}}, \\ X_2 &= \frac{\phi \pm [\phi^2 - 4a\phi(g - \lambda_1)]^{1/2}}{2a\phi}. \end{aligned} \tag{38}$$

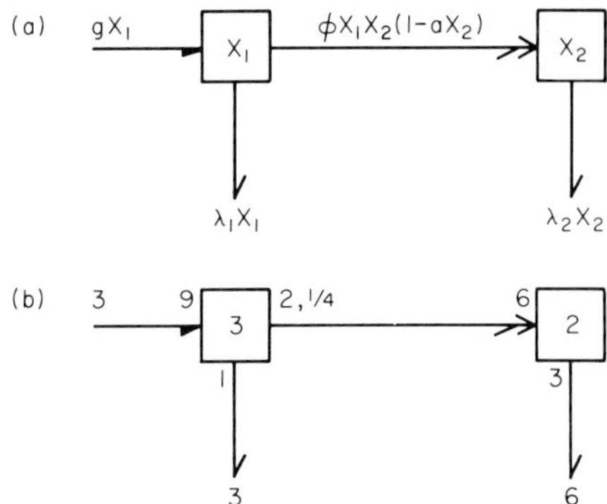

FIG. 13. Self-generating two-compartment system with a controlled nonlinear flow between compartments. (a) Functions defining flows in the system. (b) Example of the system at steady state.

In the example of this self-generating system (Fig. 13b) the two steady states are identical because the square root term in the steady-state equations [Eqs. (38)] changes to equal zero. If, in a system such as that shown in the example, compartment X_1 is displaced below its steady-state value (but not down to zero), the system will return to steady state; if, however, X_1 is displaced above its steady-state value, X_1 will expand without limit. If the two steady states for this self-generating system were not identical (and both were positive) the system would return to the steady state with the smaller value for compartment X_1 provided X_1 remains greater than zero and less than its larger steady-state value. If X_1 is displaced beyond its larger steady-state value, the compartment will expand without limit.

C. The Controlled Self-Generating Function

The concept of self-inhibition can also be applied to the input function of a self-generating system [Fig. 14 and Eq. (39)]. This one-compartment system, which might serve as a model for a population of photosynthetic organisms, is potentially feasible and very stable and has a single steady state [Eq. (40)]:

$$\dot{X} = X[g(1 - aX) - \lambda]; \tag{39}$$

$$X = \frac{g - \lambda}{ag}. \tag{40}$$

The sensitivity equations for this system are simple and indicate that, in all but one case, the direction of change of the steady-state values in response to changes in the coefficients is determined by the structure of the system (Table V).

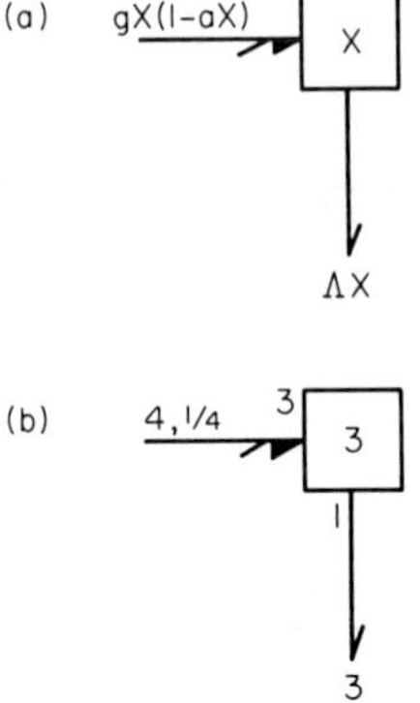

FIG. 14. One-compartment system with controlled self-generating input. (a) Functions defining flows in the system. (b) Example of the system at steady state.

TABLE V

SENSITIVITIES FOR THE CONTROLLED, SELF-GENERATING, ONE-COMPARTMENT SYSTEM[a]

Independent variable	Dependent variable	
	∂X	$\partial(\lambda X)$
∂g	+	+
∂a	—	—
$\partial \lambda$	—	?

[a] The system is given in Fig. 14a and Eqs. (39) and (40) and is expressed as the effect of an increase in each of the independent variables on the dependent variables (steady-state values for the compartment and flow).

D. FEASIBILITY OF SYSTEMS WITH CONTROLLED NONLINEAR TRANSFERS

Insofar as it could be determined by obtaining the steady-state equations for a variety of simple systems and determining that solutions do exist for a number of complex systems, it appears that all systems are potentially feasible regardless of their structure, type of input, or number of inputs. With the simple self-generating input, stability will, of course, be present only over a limited range of values for the compartment receiving the input. Complex systems will possess a large number of steady states, but this usually will not be a handicap in their use as models of natural systems, because commonly all but one of the steady states will demand negative or imaginary values in some compartments.

IV. The Use of Nonlinear Systems in Modeling

This analysis of steady state in a few nonlinear systems should make clear that the use of such systems in modeling involves a number of problems. First, not every nonlinear system is mathematically feasible in the sense that it possesses a steady state and could persist indefinitely. Thus, not every nonlinear system which might be designed on paper can serve as a model for a natural system. The self-generating, three-compartment, unbranched sequence (Fig. 5) which might at first sight appear to simulate a plant–herbivore–carnivore system in fact does not simulate that or any other persistent natural system. This in no way implies that a plant–herbivore–carnivore system cannot exist, but rather that it cannot have the mathematical relationships of Eqs. (11). Models of such a natural system must therefore utilize some different set of mathematical relationships. Altering the system in Fig. 5, for example, by replacing the self-generating input with a controlled self-generating input (Fig. 14) yields a stable and mathematically feasible system capable of serving as a model. Furthermore, even though this or any other system is mathematically feasible—i.e., by possessing a structure which permits the existence of a steady state—the actual system may be biologically infeasible because its coefficients demand zero, negative, infinite, or imaginary steady-state values in some compartments.

A second problem concerns the sensitivities of nonlinear systems. Use of a particular nonlinear system as a model must be accompanied by acceptance of a group of implicit assumptions concerning the system modeled. Adoption of the forced two-compartment system in Fig. 1a as a model of a detritus–saprovore food chain, for example, implies that in the natural system the input rate of detritus does not affect the steady-state value of the detritus compartment (Table IB). If such an assumption were not true of the actual detritus–saprovore chain, the system in Fig. 1a would be unsuited as a model for this food chain. Comparisons between the responses of the natural system modeled and the sensitivities of proposed mathematical models may provide one basis for selecting a suitable model.

A third problem scarcely touched on in this study is the dynamic behavior of nonlinear systems. If a purpose of the model is to simulate and thus predict the time behavior of a natural system, it is not enough that the model merely duplicate the steady equilibriums of the natural system. Changes in the model in response to simulated environmental changes should occur at the same rates as those in the natural system in response to actual changes. As many of the simulation chapters in this book indicate, there is no set of rules for the solution of this problem. A preliminary model of Lindeman's Cedar Bog Lake ecosystem (see Chapter 10, Vol. I) with simple nonlinear flows between compartments was found to be too sluggish in its behavior to duplicate accurately seasonal cycles of the ecosystem. Use of controlled nonlinear flows proved more satisfactory. The larger values for ϕ required by such transfers caused the system to respond more rapidly to simulated environmental changes.

A final problem and one not considered in this study is the properties of "hybrid" systems containing a variety of inputs or transfer functions, or both. This remains a field for future research since such systems are likely to find wide use as simulation models.

Acknowledgments

This research was largely done during a year's employment at the Radiation Ecology Section (Health Physics Division) of Oak Ridge National Laboratory. It was jointly sponsored by the Ford Foundation through a grant to Oak Ridge Associated Universities, and the U.S. Atomic Energy Commission under contract with the Union Carbide Corporation. I am grateful to Dr. J. S. Olson for arranging financial support for the year at Oak Ridge National Laboratory, to Dr. B. C. Patten for introducing me to systems ecology, and to other members of the Radiation Ecology Section for their help and encouragement.

REFERENCE

Gause, G. F., and Witt, A. A. (1935). *Amer. Natur.* **69**, 596.

6

Structural Properties of Food Webs

GILBERTO C. GALLOPÍN
SECTION OF ECOLOGY AND SYSTEMATICS, DIVISION OF BIOLOGICAL SCIENCES, CORNELL UNIVERSITY, ITHACA, NEW YORK

I. Introduction

A *food* or *trophic web* is an abstraction of the different paths of matter or energy transfer within a community, from primary producers to top consumer organisms. The web structure represents statements about

Present address: Fundación Bariloche, Rivadavia 986 (7°), Buenos Aires, Argentina.

feeding relationships, showing for each kind of organism its food sources and predators. Because energy and material flows are important ecological processes, and because feeding relations are a principal cause of such flows, the food web of a community reflects essential structural organization.

In emphasizing structural features, there is a tendency to regard food webs as representing fixed trophic relationships. It is important to bear in mind, however, that a web is really an instantaneous (or in some cases, average) picture of a continuously changing trophic pattern. In all natural situations the web structure changes in time with fluctuating abundance of component species as the latter move through their different life-cycle stages, interact, and die. Thus, both seasonal and shorter- and longer-term variations are to be expected in the structure of food webs, even in steady-state communities.

In any case, a food web can be represented as a collection of *points* and a collection of *links* connecting the points. The points represent kinds of organisms and the links, which may be with or without direction, define the pattern of feeding relations. The set of points may be subdivided into subsets representing *trophic levels*; each subset corresponds to a group of organisms which eats or is eaten by another such group. Quite independently of the biological meaning assigned to the points and links, a number of formal properties can be derived for such systems, properties that could, perhaps, aid in obtaining further insight into the structure and function of communities. This chapter will attempt to develop some of the properties that seem most relevant for food webs. A general acquaintance with elementary set theory will be assumed.

II. The Trophic Relation

A. Binary Relations

A community food web shows, for each pair of points, whether they are trophically related or not. Thus, the trophic relation, being a relation between two objects (points), is a binary relation. Given two sets, A and B, a *binary relation* from the set A to the set B is a *correspondence* (*mapping*, *transformation*) that relates one or more elements of B to each element of some subset of A (Dinkines, 1964). Alternatively, a *binary relation* is a subset R of the Cartesian product $A \times B$. If $R \subset A \times A$, we call R a binary relation in A. If $(a, b) \in R$, where $a \in A$ and $b \in B$, we may write aRb (Cohen and Ehrlich, 1963). If $(a, b) \notin R$, then $a\not{R}b$ (a is not in the relation R to b). In the above definition, the symbol "$\subset$" in $R \subset A \times A$ means "subset" (R is a subset of $A \times A$), the symbol "$\in$" in $(a, b) \in R$ means "element" (the ordered pair (a, b) is an element of the set R). The Cartesian product of two sets, $A \times B$, is the set of all ordered pairs (a, b) of elements $a \in A$ and $b \in B$:

$$A \times B = \{(a, b) : a \in A \quad \text{and} \quad b \in B\}.$$

Note that $A \times B = B \times A$ if and only if $A = B$.

As an example, let the set A consist of three elements and the set B of two elements,

$$A = \{a_1, a_2, a_3\}, \qquad B = \{b_1, b_2\}.$$

Let R be a binary relation from A to B that makes a_1 correspond to b_1 and b_2, and a_3 correspond to b_1 (Fig. 1). Then, $a_1 R b_1$, $a_1 R b_2$, and $a_3 R b_1$;

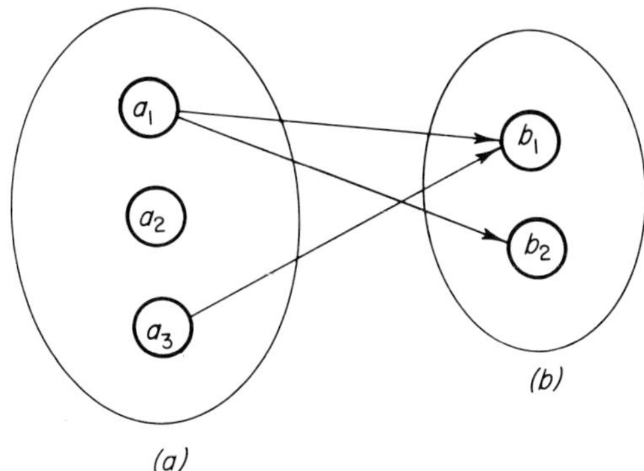

FIG. 1. Representation of a binary relation R between elements of two sets, A and B.

and further, $a_2 \not R b_1$, $a_2 \not R b_2$, and $a_3 \not R b_2$. The Cartesian product of the two sets is

$$A \times B = \{(a_1, b_1), (a_1, b_2), (a_2, b_1), (a_2, b_2), (a_3, b_1), (a_3, b_2)\},$$

and the relation R is

$$R = \{(a_1, b_1), (a_1, b_2), (a_3, b_1)\},$$

illustrating that $R \subset A \times B$.

For any relation R there exists an *inverse* (*reverse*, *converse*, *transpose*) *relation* R^* such that bR^*a if and only if aRb is true.

B. Equivalence Relations

Let R be a binary relation defined in a set A (i.e., $R \subset A \times A$). Then:

(1) R is *reflexive* if aRa for all $a \in A$.

(2) R is *symmetric* if whenever aRb, then bRa, for $a, b \in A$.

(3) R is *transitive* if whenever aRb and bRc, then aRc, for $a, b, c \in A$.

Each of these properties has a complement: R is *antireflexive* if $a \not R a$ for all $a \in A$; R is *asymmetric* if aRb precludes bRa for all $a, b \in A$; and R is *antitransitive* if aRb and bRc preclude aRc for all $a, b, c \in A$.

Note that a particular relation can have different combinations of these properties, including none of them, as in the example of Fig. 2. In this example, aRa, but $b \not R b$, $c \not R c$, and $d \not R d$; therefore R is neither reflexive nor antireflexive. Also, bRd and dRb, but bRa and $a \not R b$; R is neither symmetric nor asymmetric. Finally, cRb and bRa implies cRa, but dRb and bRa do not imply dRa; therefore R is not transitive or antitransitive.

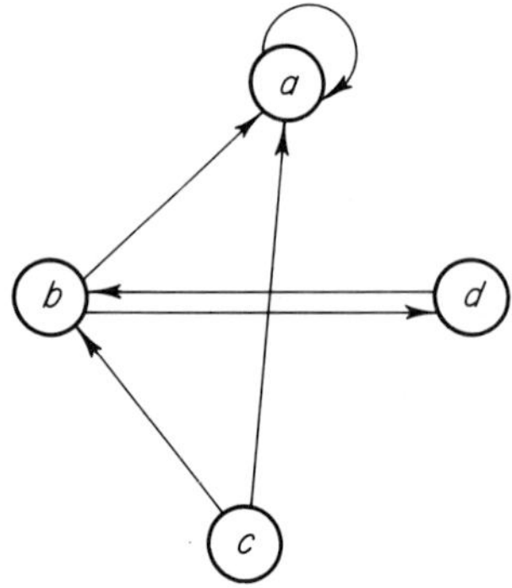

Fig. 2. Illustration of a relation that is neither reflexive nor antireflexive, symmetric nor asymmetric, and transitive nor antitransitive.

A relation R in a set A is termed an *equivalence relation* if R is reflexive, symmetric, and transitive.

C. Feeding Relations

Consider a community consisting of n populations of organisms. Since species may not be a good food web unit, as discussed later, the phrase "kind or organism," or just "organism" will be used in this discussion.

Let s_i be a population of the ith kind of organism, $i = 1, 2, \ldots, n$, and let S be the set of all populations in the community:

$$S = \{s_1, s_2, \ldots, s_n\}.$$

Feeding relations will be subsets of $S \times S$, and two types will be defined:

(1) The *trophic relation*, $T \subset S \times S$, which, in $s_i T s_j$, does not distinguish whether s_i eats s_j or s_j eats s_i, or both.

(2) The *eating relation*, $E \subset S \times S$, in which $s_i E s_j$ means s_i eats s_j without implying s_j eats s_i. The population eating and that eaten are differentiated. Note that $s_i E s_j$ implies $s_j E^* s_i$, and that the trophic relation is the set union of the eating relation and its transpose: $T = E \cup E^*$.

The trophic relations T and E have the following properties:

(1) T and E are not reflexive, except in cases where all the populations are cannibalistic. They will be antireflexive only if no cannibalism is allowed in any population.

(2) T is symmetric, but E is asymmetric unless there exist cross-eating relations between different populations. Then E is symmetric only when two populations mutually feed upon each other.

(3) T and E are not transitive because $s_i R s_j$ and $s_j R s_k$ do not imply that $s_i R s_k$, where $R = T$ or E. But T and E also are not antitransitive except if $s_i R s_k$ is not allowed whenever $s_i R s_j$ and $s_j R s_k$.

Summarizing, the only difference between T and E is that T is always symmetric and E is not, in general, symmetric. Neither T nor E are equivalence relations except in very improbable particular cases. Moreover, both cannibalism and cross-eating relations, when present, appear normally between different age groups of each population (as adults eating their young or eggs, or adults of two different populations both eating the young of the other). In formulating a food web in such cases, it appears reasonable to partition the populations into two or more age classes, each corresponding to a point in the web. If this is done, then T is antireflexive, symmetric, and not transitive, while E is antireflexive, asymmetric, and not transitive.

Thus T can be represented as a food web with "nondirected" links (Fig. 3a), and E as one with "directed" links (Fig. 3b). Note that $T^* = T$, but E^* (is eaten by) is different from E.

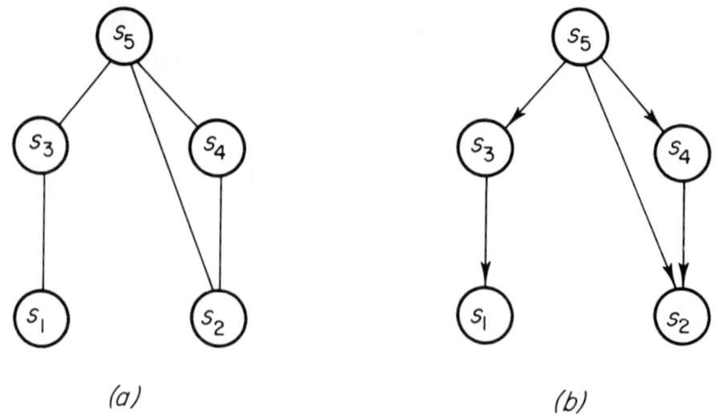

FIG. 3. (a) Illustration of the trophic relation T, and (b) the eating relation E.

III. Graph Formulation of Food Webs

A. Introduction to Graph Theory

Graph theory is a branch of topology that studies the properties of structures consisting of points variously interconnected by lines. A *graph* consists of a set X of elements called *points*, *nodes*, or *vertices*, and a family of correspondences or pairings, $R = (x_i, x_j)$, where $x_i, x_j \in X$ and $R \subset X \times X$, which indicates which vertices shall be considered to be connected (Ore, 1962). The connections are termed *links*, or *edges*, of the graph. An edge (x_i, x_j) is said to be *undirected* if the order of the two endpoints, x_i and x_j, is not considered [i.e., $(x_i, x_j) = (x_j, x_i)$]. If the order is taken into consideration, and a distinction made between the originating node and the terminating node, then the connecting edge (x_i, x_j) is said to be *directed*, with x_i the initial vertex and x_j the terminal one. When X is finite, there is a finite number of edges, and the graph is a finite graph.

A food web, consisting of a finite set S of populations and a finite set, T or E, of trophic links between some of these populations, can be represented as a finite graph. Both the undirected (S, T) and directed (S, E) graph representations are useful in some respects.

The terminology used for graph elements will follow, with some modifications, that of Busaker and Saaty (1965).

B. Problems in Representing Trophodynamics

At this point it is pertinent to discuss some of the real-world characteristics of food webs in relation to the choice of interpretation to be given to vertices, and the allocation of vertices to trophic levels. The natural tendency is to let each vertex represent a species, or other taxonomic entity when, e.g., it is difficult to sort organisms into species. But taxonomic entities may not be the most useful interpretation possible.

For example, in the case of estuarine communities, Darnell (1961) states that "as revealed by stomach analyses, most consumers appear to ingest food on the basis of ecologic rather than taxonomic association. The major food groupings as distinguished by the consumers themselves appear to include the following: phytoplankton, vascular plant material, zooplankton, microbottom animals, macrobottom animals, free-swimming fishes, and organic detritus." This is probably a fairly common situation in other communities also.

The characteristics of food items that are important to a consumer are size, abundance, accessibility, and edibility. To the extent that taxonomic entities reflect these qualities they may be suitable node-units for graph representations of food webs. But usually this is not the case as the size, abundance, etc. of taxonomic groups change greatly in ontogeny. Also, in many cases the larvae of a species belong to a completely different trophic level than the adults. Size alone can be very important in determining the trophic position of an organism, and Ivlev (1961) believes that the degree of food concentration and the size of the food organism are basic determining factors in selective feeding. Brooks and Dodson (1965) showed a striking case of size-selection of food.

Thus, even when taxonomic units reflect approximately the trophic characteristics of a population of organisms, they should be partitioned into more meaningful subunits whenever there are subsets of the same species that perform different trophic roles in a community. This introduces new complications, unquestionably, because, for example, if we represent young and adults of a given species by different vertices in a food web, we need to remember or indicate that they are strongly related in other ways than trophically. One group cannot be perturbed or eliminated without strongly affecting the other. Despite such problems, the alternative of regarding a taxonomic entity as trophically homogeneous when it is not is unacceptable.

By similar reasoning, a number of species or species' subgroups may be lumped into a single food web vertex when they have similar trophic requirements or functions. This procedure would tend to simplify food web representations. For example, Paine (1966) lumped the herbivores of intertidal trophic subwebs into broad categories (chitons, limpets, barnacles, bivalves, etc.), each category containing a number of species. The basis for lumping was the similarity of members of each group as food items for predators, although it is very probable that the classification would have changed little if it had been based on the food eaten.

Cannibalism can be represented in a food web in three different ways: If a vertex represents a species, (1) an edge (self-loop) can be drawn from the vertex back to itself, or (2) it can be ignored if it is reasoned that no new influx of energy is involved and therefore that no true trophic relation exists. If the vertices represent different intraspecific groups, (3) a trophic relation between two of these can be represented in the normal way.

A particularly important case is detritus as a food source. Darnell (1961) found that in the estuarine community of Lake Pontchartrain, organic detritus and its attendant bacteria formed the most conspicuous food item in diets of consumers. Detritus, comprising material of different biological origin in various stages of decay, is certainly less differentiated, as a result of decomposition, than its source organisms. Moreover, many detritus-feeders hardly distinguish detritus as to source, but ingest it as a heterogeneous mixture. Trophically, then, detritus can be lumped and represented by one vertex in a food web. Its sources can be indicated by appropriate edges however, if known. In representing detritus in this way, it should be remembered that a detritus vertex will have several peculiar biological characteristics, among which is the fact that it does not represent a self-perpetuating unit, as population vertices do.

Another important case is that of bacteria. Due to difficulties of taxonomic characterization, and of estimating abundance, this group is invariably lumped as an undifferentiated unit in trophic studies, and called either "bacteria" or "decomposers." This is reasonable since the bacteria, considered as food, are as indistinguishable to consumers as detritus is, and they are indeed normally eaten together. On the other hand, perhaps a useful distinction can be made on the basis of energy source, and the bacteria split into two components, autotrophic and heterotrophic. These two groups differ greatly in their trophodynamic character, the heterotrophs only being "true" decomposers.

C. Trophic Levels

The problem of allocating species or population groups to trophic levels is a complex one, as indeed is the very definition of trophic level. Odum (1959) asserts that organisms which are the same number of feeding steps (links) from primary producers belong to the same trophic level. Thus, in a graph theory frame of reference, the problem of defining trophic levels consists of partitioning the food web vertices into several disjoint subsets, each one an additional feeding step removed from the plants.

In cases where a single taxonomic entity changes trophic levels during its ontogeny, different vertices of the food web can be assigned to appropriate sets of age-groups. In the case of omnivores, the taxonomic unit may be partitioned into several vertices based on food source. For example, Cummins *et al.* (1966) allocated different percentages of the same species of a woodland stream to different trophic levels based on the proportions of different foods found in their guts. The possibility of distinguishing between "pure" and "mixed" trophic levels is also open. For example, all populations that feed on both plants and herbivores could be assigned to an "intermediate" trophic level between primary (herbivorous) and secondary (primary carnivorous) consumers.

A different approach to partitioning the organisms of a community into more or less disjoint sets ("trophic groups") is represented by Savilov's ecologic classification of benthic marine animals by mode of nutrition and degree of mobility (Nesis, 1965). This partition is not comparable to trophic levels.

In any event, many of the structural properties of food webs can be studied without partitioning the vertices into disjoint subsets. Other characteristics can only be brought out in the context of some scheme for lumping vertices. In these cases, trophic level partitions can be interpreted as arbitrary groupings based on the feeding relations, and it will be understood that members of a group may usually be related to members of other groups by nontrophic relations. In general, it will not be essential for our purposes to decide whether designated groups represent true trophic levels or not; the principal results will remain valid, and can be modified to suit the interpretation of the partition.

IV. Graph Properties of Food Webs

A. Basic Definitions

When an edge (x_i, x_j) originates and terminates on the same vertex, that edge is called a *loop*. By the preceding discussion, loops will not be considered as components of food webs. If two edges possess the same vertices as endpoints, these are termed *parallel edges*; they are *strictly parallel* when the edges are directed, and both have the same initial and terminal points. Parallel edges have no meaning in food webs. An undirected graph is said to be *simple* if it has no loops and no parallel edges; a directed graph is simple if it has no loops and no strictly parallel edges. A food web will be considered to be a simple (directed or undirected) graph.

Both for undirected and directed graphs, an edge (x_i, x_j) is said to be *incident* to the vertices x_i and x_j. Conversely, x_i and x_j are both incident to (x_i, x_j). An *isolated vertex* is one that is incident to no edge. In food webs, an isolated vertex will represent a population that is not trophically related to any others in the community. Given the changing structure of food webs, it is conceivable that vertices may be isolated at some times and not at others. A vertex that is isolated at a given time will not be considered to pertain to the food web at that time. So, a food web contains only unisolated vertices.

Let ϵ represent the edges of a graph, to be subscripted in context. According to Busaker and Saaty (1965), a finite sequence, $\epsilon_1, \epsilon_2, \ldots, \epsilon_n$, of (not necessarily distinct) edges of an undirected, finite graph is said to constitute an *edge progression* of length n if there exists an appropriate sequence of $n + 1$ (not necessarily distinct) vertices, $x_0, x_1, x_2, \ldots, x_n$, such that

$$\epsilon_i = (x_{i-1}, x_i), \qquad i = 1, 2, \ldots, n.$$

The edge progression is said to be *closed* if $x_0 = x_n$, and *open* if $x_0 \neq x_n$. An edge progression of length six is illustrated in Fig. 4. When all

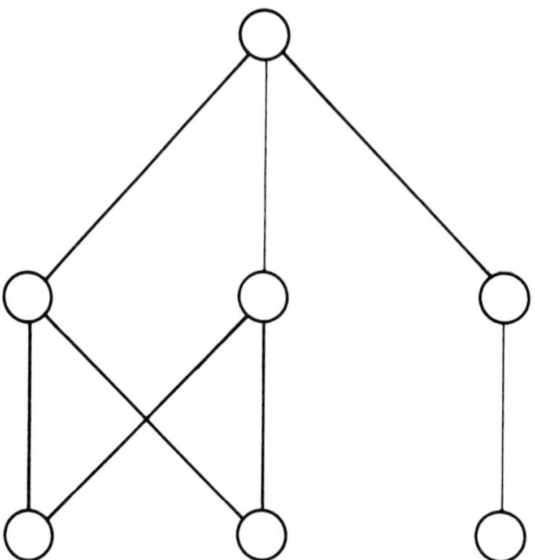

FIG. 4. An open edge progression of length six.

elements of the progression represent distinct edges, the progression is termed a *chain progression* if open, and a *circuit progression* if closed. The set of edges itself, without regard to sequencing, constitutes a *chain* if open and a *circuit* if closed. When the vertices and edges are distinct, we have *simple chain progressions* and *simple chains*, and *simple circuit progressions* and *simple circuits*.

Note that the above definitions refer to undirected graphs, i.e., in the case of food webs, to the trophic relation T. The situation for directed graphs, relation E, will be discussed later.

An important concept, both in graphs and food webs, is that of connectivity. A graph is said to be *connected* if every pair of distinct vertices is joined by at least one chain. Other graphs are *disconnected*. Figure 5

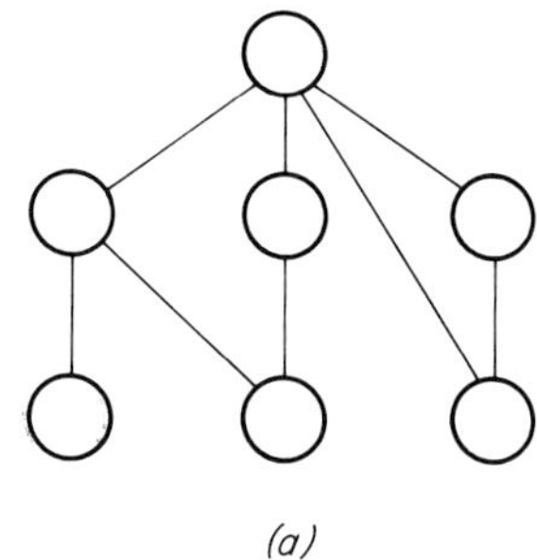

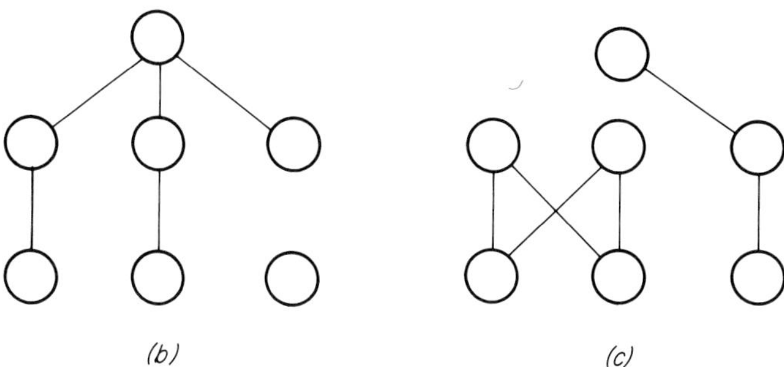

FIG. 5. Examples of connected (a) and disconnected [(b), (c)] graphs.

shows examples of both types. The relation of being connected is reflexive, symmetric, and transitive, and it is therefore an equivalence relation. It can be shown that, for equivalence relations, there exists a decomposition of the vertex set X into disjoint sets X_i such that in each subset all vertices are connected while no vertices belonging to different subsets are connected. Thus, disconnected graphs can be decomposed into a number of disjoint connected graphs. These are the *connected components* of the graph (i.e., connected within but disconnected between), and the problem of a disconnected food web reduces to one of two or more connected webs.

A *maximally connected graph* is called a *complete graph*. A graph is complete if every vertex is joined to every other vertex by an edge. A *maximally disconnected graph* is termed a *null graph*. There are no edges, or every vertex is isolated from every other. Since in food webs isolated vertices are not allowed, and in addition each disconnected segment must be self-sustaining, it is apparent that a maximally disconnected food web consists of a set of *food chains*. Each chain originates at a vertex representing a primary producer or some other source of organic import (Fig. 6).

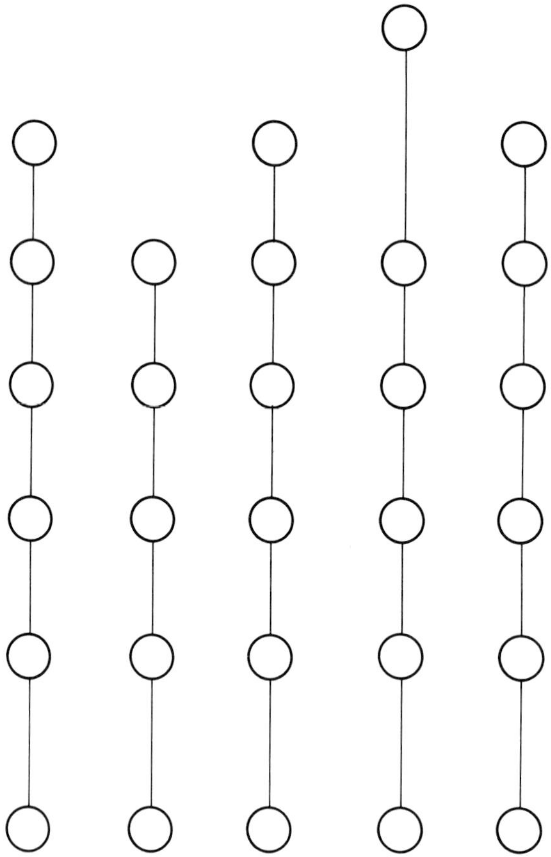

FIG. 6. Food chains: maximally disconnected food webs.

B. Characteristics of Undirected Food Webs

1. *Maximum Number of Feeding Links*

The number of combinations of n vertices taken r at a time is

$$\binom{n}{r} = \frac{n!}{r!\,(n-r)!}, \qquad 0 \leqslant r \leqslant n.$$

Note that, by definition, $\binom{n}{0} = \binom{n}{n} = 1$, and $\binom{n}{r} = 0$ when $r < 0$ and $r > n$. The maximum numbers of edges of an undirected simple graph of n vertices is the number of edges in the complete graph of n vertices,

$$\binom{n}{2} = \frac{n(n-1)}{2}.$$

Figure 7 shows the complete graphs of one, two, three, and four vertices.

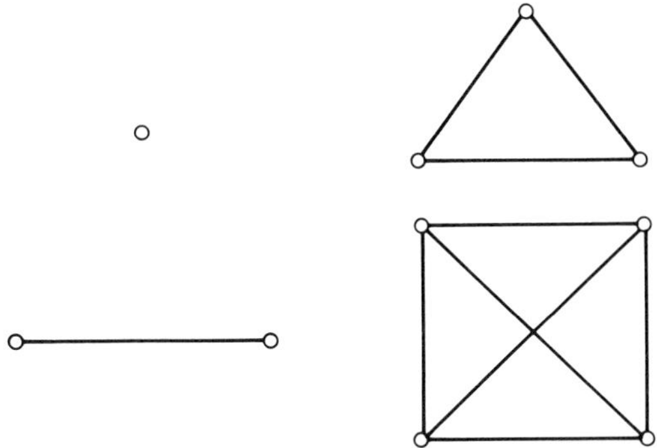

FIG. 7. Complete graphs of one, two, three, and four vertices.

The number of edges is, respectively, $\binom{1}{2} = 0$, $\binom{2}{2} = 1$, $\binom{3}{2} = 3$, $\binom{4}{2} = 6$. This result allows us to place an upper limit on the number of trophic links in any food web (directed as well as undirected, if it is assumed that x_iEx_j precludes x_jEx_i). Thus:

Conclusion 1. The absolute maximum number of links N_c, c for "complete," in a food web of n vertices is

$$N_c = \binom{n}{2}. \tag{1}$$

A graph is termed *k-partite* if its vertices can be partitioned into k disjoint sets, $X = \{X_1, X_2, ..., X_k\}$, in such a way that no edge joins two vertices of the same set (Busaker and Saaty, 1965). Each disjoint set X_i, $i = 1, 2, \ldots, k$, will be called a *trophic partition* of the food web. Clearly, a trophic level is a trophic partition because no edge is allowed among its component vertices. But the concept of trophic partition is more general than that of trophic level, as clarified shortly. Ideally, a trophic partition will include vertices which interact only by nontrophic relations, whereas interaction with members of other subsets will be only through the relations T or E. Consider a food web of n vertices combined into k trophic partitions. The maximum number of edges in this web will be lower than for the complete graph because no edges are allowed inside each partition. The maximum number of edges within the ith partition consisting of n_i vertices is $\binom{n_i}{2}$.

Conclusion 2. The maximum number of links N_p in a food web of n vertices partitioned into k trophic partitions is

$$N_p = N_{\mathrm{c}} - \sum_{i=1}^{k} \binom{n_i}{2} = \binom{n}{2} - \sum_{i=1}^{k} \binom{n_i}{2}, \tag{2}$$

where n_i is the number of vertices in partition i.

Suppose now that the trophic partitions can be ordered into a sequence in such a way as to define consecutive and nonconsecutive partitions. In addition, suppose that links are allowed only between consecutive trophic partitions, not between nonconsecutive ones. This constraint makes the trophic partitions correspond to *trophic levels* in the ecological sense that populations pertaining to one level feed only on populations of the level immediately below. The number of links between nonconsecutive trophic partitions is $\sum_{i=1}^{k} (n_i \sum_{j=i+2}^{k} n_j)$. In such a food web, the maximum number of links is

$$N_q = N_{\mathrm{p}} - \sum_{i=1}^{k} \left(n_i \sum_{j=i+2}^{k} n_j\right). \tag{3a}$$

This result can be obtained more directly by considering that the maximum number of edges between consecutive partitions composed respectively of n_i and n_{i+1} vertices is simply $n_i(n_{i+1})$.

Conclusion 3. The maximum number of links N_q in a food web of n vertices partitioned into k trophic partitions, when links are allowed only between consecutive partitions (i.e., the partitions are trophic levels), is

$$N_q = \sum_{i=1}^{k-1} n_i(n_{i+1}), \tag{3b}$$

where n_i is the number of vertices in partition i, and n_{i+1} is the number in the consecutive partition $i + 1$.

Note that election of the starting point is immaterial; the same result is obtained whether we start counting from the first or the last partition. Figure 8 shows, for a food web of six vertices and three

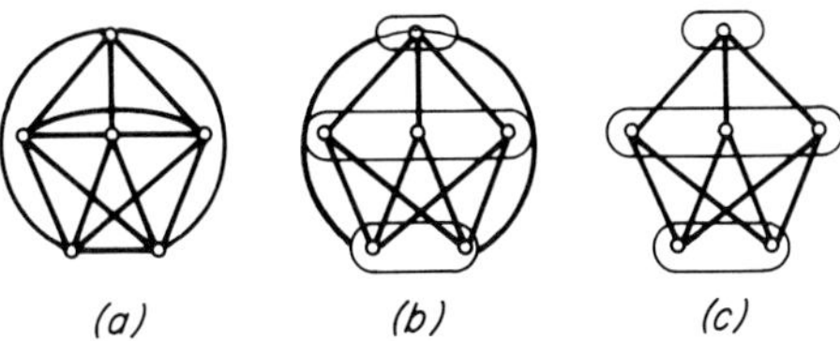

FIG. 8. A food web of six vertices and three trophic partitions, showing (a) $N_c = 15$, (b) $N_p = 11$, and (c) $N_q = 9$.

trophic partitions (containing one, three, and two vertices, respectively), the graphs for $N_c = 15$, $N_p = 11$, and $N_q = 9$.

2. *Complexity and Stability*

The foregoing conclusions are relevant to the question of stability in ecological communities. MacArthur (1955) suggested that stability of a community increases as the number of links in its food web increases. His derived stability measure is based on the amount of choice energy has in following different paths through the food web. Leigh (1965) states that the stability of a particular species increases with an increase in the number of links leading into or out of the species. In general, simple communities are recognized to be less stable than complex ones, whatever definition of stability one chooses, and complexity of the food web seems to carry some stabilizing property. Whether or not a quantitative measure of stability can be related to a given complexity, it seems important to develop a measure of food web complexity.

Without attempting to replace other formulations, it is suggested that a useful measure of complexity is given by the degree of realization of maximum possible complexity for a given food web. Let the *relative complexity* of a food web be the ratio of actual to possible number of links, "possible" number of links (i.e., *maximum complexity*) being N_c, N_p, or N_q, depending on context, and "actual" number of links defining *absolute complexity* of the web. That is, relative complexity equals absolute complexity per maximum complexity. This measure permits comparisons between different food webs independently of how many vertices they contain. So, a web with four vertices and five links will, by definition, be relatively more complex than one with 10 vertices and 20 links. Food webs with trophic partitions are to be compared by using N_p as the denominator.

Note that the requirement that a food web be connected makes the maximum absolute complexity $\binom{n}{2}$ (either a complete web, nonpartitioned, or a web subdivided into n trophic partitions with one vertex in each), and the minimum absolute complexity $n - 1$. The latter is because a graph with n vertices and less than $n - 1$ edges is disconnected, or the minimal connected graph (called a mathematical *tree*) has $n - 1$ edges (Busaker and Saaty, 1965). Figure 9 shows the initial trend of increase in absolute maximum and minimum complexity with increasing numbers of vertices.

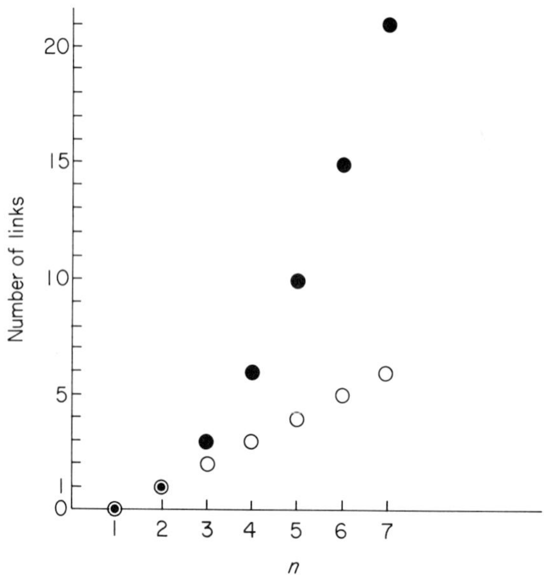

FIG. 9. Maximum $\binom{n}{2}$ and minimum $(n - 1)$ complexity of food webs with n vertices.

3. *Maximum Number of Food Chains*

Another interesting quantity related to stability is the maximum number of paths that energy can follow from producers to top consumers in a food web partitioned into trophic levels. This number is the same as the maximum possible number of different food chains from producers to top consumers.

Consider a food web partitioned into k trophic levels. For notational convenience, level 1 will be assigned to the top consumers, and level k to the primary producers. As before, n_i is the number of vertices in the ith trophic level, $i = 1, 2, \ldots, k$.

In the restrictive case in which links are allowed only between consecutive trophic levels, each food chain will have exactly $k - 1$ links. The number of steps from producers to top consumers can only be less than $k - 1$ if nonconsecutive trophic levels are permitted to be linked. On the other hand, a food chain with more than $k - 1$ links would imply some parallel edges.

Starting from the n_k producers, the number of alternative paths (of one link) to the n_{k-1} herbivores is $n_k(n_{k-1})$. For each of these possible one-link paths, there exist n_{k-2} terminal vertices at level $k - 2$. Then, the number of paths composed of two links from the level k to the level $k - 2$ is $n_k(n_{k-1})(n_{k-2})$. By repeating this procedure up to the top consumers (level 1), we obtain:

Conclusion 4. The maximum number of possible food chains (of $k - 1$ links) from the producers (here assigned level k) to the top consumers (level 1) in a food web partitioned into k trophic levels (links allowed only between consecutive levels) is

$$NF_q = n_k(n_{k-1}) \cdots (n_1) = \prod_{i=1}^{k} n_i . \tag{4}$$

The maximum number of alternative chains per producer is NF_q/n_k.

Now consider the less restrictive nonconsecutive case in which "skipping" of trophic levels is allowed, and the only constraint that remains is that no edges are allowed within trophic levels (strictly, trophic partitions). Then, the number of links in each food chain can vary from 1 to $k - 1$. In this case we have again $n_k(n_{k-1})(n_{k-2}) \cdots (n_1)$ possible paths, of length $k - 1$, from level k to level 1. But in addition, there will also be $n_k(n_{k-2})(n_{k-3}) \cdots (n_1)$ possible food chains, of length $k - 2$, from level k to level 1. Continuing this reasoning, it follows that the number of paths of any length (from 1 to $k - 1$, inclusive) will be $n_k(n_{k-1} \cdot \cdots \cdot n_1) + n_k(n_{k-2} \cdot \cdots \cdot n_1) + \cdots + n_k(n_1) = n_k(\prod_{i=1}^{k-1} n_i + \prod_{i=1}^{k-2} n_i + \cdots + \prod_{i=1}^{k-(k-1)} n_i)$. So, we can state:

Conclusion 5. The maximum number of possible food chains (consisting of any number of links between 1 and $k - 1$, inclusive) from primary producers (here, level k) to top consumers (level 1) in a food web partitioned into k trophic partitions is

$$NF_p = n_k \cdot \sum_{j=1}^{k-1} \prod_{i=1}^{k-j} n_i . \tag{5}$$

The maximum number of food chains per producer is NF_p/n_k. Figure 10 illustrates the alternative food chains of any length possible for a food web of six vertices partitioned into two producers, three herbivores, and one primary carnivore.

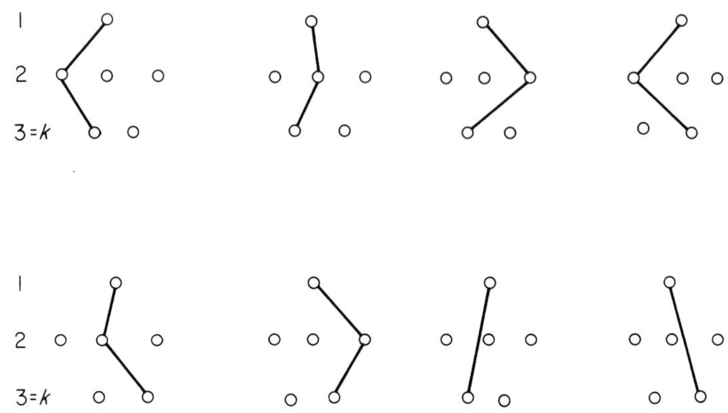

FIG. 10. Possible alternative food chains of any length (from 1 through $k - 1$) for a food web of six vertices partitioned into two producers, three herbivores, and one carnivore.

4. *Configurations of Undirected Webs*

A food web can increase its complexity in three ways:

(1) increasing the number of vertices, which has an associated increase in number of edges of at least one per new vertex;

(2) increasing the number of links without increasing the number of vertices; and

(3) oscillating between different configurations.

In the last instance it is possible to think in terms of the food web as a "resonant" structure, alternating between a set of possible configurations, each with low complexity. This dynamic process could produce a resultant food web of much greater "complexity" than each of its instantaneous configurations (Fig. 11), which would allow the system to persist with considerable stability at the same time that its component populations were undergoing significant changes in abundance, seasonally, and otherwise.

In nature, opportunistic species (MacArthur, 1960) and shifting of diets undoubtedly play an important role in such dynamics of food web

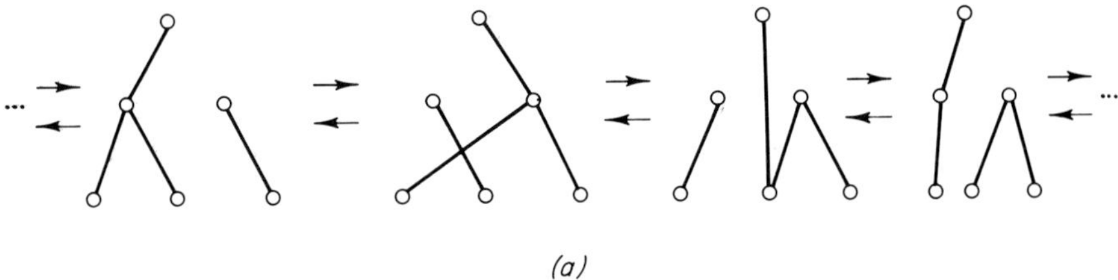

(a)

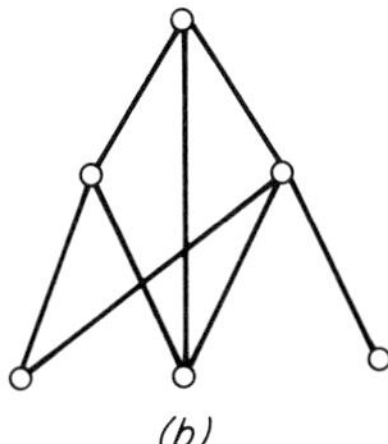

(b)

FIG. 11. (a) A series of "resonant" states of a food web, and (b) resultant or average web over a longer interval.

configurations, which would differ from, e.g., successional dynamics, in being oscillatory rather than directional in character. Succession is marked not by perpetuation of or return to a food web structure, but by a permanent change in such structure in time. In the resonant situation, the shifting between configurations should be much faster than successional changes.

In any event, it is clear that complexity of a food web, taken over a time interval longer than that necessary for transformation from one configuration to another, is usually greater than the complexity of any one of the instantaneous configurations. It should be noted that although the term "resonance" is employed here only in superficial analogy with molecular resonance, the many common features of these widely differing systems may make further comparison enlightening.

In order to study the number of possible configurations of a food web, let us first define the concept of isomorphism. Two graphs are *isomorphic* if they depict the same situation. That is, two undirected graphs are isomorphic if they have the same number of vertices, and correspondences can be established between these vertices (x_i corresponds to y_i, x_j corresponds to y_j, etc.) such that, e.g., if edge (x_i, x_j) exists (or not) in the one graph, then (y_i, y_j) exists (or not) in the other (Ore, 1963). For directed graphs, we require in addition that the endpoints of each pair of corresponding edges be ordered (initial and terminal) in the same way (Busaker and Saaty, 1965). Figure 12 shows

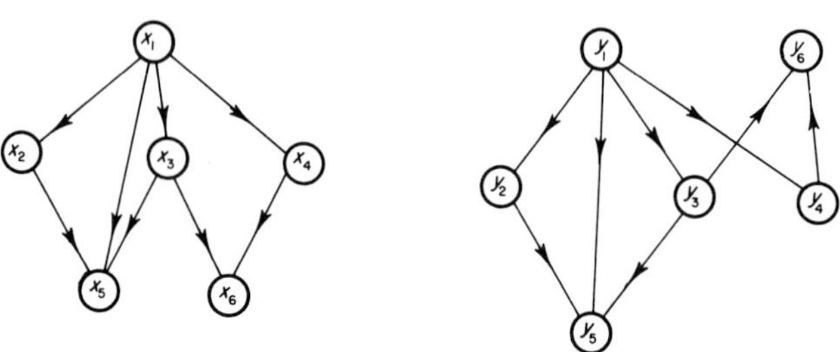

FIG. 12. Two isomorphic graphs.

an example of two isomorphic directed graphs in which x_i corresponds to y_i for $i = 1, 2, \ldots, 6$. Two graphs can be isomorphic without being equivalent if their corresponding nodes depict, e.g., different populations, or their corresponding edges pertain to different relations, or both.

The number of undirected graphs (configurations) that can be formed with n labeled vertices and m unlabeled (i.e., node attachments optional) and nonparallel edges is the number of combinations of edges of the complete graph with n vertices taken m at a time. Notice that this number includes connected and disconnected, and isomorphics graphs. So, we can state:

Conclusion 6. The number of (connected and disconnected) undirected labeled food webs that can be formed with n labeled vertices and m unlabeled links is

$$L_m = \binom{M}{n}, \tag{6}$$

where

(1) $M = N_c$ [Eq. (1)] for a nonpartitioned food web of n vertices [this case is presented by Busaker and Saaty (1965)]; the number of configurations will be denoted L_{mc}.

(2) $M = N_p$ [Eq. (2)] for a food web of n vertices partitioned into k trophic partitions; the number of configurations is L_{mp}.

(3) $M = N_q$ [Eqs. (3)] for a food web of n vertices partitioned into k trophic partitions, when links are allowed only between consecutive partitions; the number of configurations is L_{mq}.

Concerning the connectivity of these graphs, note that any (simple) graph with n vertices and more than $\binom{n-1}{2}$ edges is necessarily connected. As a proof, notice that a set of n vertices can always be partitioned into two subsets, one consisting of an isolated vertex and the other with $n - 1$ vertices. Now, the maximum number of edges of a simple graph of $n - 1$ vertices is $\binom{n-1}{2}$; if the graph has more edges, the additional ones must connect the isolated vertex with the other vertices. For the case of any number of edges (including zero, as in the null graph), we have

Conclusion 7. The number of (connected and disconnected) undirected labeled food webs that can be formed with n labeled vertices and any number of unlabeled links is

$$L = \sum_{m=0}^{M} \binom{M}{m}, \tag{7}$$

where M can be N_c, N_p, or N_q, as above, corresponding to L_c, L_p, and L_q, respectively.

For example, consider the number of different configurations for a food web of four vertices, with three trophic partitions containing one, one, and two vertices in sequence. Here, $M = N_p = \binom{4}{2} - \binom{n_1}{2} - \binom{n_2}{2} - \binom{n_3}{2} = 6 - 0 - 0 - 1 = 5$. Therefore, $L_p = \sum_{m=0}^{5} \binom{5}{m} = \binom{5}{0} + \binom{5}{1} + \binom{5}{2} + \binom{5}{3} + \binom{5}{4} + \binom{5}{5} = 1 + 5 + 10 + 10 + 5 + 1 = 32$ configurations. These are: one configuration with no links, five with one link, ten with two, ten with three, five with four, and one with five (Fig. 13).

Suppose now that we are interested only in the number of connected configurations that can be obtained from n labeled vertices and m unlabeled edges. A derivation given by Gilbert (1956) follows.

Let P be any property of connected graphs. Define $T_{n,m}$ as the (known) total number of graphs (connected and disconnected) having n labeled vertices, m unlabeled edges, and such that every connected component of the graphs has the property P. Define $C_{n,m}$ as the number of graphs in the subset of connected graphs from the set of $T_{n,m}$ graphs. Now, consider that in a graph J with $n + 1$ vertices and m edges, the vertex labeled $n + 1$ belongs to a connected component K having some number ν of vertices and some number μ of edges. The remaining part $J - K$ of J has $n - \nu$ vertices and $m - \mu$ edges. There are $\binom{n}{\nu}$ ways in which ν of the labels 1, 2,..., n can be chosen to be assigned to the graph K, then $C_{\nu+1,\mu}$ ways of picking K, and $T_{n-\nu,m-\mu}$ ways of picking $J - K$. Hence,

$$T_{n+1,\mu} = \sum_{\nu,\mu} \binom{n}{\nu} \cdot C_{\nu+1,\mu} \cdot T_{n-\nu,m-\mu}\,. \tag{8}$$

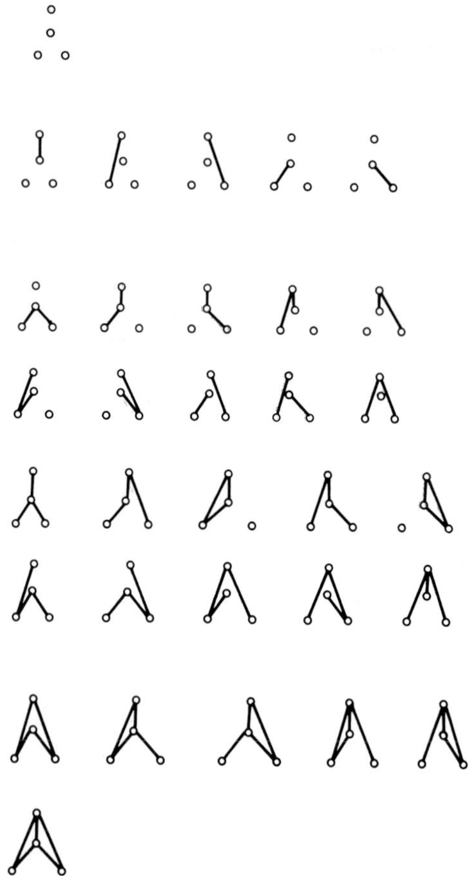

FIG. 13. The 32 possible configurations of a four-node, undirected, labeled, food web with three trophic partitions containing one, one, and two vertices in sequence.

Now, introduce the convention that $T_{0,m} = 1$ if $m = 0$, and $T_{0,m} = 0$ otherwise. Further, introduce the generating functions

$$C_n(y) = \sum_{\nu} C_{n,\nu} \cdot y^{\nu},$$

$$T_n(y) = \sum_{\nu} T_{n,\nu} \cdot y^{\nu}.$$

Then, Eq. (8) assumes the form

$$T_{n+1}(y) = \sum_{\nu} \binom{n}{\nu} \cdot C_{\nu+1}(y) \cdot T_{n-\nu}(y), \tag{9}$$

which is a recursive equation that relates $C_{n+1}(y)$ (when $\nu = n$) to $C_1(y)$, $C_2(y)$,..., $C_n(y)$ and the known $T_n(y)$. Gilbert gives also another derivation, but Eq. (9) is convenient enough for our purposes.

To use Eq. (9), $T_n(y)$, the total number of connected and disconnected graphs with n vertices and any number (y) of edges, must be specified. In doing this, note that for a graph with the property of having only nonparallel edges,

$$T_{n,m} = \binom{N_c}{m} = \binom{\binom{n}{2}}{m} = L_{mc}$$

from Eq. (6), and

$$T_n(y) = \sum_{m=0}^{\binom{n}{2}} \binom{\binom{n}{2}}{m} \cdot y^m = (1+y)^{\binom{n}{2}}. \tag{10}$$

Equation (10) is an expression of the binomial theorem (Parzen, 1960) which states that for any two real numbers, a and b, and any positive integer N, $(a+b)^N = \sum_{k=0}^{N} \binom{N}{k} \cdot a^{N-k} \cdot b^k$. The symbol "$y$" represents only a symbol to which coefficients and exponents can be attached; py^q indicates the number (p) of graphs with q edges.

Substituting Eq. (10) into (9), we count connected graphs with n labeled vertices and m unlabeled edges, none in parallel. Setting $y = 1$, we count these graphs by vertices, allowing any number of nonparallel edges. We can now state:

Conclusion 8. The number of connected, undirected, labeled, food webs that can be formed from n labeled vertices and any number of unlabeled links is $C_n(y)$, as obtained from

$$T_{n+1}(y) = (1+y)^{\binom{n+1}{2}} = \sum_{j=0}^{n} \binom{n}{j} \cdot C_{j+1}(y) \cdot T_{n-j}(y). \tag{11}$$

To illustrate, consider the following two examples:

(1) The total number of connected graphs with any number of edges is obtained by setting $y = 1$ in Eq. (11).

$$T_{n+1} = 2^{\binom{n+1}{2}} = \sum_{j=0}^{n} \binom{n}{j} \cdot C_{j+1} \cdot T_{n-j}\,.$$

For $n = 0$: $T_1 = 2^0 = 1 = \binom{0}{0} \cdot C_1 \cdot T_0\,;$

since $T_0 = 1$ [Eq. (10)], $C_1 = 1$.

$$\text{For} \quad n = 1: \quad T_2 = 2^1 = \binom{1}{0} \cdot C_1 \cdot T_1 + \binom{1}{1} \cdot C_2 \cdot T_0;$$

so $C_2 = 1$.

$$\text{For} \quad n = 2: \quad T_3 = 2^3 = \binom{2}{0} \cdot C_1 \cdot T_2 + \binom{2}{1} \cdot C_2 \cdot T_1 + \binom{2}{2} \cdot C_3 \cdot T_0;$$

so $C_3 = 4$. And so on, up to any number of vertices.

(2) The number of connected graphs with stipulated number of edges is obtained using Eq. (11) as it stands.

$$\text{For} \quad n = 0: \quad T_1(y) = (1 + y)^0 = \binom{0}{0} \cdot C_1(y) \cdot T_0(y);$$

$C_1(y) = 1$ (i.e., $1(y)^0$, signifying one graph of $n + 1 = 1$ vertex, 0 edges).

$$\text{For} \quad n = 1: \quad T_2(y) = (1 + y)^1 = \binom{1}{0} \cdot C_1(y) \cdot T_1(y) + \binom{1}{1} \cdot C_2(y) \cdot T_0(y);$$

$C_2(y) = y$ (i.e., $1(y)^1$, signifying one graph of $n + 1 = 2$ vertices and 1 edge).

$$\text{For} \quad n = 2: \quad T_3(y) = (1 + y)^3 = \binom{2}{0} \cdot C_1(y) \cdot T_2(y) + \binom{2}{1} \cdot C_2(y) \cdot T_1(y) + \binom{2}{2} \cdot C_3(y) \cdot T_0(y);$$

$C_3(y) = 3y^2 + y^3$ (three graphs with three vertices and two edges, and one graph with three vertices and three edges). And so on, up to any number of vertices. Notice that with Eq. (11), tables of $C_n(y)$ as a function of n could easily be generated by computer.

As shown before, a graph with n vertices and more than $\binom{n-1}{2}$ edges is connected. Also, a graph with n vertices and fewer than $n - 1$ edges has to be disconnected. Thus, a graph with $(n - 1)$ or more edges but $\binom{n-1}{2}$ or less edges may be connected or disconnected. When the number of edges falls in this range, the number of connected graphs can be calculated from Eq. (11). When there are more than $\binom{n-1}{2}$ edges, Eq. (6) can be used because all of the graphs will then be connected. Unfortunately, Eqs. (9) or (11) cannot be used for calculating the number of food web configurations when trophic partitions are formed.

This is because of the different possible arrangements of the vertices into trophic partitions for each number of vertices.

Incidentally, Busaker and Saaty (1965) point out that when the number of edges is very large, the total number of connected graphs (labeled) is $2^{\binom{n}{2}}$.

With respect to the problem of counting nonisomorphic graphs, Harary (1955) developed methods for obtaining the number of nonisomorphic graphs with n vertices and m edges. In food webs, however, the vertices are not freely interchangeable (because of the biological characteristics of each vertex), and the problem of isomorphism is probably less interesting than in other areas of graph applications.

C. Characteristics of Directed Food Webs

1. *Distinctions between Directed and Undirected Webs*

The eating relation E, not the general trophic relation T, is the relevant one for most purposes in formulating food webs. Consequently, it is the properties of *directed food webs* that are of greatest interest biologically. The only difference between an undirected graph and a directed one is that the endpoints of a directed edge constitute an ordered rather than an unordered pair of vertices. This difference will cause some of the properties of directed food webs to be shared with those of undirected ones, but others will be uniquely those of directed webs.

Recall that strictly parallel edges have been defined as directed edges with the same initial and terminal vertices, and it was decided that they have no meaning for food webs. Two directed edges are parallel (but not strictly parallel) if they have the same endpoints, but opposite directions. When properties of the eating relation E were discussed, parallel directed edges were also eliminated because E was decided to be asymmetric or adjustable through choice of vertices to be asymmetric. Loops also were precluded. Therefore, our discussion of directed food webs will be mainly limited to the case in which no parallel or strictly parallel edges, or loops are allowed.

In what follows, the term *arc* will be used to signify a directed edge. According to Busaker and Saaty (1965), an *arc progression* of length n is a sequence of (not necessarily) distinct arcs, $\alpha_1, \alpha_2, \ldots, \alpha_n$, such that, for an appropriate sequence of $n + 1$ vertices, $x_0, x_1, x_2, \ldots, x_n$,

$$\alpha_i = (x_{i-1}, x_i), \qquad i = 1, 2, \ldots, n.$$

An arc progression is *closed* if $x_0 = x_n$, and *open* if $x_0 \neq x_n$. An arc progression in which no arc is repeated is a *path progression* if open, and a *cycle progression* if closed. The corresponding sets of arcs, without regard to sequencing, are termed a *path* and a *cycle*. When the vertices (and therefore the arcs) are all distinct, we have *simple* path progressions, cycle progressions, paths, and cycles.

A directed graph is called *cyclic* if it contains at least one cycle, and *acyclic* otherwise. A loop is a special case of a cycle. Figure 14a

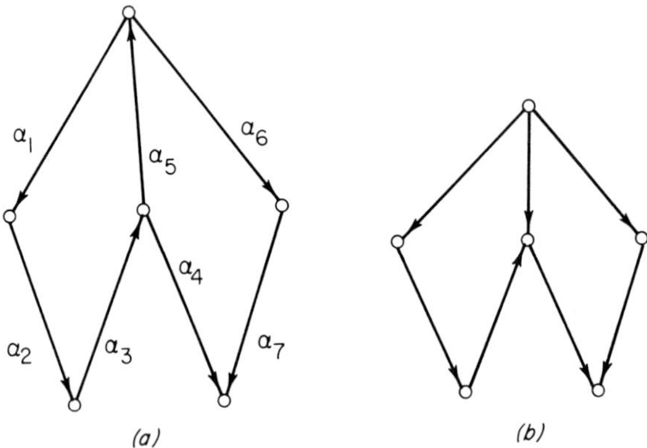

FIG. 14. (a) Cyclic graph and (b) acyclic graph.

shows a cyclic graph. The sequence α_1, α_2, α_3, α_4 forms a path progression of length 4; the sequence α_1, α_2, α_3, α_5 forms a cycle progression of length 4. Note that a sequence like α_1, α_6, α_7 is not an arc progression because the direction of the edges is not consistent. Figure 14b shows an acyclic graph. Note that while in undirected food webs circuits are allowed because the trophic relation T is symmetric, in directed food webs it is difficult to conceive the existence of cycles. The cross-eating relation already discussed at the beginning of this chapter, x_1Ex_2 and x_2Ex_1, is an example of a simple cycle. The sequence x_1Ex_2, x_2Ex_3, x_3Ex_1 is a little more complex, but unusual. A population eating a parasite of its own predator would be an example, and a scavenger feeding on the remains of its predator yet another. Thus, we can state:

Conclusion 9. Directed food webs are in general acyclic, although exceptions are possible.

2. *Maximum Length of Food Chains*

Hutchinson (1959) indicated that a theoretical upper limit exists to the maximum number of links in a food chain. Using a deliberately high estimate of the percentage of energy passing through one link that could enter the next link (20%), and assuming that each predator has twice the mass of its prey, a deliberately low value, he shows that the fifth animal in a food chain would have a population of 10^{-4} of the first. This reasoning applies equally well to whole trophic levels. Food chains in nature usually have no more than five links. These are the predator chains, and parasitic or saprophytic food chains are theoretically even shorter, as discussed by Odum (1959).

Now, in terms of our characterization of food webs as directed graphs, notice that an arc progression from top consumers to producers is a food chain. In these terms, we arrive at:

Conclusion 10. In directed food webs, in general, there are not arc progressions (food chains) more than five links long.

3. *Subwebs*

According to Conclusion 10, and ignoring some special problems raised by decomposers, a *top consumer* can be defined as a vertex having only outgoing arcs. By definition, the different top consumers of a food web are trophically unrelated to each other. Consequently, food webs can be divided into a number of sections, each terminating in a different top consumer. These can be termed *subwebs*, after Paine (1966), and as he points out, these subwebs can serve as intermediaries between the study of simple predator–prey interactions and analysis of whole food webs. Figure 15b shows the division of the food web in Fig. 15a into three subwebs. Note that, unlike trophic partitions, different subwebs may share vertices and edges in common.

4. *Configurations of Directed Webs*

Directed food webs were stated before to be acyclic and without parallel edges. A consequence of this is that all the arcs point in the same general direction, from top consumer toward producer vertices (note that the actual direction of energy flow is opposite, from producer to consumer vertices). This makes it possible to represent a food web like that of Fig. 16a as it appears in Fig. 16b. Structures such as that in Fig. 17, which cannot be so-represented, are not allowed because they imply the existence of cycles (e.g., α_1, α_2, α_3, α_4).

As a consequence of the asymmetry of the eating relation, the discussion about number of food links in undirected trophic webs remains essentially unchanged for the directed case. Therefore, Conclusions 1, 2, and 3 and the discussion about complexity hold for both undirected and directed food webs. Due to the asymmetry and the possibility of

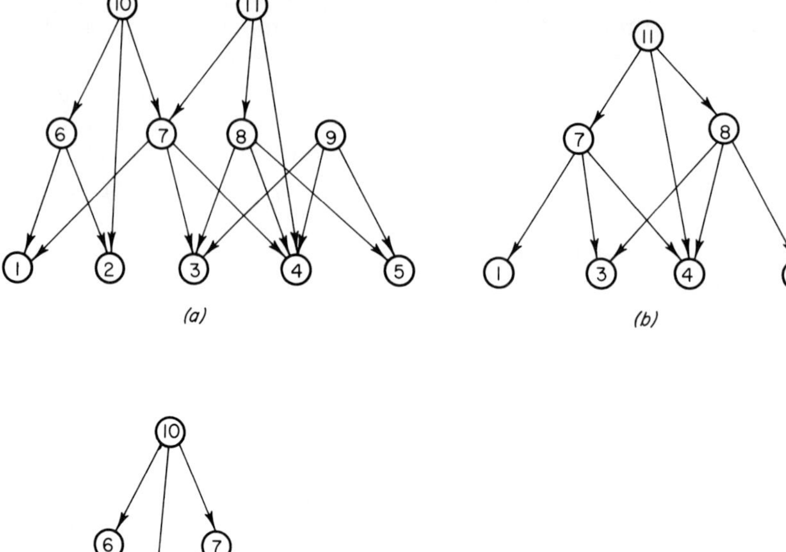

FIG. 15. (a) A food web, (b) divided up into three subwebs.

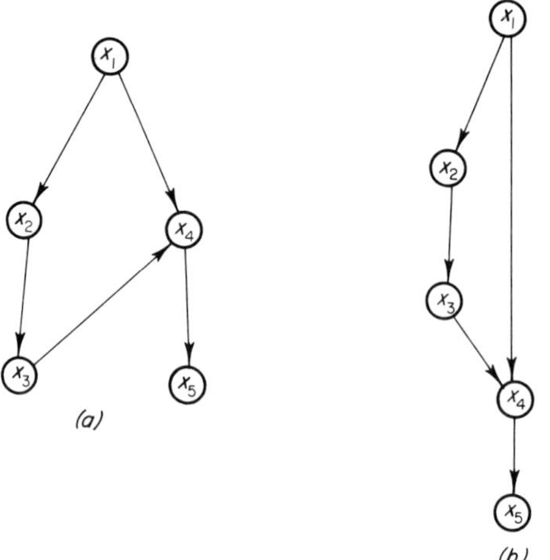

FIG. 16. Two different representations of the same food web.

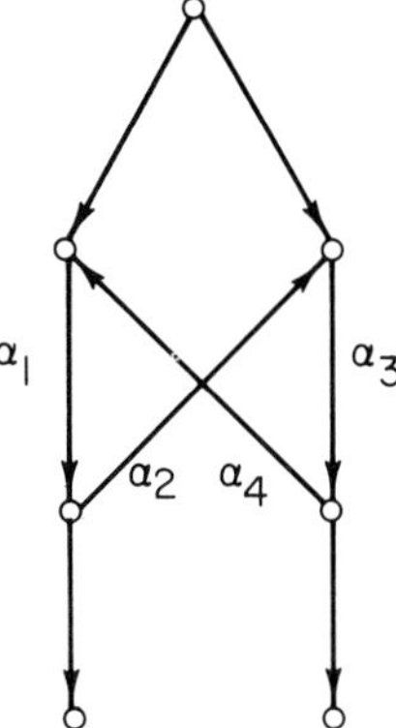

FIG. 17. A graph with a cycle, α_1, α_2, α_3, α_4, prohibited for use as a food web.

representing all the arcs in the same general direction, Conclusions 4 and 5 on the number of energy paths also remain valid for directed food webs. The statements about numbers of configurations, however, need to be modified.

When the edges of a graph are directed, the number of possible configurations with a given number of vertices clearly will increase from the undirected to the directed case. This is because in the latter instance there are two ways of inserting an arc between two nodes, (x_i, x_j) or (x_j, x_i). So there are 2^m possible combinations of directions associated with a set of m undirected edges. This leads to directed-web counterparts of Conclusions 6 and 7 for the nonpartitioned case:

Conclusion 11. The number of (connected and disconnected) directed labeled nonpartitioned food webs that can be formed with n labeled vertices and m unlabeled arcs is

$$D_{mc} = 2^m \binom{N_c}{m} = 2^m \binom{\binom{n}{2}}{m}, \tag{12}$$

where N_c is as given by Eq. (1).

Conclusion 12. The number of (connected and disconnected) directed, labeled, nonpartitioned, food webs that can be formed with n labeled vertices and any number of unlabeled arcs is

$$D = \sum_{m=0}^{\binom{n}{2}} 2^m \binom{\binom{n}{2}}{m}. \tag{13}$$

As an example, calculate D for $n = 3$. We have $D = \sum_{m=0}^{3} 2^m \binom{3}{m} = 1 + 6 + 12 + 8 = 27$, signifying 27 possible directed food webs: one with no arcs, six with one arc, twelve with two, and eight with three. All of these are shown in Fig. 18, and it should be noted that several cycles are included.

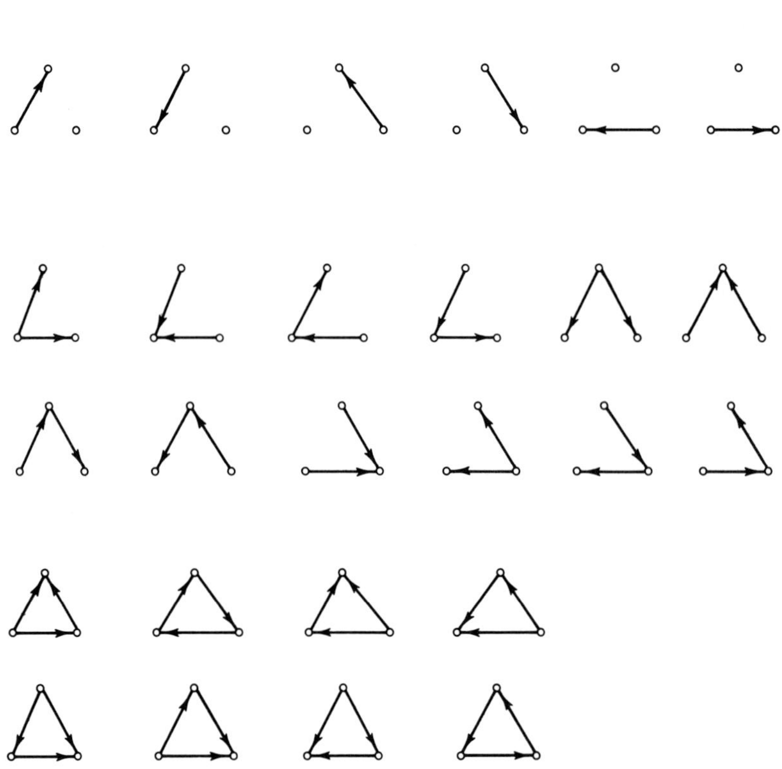

FIG. 18. The 27 possible (connected and disconnected) directed, labeled, non-partitioned, food webs which can be formed from three labeled vertices and up to three unlabeled links.

In discussing the number of configurations of directed food webs with trophic partitions, a distinction has to be made between trophic levels and other kinds of trophic partitions: (1) If trophic partitions are such that the only requirement is that no edges can exist inside the partitions, then each edge of the directed graph has two possible directions. (2) But if the trophic partitions are trophic levels, then only one direction is allowed for each edge, namely that from top consumers to producers. This holds even if "skipping" of trophic levels is allowed. In this case the number of configurations for the directed food web will be the same as in the undirected case. The following conclusions, corresponding to case 2 of Conclusions 6 and 7, can be drawn:

Conclusion 13. The number of (connected and disconnected) directed labeled food webs that can be formed with n labeled vertices and m unlabeled arcs, where the webs are partitioned into k trophic levels, is

$$D_{mp} = L_{mp}[\text{Eq (6)}] = \binom{N_p}{m}, \tag{14}$$

where N_p is as given by Eq. (2).

Conclusion 14. The number of (connected and disconnected) directed labeled food webs that can be formed with n labeled vertices and any number of unlabeled arcs, where the webs are partitioned into k trophic levels, is

$$D_p = L_p[\text{Eq. (7)}] = \sum_{m=0}^{N_p} \binom{N_p}{m}. \tag{15}$$

When the partitions are generalized trophic partitions rather than trophic levels, these conclusions are modified as follows:

Conclusion 15.

$$D'_{mp} = 2^m \binom{N_p}{m}. \tag{16}$$

Conclusion 16.

$$D_p' = \sum_{m=0}^{N_p} 2^m \binom{N_p}{m}. \tag{17}$$

There are also counterpart conclusions for case 3 of Conclusions 6 and 7:

Conclusion 17. The number of (connected and disconnected) directed labeled food webs that can be formed with n labeled vertices and m unlabeled arcs, when the webs are partitioned into k trophic levels, and when arcs are allowed only between consecutive trophic levels, is

$$D_{mq} = L_{mq}[\text{Eq. (6)}] = \binom{N_q}{m}, \tag{18}$$

where N_q is as given by Eq. (3).

Conclusion 18. The number of (connected and disconnected) directed labeled food webs that can be formed with n labeled vertices and any number of unlabeled arcs, when the webs are partitioned into k trophic levels, and when arcs are allowed only between consecutive trophic levels is

$$D_q = L_q[\text{Eq. (7)}] = \sum_{m=0}^{N_q} \binom{N_q}{m}. \tag{19}$$

When the partitions are generalized trophic partitions rather than trophic levels, these conclusions become modified to:

Conclusion 19.

$$D'_{mq} = 2^m \binom{N_q}{m}. \tag{20}$$

Conclusion 20.

$$D_q' = \sum_{m=0}^{N_q} 2^m \binom{N_q}{m}. \tag{21}$$

Considering now the number of connected configurations, the following modification of Conclusion 8 is appropriate:

Conclusion 21. The number of connected, directed, labeled, food webs that can be formed from n labeled vertices and m unlabeled arcs is 2^m times the number of connected undirected webs of n vertices and m links as given by Eq. (11). For instance, in the second example following Conclusion 8, we had $C_3(y) = 3y^2 + y^3$ (three undirected graphs with three vertices and two edges, and one graph with three vertices and three edges). For the directed case we have $CD_3(y) = 3y^2(2^2) + y^3(2^3)$, where in the first term $m = 2$ and in the second $m = 3$. Thus, $CD_3(y) = 12y^2 + 8y^3$ (12 directed graphs with three vertices and two arcs, and eight graphs with three vertices and three arcs).

Gilbert (1956) has contemplated the case of connected directed graphs by putting $T_n(y) = (1 + y)^{n(n-1)}$, but this modification includes parallel directed edges, although not strictly parallel ones.

Lindeman (1942) recognized that higher trophic levels are less distinctly recognizable than lower levels due to greater diversification of diets. Consequently, often producer and herbivore levels can be easily characterized, but predators and especially omnivores become more difficult. Thus, we have the common situation where part of the system is more blurred than the rest, i.e., knowing which populations constitute some of the trophic partitions, but not all.

If one is interested in the maximum number of arcs, or the number of configurations of such a system, the available information can be used. The procedure consists of allocating the populations known to pertain to some trophic partition, say producers or herbivores, into a number of partitions. All the dubious higher consumers that cannot be grouped into trophic partitions are allocated to one trophic partition each. Then the formulae for the *k*-partitioned food webs can be used. This approach will allow links between each of the higher consumers and any of the other vertices in the web, but still edges within the known partitions will not be allowed.

D. Edge–Vertex Relationships

1. *Structural Significance of Node Degrees*

We will now consider some properties of the vertices in relation to the number of edges. The number of edges incident on a vertex x_i is the *degree* of x_i, denoted $\delta(x_i)$. The *positive degree* of a vertex $\delta^+(x_i)$ is the number of outgoing directed edges, and the *negative degree* $\delta^-(x_i)$ is the number of incoming directed edges. Clearly,

$$\delta(x_i) = \delta^+(x_i) + \delta^-(x_i).$$

Following Busaker and Saaty (1965), for any graph

$$\sum_{i=1}^{n} \delta(x_i) = 2N_\epsilon ,$$

where N_ϵ is the number of undirected edges. For directed graphs,

$$\sum_{i=1}^{n} \delta^+(x_i) = \sum_{i=1}^{n} \delta^-(x_i) = N_\alpha ,$$

where N_α is the number of directed edges. Thus,

$$\sum_{i=1}^{n} \delta(x_i) = 2N_\alpha .$$

Food webs have the following general node–vertex characteristics:

(1) Producer vertices have $\delta^+(x_i) = 0$.

(2) Top consumers have $\delta^-(x_i) = 0$ (excepting dead consumers going to decomposers).

(3) Nonproducer and nontop-consumer vertices have $\delta^+(x_i) > 0$ and $\delta^-(x_i) > 0$.

(4) $\delta^+(x_i)$ generally increases in the direction from producers to top consumers (Paine, 1963).

From a purely structural point of view, the most important vertices in a food web are those with the greatest number of incident edges, i.e., those with the greatest numbers of node degrees. Biologically, these represent generalized predators and/or universal prey. Elimination or introduction of such vertices produce more important changes in the food web structure than adding or substracting specialized vertices. In many communities top consumers have the highest node degrees, and they are known to be important in community function. Paine (1966) provides an example of dramatic change induced in the composition of an intertidal community by removing the top predator of a subweb. The number of species in the area decreased from 15 to 8, and the total standing crop increased noticeably.

2. *Trophic Distance*

As another property of food web vertices, it is possible to define a *trophic distance* between two vertices in a given trophic partition. The trophic distance may characterize, in relation to food, the potential for competition or other nonfeeding interspecific interactions between members of the same trophic partition.

Let $S^+(x_i, x_j)$ denote the *prey similarity,* of two vertices, x_i and x_j, within the same trophic partition:

$$S^+(x_i, x_j) = \delta^+(x_i, x_j)/[\delta^+(x_i) + \delta^+(x_j) - \delta^+(x_i, x_j)],$$

where $\delta^+(x_i)$ is the number of outgoing arcs (and therefore the number of prey vertices) of node x_i, $\delta^+(x_j)$ is the number of prey vertices of node x_j, and $\delta^+(x_i, x_j)$ is here defined as the number of prey vertices common to both x_i and x_j. Then $S^+(x_i, x_j)$ ranges from 0 $[\delta^+(x_i, x_j) = 0]$ to 1 $[\delta^+(x_i) = \delta^+(x_j) = \delta^+(x_i, x_j)]$. This index is formally analogous to one developed by Jaccard (1912) and used by plant ecologists to compare floras. The *trophic distance with respect to prey* between vertices x_i and x_j can be defined as

$$D^+(x_i, x_j) = 1/S^+(x_i, x_j), \qquad (1 \leqslant D^+(x_i, x_j) < \infty). \tag{22}$$

In a similar way, *predator similarity* of two vertices can be defined as

$$S^-(x_i, x_j) = \delta^-(x_i, x_j)/[\delta^-(x_i) + \delta^-(x_j) - \delta^-(x_i, x_j)].$$

The corresponding *trophic distance with respect to predators* is

$$D^-(x_i, x_j) = 1/S^-(x_i, x_j) \qquad (1 \leqslant D(x_i, x_j) < \infty). \tag{23}$$

A *total trophic distance* between two vertices x_i and x_j is defined by

$$D(x_i, x_j) = 2/[S^+(x_i, x_j) + S^-(x_i, x_j)] \qquad (1 \leqslant D(x_i, x_j) < \infty). \tag{24}$$

As an example, consider the food web of Fig. 19. The total trophic

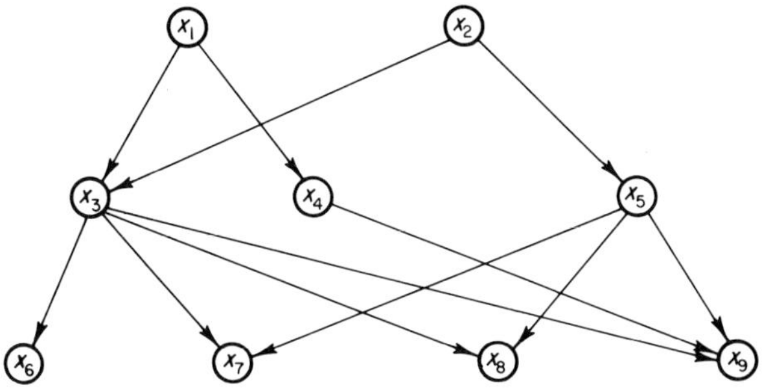

FIG. 19. A food web for comparing trophic distances between x_3 and x_4, x_3 and x_5, and x_4 and x_5.

distances between x_3 and x_4, and x_3 and x_5, and x_4 and x_5 are computed as follows:

$$\begin{aligned}
\delta^+(x_3) &= 4 & \delta^-(x_3) &= 2\\
\delta^+(x_4) &= 1 & \delta^-(x_4) &= 1\\
\delta^+(x_5) &= 3 & \delta^-(x_5) &= 1\\
\delta^+(x_3, x_4) &= 1 & \delta^-(x_3, x_4) &= 1\\
\delta^+(x_3, x_5) &= 3 & \delta^-(x_3, x_5) &= 1\\
\delta^+(x_4, x_5) &= 1 & \delta^-(x_4, x_5) &= 0\\
S^+(x_3, x_4) &= 1/(4 + 1 - 1) = \tfrac{1}{4} & S^-(x_3, x_4) &= 1/(2 + 1 - 1) = \tfrac{1}{2}\\
S^+(x_3, x_5) &= 3/(4 + 3 - 3) = \tfrac{3}{4} & S^-(x_3, x_5) &= 1/(2 + 1 - 1) = \tfrac{1}{2}\\
S^+(x_4, x_5) &= 1/(1 + 3 - 1) = \tfrac{1}{3} & S^-(x_4, x_5) &= 0/(1 + 1 - 0) = 0
\end{aligned}$$

$$\begin{aligned}
D(x_3, x_4) &= 2/(\tfrac{1}{4} + \tfrac{1}{2}) = 2.67\\
D(x_3, x_5) &= 2/(\tfrac{3}{4} + \tfrac{1}{2}) = 1.60\\
D(x_4, x_5) &= 2/(\tfrac{1}{3} + 0) = 6.00.
\end{aligned}$$

Vertices x_3 and x_5 are almost indistinguishable in terms of total trophic distance, whereas vertices x_4 and x_5 are very different indeed.

Based on the above discussion, we can state:

Conclusion 22. Two vertices of a food web whose total trophic distance is one are mutually interchangeable (identical) from a structural point of view.

It is noted that the concepts of trophic distance, and predator and prey similarity are qualitative in nature since they do not consider the "strength" of trophic relations, or how much energy flow occurs through each link. A quantitative extension of these ideas is possible, though, through use of weighting factors to represent quantitative considerations. In any event, the total trophic distance is a measure of structural interchangeability of vertices, and a community or subweb with predominantly small trophic distances would be expected to be quite different from one with predominantly large ones.

In a "most complete" food web with trophic partitions, the trophic distance between any two members of a trophic partition is one. This is because each vertex in a trophic partition is connected to all vertices outside the partition or, when links are not allowed between non-consecutive partitions, each vertex of a partition is connected to all other vertices of the consecutive partitions. Since all the vertices of a partition are identically connected to vertices outside the partition, their trophic distances are minimal, i.e., one.

From this it can be inferred that food webs of high relative complexity have low trophic distances, and conversely, and a way is open to estimating relative complexity without knowing all the details of a food web's structure. Knowledge of a sample of prey–predator relations is sufficient to obtain trophic distances of some members of each trophic partition. Mean distances over the set of partitions should suffice to give a first measure of relative complexity. The measure can, of course, be refined with further information.

Optimal relative complexity is probably less than the maximum, and the optimum average trophic distance in a food web is probably somewhat greater than one. This can be reasoned from the fact of the mean trophic distance in a web of maximum complexity being one, meaning that all the populations of a trophic partition have the same predators and prey (not precluding, in any way, different energy flow characteristics, however). Thus, at least qualitatively, the vertices of a partition have a degree of niche overlap, which would lead to competitive interactions and consequent reduction in number of feeding connections. This tendency would, on the other hand, be compensated by diversification of other aspects of niches than trophic relations. The resulting balance would seem to be in favor of mean trophic distances greater than one, and therefore relative complexities somewhat less than maximum.

E. Matrix Representations of Graphs

Geometric graphs can be represented in other ways than as systems of points and lines, and one of these ways is by matrices. This subject is discussed more extensively by Busaker and Saaty (1965).

1. *The Incidency Matrix*

The *incidency matrix* shows whether a given edge is incident with a given vertex or not. If a graph has n vertices and m edges, the incidency matrix is an $n \times m$ matrix whose rows correspond to the vertices and columns to the edges. For undirected graphs, the matrix element of the ith row and jth column is 0 if the jth edge is not incident on the ith vertex, and 1 if it is. For directed graphs the entry is 0 if the jth edge is not incident with the ith vertex, 1 if the edge is incident and directed away from the vertex, and -1 if it is incident and directed toward the vertex. As examples, the incidency matrices corresponding to Fig. 20a and 20b are as follows.

	ϵ_1	ϵ_2	ϵ_3	ϵ_4	ϵ_5	ϵ_6	ϵ_7
x_1	1	0	0	0	0	0	0
x_2	0	1	1	0	0	0	0
x_3	0	0	0	1	0	0	1
x_4	1	1	0	0	0	1	0
x_5	0	0	1	1	1	0	0
x_6	0	0	0	0	1	1	1

(Fig. 20a)

	α_1	α_2	α_3	α_4	α_5	α_6	α_7
x_1	0	0	0	0	0	0	-1
x_2	-1	0	0	0	0	0	0
x_3	0	-1	-1	0	0	-1	0
x_4	1	1	0	0	-1	0	0
x_5	0	0	1	-1	0	0	0
x_6	0	0	0	1	1	1	1

(Fig. 20b)

Observe that in both cases there are two and only two nonzero entries in each column, and that in the directed case these are of opposite sign.

2. *The Adjacency Matrix*

Elements of the *adjacency matrix* denote the number of edges incident with vertices x_i and x_j in undirected graphs, and in directed graphs they indicate the number of edges directed from x_i (rows) to x_j (columns). The entries are restricted to nonnegative integers. The adjacency matrices for Figs. 20a and 20b are

	x_1	x_2	x_3	x_4	x_5	x_6
x_1	0	0	0	1	0	0
x_2	0	0	0	1	1	0
x_3	0	0	0	0	1	1
x_4	1	1	0	0	0	1
x_5	0	1	1	0	0	1
x_6	0	0	1	1	1	0

(Fig. 20a)

	x_1	x_2	x_3	x_4	x_5	x_6
x_1	0	0	0	0	0	0
x_2	0	0	0	0	0	0
x_3	0	0	0	0	0	0
x_4	0	1	1	0	0	0
x_5	0	0	1	0	0	0
x_6	1	0	1	1	1	0

(Fig. 20b)

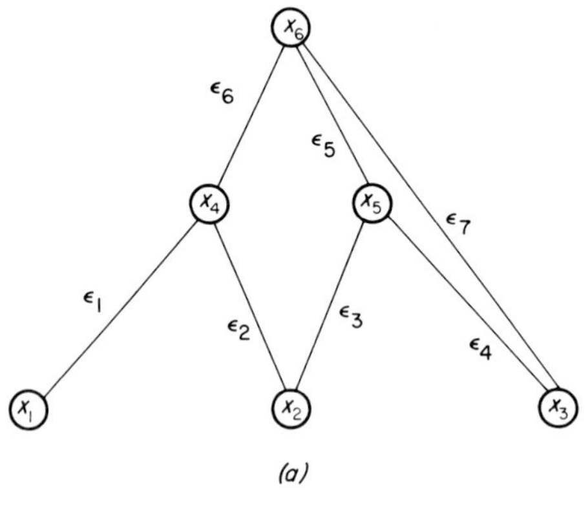

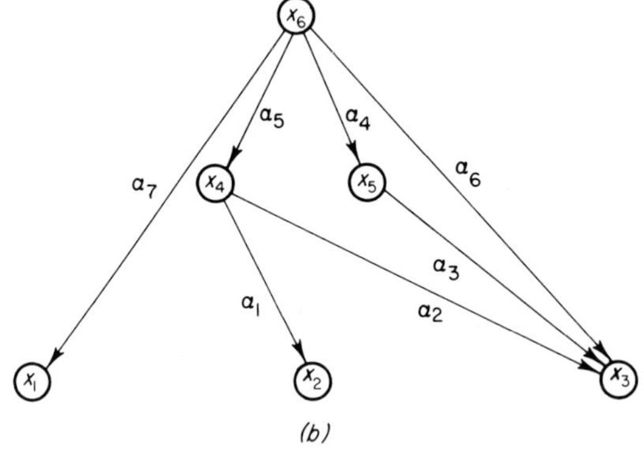

FIG. 20. Example (a) undirected and (b) directed graphs.

Notice that the undirected adjacency matrix is symmetrical about the principal diagonal. Note also that the directed matrix is of the same general form as compartment model coefficient matrices, the entries only being different-valued. Denoting adjacency matrices for directed graphs by A, it can be proved that the matrix A^n gives the number of arc progressions of length n between any two vertices of the directed graph. For example in the graph of Fig. 20b the element of the 6th row, 3rd column of A^2 is 2, indicating the two arc progressions of length 2: α_5, α_2 and α_4, α_3 from x_6 to x_3 shown in the figure.

Properties of the binary relation R represented by a graph can be obtained from the adjacency matrix:

(1) If R is reflexive, A has only 1's in the main diagonal.

(2) If R is antireflexive, A has only 0's in the main diagonal.

(3) If R is symmetric, A is symmetric (i.e., $a_{ij} = a_{ji}$).

(4) If R is asymmetric, A^2 will have the same entries in the main diagonal as A.

(5) If R is transitive, a_{ij} will be positive whenever the corresponding element of A^2 is positive.

V. Status in Food Webs

It is of interest here to consider briefly the concept of status introduced by Harary (1961). According to Harary, "the *status* of a person A in an organization is the number of his immediate subordinates plus twice the number of their immediate subordinates (who are not immediate subordinates of A) plus three times the number of their immediate subordinates (not already included), etc." The *contrastatus* of A is the status of A in the organization chart obtained by reversing the direction of every directed line in the chart. *Net status* is the status minus the contrastatus.

This is an interesting concept. Applied to food webs it yields a measure, in the structural sense, of the relative importance of an organism in a community. Harary would distinguish "powerful" and "powerless" organisms based on these concepts, and in the framework of ecology they might reflect the degree of structural perturbation on a food web if a given vertex were to be eliminated.

VI. Summary

This chapter has considered some of the structural properties of food webs and the properties of trophic relations. The food web has been viewed as an abstract graph-theoretic system, and an attempt has been made at combining mathematical properties with biological interpretations. The attempt should be considered preliminary.

First emphasis has been on obtaining "boundary conditions" (Slobodkin, 1961) for the structure of food webs. But this is only part of the story. The very complex problem of dynamics of biological interactions in a community, interactions which produce living food webs, and the aspect of energy and matter flows in these webs, have not been considered except incidentally. These subjects are treated in detail throughout this book, however, particularly in the form of compartment models.

The theoretical approach of this chapter is compatible with these more dynamic models if we recognize the correspondences between "vertices" and "compartments," and between "edges" and mathematically defined "flows." Then, the present treatment can be viewed as forming the base of a hierarchy of models leading from static structural features of ecosystems toward dynamic functional ones (e.g., Patten, 1966).

Mathematicians and engineers have developed a whole body of theory concerned with optimization of flows through networks. "To understand ecosystems," it has been said, "will be to understand networks" (Patten and Witkamp, 1967). Therefore, application of these theories to food webs will surely yield significant results, although in many instances the existing theory will have to be modified (for example, in usual formulations the flow is assumed to be conserved from source to sink). Since the structure of the network through which energy flows in ecosystems corresponds to the food web, the present study may contribute to development of a general theory of flows in biological networks. The principal results of this study are summarized below.

A food web can be represented as a set of vertices and a set of links. The links may be directed or undirected. An undirected link represents the *trophic relation* T between two vertices, a relation which is antireflexive, symmetric, and not transitive. The relational statement $x_i T x_j$ means "x_i eats or is eaten by x_j." A directed link represents the *eating relation* E, which is antireflexive, asymmetric, and not transitive. The relational statement $x_i E x_j$ means "x_i eats x_j," and the inverse relation $x_i E^* x_j$ means "x_i is eaten by x_j." In the case of T, T^* and T have the same meaning due to the symmetric property.

A food web can be divided into *trophic partitions*, which are disjoint subsets of the set of food web vertices such that no links are allowed between vertices of the same partition. Consecutive trophic partitions can be defined also, and when links are allowed only between consecutive partitions, these then became *trophic levels*.

A typical *undirected food web* can be characterized as follows:

(1) It is simple (loops and parallel edges not allowed).

(2) It is connected (no isolated vertices, i.e., vertices unlinked to the web).

(3) Its maximum number of links is N_c, N_p, or N_q [Eqs. (1), (2), and (3), respectively], for unpartitioned, unconstrained partitioned, and constrained partitioned cases, respectively.

(4) The maximum number of alternative food chains from producers to top consumers is NF_p [Eq. (5)] for the (unconstrained) partitioned case, and NF_q [Eq. (4)] for the (constrained) partitioned case in which links are only allowed between consecutive trophic partitions.

(5) The total number of (connected and disconnected) possible labeled configurations of a food web is L_c, L_p, and L_q [Eq. (7)], respectively, for the three cases above.

(6) The number of possible connected labeled configurations of a food web is $C_n(y)$, as obtained from Eq. (11).

A typical *directed food web* has the following properties:

(1) It lacks loops, and parallel and strictly parallel links.

(2) It is acyclic.

(3) There are no directed edge progressions (*food chains*) with more than five links.

(4) There is a set of vertices with only incoming edges (*primary producers*).

(5) There is a set of vertices with only outgoing edges (*top consumers*).

(6) The maximum number of links is the same as for the undirected case.

(7) The number of alternative food chains is the same as in the undirected case.

(8) The number of possible (connected and disconnected) labeled configurations of a food web is D, D_p, D_p', D_q, and D_q' [Eqs. (13), (15), (17), (19), and (21), respectively] for the following respective cases: (a) unpartitioned, (b) division into trophic levels, (c) division into other trophic partitions, (d) division into trophic levels with links only between consecutive levels, and (e) division into other trophic partitions with links only between consecutive partitions.

(9) The number of possible connected labeled food webs is $CD_n(y)$ computed as $2^m \cdot C_n(y)$, where m is the number of unlabeled links, and $C_n(y)$ is obtained from Eq. (11).

(10) The degree of the vertices (number of attached edges) in a food web generally increases in the direction from producers to top consumers.

In addition, it was noted that properties such as *relative complexity*, *trophic distance*, and *status* could be defined in terms of relationships between numbers of links and vertices.

Acknowledgments

The author is grateful to Dr. B. C. Patten, without whose encouragement and help this chapter would not have been written, and to Dr. R. H. Whittaker, Dr. D. J. Hall, and Mr. M. L. Kreithen for useful comments and suggestions offered about the manuscript.

REFERENCES

Brooks, J. L., and Dodson, S. I. (1965). *Science* **150**, 1.

Busaker, R. G., and Saaty, T. L. (1965). "Finite Graphs and Networks." McGraw-Hill, New York.

Cohen, L. W., and Ehrlich, G. (1963). "The Structure of the Real Number System." Van Nostrand, Princeton, New Jersey.

Cummins, K. W., Coffman, W. P., and Roff, P. A. (1966). *Verh. Int. Verein. Limnol.* **16**, 627.

Darnell, R. M. (1961). *Ecology* **42**, 553.

Dinkines, F. (1964). "Elementary Concepts of Modern Mathematics." Meredith, New York.

Gilbert, E. N. (1956). *Can. J. Math.* **8**, 405.

Harary, F. (1955). *Trans. Amer. Math. Soc.* **78**, 445.

Harary, F. (1961). *Gen. Systems* **6**, 41.

Hutchinson, G. E. (1959). *Amer. Natur.* **93**, 145.

Ivlev, V. S. (1961). "Experimental Ecology of the Feeding of Fishes." Yale Univ. Press, New Haven, Connecticut.

Jaccard, P. (1912). *New Phytol.* **11**, 37.

Leigh, E. G. (1965). *Proc. Nat. Acad. Sci.* **53**, 777.

Lindeman, R. L. (1942). *Ecology* **23**, 399.

MacArthur, R. H. (1955). *Ecology* **36**, 533.

MacArthur, R. H. (1960). *Amer. Natur.* **94**, 25.

Nesis, K. N. (1965). *Oceanology* **5**, 96.

Odum, E. P. (1959). "Fundamentals of Ecology." 2nd ed. Saunders, Philadelphia, Pennsylvania.

Ore, O. (1962). *Amer. Math. Soc. Colloquium Publ.* **38**, 1.

Ore, O. (1963). "Graphs and Their Uses." Random House, New York.

Paine, R. T. (1963). *Ecology* **44**, 63.

Paine, R. T. (1966). *Amer. Natur.* **100**, 65.

Parzen, E. (1960). "Modern Probability Theory and its Applications." Wiley, New York.

Patten, B. C. (1966). *Bio. Sci.* **16**, 593.

Patten, B. C., and Witkamp, M. (1967). *Ecology* **48**, 813.

Slobodkin, L. B. (1961). *Amer. Natur.* **95**, 147.

7

Niche Quantification and the Concept of Niche Pattern

HERMAN HENRY SHUGART, JR.* AND BERNARD C. PATTEN
DEPARTMENT OF ZOOLOGY AND INSTITUTE OF ECOLOGY
UNIVERSITY OF GEORGIA, ATHENS, GEORGIA

University of Georgia, *Contributions in Systems Ecology*, No. 10.

* Present Address: Ecological Sciences Division, Oak Ridge National Laboratory, Oak Ridge, Tennessee.

I. Introduction

Ecological niche theory developed primarily in animal ecology as a heuristic concept. Grinnell (1904, 1917) introduced the term in the context of bird habitat preferences and later (Grinnell, 1928) defined the niche more broadly as, "... the ultimate distributional unit within which each species is held by its structural and instinctive limitations." As with many ecological concepts, the niche has been variously defined. An animal's niche has been related to its trophic position (Elton, 1927), its competitive interactions with other species (Lotka, 1932; Gause, 1934; Hardin, 1960), and its feeding preferences (Weatherly, 1963).

Much recent work in niche theory (Hutchinson, 1965; Levins, 1968; Vandermeer, 1970) has used the competition model (Volterra, 1926; Lotka, 1932) as a starting point. The Lotka–Volterra model expresses a species' relationship to a given set of environmental conditions with a constant parameter (K, carrying capacity), and the relationships of the species with other similar species as a set of constant parameters (α's, competition coefficients). Although these parameters are probably not constant, typical studies involve developing and manipulating matrices of such constants.

The present study approaches the problem of the ecological niche from a different point of view, namely that of habitat selection. The primary goal will be to develop and utilize a quantifiable geometric model of a species' niche. This niche model will be used to explore the habitat selection and resource division of 20 ecologically similar bird species, the winter fringillids of northern Georgia. Although the model will be applied in the context of bird habitat selection, it is general and can be used for a variety of organisms and with a variety of environmental variables.

II. The Model

Hutchinson (1944, 1957, 1965) formalized the niche concept by considering the niche as a hypervolume in an n-dimensional Euclidean space with each axis of this hyperspace corresponding to some relevant environmental variable. Usually the niche space is constructed by assuming a Cartesian coordinate system of relevant variables (Hutchinson, 1965). This assumption is not necessary for the niche model presented in this paper.

In this and all subsequent sections, let:

A = any nonsingular $n \times n$ matrix;

A^{-1} = the inverse of A;

A^{T} = the transpose of A;

$E(argument)$ = the expectation operator;

var$(argument)$ = the variance operator;

Σ = the variance–covariance matrix for the n variables used in a niche study; a typical element of Σ, O_{ij}, is the covariance between variable i and variable j; O_{ii} is the variance of variable i;

$\mathbf{x}$ = a habitat difference vector; a typical element of $\mathbf{x}$, x_i, is obtained by subtracting the value of variable i for a given point in a Euclidean hyperspace from the value of variable i for some other point in the space;

$\mathbf{y}$ = a habitat difference vector obtained similarly to $\mathbf{x}$.

An infinite number of coordinate systems are possible within a given Euclidean space. Since many of the problems in niche theory are geometric problems such as measuring niche overlaps and determining niche volumes, it is imperative to measure distances in the space to quantify these abstractions. In a Euclidean two-space (plane) with orthogonal (right-angle) coordinate axes having no differences in scale, the distance between two points can be determined by the Pythagorean theorem (Fig. 1a).

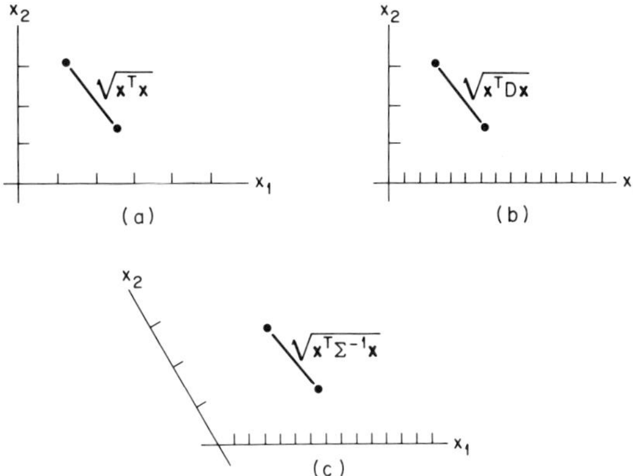

FIG. 1. Pythagorean theorem in various coordinate systems used in a Euclidean space. (a) Orthogonal, equal scale axes. (b) Orthogonal, unequal scale axes. (c) Non-orthogonal, unequal scale axes.

Let (a_1, b_1) be the coordinates of point P, and (a_2, b_2) be the coordinates of point Q. Then, the distance d from P to Q is

$$d = [(a_1 - a_2)^2 + (b_1 - b_2)]^{1/2}.$$

Generalizing to Euclidean n-space with equal-scale, orthogonal coordinates, let $a = a_1 - a_2$, $b = b_1 - b_2, \ldots, n = n_1 - n_2$, where $(a_1, b_1, \ldots, n_1)$ and $(a_2, b_2, \ldots, n_2)$ are coordinates for points P and Q, respectively; then $d = (a^2 + b^2 + \cdots + n^2)^{1/2}$ (Friedrichs, 1965). If the difference vector, $\mathbf{x}$, has elements $a, b, \ldots, n$, then $d^2 = \mathbf{x}^\mathsf{T}\mathbf{x}$ by matrix multiplication.

If the axes are orthogonal but have different scales (Fig. 1b), it is necessary to multiply the elements of $\mathbf{x}$ by appropriate scaling factors. This can be done by multiplying by an $n \times n$ diagonal scaling matrix D, such that $d^2 = \mathbf{x}^\mathsf{T}D\mathbf{x}$.

If the axes are of unequal scale and nonorthogonal (Fig. 1c), then $d^2 = \mathbf{x}^\mathsf{T}\Sigma^{-1}\mathbf{x}$, where Σ^{-1} is the inverse of the variance–covariance matrix associated with the variables used to form the coordinate axes. The use of this generalized distance measure developed in anthropometric statistics, and detailed proofs and discussion may be found in Mahalanobis (1949) and Mukherjee and Bandyopadhyay (1965). The distance is often referred to as a "generalized Mahalanobis distance" or "V" (Rao, 1952).

To demonstrate the theory and assumptions behind this distance measure, assume an n-dimensional Euclidean space with two points of interest in the space. Since any such space has an infinite number of possible coordinate systems, let $\mathbf{x}$ be a difference vector between the coordinates of the two points in a coordinate system with orthogonal, equal-scale axes. Let the variables forming the axes be standardized normal variables that are statistically independent. In the normal case, independence implies a correlation of zero between all variables. With this coordinate system $d^2 = \mathbf{x}^\mathsf{T}\mathbf{x}$ by the Pythagorean theorem. If there is another coordinate system with nonorthogonal (correlated to one degree or another) and unequal-scale axes, then d^2 can be obtained by relating this coordinate system to the orthogonal system. Let $\mathbf{y}$ be the difference vector in the new coordinate system, with $\mathbf{y} = A\mathbf{x}$, where A is any nonsingular (all rows (columns) linearly independent) matrix of rank n. A must be nonsingular or all the axes of the orthogonal coordinate system will not be represented in the new coordinate system. If E is the expectation operator and var is the variance operator, then $E(\mathbf{y}) = 0$ and $\text{var}(\mathbf{y}) = A\ \text{var}(\mathbf{x})\ A^\mathsf{T}$. But $\text{var}(\mathbf{x}) = I$, the identity matrix, by the standardized-normal assumption, so $\text{var}(\mathbf{y}) = AA^\mathsf{T}$, which is Σ, the variance–covariance matrix, by definition. Now $\mathbf{x} = A^{-1}\mathbf{y}$ and $\mathbf{x}^{\mathsf{T}} = \mathbf{y}^{\mathsf{T}}(A^{-1})^{\mathsf{T}} = \mathbf{y}^{\mathsf{T}}(A^{\mathsf{T}})^{-1}$, so that the distance $d^2 = \mathbf{x}^{\mathsf{T}}\mathbf{x}$ becomes $d^2 = \mathbf{y}^{\mathsf{T}}(A^{\mathsf{T}})^{-1}A^{-1}\mathbf{y}$ in the new coordinate system. By the commutativity of a matrix with its inverse, $d^2 = \mathbf{y}^{\mathsf{T}}(AA^{\mathsf{T}})^{-1}\mathbf{y}$ or $d^2 = \mathbf{y}^{\mathsf{T}}\Sigma^{-1}\mathbf{y}$, Σ the variance–covariance matrix. Due to the properties of the Σ matrix in the normal case, it is possible to reverse the above development to relate distance to an orthogonal axis given the variance–covariance matrix. This can be accomplished because A, such that $\Sigma = AA^{\mathsf{T}}$, exists if Σ is normally distributed. Thus, it is sufficient to assume that a set of variables has a multivariate normal distribution to be able to use the formula $d^2 = \mathbf{y}^{\mathsf{T}}\Sigma^{-1}\mathbf{y}$ to obtain distance in a generalized hyperspace. In practice Σ is estimated by S, the sample variance–covariance matrix. Calculating distances under the Σ^{-1} or S^{-1} metric allows one to account for the interrelations among environmental variables and circumvents the restrictive assumption of independence needed to construct an orthogonal coordinate system.

Having established a mathematical development of a niche hyperspace model, it is now logical to test the model on biological data. Levins (1968) points out that any usable model of the niche must allow:

(1) a measure of niche width,

(2) a measure of niche overlap between species,

(3) a method of using the niche model to explain species diversity, and

(4) a measure of niche dimension.

Each of these categories will be discussed below.

III. Methods and Materials

A. The Species

In winter, migratory and resident populations of the bird family Fringillidae* (American Ornithologists' Union, 1957) represent a diverse, highly mobile vertebrate fauna. In northern Georgia, winter fringillids

* There is some debate at this time concerning higher taxa classification of birds. The birds included in the study are referred to as fringillids in that the species studied included all species of family Fringillidae (American Ornithologists' Union, 1957) normally found in winter in northern Georgia. Using an alternative classification (Mayr and Greenway, 1956), the study includes species of family Emberizidae, subfamilies Richmondinae and Emberizinae, and family Fringillidae, subfamily Carduelinae, typical for the region in winter.

TABLE I

COMMON NAME, SCIENTIFIC NAME[a]

Species	Abundance	Habitat	Mnemonic
Cardinal *Richmondena cardinalis* (Linnaeus)	Common	Edges, forests, thickets	CARD
Evening grosbeak *Hesperiphona vespertina* (Cooper)	Irregular	Deciduous forests	GROS
Purple finch *Carpodacus purpureus* (Gmelin)	Irregular	Deciduous forests	PURP
Pine siskin *Spinus pinus* (Wilson)	Irregular	Edges, forests	PINE
American goldfinch *Spinus tristis* (Linnaeus)	Common	Open areas, edges, forests	GOLD
Rufous-sided towhee *Pipilo erythrophthalmus* (Linnaeus)	Common	Edges, thickets	TOWH
Savannah sparrow *Passerculus sandwichensis* (Gmelin)	Common	Fields	SAVA
Grasshopper sparrow *Ammodramus savannarum* (Gmelin)	Uncommon	Fields	GRAS
Henslow's sparrow *Passerherbulus henslowii* (Audubon)	Uncommon	Thicker fields	HENS

Vesper sparrow *Pooecetes gramineus* (Gmelin)	Uncommon	Open fields, edges	VESP
Bachman's sparrow *Aimophila aestivalis* (Lichtenstein)	Rare	Pine forest edges	PW-S
Slate-colored junco *Junco hyemalis* (Linnaeus)	Common	Edges	JUNC
Chipping sparrow *Spizella passerina* (Bechstein)	Common	Edges	CHIP
Field sparrow *Spizella pusilla* (Wilson)	Common	Brushy fields, edges	FELD
White-crowned sparrow *Zonotrichia leucophrys* (Forster)	Uncommon	Open fields	WC-S
White-throated sparrow *Zonotrichia albicollis* (Gmelin)	Common	Edges, forests	WT-S
Fox sparrow *Passerella iliaca* (Merrem)	Common	Edges, thickets	FOX
Lincoln's sparrow *Melospiza lincolnii* (Audubon)	Rare	Thick brushy edges	LINC
Swamp sparrow *Melospiza georgiana* (Latham)	Uncommon	Edges, wet places	SWAM
Song sparrow *Melospiza melodia* (Wilson)	Common	Fields, edges	SONG

[a] American Ornithologists' Union (1957).

occupy a wide variety of habitats and there is evidence for strong selective pressure toward diversification in winter populations (Fretwell, 1969). Unlike most breeding bird populations (Tramer, 1969) winter fringillids tend to be most diverse and dense in early (old-field) and intermediate (edgelike) successional stages. A list of the 20 species of fringillids included in the study and their general winter habitat preferences is presented in Table I.

B. General Sampling Methods

Observations of environmental variables associated with 1756 individual birds were collected during January and February, 1970. For each individual, 69 structural measures of vegetation were recorded. These 69 variables were reduced to 44 variables (Table II) chosen to

TABLE II

Description of the 44 Variables Used to Describe Bird Habitats

Mnemonic	Variable	Methods
CANP	Canopy closure	Index of canopy thickness obtained from point samples (James and Shugart, 1971)
UNDP	Thickness of the understory	Index of canopy thickness obtained from point samples. Modified from James and Shugart (1971)
GRCP	Thickness of the ground-level vegetation	Plotless point samples with ocular tube (Winkworth and Goodall, 1962)
CANE	Evergreenness of canopy vegetation	Expressed as a percentage of CANP based on point samples
UNDE	Evergreenness of understory vegetation	Expressed as a percentage of UNDP based on point samples
GRCE*	Evergreenness of ground-level vegetation	Expressed as a percentage of GRCP based on point samples
TALL	Height in meters of the tallest vegetation within 10 m of a point	Obtained by simple triangulation (James and Shugart, 1971)
MTHT	Canopy height in meters	Height of vegetation directly over sample point obtained by triangulation (James and Shugart, 1971)
ZBIT	Lack of tree-size vegetation in 90° sectors around a point	Bitterlich reading (BAF = 10) taken in four equal (90°) sectors. ZBIT is the number of zero Bitterlich readings in each of the four sectors (Grosenbaugh, 1952)

Table continued

TABLE II (*continued*)

Mnemonic	Variable	Methods
BASA	Basal area of trees around a point (estimated square feet of cross section of trees at breast height per acre)	Bitterlich readings (BAF = 10) taken in four equal sectors, summed and multiplied by ten (Grosenbaugh, 1952)
HBAR	Evenness of distribution of trees around a point	Bitterlich (BAF = 10) taken in four equal sectors, denote each sector, s_i , $i = 1,\ldots, 4$. $p_i = s_i/\Sigma s_i$; HBAR $= -\Sigma p_i \log_e(p_i)$ (Grosenbaugh, 1952)
ADOT	Average diameter of trees in centimeters	Quarter method (Cottam and Curtis, 1956) on vegetation $\geqslant 3$ inches DBH (diameter at breast height)
ADIT	Average distance of trees from point in meters	Quarter method (Cottam and Curtis, 1956) on vegetation $\geqslant 3$ inches DBH
BDOT	Maximum of four tree diameters used to obtain ADOT	
BDIT	Maximum of four distances used to obtain ADIT	
LDOT*	Minimum of four tree diameters used to obtain ADOT	
LDIT	Minimum of four distances used to obtain ADIT	
ADOS*	Average diameter of shrub-level ($\leqslant 3$ inches DBH, $\geqslant 1$ m in height) vegetation in millimeters $\times 2.0$	Quarter method (Cottam and Curtis, 1956) excluding vegetation $\geqslant 3$ inches DBH and $\leqslant 1$ m in height
ADIS*	Average distance of shrub-level vegetation in meters	Quarter method (Cottam and Curtis, 1956) excluding vegetation $\geqslant 3$ inches DBH and $\leqslant 1$ m in height
BDOS*	Maximum of four diameters used to obtain ADOS	
BDIS*	Maximum of four distances used to obtain ADIS	
LDOS*	Minimum of four diameters used to obtain ADOS	
LDIS*	Minimum of four distances used to obtain ADIS	
TREE*	Estimated number of tree species within 0.10 acre circle	0.10 acre circle sampling method (James and Shugart, 1971)
SHRB*	Estimated number of species of understory ($\leqslant 3$ inches DBH, $\geqslant 1$ m in height) plants within 4 m of point	Modified from 0.10 acre sampling method (James and Shugart, 1971)

Table continued

TABLE II (*continued*)

Mnemonic	Variable	Methods
GRCO*	Estimated number of species of ground-level ($\leqslant 1$ m in height) vegetation within 1 m of point	1 m radius measured
HITS*	Thickness of understory vegetation at breast height	Average of two 0.01 acre arm-length rectangles centered on the sample point (Penfound and Rice, 1957)
BHIT	Maximum of two values used to obtain HITS	
LHIT	Minimum of two values used to obtain HITS	
ST-1	Average number of contacts of the vegetation touching a 1 m long rod sliding horizontally through a bar perpendicular to the ground at heights of 0.25, 0.50, 0.75, and 1.0 m	Wilson (1959)
ST-2	As ST-1 with readings at 1.25, 1.50, 1.75, and 2.0 m	Wilson (1959)
ST-3*	As ST-1 with readings at 2.25, 2.50, 2.75, and 3.0 m	Wilson (1959)
AR-1	Average area of 1 m rod covered by vegetation contacts as in ST-1	
AR-2	Average area of 1 m rod covered by vegetation contacts as in ST-2	
AR-3	Average area of 1 m rod covered by vegetation contacts as in ST-3	
NO-1	Average number of contacts from ST-1 which were identifiable seed heads at point of contact	
NO-2	As NO-1, using contacts from ST-2	
NO-3*	As NO-1, using contacts from ST-3	
BW-1*	AR-1 divided by ST-1	
BW-2	AR-2 divided by ST-2	
BW-3*	AR-3 divided by ST-3	
SE-1*	ST-1 divided by NO-1	
SE-2	ST-2 divided by NO-2	
SE-3	ST-3 divided by NO-3	

* Denotes variables (singular variables) which were later found to be predictable from linear combination of other variables.

quantify the structure of the environment associated with a given individual bird. Table II contains a mnemonic for each of these 44 variables, what each variable measures, and methodological descriptions. Individual records were obtained by cruising a spectrum of typical north Georgia habitats, and sampling individual fringillids by the wandering quarter method (Cottam and Curtis, 1956). Care was taken to sample in all representative habitat types for the region.

Since winter fringillids tend to flock, individual data vectors taken from birds in flocks were retained as a single record in the data analysis. This served as a statistical constraint against entering a flock bias into the data set used to construct S^{-1}, and also allowed testing for differences in habitats in which a given species flocked, did not flock, or was absent.

Eighteen large-area plots in uniform habitats, varying in size from 5 to 30 acres, were sampled to obtain actual densities of the 20 species in different habitats. In each of these large-area plots, bird species densities were recorded on several dates during the study period.

IV. Results and Discussion

A. Individual-Level Responses of Fringillids to Habitat Variables

It is axiomatic that birds are not randomly distributed. In fact, most bird watchers can predict which bird species will occur in a given habitat at a particular season, and species habitat preferences are often used as criteria in field identification (Peterson, 1947). Birds appear to select habitats by factors not immediately connected with their survival (Svardson, 1949), so that their habitat selection is based on proximate rather than ultimate factors (Lack, 1949). Ultimate factors are directly connected with an individual's survival (e.g., food, protection from predators); proximate factors are indirectly connected to an individual's survival, but are dependable and easily perceived predictors of ultimate factors (e.g., a given vegetative structure indicating a high likelihood of finding food or shelter). One might expect highly mobile animals such as birds to use proximate factors to sort favorable from unfavorable habitats, and it is generally agreed that bird habitat selection is based largely on gross visual aspects of the vegetational physiognomy (Hilden, 1965).

To test whether any of the 20 bird species "selected" for any of the 44 habitat variables, the data were analyzed by a one-way analysis of variance (Steel and Torrie, 1960) after being grouped in three levels:

(1) study plots in which a given species did not occur,
(2) study plots in which the species occurred in low numbers (1–5 individuals),
(3) study plots in which the species occurred in high numbers ($\geqslant 6$ individuals).

A study plot is a microhabitat description. Study plots were formed by collecting all individual data sets which were identical (same date, time, and same value for each of the 44 variables), and noting the number of individuals of each species associated with the data set.

These results are found in Table III. Due to the size (20×44) of the matrix (Table III), only the significance of the analysis is indicated. A fully expanded table of F values and associated degrees of freedom is found in the Appendix.

Some fringillids select various habitat variables to a significant degree while others do not, but no obvious general patterns are discernible in Table III. Some abundant birds (Cardinal, Towhee, Savannah Sparrow) select many of the habitat variables; other equally abundant species (Goldfinch, Chipping Sparrow, Field Sparrow) select virtually none. Since Table III represents an initial univariate pass over a multivariate data set, the results should be viewed species by species. Thus, it is appropriate to look at the individual species responses, rather than at responses of the entire group.

B. Rufous-Sided Towhee, Cardinal, Savannah Sparrow, and Purple Finch—Four Different Strategies of Selecting Habitats

With univariate analysis (Table III) one cannot determine the effects of correlations between the different variables. For example, a species might select its habitats solely on the basis of canopy closure, but show significant selection for other variables (e.g., tree height) which are highly related to canopy closure. Discriminant function analysis (Rao, 1952) can be used to account for these correlations and to provide an ordering of the variables according to their importance.

The discriminant vector, $\mathbf{a}^{\mathsf{T}}$, is a vector of arbitrary weights such that the F statistic for testing the univariate hypothesis,

$$H_0: \qquad u_{1z} = u_{2z} = u_{3z} = \cdots = u_{nz}\,,$$

where $u_{iz} = \mathbf{a}^{\mathsf{T}}\mathbf{u}_i$; $\mathbf{u}_i$, being a mean habitat vector for the ith group, is a maximum. In the present case, there are three groups—plots in which a given species was present, absent, or abundant. Detailed mathematical discussion of the discriminant function and its derivation may be found elsewhere (Rao, 1952). The discriminant vector $\mathbf{a}^T$ is determined using matrix differentiation and then maximizing the above-mentioned F statistic. By multiplying the discriminant vector and a given habitat vector, one obtains a single-valued function, the discriminant function (Rao, 1952). The discriminant function has been used in studies with breeding birds to reduce multivariate habitat data to a single function (Cody, 1968; Hespenheide, 1971), but its best use is in determining differences between groups. The absolute value of the correlation between a given variable and the discriminant function, which supplies the maximum separation among the groups, indicates how well the variable separates the groups. Selection of habitat variables for this analysis refers not to behavioral selection (Klopfer and MacArthur 1961; Klopfer, 1967), but to the power of these variables to classify habitats in which the species occurred. Most of the variables used in this study measure the structure of the vegetation and quantify proximate environmental factors. But a few of the variables measure food availability (NO-1, NO-2, NO-3, SE-1, SE-2, SE-3), and can probably be considered measures of ultimate factors.

Necessary information for a discriminant function analysis of this sort are: univariate F values, indicating the degree of selection for a variable (Appendix); correlations between each variable and the discriminant function, indicating the power of variables to separate habitats in which a species is absent, present, or abundant; and two mean vectors. The mean vectors, one for habitats in which the species were present or abundant, and the other for habitats in which the species were absent, are needed for interpretation of the correlation with the discriminant function.

1. *Cardinal*

Although the Cardinal is an extremely common and seemingly ubiquitous bird in the region, it seems to be rigid in its habitat selection, with significance in 21 of the 44 variables used in this study (Table III). Most of the 44 variables are concerned with trees and understory vegetation, so it is important to determine which of these conceptually similar variables are most strongly selected. Thus, a discriminant function analysis is appropriate (Table IV). To assure that no variable of possible significance was omitted, any variable with an F value greater than 1.0 ($\alpha = 0.50$) was included. The best single variable (highest correlation with discriminant function) is UNDP, which measures general thickness of the understory (Table II). Other strong variables are TREE

TABLE III

HABITAT SELECTION MATRIX INDICATING THE RESULTS OF 880 UNIVARIATE ANALYSES OF VARIANCE TO DETERMINE SIGNIFICANT SELECTION OF HABITAT VARIABLES FROM RANDOM FOR EACH OF THE 20 SPECIES FOR EACH OF THE 44 VARIABLES[a]

Variable	Species																			
	CARD	GROS	PURP	PINE	GOLD	TOWH	SAVA	GRAS	HENS	VESP	PW-S	JUNC	CHIP	FELD	WC-S	WT-S	FOX	LINC	SWAM	SONG
CANP		**				**	***	*	*		*				**	**				**
UNDP	***					***	***	**	*	*						**				*
GRCP		*	***						***											*
CANE							*				***	***	***							
UNDE											***	***	***							
GRCE							*		***											
TALL	**	*	**			***	***	***	**	*		***	*		**	***				**
MTHT	**	***	***		*		***	*								***				*
ZBIT	**	***	***				***	**	**						*	***				**
BASA	**	***	***				***	*								***				*
HBAR						***	***	*				**	**	**						
ADOT	*	*	*				***	*	**			***				***				*
ADIT	**					*	***	***	***	*		**			**	**				
BDOT	*	***	**				***	*	*			**				**				**
BDIT	**					*	***	***	**			**			**	**				
LDOT							*					***				***				
LDIT	*					*	***	***	***	*		**			**	**				
ADOS						**	*								**					

ADIS	**					*	***	***		***		*								*
BDOS						***													***	
BDIS	**					*	***	***		**		*			*					*
LDOS						***													***	
LDIS	**					*	***	***	*	***										*
TREE	***	***	***				***	*								***				*
SHRB	**			**		*	***	**								**		**	*	
GRCO			*						***					**	**					***
HITS	**			*		**	**	***										***	**	
BHIT	**					**	**	***										***	**	
LHIT	**			*		***	*											***	**	
ST-1									***			**			**			***	*	
ST-2															**		*			
ST-3	*		***																	
AR-1									***									***		
AR-2															**		**	*		
AR-3	*		**			*														
NO-1							*		***			*						*	*	*
NO-2															*		**			*
NO-3			***			*														
BW-1									***											
BW-2						*	**	*									***	*	**	***
BW-3	**				**	***					***									
SE-1							***		*											
SE-2						**									***		***		**	***
SE-3				***		***													***	

[a] Asterisks give level of significance: *, $\alpha = 0.05$; **, $\alpha = 0.01$; ***, $\alpha = 0.001$.

TABLE IV

UNIVARIATE *F*, CORRELATION WITH DISCRIMINANT FUNCTION AND MEAN VECTORS FOR 29 VARIABLES SELECTED WITH *F* VALUES GREATER THAN 1.0 BY THE CARDINAL

Variable	Univariate *F*	Correlation with discriminant function	Mean values of plots in which species occurred[a]	Mean values of random sample[a]
CANP	2.67	0.298	27.98	19.98
UNDP	8.44	0.575	51.84	32.46
GRCP	2.56	−0.009	67.50	77.78
TALL	6.15	0.482	15.84	11.43
MTHT	5.38	0.244	9.37	6.32
ZBIT	4.73	−0.340	1.81	2.51
BASA	4.53	0.299	57.63	35.02
HBAR	2.69	−0.348	0.94	1.09
ADOT	3.14	0.347	0.31	0.24
ADIT	5.52	−0.424	140.90	361.70
BDOT	3.91	0.324	0.47	0.34
BDIT	5.35	−0.401	171.80	401.50
LDIT	4.35	−0.390	109.70	308.90
ADIS	6.37	−0.478	4.16	24.10
BDIS	6.29	−0.481	7.23	27.90
LDIS	5.45	0.436	1.66	20.21
TREE	7.32	0.517	3.47	2.03
SHRB	6.01	0.513	5.53	4.03
GRCO	1.54	−0.177	6.76	7.83
HITS	4.61	0.443	56.17	34.05
BHIT	4.21	0.412	69.39	43.42
LHIT	4.37	0.443	42.95	24.67
ST-2	1.75	0.121	1.47	0.86
ST-3	3.49	0.281	0.29	0.09
AR-3	4.42	0.427	0.31	0.11
NO-3	2.25	0.346	0.04	0.01
BW-3	4.67	0.465	0.32	0.16
SE-1	2.18	−0.146	0.14	0.20
SE-2	2.09	0.163	0.09	0.09

[a] The last two columns form the mean habitat vectors for Cardinal, and for a random sample, respectively.

(estimating tree species present and indicating preference for diverse forests) and SHRB (estimating shrub species present). Most of the correlations are weak, with absolute values of 0.575 or less, indicating that the Cardinal selects its habitats using a variety of factors, most of which are related to the understory and canopy vegetation. Mean values for the Cardinal (Table IV) indicate a general preference for edges, and edgelike habitats. Generally similar results have been obtained for Cardinals in the breeding season (Dow, 1969a, b). The Cardinal does not appear to select habitats according to food availability and fits the expected pattern of selecting habitats on the basis of proximate rather than ultimate factors (Lack, 1933; Svardson, 1949).

2. *Rufous-Sided Towhee*

The Towhee, like the Cardinal, is a common bird in northern Georgia. The two species seem to prefer somewhat similar habitats and are often found together. Upon analysis (Table V), the Towhee also selects for various factors relating to the understory. The strongest such factor is LDOS (size of the smallest nearby shrub) reflecting the tendency of Towhees to be found near cover. Unlike the Cardinal, the Towhee selects for food availability, particularly SE-3. BDOS (size of largest nearby shrub) is also a strong variable. Inspection of the mean vectors (Table V) indicates that Towhee microhabitats are best typified by large shrubs (LDOS, BDOS) and availability of seeds in the upper layers of the understory (SE-3).

There are two very different habitat selection strategies represented by the Cardinal and the Towhee. The Cardinal apparently selects its habitats on the basis of understory and canopy factors (proximate factors); the Towhee selects on the availability of cover and food (ultimate factors). The two species frequently occur in the same habitats, but apparently for different reasons.

The importance of proximate factors (Lack, 1933; Svardson, 1949) in bird habitat selection is not clear. Of the 20 species included in this study, 11 species select for food availability to one degree or another (Table III). The Towhee represents a rather extreme example of ultimate factor selection.

3. *Savannah Sparrow*

The Savannah Sparrow, like the Cardinal and to some extent the Towhee, selects its habitats primarily according to canopy and understory factors (Table VI). This is initially misleading for the Savannah Sparrow is an open-country bird. The Savannah Sparrow significantly selects on canopy and understory because its habitats are typically void of trees and thick undergrowth as indicated by the mean vectors (Table VI). Factors that relate to food (SE-1, SE-2) are not of great consequence (Table VI), indicating that this species selects its habitats on the basis of proximate factors. The Savannah Sparrow differs from the Cardinal in that its habitats are typified by absence rather than presence of these proximate factors.

TABLE V

UNIVARIATE F, CORRELATION WITH DISCRIMINANT FUNCTION AND MEAN VECTORS FOR 29 VARIABLES SELECTED WITH F VALUES GREATER THAN 1.0 BY THE RUFOUS-SIDED TOWHEE

Variable	Univariate F	Correlation with discriminant function	Mean values of plots in which species occurred[a]	Mean values of random sample[a]
CANP	4.51	0.129	34.13	19.98
UNDP	7.42	0.253	58.70	32.46
TALL	7.22	0.234	17.70	11.43
MTHT	1.61	−0.021	7.83	6.32
ZBIT	1.86	−0.072	1.96	2.51
HBAR	15.00	−0.349	0.64	1.09
ADIT	4.28	−0.138	118.30	361.70
BDIT	3.01	−0.111	188.60	401.50
LDIT	4.70	−0.148	48.28	308.90
ADOS	6.38	0.321	2.86	2.21
ADIS	4.08	−0.181	3.15	24.10
BDOS	16.82	0.534	3.73	3.16
BDIS	3.24	−0.111	5.58	27.90
LDOS	21.18	0.595	1.87	1.51
LDIS	3.24	−0.159	1.32	20.21
TREE	1.60	0.149	2.96	2.03
SHRB	3.49	0.223	5.70	4.03
HITS	7.40	0.276	73.93	34.05
BHIT	5.48	0.244	85.35	43.42
LHIT	8.95	0.294	62.52	24.67
ST-3	1.70	0.163	0.13	0.09
AR-1	1.41	0.007	4.09	2.75
AR-3	3.07	0.220	0.21	0.11
NO-3	3.02	0.213	0.03	0.01
BW-1	1.32	−0.162	1.10	0.84
BW-2	3.10	0.136	0.84	0.05
BW-3	7.32	0.345	0.35	0.16
SE-2	4.89	0.195	0.24	0.09
SE-3	18.17	0.550	0.03	0.01

[a] The last two columns form the mean habitat vectors for the Rufous-Sided Towhee and a random sample, respectively.

4. *Purple Finch*

Whereas the previously discussed species selected for a large number of variables (Tables IV, V, and VI) with rather low correlations of any single variable with the discriminant function, the habitat selection of the Purple Finch is focused on a single variable, BASA (r = −0.809,

TABLE VI

UNIVARIATE F, CORRELATION WITH DISCRIMINANT FUNCTION AND MEAN VECTORS FOR 29 VARIABLES SELECTED WITH F VALUES GREATER THAN 1.0 BY THE SAVANNAH SPARROW

Variable	Univariate F	Correlation with discriminant function	Mean values of plots in which species occurred[a]	Mean values of random sample[a]
CANP	13.51	0.425	0.00	19.98
UNDP	15.61	0.427	2.32	32.46
CANE	3.74	−0.223	0.00	15.45
UNDE	2.33	0.176	0.00	6.73
GRCE	4.31	−0.165	1.20	1.48
TALL	25.08	0.578	1.18	11.43
MTHT	9.61	0.358	0.61	6.32
ZBIT	14.66	−0.442	4.00	2.51
BASA	7.03	−0.298	0.00	35.02
HBAR	6.67	0.280	1.39	1.09
ADOT	10.44	0.374	0.11	0.24
ADIT	27.67	−0.608	928.20	361.70
BDOT	12.18	0.404	0.11	0.34
BDIT	20.08	−0.518	928.40	401.50
LDOT	3.67	0.221	0.11	0.17
LDIT	34.07	−0.675	928.00	308.90
ADOS	5.44	0.266	1.38	2.21
ADIS	44.51	−0.768	80.88	24.10
BDIS	32.93	−0.658	81.89	27.90
LDIS	46.69	−0.789	71.19	20.21
TREE	10.38	0.372	0.00	2.03
SHRB	19.34	0.506	0.57	4.03
HITS	5.34	0.266	2.23	34.05
BHIT	5.91	0.280	3.54	43.42
LHIT	3.67	0.221	0.93	24.67
NO-1	3.94	−0.157	0.89	0.64
BW-2	5.46	0.270	0.07	0.05
SE-1	11.56	−0.223	0.31	0.20
SE-2	2.29	0.175	0.00	0.09

[a] The last two columns form the mean habitat vectors for the Savannah Sparrow and a random sample, respectively.

Table VII). BASA is a general indicator of forest. Since the next strongest variables are TREE and MTHT, the Purple Finch apparently selects for forest, particularly more diverse (hardwood or mixed-hardwood) forest as indicated by the mean vectors (Table VII). There is also a rather weak selection of food availability in the upper portion of the understory (NO-3 and SE-3, Table VII).

TABLE VII

UNIVARIATE *F*, CORRELATION WITH DISCRIMINANT FUNCTION AND MEAN VECTORS FOR 18 VARIABLES SELECTED WITH *F* VALUES GREATER THAN 1.0 BY THE PURPLE FINCH

Variable	Univariate *F*	Correlation with discriminant function	Mean values of plots in which species occurred[a]	Mean values of random sample[a]
CANP	1.27	−0.178	36.67	19.98
GRCP	8.64	0.512	17.50	77.78
TALL	4.87	−0.382	22.17	11.43
MTHT	9.72	−0.531	19.17	6.32
ZBIT	7.65	0.482	0.00	2.51
BASA	21.85	−0.809	176.70	35.02
ADOT	3.13	−0.281	0.44	0.24
ADIT	2.24	0.259	6.23	361.70
BDOT	5.89	−0.422	0.77	0.34
BDIT	2.42	0.269	9.17	401.50
LDIT	1.62	0.220	2.58	308.90
TREE	9.81	−0.545	7.00	2.03
SHRB	2.69	−0.262	6.33	4.03
GRCO	4.22	0.341	3.17	7.83
ST-3	9.42	−0.473	0.95	0.09
AR-3	6.08	−0.399	0.79	0.11
NO-3	6.97	−0.370	0.13	0.01
SE-3	1.35	0.198	0.03	0.01

[a] The last two columns form the mean habitat vectors for the Purple Finch and a random sample, respectively.

In the four species discussed above, there appear to be four different strategies of habitat evaluation. The four examples represent extreme cases but in general the strategies of most of the 20 species (Table III) can be broken down into categories of:

(1) proximate or ultimate selection,
(2) single or multiple factor selection,
(3) presence or absence selection.

Additionally, there are species such as the Field Sparrow and the American Goldfinch (FELD and GOLD, Table III) which appear to select none of the variables to any great extent. The lack of significance of selection in these two species is not due to small sample size (the largest sample is that of the Field Sparrow), but reflects great variability in the habitat preference of these two species. Thus, an additional strategy relating to generality must be included:

(4) No selection.

More will be said about this category below.

C. Species-Pair Interactions on Various Habitat Variables

Since some fringillids select various habitat variables as indicated by one-way analysis of variance (Table III), it is logical to ask if the species affect one anothers' habitat selections. With 44 variables and 20 species, there are 8360 (190 species-pairs times 44 variables) 3 × 3 factorial-design analyses of variance possible. Clearly, an interaction analysis is feasible only for selected variables and selected species-pairs. Since the data were collected with a design directed toward mapping distances in a Euclidean hyperspace, the data set lacks the resolution necessary to make definitive statements concerning the causes of any significant interactions. It is the purpose of this section to document the existence of these interactions, and suggest their possible causes.

Of the 8360 possible paired interactions in this study, about 8000 cannot be tested due to the fact that the species-pairs never occurred together on the same plot (for definition of plot see page 294). This is due to the low likelihood that two individual birds, even if they are very close to one another, will have identical habitat vectors, and is an artifact of the definition of a plot rather than a testimonial for efficiency of resource division or competitive exclusion.

Morse (1970) studied several aspects of mixed-species flocks and found that certain pairs of flocking species altered their foraging behavior in one anothers' presence. Such behavioral changes constitute "ecological shifts" as defined by MacArthur and Wilson (1967). Examining the interactions of the Cardinal and Rufous-sided Towhee (Table VIII for typical ANOVA table), ecological shifts in habitat parameters are found for variables TALL and BW-3. These interactions indicate a tendency for Cardinals and Towhees to be found together in habitats in which neither is particularly likely to be found separately. Significant interactions were also found for Chipping Sparrows and Juncos on variable CANE ($F = 13.72$, $\alpha = 0.001$), and variable UNDE ($F = 26.52$, $\alpha = 0.001$); Swamp and Song Sparrows on variable SE-3 ($F = 5.52$, $\alpha = 0.01$); Fox and Song Sparrows on variable NO-2 ($F = 17.08$, $\alpha = 0.001$).

Upon inspection, all of these interactions demonstrated a tendency for

TABLE VIII

RESULTS OF 3 × 3 FACTORIAL DESIGN ANALYSIS OF VARIANCE TESTING FOR INTERACTIONS BETWEEN CARDINAL AND RUFOUS-SIDED TOWHEE ON 14 MUTUALLY SELECTED VARIABLES

Variable	F Value for CARD × TOWH interaction with three degrees of freedom
UNDP	2.30
TALL	10.33[a]
ADIT	0.63
BDIT	1.67
LDIT	0.11
ADIS	0.44
BDIS	0.32
LDIS	0.33
SHRB	1.97
HITS	1.28
BHIT	1.19
LHIT	1.18
AR-3	0.66
BW-3	2.76[b]

[a] Significant at the $\alpha = 0.001$ level.
[b] Significant at the $\alpha = 0.05$ level.

the species to be found together under what would normally be considered "marginal" conditions for both of the species. This suggests a tendency for these species to flock under marginal habitat conditions, behavior that has been noted by Morse (1970), who discusses its possible adaptive significance in mixed-species flocks. Such interactions could also be caused by pairs of species overlapping in their habitat preferences only in marginal areas.

It is important to note that in none of the cases were the interactions of a negative sort and that the species tested for interaction were species which were thought to be likely competitors. The indication is that if there is a behavioral analog to competitive exclusion in the microhabitats of these birds, it is much more subtle than its converse, the tendency for ecologically similar species to occur together in higher numbers than one would expect from habitat conditions alone.

D. ELIMINATION OF SINGULARITIES FROM THE S^{-1} METRIC

Before the S^{-1} matrix can be used to calculate hyperspace distances, any singularities in the matrix must be eliminated. For these purposes, a singular matrix contains a row (or column) which is linearly dependent or nearly so. Linear dependence is defined in a less restrictive manner than in linear algebra, because it is important to eliminate any variables which are so highly intercorrelated with other variables that the residual variance is extremely small. These variables would not be strictly linearly dependent unless this residual is zero, which is highly unlikely.

Singular variables were eliminated by checking for near-zero leading elements in a forward Doolittle computer iteration written by J. E. Dunn (University of Arkansas Computer Center). Of the 44 original variables, 17 were singular (Table II). With these variables eliminated, it was then possible to use the S^{-1} matrix as a niche metric.

V. Niche Theory

Hutchinson (1965) recognized two aspects of a species' niche, its *fundamental niche* and its *realized niche*. The fundamental niche is the hypervolume of the hyperspace in which the species can potentially survive, and the realized niche is the hypervolume actually occupied by the species in a given environment. The fundamental niche is a measure of the species' potential to live under various conditions; the realized niche is dependent upon environmental conditions in a given geographical region.

A. A Realized Niche Measure

A niche measurement which captures the main elements of Hutchinson's (1965) realized niche concept can be obtained straightforwardly in the Euclidean space under the S^{-1} metric. The origin of the coordinate system in this space has as its origin the vector of the regional means for each variable. Since this hyperspace is theoretically grained with a normal probability density function, points (habitat coordinate vectors) at a distance from the origin of the coordinate system (mean vector for the region) are less frequent in the region than points that are close to the origin. Figure 2 diagrams the effect of the realized niche component in a hypothetical one-space for species with equal niche sizes (i.e., $a_1 - a_2 = b_1 - b_2$). Thus, one can obtain a measure of the likelihood of a given species encountering favored habitats in a given region as the inverse distance of the species' average habitat vector (niche center) from the origin. Table IX contains these distances for the 20 species treated in this study.

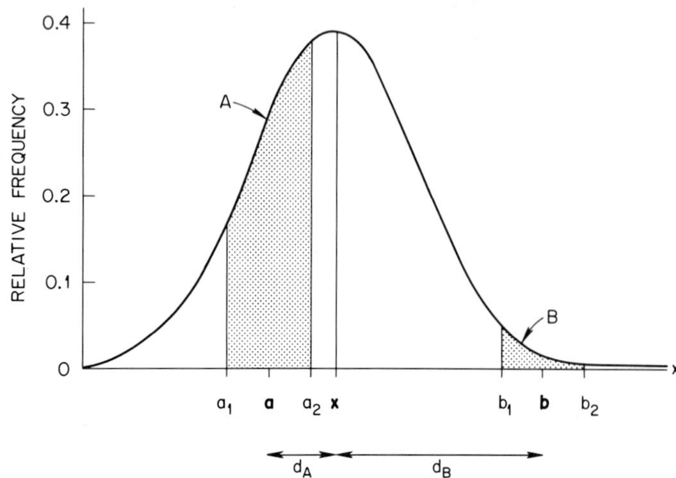

FIG. 2. Effect of realized niche component on microhabitats available to two species in a one-space. x is an environmental variable, $\bar{x}$ is the mean of the variable in a given region and is the origin of the one-space. $a_1 - a_2 = b_1 - b_2$. The realized niche measure for species A (d_a) is less than that of species B (d_b) indicating that species A has more available habitat in the region.

B. RARITY AND THE REALIZED NICHE MEASURE

The two rarest fringillids in northern Georgia, the Lincoln's Sparrow (LINC) and the Bachman's Sparrow (PW-S) show extreme reactions in distance from the origin (Table IX). The Lincoln's Sparrow apparently is rare primarily due to a lack of suitable habitats ($d = 0.305$) in the region. This fits the observations of C.W. Helms and S.A. Gauthreaux (personal communication) that the Lincoln's Sparrow is more abundant during fall migration than during the winter. Although the evidence is anecdotal, the migratory habitats of the species (old fields, backyard bird feeders, and a variety of other habitats) are marginal when compared to the winter habitat. This would indicate a considerably larger number of Lincoln's Sparrows move through the area than overwinter in it, and strengthens the premise that there is a shortage of prime habitat for the species in the region.

Mean Bachman's Sparrow habitat is quite close to the mean available habitat in the Clarke County area and the rareness of the Bachman's Sparrow cannot be attributed to any lack of available winter habitat ($d = 0.066$). This does not preclude the possibility that the species may be limited by available habitat during other periods or regions of its annual cycle.

TABLE IX

DISTANCE FROM THE ORIGIN (*d*) OF THE NICHE CENTERS OF EACH OF THE 20 FRINGILLIDS. *d* CAN BE CONSIDERED A MEASURE OF THE REALIZED NICHE OF EACH SPECIES

Species	Distance from origin
CARD	0.165
GROS	0.148
PURP	0.523
PINE	0.737
GOLD	0.101
TOWH	0.150
SAVA	0.107
GRAS	0.160
HENS	0.153
VESP	0.109
JUNC	0.106
PW-S	0.066
CHIP	0.067
FELD	0.094
WC-S	0.118
WT-S	0.109
FOX	0.140
LINC	0.305
SWAM	0.152
SONG	0.106

There is no clear-cut relation between the realized niche measure (*d*) and species rarity. Species with large distances (Table IX) such as the Purple Finch (PURP), Pine Siskin (PINE) and Lincoln's Sparrow (LINC) are typically less abundant than related or ecologically similar species. From casual field observations, these three species appear to be limited by suitable habitat. However, there are rare species which seem to have no shortage of suitable habitat (Bachman's Sparrow, PW-S; Henslow's Sparrow, HENS) and which are probably limited by other factors not included in this 27-space, perhaps in different seasons or different geographic locations.

C. FUNDAMENTAL NICHE MEASURES

There are two criteria which any quantification of Hutchinson's (1965) fundamental niche must meet:

(1) The quantification must provide some method of testing for niche overlaps among species' niches.

(2) The quantification must express the size of a species' niche hypervolume in the simplest possible manner with minimum loss of information.

The first criterion will be called the *resource division problem*; the second, the *niche width problem*.

1. *Resource Division: The Distances from Niche Centers*

Niche overlap implies that there is some degree of similarity in the various species' requirements; lack of overlap is indicative of efficient resource division (Cody, 1968). To evaluate resource division one needs a measure of the distance between any two species' niches, and a measure of the dispersion around the niche center for each species. The distance between two niche centers (mean habitat vector for each species) can be calculated under the S^{-1} metric using the formula $d = \mathbf{x}^{\mathrm{T}} S^{-1} \mathbf{x}$, where $\mathbf{x}$ is the difference vector obtained by subtracting the mean habitat vectors of two species. Since a resource division problem is concerned with

TABLE

MAPPING MATRIX OF DISTANCES BETWEEN EACH OF ALL

	CARD	GROS	PURP	PINE	GOLD	TOWH	SAVA	GRAS	HENS
CARD	—								
GROS	0.116[a]	—							
PURP	0.350[a]	0.352[a]	—						
PINE	0.731[a]	0.827[a]	1.178[a]	—					
GOLD	0.059[a]	0.041	0.352[a]	0.735[a]	—				
TOWH	0.131[a]	0.128[a]	0.498[a]	0.410[a]	0.079[a]	—			
SAVA	0.055[a]	0.061[a]	0.378[a]	0.743[a]	0.021	0.092[a]	—		
GRAS	0.103$_3$	0.134[a]	0.364[a]	0.758[a]	0.065[a]	0.139[a]	0.047[a]	—	
HENS	0.100[a]	0.076[a]	0.442[a]	0.788[a]	0.056[a]	0.116[a]	0.042	0.119[a]	—
VESP	0.084[a]	0.075	0.422[a]	0.753[a]	0.030	0.096[a]	0.007	0.044	0.073
JUNC	0.061[a]	0.052[a]	0.392[a]	0.768[a]	0.021	0.082[a]	0.026[a]	0.073[a]	0.067[a]
PW-S	0.193	0.147	0.537[a]	0.744[a]	0.107	0.146	0.132	0.183	0.165
CHIP	0.104[a]	0.084[a]	0.444[a]	0.745[a]	0.049[a]	0.084[a]	0.051[a]	0.103[a]	0.088[a]
FELD	0.054[a]	0.056[a]	0.373[a]	0.733[a]	0.013	0.071[a]	0.016[a]	0.060[a]	0.050[a]
WC-S	0.072[a]	0.079[a]	0.370[a]	0.656[a]	0.031	0.071[a]	0.027	0.067[a]	0.060
WT-S	0.060[a]	0.046[a]	0.399[a]	0.734[a]	0.015	0.068[a]	0.029[a]	0.082[a]	0.063[a]
FOX	0.070[a]	0.071[a]	0.395[a]	0.829[a]	0.034	0.117[a]	0.042[a]	0.107[a]	0.083[a]
LINC	0.311[a]	0.283[a]	0.551[a]	0.932[a]	0.252[a]	0.316[a]	0.257[a]	0.308[a]	0.283[a]
SWAM	0.094[a]	0.113[a]	0.432[a]	0.495[a]	0.071[a]	0.040[a]	0.078[a]	0.147[a]	0.093[a]
SONG	0.061[a]	0.062[a]	0.386[a]	0.753[a]	0.015	0.080[a]	0.019[a]	0.068[a]	0.057[a]

[a] Indicates $\alpha = 0.01$ level.

testing for overlap, it is important to use a dispersion measure which allows statistical evaluation. The Hotelling T^2 statistic (Morrison, 1967) provides such an evaluation. The dispersion of each species' niche about its niche center is measured and accounted. Thus, two species with narrow niches (slight dispersion around their respective niche centers) might not overlap even though their niche centers are quite close in the hyperspace, or conversely two species with wide niches (great dispersion) might overlap even though their niche centers are quite distant in the hyperspace.

To determine the degree of overlap among the various bird species, the distance between the centers of each pair of species' niche hypervolumes was calculated in the hyperspace under the S^{-1} metric (Table X). These distances were tested for significance using the Hotelling T^2 statistic (Morrison, 1967). About 70% of these distances are statistically significant at the $\alpha = 0.01$ level. The lack of significance for the remaining 30% of the distances can most easily be attributed to small sample sizes of rarer species. Thus, the niche hypervolumes are different, there is

X

Possible Pairs of Niche Centers for the 20 Species

VESP	JUNC	PW-S	CHIP	FELD	WC-S	WT-S	FOX	LINC	SWAM	SONG
—										
0.33	—									
0.136	0.098	—								
0.059	0.023	0.068	—							
0.024	0.007	0.113	0.029	—						
0.036	0.036[a]	0.142	0.062[a]	0.018	—					
0.035	0.014	0.125	0.041[a]	0.012	0.034	—				
0.054	0.039	0.166	0.069[a]	0.032	0.039	0.033	—			
0.268[a]	0.250[a]	0.354[a]	0.274[a]	0.245[a]	0.260[a]	0.238[a]	0.247[a]	—		
0.094[a]	0.078[a]	0.169	0.102[a]	0.063[a]	0.056[a]	0.066[a]	0.077[a]	0.280[a]	—	
0.027	0.019	0.131	0.042[a]	0.010[a]	0.019[a]	0.014	0.014	0.246[a]	0.064[a]	—

little overlap in the species' niches, and the resource division of these species can be explained by habitat variables.

Since the niches do not overlap to any great extent, it is not possible to use this data set to test the competitive exclusion principle. It is interesting to note that many of the distances in Table XI are not what one would intuitively expect from anecdotal field observations. This is

TABLE XI

CONSTANTS FOR RELATIVE DENSITY FUNCTIONS QUANTIFYING NICHE WIDTH OF 13 FRINGILLIDS[a]

Species	c_1	r
CARD	4.70	−1015.20
GROS	2.15	−1161.50
PURP	1.83	−35.81
PINE	1.75	−462.77
GOLD	3.43	−2087.44
TOWH	3.63	−268.00
SAVA	4.17	−547.38
JUNC	3.65	−53.89
CHIP	2.90	−816.23
FELD	24.85	−1169.59
WT-S	14.01	−1017.16
SWAM	1.72	−136.90
SONG	4.95	−193.30

[a] Functions are of the form $x(\delta) = c_1 e^{r\delta}$ where x = likelihood of finding an individual of the species at a given location having some associated habitat vector; δ = distance from the center of the species' niche; r = a rate ($r \leqslant 0$), and c_1 is the constant of integration.

due in part to the difficulty of conceptualizing a 27-dimensional space, particularly one in which the axes are not at right angles. But it is also due to the tendency of field observers to weigh certain environmental factors more heavily than others, for example, in the case of the variables used in this study, those concerned with trees and larger underbrush.

2. *The Problem of Niche Width*

The Hotelling T^2 statistic provides a measure of niche width as dispersion around a niche center, but this dispersion measure is too complex to be used as a niche width measure in any large hyperspace. If one were using a one-space for the niche hyperspace, the dispersion measure about a species' niche center would simply be the variance about the mean. Unfortunately, in the n-space case the multivariate analog of this variance is an $n \times n$ matrix, the variance–covariance matrix associated with a given species. Thus, the Hotelling T^2 statistic allows for testing the overlaps of niches, but it does not provide a simplified expression of niche width; rather, it provides niches as hypervolumes.

Maguire (1967) measured protozoan niche widths using nonparametric rank-order statistics, and studied the changes in niche "shapes" in successional and climax species. Maguire's method provides no determination of singular variables, i.e., variables which can be predicted by other variables, but does not depend upon any assumption of normal distributions. Other niche-width indices have been developed to aid in studying competitive interactions between various species. The best known of these is Levins' (1968) derivation from information theory, $W = -\sum r_p \log r_p$, where W is the niche width index and r_p is the proportion of the species' abundance occurring in environment p. The W formulation includes components of both the realized and fundamental niche as defined herein, and is thus probably most valuable as a general measure of a species' adaptation to the environmental conditions in a given area. McNaughton and Wolf (1970) have produced an index that is a possible competitor of W (Levins, 1968), but it lacks mathematical rigor (Shugart, 1970).

One simple measure of niche width can be developed from differential equations describing relative density functions along a generalized axis measuring distance from the center of a species' niche. Assume that as one moves outward in any direction in hyperspace from the center of a species' niche, the likelihood of finding an individual bird in the habitat described is a continuous, differentiable, single-valued function. Assume further, that although this function may be highly complex, it can be approximated by the differential equation

$$dx/d\delta = rx,$$

where

x = some relative density function which expresses the likelihood of finding an individual of the species at a given location having some associated habitat vector,

δ = distance from the center of the species' niche, and

r = a rate ($r \leqslant 0$), in effect defining niche width.

The general solution of this equation has the form $x(\delta) = c_1 e^{r\delta}$, where c_1 is the constant of integration.

One way to estimate c_1 and r might be to: (1) divide the distance from the species' niche center into very small intervals; (2) determine how frequently individuals are sampled in each interval; (3) log transform this number of individuals; (4) find the slope and intercept by linear regression techniques; and (5) exponentiate the result. The value of r is an inverse measure of the species' niche width, and the value of c_1 is a measure of the species' relative abundance in optimal habitat ($\delta = 0$). Since the r measure ($r \leqslant 0$) describes a negative exponential curve form, as $|r|$ increases the width of the species' niche decreases.

The Slate-colored Junco (JUNC, $r = -35.81$) is the species with the widest niche, and the Goldfinch (GOLD, $r = -2087.44$) has the narrowest niche (Table X). Species such as the Field Sparrow (FELD, $c_1 = 24.85$) and the White-throated Sparrow (WT-S, $c_1 = 14.01$) can be found in unusually high numbers in optimal habitat, and seem to be unique in this regard (Table XI).

3. *Comparison of Levins' Niche Width Index and the d, c_1, and r Measures*

Since relative abundances of each of the species on 18 study plots were collected during the course of the study, it was possible to calculate Levins' (1968) niche breadth index for comparative purposes. Since the W index contains aspects of both fundamental niche and realized niche, it was logical to investigate the ability of d, c_1, and r to predict W in a stepwise multiple regression technique (Steel and Torrie, 1960). The results of this analysis are summarized below:

Step 1. $W = 1.36025$
coefficient of determination $= 0.000$,

Step 2. $W = 2.33514d + 1.82084$
coefficient of determination $= 0.260$,

Step 3. $W = -0.00025r - 2.09759d + 1.59988$
coefficient of determination $= 0.285$,

Step 4. $W = 0.01791c_1 - 0.00021r - 1.94329d + 1.49947$
coefficient of determination $= 0.299$.

There is little predictability of Levins' W from c_1, r, and d; and any relation between the W index and the present indices of niche width is in the realized niche component of both measures, as indicated by the ordering of the three variables in the above expressions for W.

D. Using Niche Widths and Distances to Predict Relative Abundance of Species

Having obtained a measure of the realized niche (d) and two measures associated with niche width (r, c_1), it is logical to test these measures with actual field data. A useful data set for this test is the Athens, Georgia Christmas Bird Counts (Peake, 1966, 1967, 1968; Kilgo, 1969, 1970). Anyone who has worked with Christmas Count data is aware of the variability of this data set. Counts vary with the yearly changes in bird populations, seasonal weather conditions affecting bird observability, the ardor of the bird counters, and a great variety of other factors including outright sample bias and a tendency of some Christmas Counts to list an inordinate number of rare species. With all these shortcomings, the Christmas Counts in general represent unique and often valuable information concerning winter bird populations. The Athens, Georgia count is a more representative count than average because most of the observers have been exposed to some statistical sampling theory. Additionally, the counts are typically less variable if relative abundances are averaged over several counts.

Since the realized niche measure is based on the Athens area, and the fundamental niche measures should be free of regional effects save geographical changes in niche dimensions, the three measures were used to predict the relative abundances of the 13 species averaged over the last five years. Using multiple and partial linear regression (Steel and Torrie, 1960) the following equation was obtained:

$$y = 0.67370c_1 + 0.00038r - 8.15970d + 6.07907,$$

where

y = relative abundance,
c_1 = relative abundance in optimal habitat,
r = niche width, and
d = realized niche.

The coefficient of determination for the above equation was 0.601, indicating an unusually high level of predictability considering the typical variability of Christmas Count data. Not only is the regression line able to predict species relative abundance, but it fits the hyperspace representation of species' niches, that is, as c_1 (relative abundance in optimal habitat) increases, the species' abundance increases, an increase in niche width (decrease in $|r|$, the loss rate for a negative exponential) or a decrease in d (the realized niche measure) both increase relative abundance. An extremely common species could be expected to have a wide niche (small $|r|$), to be abundant in optimal habitat (large c_1), and to be well adapted to the region (small d). Thus, the niche measures in the hyperspace under the S^{-1} metric are both predictive and logical in their prediction.

VI. Niche Patterns

A potential value of the species' niche concept is in providing insight into the structure and function of biotic communities. The niches of species comprising a given community can be represented as a hyperspace. Each axis of this hyperspace would be a variable describing the species in a community. Such a hyperspace will be referred to as a *niche pattern* for a given community.

Using r, c_1, and d as general measures of niche width and position, it is possible to look at the general niche patterns of the fringillid community. Using a three-dimensional graph (Fig. 3) with the r axis

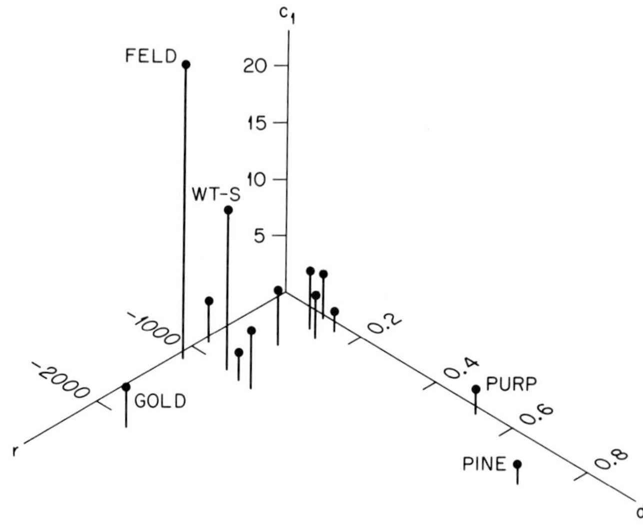

FIG. 3. Three-dimensional graph of r, c_1, and d components of 13 species' niche descriptions.

measuring niche width (niches become narrower as $|r|$ increases), the c_1 axis measuring relative density in optimal habitat (relative density increases with higher values of c_1), and the d axis measuring the realized niche (a species' adaptation to the region decreases as d increases), the species' niches can now be viewed holistically. Most of the species (unlabeled dots, Fig. 3) form a cluster, indicating that they have large realized and fundamental niches (small d's and small $|r|$'s) and do not occur in inordinately large numbers in optimal habitats (small c_1's).

Two species, the White-throated Sparrow (WT-S) and the Field Sparrow (FELD), differ from the other fringillids in that they can be found in large numbers in optimal habitat (large c_1 values, Fig. 3). This accounts in part for their high abundance in the area, an abundance which cannot be accounted for by their values of d and r alone.

Two other species, the Pine Siskin (PINE) and the Purple Finch (PURP), have high d values (Fig. 3) indicating a lack of suitable habitat, and the Goldfinch (GOLD) has an unusually narrow niche (large $|r|$). The feeding behavior of these three species is typical of species operating in a coarse-grained environment (MacArthur and Wilson, 1967; Cody, 1968); the relative positions of all three species in Fig. 3 support this conclusion. The Purple Finch and the Pine Siskin have large d's indicating that there is a shortage of available habitat. The Goldfinch, although it has a small d, has a narrow niche (large $|r|$) and spends considerable time searching for suitable habitats.

With the exceptions mentioned above, the niche patterns of most of the more abundant species treated in this study are quite similar. Viewed as responses to specific habitat variables (see previous sections), there is no readily apparent pattern in the habitat selection of these birds, yet most of the species fall into about the same pattern when their niches are viewed holistically in a niche pattern diagram. The significance of this pattern of similarity of general niche measurements will be discussed below.

A. Theoretical Implications of Niche Quantification

1. *The Significance of Niche Position*

There are several theoretical questions arising from this niche quantification method. One such question is "What are the expected characteristics of species whose niches fall in different positions in niche pattern diagrams such as Fig. 3?" Of the eight possible positions for all combinations of high or low values of c_1, $|r|$, and d (Fig. 4), four positions (5, 6, 7, and 8) are represented by species included in this study. Expected characteristics of species at these positions (Fig. 4) might be:

Position 1 (high d, high $|r|$, high c_1): Species should be limited in habitat in the region (high d), have a narrow niche (high $|r|$), and be

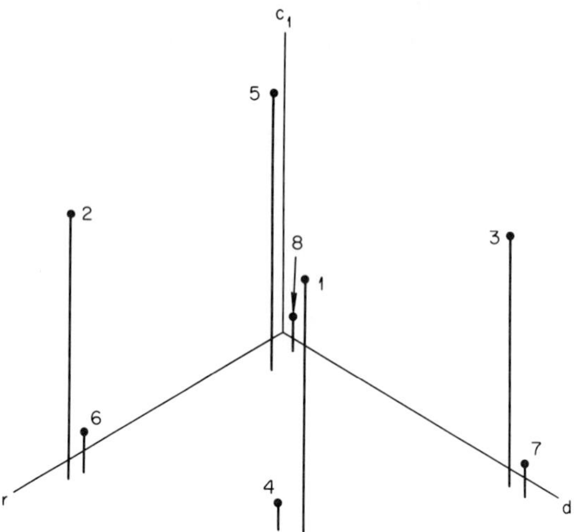

FIG. 4. Positions of hypothetical species' niches on niche pattern diagram. Positions correspond to high or low values of r, c_1, and d in the following pattern.

found in high abundances in favored habitat (high c_1). Species could be expected to be rare over large areas, but locally abundant where conditions permit. Species' habitats are so rare that even with a high degree of mobility, individuals probably could not find new habitats. These species would most likely be found in microhabitats that persist over long periods of time, and the species could be expected to become extinct under any conditions altering these specific sites. Possible examples: the Ozark Cave Salamander, *Typhlotriton spelaeus*, other troglodytes except bats.

Position 2 (low d, high $|r|$, high c_1): Species should be well adapted to the region (low d) with a narrow niche (high $|r|$), and found in relatively high numbers in optimal habitat (high c_1). Species' habitats are uncommon, and mobility allowing individuals to move to favorable microhabitats should be expected. These species would be expected to be widely dispersed over large geographic areas, with clumped distributions at favorable sites. Because of their high mobility and high abundance at specific sites, they might be expected to colonize new habitats or islands (MacArthur and Wilson, 1967), but the narrow niches typical of these species would probably limit the success of colonization at most sites. Possible example: House Sparrow, *Passer domesticus*, which has a narrow niche (almost always found in immediate proximity to human buildings), is capable of high abundance, and appears to be well adapted to many North American areas (towns).

Position 3: Species should be poorly adapted to the region (high d), with wide niches (small $|r|$) and high abundance in favorable sites (high c_1). Species' distributions and mobility should be like that of species in Position 2. Species with niches in Position 3 should be superior island colonizers because of their intrinsic generality (small $|r|$). Possible examples: Most species which appear preadapted for island colonizing (MacArthur and Wilson, 1967), many successional species.

Position 4: Species with narrow niches (large $|r|$), poorly adapted to the region (large d), and found in low abundance in optimal habitats (low c_1). Probably no species which is a permanent component of a given community has a niche at Position 4. Possible examples: Rare bird species found for short periods of time far from their normal range.

Position 5: Species with wide niches (small $|r|$), well adapted to the region (small d), and found in high numbers at optimal sites (high c_1). Species with these characteristics could be expected to dominate the community. Examples: Field Sparrow (FELD, Fig. 3) and White-throated Sparrow (WT-S, Fig. 3).

Position 6: Species with narrow niches (large $|r|$), well adapted to the region (small d), and found in low numbers at optimal sites (low c_1). Like species in Positions 2 and 3, these species should be highly mobile, and adapted for finding favorable sites. These species would typically be less common over wide geographical areas, but found in small numbers locally. Example: The Goldfinch (GOLD, Fig. 3) is a species between Positions 6 and 2.

Position 7: Species with wide niches, poorly adapted to the region, and found in low numbers at optimal sites. Mobility characteristics and distribution as in Positions 2, 3, and 6. Examples: Purple Finch (PURP, Fig. 3), and Pine Siskin (PINE, Fig. 3).

Position 8: Species with wide niches, well adapted to the region, and found in low numbers at optimal sites. The typical species in the present study are near Position 8. These species are relatively common and widely distributed. Typical examples: Song Sparrow, Fox Sparrow.

Since there are no other studies using the same methodology, it is not possible at present to determine what niche patterns, if any, exist in other faunal or floral assemblages. One might expect diverse communities in unchanging environments to have a large number of Position 6 species (Maguire, 1967), and less diverse communities to have Position 5 species. One can only speculate about what general niche patterns might be associated with given communities, and the description of general niche patterns as an index to community designs would appear to be a fruitful field of investigation.

2. *Hypothetical Niche Patterns*

As a demonstration of the possible use of niche pattern diagrams, one can speculate about possible niche patterns in various communities. Figure 5, for example, illustrates possible niche-pattern differences in

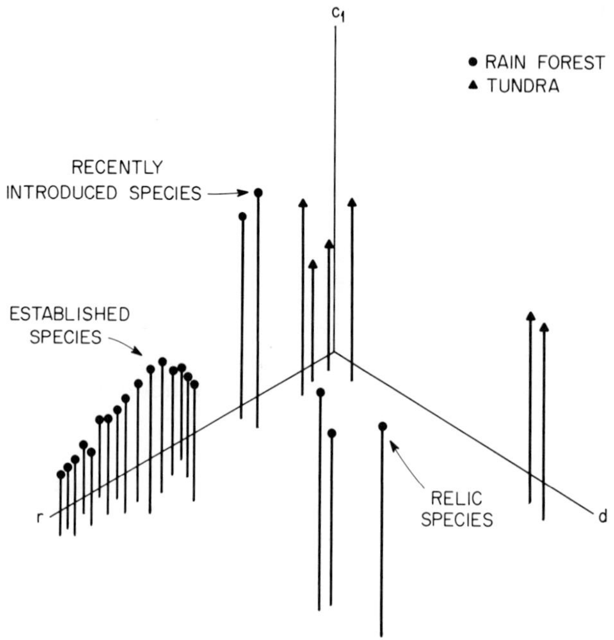

FIG. 5. Hypothetical niche pattern diagram for consumers in tundra and tropical rainforest.

consumers of arctic tundra and tropical rainforest. In the tundra, one might expect the observed tendencies for many of the consumers to either gather in herds (e.g., reindeer, musk oxen) or to demonstrate abrupt increases in population density (e.g., voles) to produce high c_1 values in a niche pattern (Fig. 5). One might also expect the species' niches to be wide (small $|r|$) because of the environmental variability, particularly in seasonal temperature changes. If this is the case, then any differences in species' relative abundance would be due to differences in the d component (likelihood of a given species finding suitable habitat).

In the tropical rainforest (Fig. 5), species niches are felt by many observers to be rather narrow (large $|r|$), and organisms often are widely dispersed over large areas (small c_1). Assuming that most of the rainforest consumers have had time to adapt to the environment (small d), most of the species' niches should be near Position 6 on a niche pattern diagram (Fig. 4). Relic species in such an assemblage should be expected to occur at Position 1. Recently introduced species to the region would probably differ most radically in niche width (they should have wider niches or small $|r|$).

Floral or faunal components of islands (Fig. 6), if the island is near a

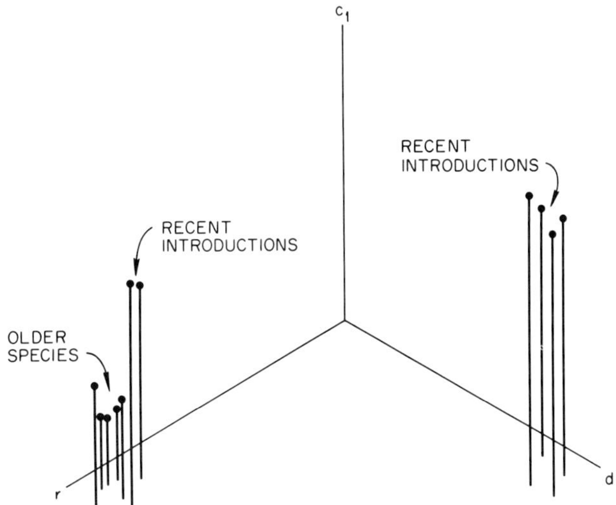

FIG. 6. Hypothetical niche pattern diagram for floral or faunal components of an island near a mainland source of species.

mainland source of species, should have a recently introduced component of colonizing species (MacArthur and Wilson, 1967). Older, established species could probably be expected in Position 6 (Fig. 6). A species equilibrium curve (MacArthur and Wilson, 1967) could then be expressed as a time-varying trajectory through the niche pattern space, moving the positions of introduced species toward the positions of the established species.

Another niche pattern with time-varying attributes might be desert plants (Fig. 7). Species such as cacti which are ecologically active during most of the year and widely dispersed would probably be found at Position 8. Annual plants which bloom in great numbers when the time

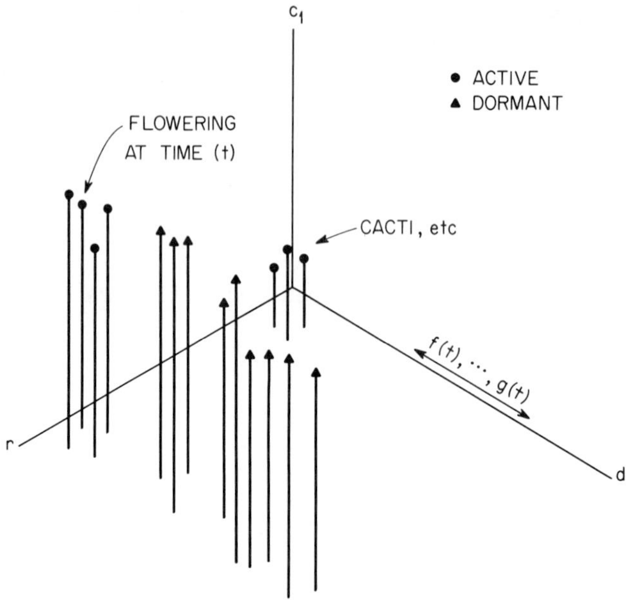

Fig. 7. Hypothetical niche pattern diagram for plant components of deserts.

of the year and environmental conditions are right (large $|r|$ and high c_1), should be viewed in a dynamic perspective. As time and conditions vary, the position of these species along the d axis should change. When the d value is small enough for a given species, it flowers, seeds, and dies back during a short period of time. Thus, the annual plants flowering in a desert at a given time are probably in Position 2. The annual plants in seed-form at a given time are in motion between Positions 2 and 1.

3. *Possible Changes in Niche Position Through Time*

It was not possible to connect the niche quantifiers used in the present study to niche measurements (Levins, 1968) relating strongly to competition. In the case of the present study, no species which were possible competitors overlapped to any appreciable extent, so it was not possible to generalize the niche concept in the context of this methodology to include competition. The great advantage in including competition as an integral part of niche theory is in the ease with which competition can be related to natural selection and evolution (MacArthur and Wilson, 1967; Levins, 1968). Since the present methodology does not include competition, it is appropriate to indicate the ways that the niche quantifiers might be linked to evolutionary theory.

One method of linking the r, d, and c_1 niche quantifiers to evolution is to determine in what directions positions of species' niches in a niche pattern diagram (Figs. 3, 4, 5, 6, and 7) could be expected to move over long periods of time. In these niche pattern diagrams there are six basic directional components (positive and negative along each of the coordinate axes, Fig. 8) for changing positions. Of the six directions, one can speculate with some assurance concerning four of them. These four components are:

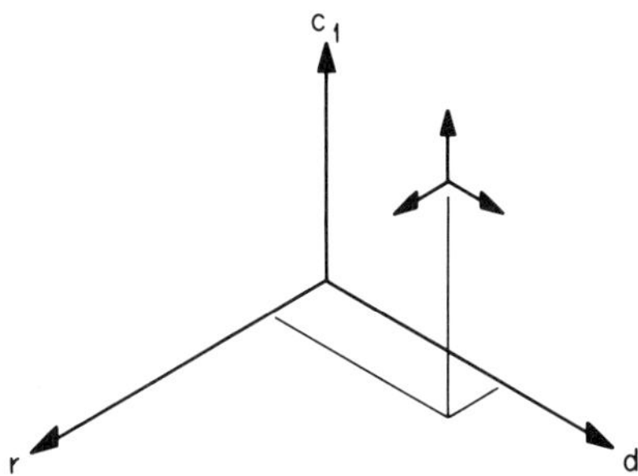

FIG. 8. Directional components for changing the position of a species' niche in a niche pattern (c_1, d, r) diagram.

(1) Increasing along the d axis—loss of adaptation to the region. The magnitude of this component is most likely a function of the rate at which an environment changes. Evolutionary strategies in changing environments have been discussed elsewhere (Levins, 1968), but, in general, a rapidly changing, unpredictable environment should make it difficult for a given species to adapt rapidly enough to minimize d.

(2) Decreasing along the d axis—regional adaptation. This component is a function of natural selection, and environmental constancy. In environments which are unchanging over long periods of time, natural selection should minimize distances along the d axis.

(3) Increasing along the r axis—species niche width reduction. Competition is probably highly related to this component, and competitive interaction would most logically reduce species' niche widths.

(4) Decreasing along the r axis—species niche width increase. This is probably due to a tendency for species to expand their niches, and to utilize unused parts of an environment. Crowell's (1961) study of mainland and island birds provides an example of this component in operation. The so-called "richness' component (Tramer, 1969) of the Shannon diversity formula, probably increases as the number of species in a community with narrow niches (large $|r|$'s) increases.

The other two directional components, movement up or down the c_1 axis, are difficult to identify. As species develop internal density controls (e.g., territoriality) over time, one would expect c_1 values to be reduced. However, other traits thought to be highly evolved over long periods of time (e.g., social aggregations, colonies of organisms) might actually tend to elevate the c_1 values. As "richness" is related to the r axis, the "equitability" component of Shannon's formula (Tramer, 1969) is related to the c_1 axis, which measures the potential for a given species to dominate at a particular site.

B. Applied Uses of Niche Quantification

The realized niche measure (d) might be useful in an ecosystem management context to estimate the effect of environmental changes on species. Environmental alterations which increase the distance between a species' mean and the regional mean will decrease the abundance of the species. Unfortunately, the converse is not necessarily true. Decreasing the distance will increase the abundance of a species only if the factors which limit the species are included in the hyperspace, and if internal density regulating mechanisms will not limit the abundance of the species regardless of the habitat improvement. Thus, the method will identify a "bad" management program relative to a given species, but it cannot extrapolate beyond the data set to positively identify a "good" management technique. For example, if one eliminated the moist, thickly overgrown tangles in the area, the Lincoln's Sparrow (an apparently habitat-limited species) would probably no longer be found in the region. Conversely, increasing the typical habitat of the Lincoln's Sparrow would increase the species' abundance only if the species is not limited by other unknown factors, or in another region at another time of the year (e.g., limited on the northern breeding grounds).

In an era of environmental alterations caused by man, a technique which provides a method of determining detrimental changes should have value. The method could not have predicted the spread of the House Sparrow, *Passer domesticus*, over North America (it would only indicate that such an invasion was possible), but it could have predicted the demise of the Ivory-billed Woodpecker, *Campephilus principalis*, with harvesting of the southern hardwood forests.

Additionally this methodology could be used to assess the likelihood that an introduced species might "take" in a given region. Given niche parameters (d, r, and c_1), and a regression equation such as the one developed in the previous section, one can obtain the expected abundance of an introduced species. For example, in the present study a species with a small $|r|$, high c_1, and small d could be expected to survive in the region. If a species differs from the established niche pattern (Fig. 3) in the direction of c_1, the hypothetical species could become overabundant. The most recent "introduced" species in the winter fringillid assemblage, the Evening Grosbeak, which has recently extended its range into the region, fits the established niche pattern (Fig. 3) with a low negative r, small d, and small c_1. Introductions of species with high c_1 values probably should be avoided. Such information might be useful in conjunction with game species introduction programs, and equally useful in determining which foreign species should be most carefully quarantined from certain regions. To be used to best advantage, one should determine r, d, and c_1 values in all seasons. The method would probably be most workable with non-migratory species, as migrants may well be limited in other geographical areas.

An additional use of the niche quantification methodology, having both applied and theoretical implications, is in the field of ecosystem modeling. One section of this paper treated the development of a very simple differential equation to model a species' relative abundance response to a trajectory through the Euclidean space under the S^{-1} metric. This same technique could possibly be used to quantify multivariate constraints (Ashby, 1965) on dynamic ecosystem models. Since the niche quantification method can be used to describe phenomena not restricted to species niches, it is possible to use hyperspace distances and conditions to set explicitly varying coefficients in dynamic models of various biological systems.

VII. Summary

This study presents a general methodology for niche quantification, and uses quantified niche measures to develop the niche pattern diagram. The niche pattern diagram is potentially useful for theoretical studies of community structure and function, and for a number of applied uses including species introduction problems, and assessing possible results of environmental changes.

Niches of winter fringillids in northern Georgia were quantified using variables associated with habitat selection. A measure of species adaptability to the study region ("realized niche" measure) was calculated. The species' niche was viewed as a density configuration in a Euclidean hyperspace, and this density function was approximated with an ordinary linear differential equation to quantify niche width. Using these niche quantifiers it was possible to predict species' relative abundance over large areas. Although their habitat selection strategies were varied, the overall patterns of 13 common fringillid species' niches were similar with a few logical exceptions.

The niche pattern diagram was also discussed in connection with hypothetical communities, as a possible method for analyzing community design, for predicting consequences of environmental changes on community components, and for predicting the results of species introductions.

Acknowledgments

We thank Dr. R. E. Bargmann for statistical consultation and invaluable assistance in developing a geometric niche model; Mrs. Thelma Richardson for her statistical and programming consultation; and several graduate students, including R. Mural, G. I. Child, K. P. Able, J. J. Mahoney, and J. A. Holbrook, for time spent interacting on the subject of niche theory.

REFERENCES

American Ornithologists' Union (1957). "Check-list of North American Birds." Lord Baltimore Press, Baltimore, Maryland.

Ashby, W. R. (1965). *In* "Progress in Biocybernetics" (N. Wiener and J. P. Schade, eds.). Elsevier, Amsterdam.

Cody, M. L. (1968). *Amer. Natur.* **102**, 107–147.

Cottam, G., and Curtis, J. T. (1956). *Ecology* **37**, 451–460.

Crowell, K. (1961). *Proc. Nat. Acad. Sci.* **47**, 240–243.

Dow, D. D. (1969a). *Can. J. Zool.* **47**, 103–114.

Dow, D. D. (1969b). *Can. J. Zool.* **47**, 409–419.

Elton, C. S. (1927). "Animal Ecology." Sidgwick and Jackson, London.

Fretwell, S. (1969). *Evolution* **23**, 406–420.

Friedrichs, K. O. (1965). "From Pythagorus to Einstein." Random House, New York.

Gause, G. F. (1934). "The Struggle for Existence." Williams and Wilkins, Baltimore, Maryland.

Grinnell, J. (1904). *Auk* **21**, 364–382.

Grinnell, J. (1917). *Auk* **34**, 427–433.

Grinnell, J. (1928). Univ. Calif. Chron., Oct. 1968, pp. 429–450.

Grosenbaugh, L. R. (1952). *J. Forestry* **50**, 32–37.

Hardin, G. (1960). *Science* **131**, 1292–1297.

Hespenheide, H. A. (1971). *Auk* **88**, 61–74.

Hilden, O. (1965). *Ann. Zool. Fenn.* **2**, 53–75.

Hutchinson, G. E. (1944). *Ecology* **25**, 3–26.

Hutchinson, G. E. (1957). *Cold Spring Harbor Symp. Quant. Biol.* **22**, 415–427.

Hutchinson, G. E. (1965). "The Ecological Theater and the Evolutionary Play." Yale Univ. Press, New Haven, Connecticut.

James, F. C., and Shugart, H. H. (1971). *Audubon Field Notes* **24**, 727–736.

Kilgo, J. (1969). *Audubon Field Notes* **23**, 231.

Kilgo, J. (1970). *Audubon Field Notes* **24**, 231.
Klopfer, P. H. (1967). *Wilson Bull.* **79**, 290–300.
Klopfer, P. H., and MacArthur, R. H. (1961). *Amer. Natur.* **95**, 223–226.
Lack, D. (1933). *J. Anim. Ecol.* **2**, 239–262.
Lack, D. (1949). *In* "Genetics, Paleontology, and Evolution" (G. L. Jepsen, E. Mayr, and G. G. Simpson, eds.). Princeton Univ. Press, Princeton, New Jersey.
Levins, R. (1968). "Evolution in Changing Environments." Princeton Univ. Press, Princeton, New Jersey.
Lotka, A. J. (1932). *J. Washington Acad. Sci.* **22**, 461–469.
MacArthur, R. H., and Wilson, E. O. (1967). "The Theory of Island Biogeography." Princeton Univ. Press, Princeton, New Jersey.
Maguire, B., Jr. (1967). *Amer. Natur.* **101**, 515–526.
Mahalanobis, P. C. (1949). *Sankhy* **9**, 237–239.
Mayr, E., and Greenway, J. C. (1956). Breviora 58. Museum of Comparative Zoology, Cambridge, Massachusetts.
McNaughton, S. J., and Wolf, L. L. (1970). Dominance and the niche in ecological systems. *Science* **167**, 131–139.
Morrison, D. F. (1967). "Multivariate Statistical Methods." McGraw-Hill, New York.
Morse, D. H. (1970). *Ecol. Monogr.* **40**, 119–168.
Mukherjee, R., and Bandyopadhyay, S. (1965). *In* "Contributions to Statistics" (C. R. Rao, ed.). Eka Press, Calcutta.
Peake, R. (1966). *Audubon Field Notes* **20**, 198.
Peake, R. (1967). *Audubon Field Notes* **32**, 186.
Peake, R. (1968). *Audubon Field Notes* **22**, 205–206.
Penfound, W. T., and Rice, E. L. (1957). *Ecology* **38**, 660–661.
Peterson, R. T. (1947). "A Field Guide to the Birds." Houghton Mifflin, Boston, Massachusetts.
Rao, C. R. (1952). "Advanced Statistical Methods in Biometric Research." Wiley, New York.
Shugart, H. H. (1970). *Science* **170**, 1335.
Steel, R. G. D., and Torrie, J. H. (1960). "Principles and Procedures of Statistics." McGraw-Hill, New York.
Svardson, G. (1949). *Oikos* **1**, 157–174.
Tramer, E. J. (1969). *Ecology* **50**, 927–929.
Vandermeer, J. H. (1970). *Amer. Natur.* **104**, 73–83.
Volterra, V. (1926). *Mem. R. Acad. Naz. Lincei, Ser. 6* **2**, 31–113.
Weatherly, A. H. (1963). *Nature* **197**, 14–17.
Wilson, J. W. (1959). *New Phytol.* **58**, 92–101.
Winkworth, R. E., and Goodall, D. W. (1962). *Ecology* **43**, 342–343.

Appendix. Degrees of Freedom, and

F values less than

	CARD	GROS	PURP	PINE	GOLD	TOWH
Degrees of freedom	(2250)	(2250)	(2250)	(1251)	(2250)	(2250)
CANP	2.67	5.17	1.27		2.49	4.51
UNDP	8.44					7.42
GRCP	2.56	3.89	8.64		1.17	
CANE						1.16
UNDE						
GRCE	1.29					
TALL	6.15	3.52	4.87	2.01		7.22
MTHT	5.38	10.25	9.72	2.25	3.11	1.61
ZBIT	4.73	7.87	7.65	1.86	2.46	1.86
BASA	4.53	15.39	21.85		1.32	
HBAR	2.69					15.00
ADOT	3.14	3.52	3.13			
ADIT	5.52	2.25	2.24	1.25	1.18	4.28
BDOT	3.91	11.55	5.89			
BDIT	5.35	2.43	2.42	1.36	1.18	3.01
LDOT						
LDIT	4.35	1.63	1.62			4.70
ADOS	1.22					6.38
ADIS	6.37					4.08
BDOS						16.82
BDIS	6.29					3.25
LDOS						21.18
LDIS	5.45			74.36		3.24
TREE	7.32	9.12	9.81	2.66	2.02	1.60
SHRB	6.01	1.33	2.69	8.89		3.49
GRCO	1.54	1.80	4.22		1.28	
HITS	4.61			4.07		7.40
BHIT	4.21			2.69		5.48
LHIT	4.37			5.54		8.95
ST-1		1.09				1.05
ST-2	1.75					
ST-3	3.49		9.42			1.70
AR-1						1.41
AR-2	1.41					
AR-3	4.42		6.08			3.07
NO-1		1.53				
NO-2				3.15		
NO-3	2.55		6.97	2.71		3.02
BW-1						1.32
BW-2	1.53			1.41	1.69	3.10
BW-3	4.67			1.02	4.75	7.23
SE-1	2.18	1.79				
SE-2	2.09			4.53	1.22	4.89
SE-3			1.35	48.55		18.17

F Values Associated with Table III

1.0 are deleted.

SAVA	GRAS	HENS	VESP	PW-S	JUNC	CHIP	FELD
(2250)	(1251)	(1251)	(2250)	(1251)	(2250)	(2250)	(2250)
13.51	6.35	4.69	2.29	5.44	2.78	1.65	
15.61	8.40	4.68	3.05				
	1.11	15.51	2.48		1.61		1.01
3.74	1.85	1.37		14.71	10.88	12.48	
2.33	1.16			44.96	13.68	20.00	1.16
4.31	1.82	38.03			1.47		
25.08	12.89	7.16	3.06	1.17	8.37	3.26	
9.61	5.41	3.50	1.29	1.23		1.24	1.32
14.66	7.52	5.55	1.78				
7.03	3.56	2.64	1.09				2.89
6.67	3.83	2.84			5.87	5.88	4.01
10.44	5.65	4.18	1.85		7.96		1.28
27.67	17.42	12.73	3.82	1.20	6.07	2.77	
12.18	6.46	4.77	1.81		5.66		
20.08	13.42	9.85	2.89	1.27	5.19	2.91	
3.67	2.27	1.69			9.26		2.05
34.07	20.23	14.75	4.55		5.72	2.07	
5.44	1.95		1.51		1.05		
44.51	24.64	3.10	8.27		3.18		
1.65							
32.93	18.86	2.04	6.37		3.00		1.31
46.69	27.76	3.94	9.31		2.87		
10.38	5.13	3.79	1.60				1.13
19.34	13.20	2.09	2.79				
		11.73	1.91	1.21	2.98		3.13
5.34	33.82	2.07					1.24
5.91	37.32	2.20					
3.67	2.35	1.54			2.48	1.11	2.24
		11.68			4.09		1.24
1.50					1.04		
		15.12			2.49		1.00
1.78	1.00				1.11		
3.94	1.98	64.42			3.64		
1.26					1.16		
	1.00	52.91			1.72	1.04	
5.46	4.11	3.05			2.57		1.65
1.90				6.40		2.06	
11.56	1.36	6.21	2.54	1.30	1.60	2.99	
2.29	1.46	1.09			2.26		

PART IV

APPLICATIONS AND PROSPECTS

This section presents a sampling of systems ecology applications. Taken altogether, it provides a reasonably balanced and accurate picture of the practical capability of ecological systems analysis and simulation at the present time. Performance does not come up to publicity (see Preface), but prospects for rapid improvement are good given a willingness to let pragmatism guide sound scientific development without demanding unrealistic short-term successes.

Chapter 8 is on fisheries management. Departing from conventional attitudes in this field, Dr. Saila defines a fishery as "a holistic complex including fishes and other organisms, man, machines, as well as the environment...." Fisheries management, in effect, must thus proceed by consideration of the whole. Several examples from the author's experience in applying this principle are provided. The first is an extensive planning model for a haddock fishery that illustrates well the holistic approach. The second example is an application of Bayes' theorem in decision making, illustrated by a case of a fishing cooperative wanting to assess the wisdom of expanding operations. The final example is a lobster simulation study. A traditional population biology approach is taken, demonstrating the ease of slipping out of the holistic mode in practice, even when the principle is expressly adopted as a point of operating philosophy. There will, it seems, be psychological tendencies as well as technical problems to overcome in developing the system concept for practical ends.

Chapter 9, by Dr. G. J. Paulik, is a review of the current status of digital computer simulation modeling in applied ecology. The view is taken that mathematical modeling is inherently good, and that, coupled to simulation, modeling reaches a high degree of utility for the resource ecologist. The need to continue missionary work to so convince ecologists is noted. The uses of simulation in solving statistical ecology problems are considered, followed by a review of resource management models from the fisheries field. Examples with a biological emphasis are presented, and then several with bioeconomic goals such as forecasting salmon runs, evaluating consequences of gear restrictions, etc. A section is devoted to the uses of simulation models in training applied ecologists. Simulation gaming, in particular, comes through as one of the most potentially effective means of imbuing students with a holistic insight by demonstrating system-wide consequences of management decisions. Paulik's attention to training represents the only significant commentary in this and the preceding volume on systems ecology education. The omission is significant, not because training is unimportant, but because it is the most grossly neglected area of systems ecology today. The kinds of applied capabilities envisioned for systems ecology begin with the painstaking procedures of the day-to-day classroom, and if the classroom does not receive attention in the form of curriculum development, then in the long run ecologists may not have much to offer to the solution of applied ecology problems. The chapter concludes with observations on the future of simulation modeling in resource science, noting that it will be the "flesh-and-blood" ecologists who end up keeping the "electronic" ecologists honest.

Chapter 10 is devoted to uses of systems concepts and methods in courtroom environmental defense. The different ways that systems studies can be used in court action are brought out. The two principal ways are methods serving as a vehicle for integrating data and processes, and concepts serving to present a holistic view of the environmental system before the courts. Techniques by which scientific evidence may be developed in a systems format are examined, illustrating the level of scientific and technical preparation necessary for an individual to testify effectively. The cases discussed come from several notable experiences of the author, and include the now famous Wisconsin DDT hearings, action against the AEC's Project Rulison in Colorado, and several others. What distinguishes the presentation and makes for fascinating reading is the use of trial transcript material to illustrate how the scientific record is written in court.

Chapter 11 considers systems ecology and the future of human society. Its point of departure is the new environmentalism that has swept the U.S. in the past few years. The major premises of this movement are accepted. Man is tied to a global ecosystem, and planning for survival and quality existence means finding ways to handle the complexity. The appropriateness of systems ecology for attacking such problems reflects the similarity of its methods to the nature of the global ecosystem, and the ability of its concepts to provide a single integrated framework. "These properties," it is said, "are not those of ecology but of systems methods; ecology is only one of several integrative disciplines." And for Dr. Foin and the Davis Environmental Systems Group that he here represents, the methods of systems ecology come down to simulation modeling.

Much of the chapter is devoted to the CALSIM (California Simulation) project, a team study to explore the feasibility of developing a model of social, economic, and ecologic aspects of the state of California. A number of intriguing findings are reported, e.g., some social consequences of population growth. In respect to education taxation, 1 % growth dilutes tax effectiveness by 25 %, 2 % growth by half. On "law-and order," in comparison to a stable population, 1 % growth leads to 70 % more personal and 153 % more property crimes. For 2 % growth, the corresponding figures are 398 % and 537 %, respectively. The principal conclusion from the CALSIM effort relates to inadequate data bases in simulation modeling. That "no simulation model can be better than the data used to construct it" is passively accepted as the basis for concluding that, for the foreseeable future, detailed whole system models which reliably predict the time course of human societies are probably not feasible. Fortunately, such a conclusion overlooks the distinct possibility, being entertained by others of us at the present time, that with the aid of a little theory models may very well have the potential to be far better than the data that go into them. In fact, the subject of modeling in data supplementation is one that can be expected to see considerable exploration in the future.

Chapter 12 concludes the book with a broad brush scenario on next-generation "ecomodeling." The brief history and current status of ecological modeling are reviewed and the growth of interest and activity documented. The broadening scope of mathematics being brought in is indicated, as well as the trend away from individual efforts to programs involving interdisciplinary teams. Academically motivated models are contrasted with applied models, one essential difference being that the latter must provide convenient access to parameters and inputs that are subject to manipulation in management procedures. In developing his own prescriptions for ecomodeling, Clymer puts forth the notion of *hierarchical modeling* which he says is consistent with the hierarchical nature of natural systems. Several examples of the hierarchical approach are provided. Then, a few "next-generation" ecomodels are developed in outline form. Of particular interest is the extension of ecology to the spheres of human personality and interactions, and to urban health care systems. The author regards ecology as a valid perspective on all large scale man/nature systems, and the latter as the most compelling motive for the advance of ecology.

The final section, "Closure," speaks eloquently of the challenges for systems ecology, of the strengths and limitations of ecomodeling, and of its satisfactions and frustrations. A fitting inspirational note is struck to provide a graceful exit from the collection of ideas that is these first two volumes of *Systems Analysis and Simulation in Ecology*.

8

Systems Analysis Applied to Some Fisheries Problems

SAUL B. SAILA
GRADUATE SCHOOL OF OCEANOGRAPHY
UNIVERSITY OF RHODE ISLAND, KINGSTON, RHODE ISLAND

I. Fisheries Problems in Perspective

Fisheries science has sometimes been defined as a branch of quantitative ecology, concerned with defining factors which control the yields of fisheries and with means for maintaining these yields at the highest levels on a sustained basis. According to Cushing (1968) there are two areas of fisheries science: (1) the study of the natural history of stocks, and

(2) the study of the dynamics of the stocks. The broad definition utilized herein is that the subject includes the systematic application of various kinds of information to a fishery with a view toward ultimate prediction and control. In this context it appears that we are concerned with a holistic complex including fishes and other organisms, man, machines, as well as the environment—a system in which various inputs or outputs can initiate events which are removed in both time and space.

One of the important aspects of fisheries science involves the quantitative relationships between stocks of fish and man as an exploiter of these stocks. For the most part fisheries problems (especially those dealing with commercially exploited stocks) are complex problems in applied science and decision making. This is the case because man is usually interested in the fishery resource from the standpoint of man and not the resource. The only matter of real significance to a population (stock) of fish is that it not be depleted (by any means) to a value below which it can perpetuate itself. All facets of man's interest in the resource (influenced by the biology of the fish) have their basis in economic and social aspects of man's behavior. What is done with a marine resource (fishery) seems to be governed by the desires of man and appears to have little connection with the resource per se. For economically important marine fishes, the only way man affects a population significantly at present is by fishing. This can be varied in efficiency, amount, method, and area fished. On a world-wide basis pollution is still negligible, artificial enrichment or fertilization is uneconomical, epidemics are uncontrollable, and widespread stocking of marine fishes from hatcheries seems impractical. In light of the above and in order to better understand and to obtain workable solutions to natural fisheries problems a sophisticated theory of fishery population dynamics has been developed (Baranov, 1918; Beverton and Holt, 1957; and Ricker, 1958). The work of the above-mentioned authors and several others has consisted primarily in developing deterministic models of population growth and death processes. These have been combined to form yield equations which can be solved in closed analytical form under various simplifying assumptions. Only relatively recently (Royce *et al.*, 1963; Paulik, 1969; and Watt, 1966, 1968) have systems analysis techniques been suggested for and applied to fisheries problems. This is somewhat surprising because systems analysis techniques have been applied to engineering and military problems for some time with good success. In general when one applies information toward a defined goal (prediction) one is dealing with a problem in systems analysis and not in empirical science. The general objectives of systems analysis are to analyze specific situations in order to ascertain how adequately these situations (systems) meet certain specific problems. On the other hand empirical sciences attempt to derive or describe general laws. It seems clear that the methods of systems analysis are potentially valuable for a better understanding of fisheries problems. The material in this chapter is presented with a view toward promoting more widespread interest in the application of systems analysis techniques to a broad range of problems in fisheries science. The author is convinced that these will be useful in both research and management aspects of the science.

A. Fisheries Development

Although problems relating to the management of world fisheries resources are complex, they are very important to human welfare and justify considerable effort as well as expense in an attempt to gain better understanding of them. In 1850 the world catch of fish and shellfish (excluding whales) was about two million tons. By 1900 it had increased to four million tons, in 1930 to 10 million tons and to 20.2 million tons in 1950. The world catch of fish has doubled in the last decade to more than 50 million tons in 1965. It is interesting to note that most of the expansion in world fisheries has been derived from relatively few species. For example, the yield of a single species, the anchovy (*Engraulis ringens*) which is found off the coasts of Chile and Peru has increased from 50,000 tons in 1954 to approximately 10 million tons in 1964. This species has reached a condition of over exploitation by man in about one decade. On the other hand some of the sardine-like fish off the African coasts and some bathypelagic fishes in much of the world ocean are still underexploited. Other fishes, such as the haddock and other bottom living fishes in the Northwest and Northeast Atlantic are now being exploited at nearly maximum rates. A previously important marine resource, the Antarctic blue whale, has been driven virtually to extinction. From the above it is evident that some species of marine organisms are already over exploited and that the world fishery is developing much more rapidly now than it has in the past.

From recent reports of the President's Scientific Advisory Committee (PSAC) and other agencies it is also evident that the world ocean will be increasingly utilized as a source of animal protein in an attempt to alleviate already serious socioeconomic consequences of protein malnutrition. Among recent developments in further utilization of marine resources are increased production of fish meal for animal feeds and fish protein concentrate, a refined fish meal for beneficial use in human nutrition. It has been estimated (Schaefer, 1965) that the world ocean is capable of producing on the order of 200 million tons of fish annually on a sustained basis. Others have been considerably more optimistic. Regardless, a major effort will be necessary on the part of fishery scientists to achieve the goal of considerably increased yields while maintaining stocks in a condition which permits sustained yields at economically acceptable costs.

The complexity of fisheries science is evident when one considers the numerous and diverse kinds of information required for prediction and control. Detailed biological information, including physiology as related to growth and metabolism, behavioral information as related to migration and movement, and embryological studies as related to reproduction are only some of the biological requirements. In addition, a knowledge of the environment in terms of its diverse and dynamic physical and chemical properties as well as the interests and desires of man and the performance of his equipment must be considered. In view of the above complexities many fisheries problems in the past have been "solved" in one way or another even if the "solution" has been to ignore the problem and hope it would go away! More often, however, fishery scientists have sought for better ways of doing things and these have resulted primarily in development of deterministic population models. Knowledge, usually the lack of it, is a limiting factor during model construction. It is a critical limiting factor when one attempts to formulate a complex problem in analytical terms, because despite the sophistication of today's mathematics there are many complex problems which cannot be stated so that they can be solved analytically. This statement refers to the state of mathematics, the complexity of the problems we want to solve, and the irregularities with which we often compound them.

II. Systems Analysis and Simulation Models—General

It seems obvious that many natural fisheries problems cannot be solved experimentally by manipulating the fishery (system) because of time and cost considerations as well as resistance to interference with ongoing operations by fishermen. When the system (fishery) cannot be studied directly, it can be studied by means of models. Some analytical models of fish population dynamics have already been referenced. The term "model" as used here refers to a formalized theory or stylistic interpretation of a body of propositions that a theory represents. Just as there are many theories concerning the workings of a particular system so there can be many models that formalize a theory. It seems that fisheries problems should be considered for solution by numerical models where there is insufficient knowledge to construct an analytical model or an analytical model is not available. Numerical procedures are designed to produce usable solutions by replacing complex problems with simpler ones which can be solved and which approximate (more or less) solutions to the original problems. An example of such a procedure is the numerical integration of a mathematical function for which an integration formula does not exist. One method of numerical integration is to fit a series of rectangles of small width into the area beneath the function. In this case an analytical solution to the integration problem is replaced by a numerical one. It should be pointed out, however, that use of numerical procedures instead of analytical procedures places additional burdens on problem solvers because procedures for estimating a solution (model building) and errors inherent in the estimates (model use) must be considered in addition to the basic problem itself.

As indicated previously a system (fishery) is usually studied in order to predict its behavior. We may want to know how much exploitation is possible without adversely affecting sustained yields; we may want to know how environmental variables or abundance of the parent stock affect recruitment; how to maximize economic returns; how a new marketing procedure will affect consumer preference; or how a new vessel will affect a fisherman's return on investment. These questions, and many others can only be answered by models composed of mechanisms that are able to reproduce relevant aspects of system performance and behavior.

Systems can be broadly defined as bounded sectors of reality. Two objects are considered to act as a system when they are integrated to the extent that the performance of one affects that of the other. A study of one cannot be made in isolation from the other without losing effects caused by their interaction. Questions about the performance of an integrated system cannot be answered by studying its components separately. They must be studied together. More and more we find that systems analysis studies are being made of complex phenomena composed of large numbers of objects, each interacting with the other according to complex performance rules. This is due to an increasing recognition that systems must be treated as a whole and not as a sum of their component parts.

"Simulation" is a term commonly applied to the use of models to study systems. In our case the term is restricted to the use of numerical models. A simple example of a simulation model would be a flow chart or diagram representing the passage of an anadromous fish species through a particular kind of fishway in a given river. The overall system logic describes the movement of fish through the fishway by logical relationships and paths. Particular parts of the model can be symbolic if they describe certain mathematical or statistical procedures for determining such things as the amount of time spent in actual ascent, or the size of the queue under changing flow rates and swimming speeds. In general, simulation is considered a part of systems analysis. It normally involves use of a numerical model to study the behavior of a system as it operates over time. Most models of this nature are implemented on digital computers since we are usually interested in advancing a system through time in discrete steps. Simulation is an experimental and numerical technique and is usually more expensive to use than analytical solutions to fisheries problems. Despite the expense, the technique is becoming widely used in some disciplines because analytical solutions are not available. Indeed, simulation may sometimes be preferred over real-world experiments. This is true when there are difficulties in controlling certain parameter changes or in keeping ambient conditions reasonably constant. Simulation can often aid in understanding a system which contains a large amount of statistical variation.

A. Simulation as a Systems Analysis Tool

Systems analysis looks at measures of system performance as both measures of average behavior as well as measures of dynamic response. Measures of average performance (means, standard deviations, coefficients of variation) are fairly common outputs of systems studies. These measures usually allow systems to be compared statistically; i.e., system *A* is preferred to system *B* if the values of its average performance are "better." The assessment of "better" takes into account average performance measures and their variation over time.

In systems analysis it may also be desirable to consider comparisons other than static comparisons. So-called system dynamics measure the way a system responds to shocks and disturbances. Typical dynamic performance measures include the sample correlogram and the sample spectrum. These measures portray the time-dependent behavior of the system. Thus they allow one to discriminate between systems that have identical average performance but different behavioral characteristics. Using system dynamics it is possible, for example, to select that system which responds most favorably to fluctuating prices from several systems, which, on the average, perform similarly. Simulation is one of the few tools available for estimating system dynamics.

Assuming a decision is made to undertake a simulation study of a fishery or fisheries problem there is a great deal of technical information which must be obtained and studied before the study can be done. Simulation programs can be written with relatively little formal training in computer science. The output from many programs may appear to reproduce the behavior of the object system. However, the model must not only resemble the system under study, it must also act like it. In fisheries science it seems highly desirable that a group or "team" approach be utilized in developing systems models of fisheries problems. The reason a number of people with various kinds of "discipline orientation" are necessary is that the problem and its solution demand knowledge of numerous disciplines including: biological sciences, oceanography, economics, various kinds of technology, business administration, and some programming and operations-research skills. This is a large order for any one individual.

Just as theories rest on various assumptions, so do simulation models. As assumptions put bounds on a theory, they structure a model and make it usable. The choice of assumptions for a simulation model depends on many things: the nature of the model, the environment in which it exists, and the use to which the model will be put, are examples. The more highly structured a model, the more numerous its assumptions.The narrower a model becomes, the more use it must make of assumptions to limit the world in which it is embedded and to create a suitable climate for precision. It is highly important that assumptions for simulation models be clearly stated and not hidden in the programming. The assumptions largely determine the purposes of a model and also determine the credibility that then is ascribed to its predictions. In general it seems that simulation models have been used and very probably will be used in the future both as explanatory (descriptive) as well as predictive devices. That is, the researcher is concerned with both system response and system dynamics. Two important questions are involved in systems analysis with simulation models. They are: (1) What use will be made of the model, or what questions will be asked; and (2) What are the requirements of accuracy and precision? Answers to these questions largely determine the structure of the models regardless of the discipline to which they are applied.

B. Elementary Systems Analysis Modeling Concepts

It has already been stated that systems can be considered as bounded sectors of reality. One of the early considerations in a simulation study is the boundary of the world to be included in the model. The fact that models are selective requires that system boundaries be defined and assumptions be made concerning the way the enclosed system interacts with the world which lies outside its boundaries.

When a system has been defined, the objectives of the simulation study determine what the model of the system will look like. It is apparent that there can be many models of the same system. Models differ from each other because: (1) the theories they formalize may be different, and (2) they employ different technical mechanisms.

An example of a system which can have many models, i.e., theories about how it operates, is a marine fishery. For a particular fishery there can be biological, economic, or operations models. Each model is different, viewing the system from a different point of view, and yet all models may include similar elements. This is so because variations in assumptions about how subsystems interact or differences in degrees of detail specifying systems structure are possible.

Systems are distinguished from one another by their static and dynamic structures. The entities (objects such as fish, people, boats, markets, etc.) that make up a system, along with their associated attributes (characteristics such as size, age, weight, economic status, etc.) and membership relations (connections between entities, such as spawner-recruit relations or individual boat, fleet member, or cooperative interactions) define its static structure. The activities in which these entities engage define its dynamic structure. It is possible to have systems with identical static structures but different dynamic structures and vice versa.

A system is said to be in a certain state when its entities have properties unique to that state. When looked at from a point in time, a system model is in one of a large number of states. Systems simulation is the manipulation of a model to reproduce the operations of a system as it moves through time. As such, a systems analyst is concerned with techniques which move a system from state to state, and with techniques which permit inferences from these movements.

A simulation model is essentially a system-state generator. Given an initial system state, it moves a system to new states using information contained in the system, extracted from previous state changes, and communicated from outside the system boundaries.

A systems model's structure is affected by a whole host of factors. These include:

(1) the purpose of the model,
(2) the accuracy and precision required of output,
(3) the detail required to achieve required precision,
(4) the assumptions required, and
(5) the availability of data.

System analysis modeling is considered to be an iterative process. Several stages are involved in the modeling sequence. These include:

(1) problem definition in general terms, definition of gross system boundaries, and general output requirement definition;

(2) statement of initial assumptions, structure definition, construction of minimum system model, and preliminary assessment of assumptions;

(3) determination of input data requirements and availability and modification of model structure if unavailability is high;

(4) determination of output possibilities; and

(5) precise specification of final model including: programming language, model plans, and reexamination of previous steps.

It usually costs a great deal of money to program a systems model and run it on a computer. Generally speaking, it seems that models directed at specific goals are more efficient than general purpose models.

In summary, it can be said that a system simulation model is a flexible tool, enabling a situation (such as a marine fishery) to be studied in more detail than heretofore possible. It is (at least for fisheries science) relatively new and untried. It offers the promise of relatively great rewards when properly applied. Some specific examples follow an initial conceptual scheme development.

C. Fisheries Conceptual Scheme

A systems approach seems desirable to permit considering simultaneously not only biological properties of populations of fishes and their behavioral responses to various stimuli, but also the complex of environmental, social, economic, and equipment variables which affect the outcome and ultimate success of a fishery. Even beyond this, the judgment of experienced fishermen (or possibly the operators of a command and control center) must be effectively integrated with the techniques and results of systems analysis. Few people involved with fisheries problems would argue that there is an effective substitute for judgments made by experienced people. Decisions on various fishing operations can be based on both experience and systems analysis as illustrated in Section IV.

Applications of experimental design and simulation modeling techniques are proposed as a conceptual plan for optimizing the exploitation efficiency and yield of a fishery. The model is conceived as a symbolic scheme exhibiting the interrelatedness as closely as possible of the functions of the total system and enabling continually improved prediction of its output response characteristics for given input configurations. The model must be operational in the experimental sense and is not construed as a strictly mathematical scheme, or an ensemble of a priori concepts to be arbitrarily imposed on system operations. A preliminary plan of the proposed system is shown in Fig. 1 as a series of symbols depicting postulated man–machine combinations and interactions, all combining to produce a desired objective. The purpose of the experimental design is to assist in defining objectives, to test the system to measure objectives, and to determine contribution of various systems blocks to the objectives.

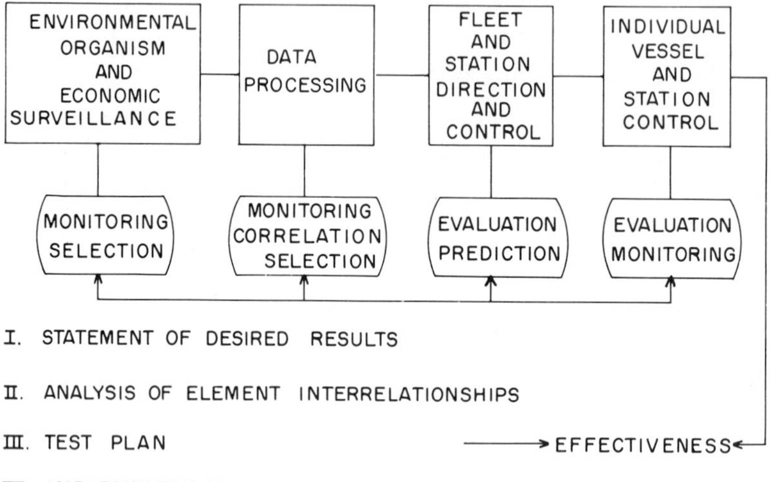

Fig. 1. An indication of the steps involved in the application of experimental design to a complex system. In particular, a symbolic model is shown for a suggested fishery exploitation system.

The objectives of an experimental design include means for measuring improvement of the proposed system over existing or past fishery exploitation methods. By improvement is meant (1) rational distribution of fishing effort to maximize the sustained catch and income from desired species per unit of effort, and (2) improved prediction ability for determining the concentration and composition of catchable stocks in the area under consideration. Criteria of improvement in both of the above cases could be quantified by measuring certain outputs, such as catch per unit investment spent on effort, and index values might be assigned to detection and identification probabilities. Obviously, the absolute magnitude of improvement would vary seasonally with reference to a particular geographic area and species. However, it does not seem prohibitively difficult to obtain indications of the efficiency of the integrated system at all times. Comparisons of current system performance with available catch, effort, and monetary return data from previous years would be used for establishing a base and making continued evaluations.

Briefly, the model suggested is considered to be a symbolic representation of a system—in this case the environment, the organisms, man, and the equipment he employs for surveillance and capture. The concept of a model is predicated on the assumption that it is possible to abstract from a complex system certain pertinent and discernible relationships and to mathematize and quantify these relationships with a view toward describing the behavior of the system under all conditions. Initial steps of modeling consist of deriving concepts that describe the purpose, factors operating, and pertinent parameters or state variables, all of which go toward erecting a frame of reference in which the model is to be operational. This phase is the one in which planned experiments are involved to define the above. Obviously, this has not yet been done in the scheme suggested nor systematically in any existing fishery. However, this work is entirely possible with existing equipment given adequate funds and time. When these are available the model could establish a theoretical experimental situation within which the entire system could be evaluated with respect to the previously defined criteria of effectiveness.

The preliminary flow chart, Fig. 2, gives some indication by means of

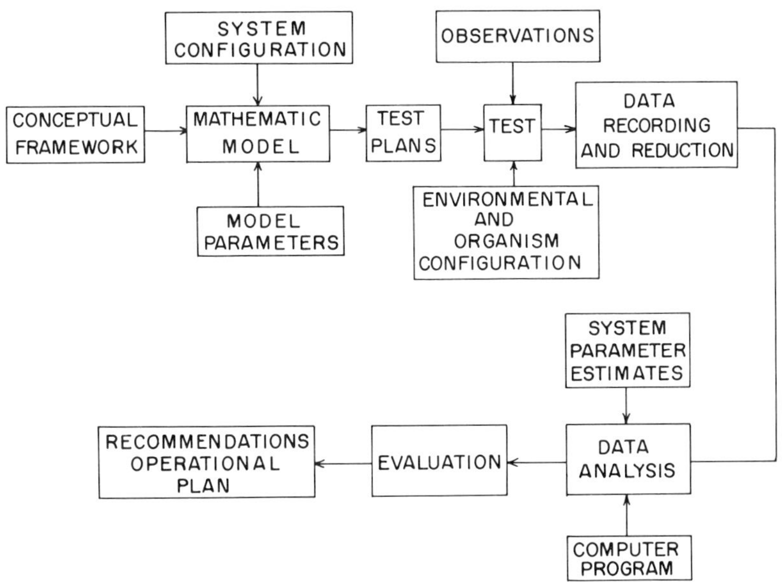

FIG. 2. A generalized flow chart of the steps involved in system modeling for a fishery exploitation system.

a block diagram of a first approximation toward the distinctive logical and sequential steps in system modeling. It is submitted that test data from carefully planned experiments could provide the necessary quantitative empirical data to fill out the model and to evaluate it. Some of these are already available for certain fisheries. The model would then be continually improved and modified with new test data but would be regularly utilized to describe and predict the response of the system to the various inputs which are time-variable.

The conceptual approach suggested above for the analysis of a complex system, such as a marine fishery, involves certain steps. The approach is based on assuming the existence of systems and that these systems can be conceived to consist of elements and element pair relations. The analysis of any complex system seems to involve:

(1) preliminary selection and classification of the system elements affecting the predesignated problem,

(2) initial determination of the existence of direct relations between element pairs,

(3) restatement of system elements and direct relations to achieve greater consistency,

(4) mathematical modeling of those elements and relations which lend themselves to it,

(5) model evaluation in terms of completeness or adequacy,

(6) description of relations not represented by the mathematical models, and

(7) judgment integration of the model and description.

It should be appreciated that any conceptual representation of a system, since it is not the system itself, can represent only some of the system's attributes and interrelationships. For this representation to be meaningful, it must structure all those attributes and interrelationships which are relevant to the purpose for which the system is analyzed. In addition, the representation must be fairly easy to manipulate. Finally, the judgment of the fisheries manager or operator must be integrated with the techniques of systems analysis. The integration of judgment with results of systems analysis is believed to embody the concepts of a priori, a posteriori, and conditional probability.

The promise of such an approach seems technically and, perhaps soon, economically feasible as well. However, details such as particular oceanographic equipment configurations for data gathering, the electronic and data handling components that go into data acquisition, interrogation sequences, and computer configurations are still rapidly developing, and it would seem unreasonable to specify these more precisely at present. These elements, among others, must be structured for the specific problem and its objectives.

III. Application of Systems Analysis—The Georges Bank Haddock Fishery

A. Introduction

This section is based on a study, financed by contract with the National Council on Marine Resources and Engineering Development, Executive Office of the President. The Council's approval to utilize portions of the work is acknowledged. However, the findings, recommendations, and opinions expressed in this section do not necessarily represent viewpoints of the Council. The entire study is entitled, "A Systems Analysis of Specified Trawler Operations" and was performed by Marine Technology, Inc., a division of Litton Industries, for whom the author served as biological consultant. The reports of this study have been released through the Clearinghouse for Federal Scientific and Technical Information and are available as: PB 178 661, "A Systems Analysis of Specified Trawler Operations, Vol. I, Executive Summary"; and PB 178 662, "A Systems Analysis of Specified Trawler Operations, Vol. II, Detailed System Description."

Although the objectives of this study involved construction and testing of a generalized fishing systems model, the specific data base was derived from the Boston otter trawl operations on the Georges Bank haddock fishery. A fishing system as used herein is defined as a collection of equipment used for the purpose of commercial fishing. At a minimum, the fishing system consists of a fishing platform, propulsion machinery, and fish capture and handling gear. At the other extreme it could include a whole fleet of vessels.

The overall orientation of the model to be described is economic/financial. However, biological and technological considerations are included but not at the level suggested in the conceptual scheme of the previous section. The group involved with this study included professional people from several disciplines, such as engineering, economics, biology, and finance. After a brief consideration of all elements of the systems flowchart, a detailed description of the biological submodel is provided in order to illustrate the nature of the program, and the flexibility of the submodels which may be incorporated into any systems model.

B. Flow Chart Description

A graphic representation of the systems model is shown in Fig. 3. It illustrates in a step-wise fashion the elements of the planning model. It begins with opportunity identification and proceeds through the elements of the model to analysis procedures which include detailed reports.

1. *Opportunity Identification*

A general and specific opportunity system function are illustrated in Fig. 3. General opportunity may be defined as the totality of investments or alternative courses of action available to the entrepreneur or policy planner. The entrepreneur or policy planner is obliged to search the general opportunity set by intuitive, empirical, or analytical methods. In terms of this problem, the search of the general set of opportunities could be considered terminated by the decision to engage in a fishing enterprise or in determining policies for the fishing industry. The above-mentioned decision, for the purposes of the model can be an individual or group-level decision or a national policy decision.

Specific opportunity identification occurs when the decision to enter into fishing activity has been made. The specific opportunity level operations should lead the fishing operator to "go into business" in a fishery at a given location with the requisite physical equipment. In this case we are concerned with the Boston haddock fleet exploiting the Georges Bank haddock fishery.

There are at least two groups of contrasts and their effects on fishing which must be considered. They are physical resources and the socioeconomic situation. Among the physical resources requiring assessment are:

(1) availability of stocks of desirable species,
(2) availability of capital,
(3) availability of suitable labor, and
(4) availability of machinery and technology.

Among the socioeconomic considerations are:

(1) the need for goods and services planned,
(2) public restraints, and
(3) type and strength of potential competition.

The above lists are only indicative of the elements which should be considered by prospective fishermen or policy planners. After some form of analysis of these general considerations it is necessary to hypothesize alternative methods of accomplishing the specified system objectives.

2. *Alternative Systems Design Specifications*

In this study both entrepreneurial and environmental aspects of commercial fishing were considered. In addition interactions and stability were evaluated. A large amount of information on conventional side trawlers, stern trawlers, transport boats, and factory trawlers was collected and organized for use in developing specifications for alternative systems. This information included: boat description and function; general characteristics; plan and profile views; initial construction cost function; daily operating cost function, where applicable; on-board processing equipment costs, where applicable; daily operating costs of trawlers when operated on a non-lay basis, excluding fuel and crew cost; daily fuel cost for diesel powered trawlers; payload capacity of fishing boats; main engine maintenance and repair costs; estimated cost of trawler fish-capture gear; a method of estimating the approximate scrap value of a trawler; and an approach for estimating the standard fishing day coefficient. From the above-mentioned information, a systems analyst can design alternative systems in terms of initial and operating costs. The following list of variables were identified as having significant implications in the analysis of fishing systems. The capitalized parenthetical name is a symbolic mnemonic description of the variable (up to and including six alphanumeric characters).

a. *Fishing System Operational Variables. Standard Fishing Day Coefficient* (COSFDY): an empirical measure that relates performance of one fishing system to another. Thus, boats of the Boston trawler fleet of between 200 and 249 gross registered tons and with 400 to 765 installed horsepower (main engine) have a standard fishing day coefficient of 1.00. Appoximately 50% of the Boston offshore trawler fleet falls within this group. This group is also referred to as Size Class 3a in Bureau of Commercial Fisheries literature. The BCF definition of this variable was modified by including installed horsepower as a parameter used in the estimation of the standard fishing day coefficient.

The standard fishing day coefficient can also be considered a system performance index that expresses operational efficiency. Thus, if the analyst wishes to evaluate cost benefits of equipment or improved on-board fish-handling methods which will result in catching more fish, it will be necessary to estimate a new standard day coefficient that would reflect improvements to the fishing system. The incremental cost of installing and operating the improved system must then be reflected by modification to selected cost variables.

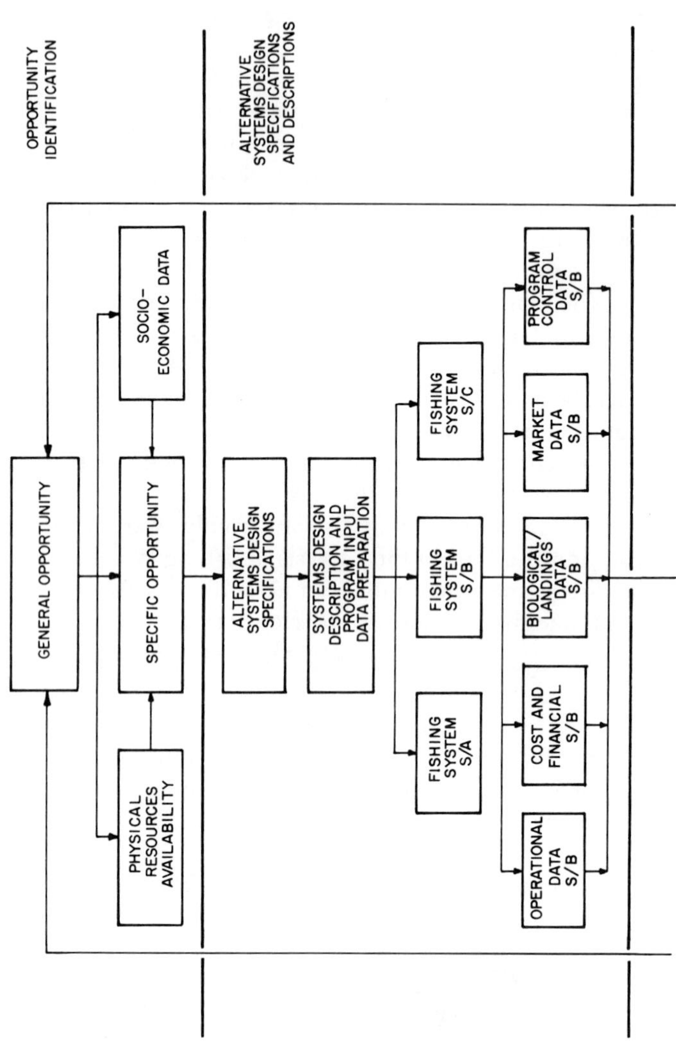
OPPORTUNITY IDENTIFICATION
ALTERNATIVE SYSTEMS DESIGN SPECIFICATIONS AND DESCRIPTIONS
GENERAL OPPORTUNITY
SOCIO-ECONOMIC DATA
SPECIFIC OPPORTUNITY
PHYSICAL RESOURCES AVAILABILITY
ALTERNATIVE SYSTEMS DESIGN SPECIFICATIONS
SYSTEMS DESIGN DESCRIPTION AND PROGRAM INPUT DATA PREPARATION
FISHING SYSTEM S/C
FISHING SYSTEM S/B
FISHING SYSTEM S/A
PROGRAM CONTROL DATA S/B
MARKET DATA S/B
BIOLOGICAL / LANDINGS DATA S/B
COST AND FINANCIAL S/B
OPERATIONAL DATA S/B

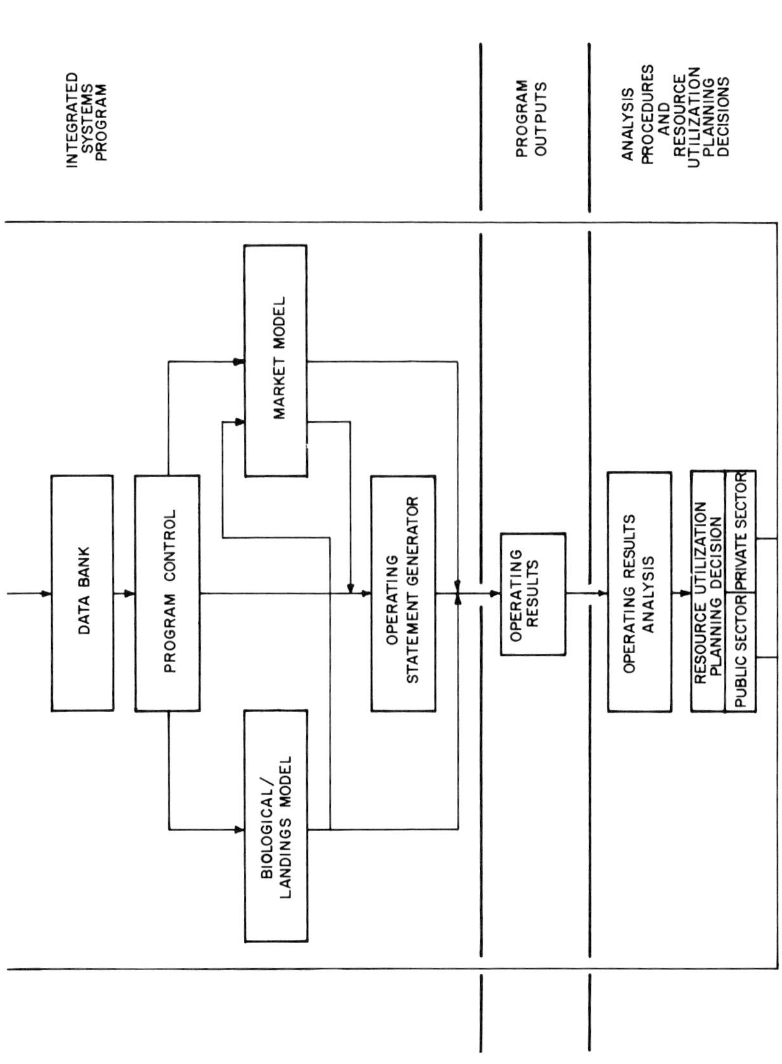

FIG. 3. Generalized resource utilization flow chart within which the Georges Bank haddock fishery model was developed.

If the analyst chooses not to specify a given standard day coefficient, he may specify a range of values and evaluate the systems improvement in terms of *return on investment* (ROI) by specifying a desired acceptable rate and selecting the standard day coefficient that is appropriate to the specified return on investment.

Crew Size (CREW): the number of individuals aboard the fishing system engaged in fishing or operating the fishing system. This value may be specified for a single boat system or a system involving a group of boats.

Calendar Days per Trip Coefficient (COTRIP): the average number of days per trip (port to port) for a given fishing system. Thus (DAYS)/(COTRIP) is equal to the number of trips made during the specified period, and is rounded to an integer value.

Calendar Days at Sea Fishing (DAYS): the number of calendar days expended fishing by a given fishing system during a given period in the waters of an identifiable fishery.

Catch Distribution Factor (CDIST): the catch distributed by fish product by year based on the reference catch. Catch is distributed on the basis of the model period annual landings in metric tons of the reference fish product. In this study the empirical data was for haddock, thus making haddock the reference fish product.

U.S. Catch Distribution (USDIST) :the U.S. Catch Distribution if different from CDIST. If the REPDEL option is used, USDIST is required to specify the species caught by year. The REPDEL option refers to the specific elements of the U.S. catch and is explained more fully in later portions of this listing.

b. *Fishing System Cost and Financial Variables. Fishing System Platform Cost* (CPLAT): the initial or replacement cost of the hull and machinery components of a fishing system.

Fishing System Gear, Equipment, and Net Cost (CFAPP): the initial or replacement cost of fish-handling and capture gear.

Cost of On-Board Processing Equipment (COBPRO): the cost of on-board fish product processing equipment such as filleting, packaging, freezing, and by-product manufacturing equipment and accessory machinery.

Subsidy (SUBSID): that portion of the cost of the fishing system paid for by others, primarily government subsidy monies.

Scrap Value of the Fishing System (SCRAP): that minimum value assignable to the fishing system when brought to scrap conversion. This value is used as a lower boundary for the fishing system disposal value in the present value computation.

Joint Trip Fixed Cost (COVTJC): a coefficient used to allocate those fixed costs which are trip-dependent and are deducted from gross revenue before that boat or crew share is allocated, such as the additional compensation to the engineers, mate, sounding machine cost, shore guards, radar, fish handlers, and icing-up.

Joint Variable Costs (COJVAC): a coefficient used to allocate those joint costs that are revenue-dependent and that are deducted from gross revenue before the boat or crew share is allocated (e.g., wharfage charges, scales charges, exchange fee, welfare fund, and ice for three months).

Crew Share Coefficient, Lay Distribution (COCLAY): that portion of the revenue after joint costs have been deducted that is distributed by agreement, lay, to the fishing system crew.

Wages per Day per Man (RATWAG): wages paid to fishing systems crews including fringe benefits when no lay agreement is used.

Other Variable Costs to Gross Income (OVCPCT): other variable operating costs, excluding labor, incurred against the gross revenue for the period being studied when a non-lay financial statement is to be generated, such as fuel, lubricants, etc.

Daily System Operating Cost (DAYSOC): a daily operating cost incurred for services, equipment lease, and carry boat costs, that is charged against the fishing system.

Rate of P & I Insurance per Crew Member (RATCIN); the basic annual cost per crew member for personal and indemnity insurance.

Change in P & I Insurance Rate per Year (DELCIN): the annual increase or decrease in personal and indemnity insurance per crew member, RATCIN.

Hull Insurance Rate on Market Value (RATIMV): the hull and machinery insurance rate based on fair market value.

Hull Insurance Rate on Replacement Cost (RATIRC): the hull and machinery insurance rate based on replacement cost.

Repair and Maintenance Rate on Fixed Assets (RATRAM): the rate of repair and maintenance per year for the specified fishing system and as associated to the cost of the fishing system, CPLATT + CFAPP.

Annual Change of Repair and Maintenance Rate of Fixed Assets (DELRAM): this variable allows for the increase of repair and maintenance costs as a function of fixed asset age.

Inflation–Deflation Rate (RATINF): a rate of inflation or deflation that is indicative of either increasing or decreasing operating cost as a function of time.

Administration Rate (RATADM): the rate of application of administrative expenses to the enterprise.

Working Capital Factor (COWCAP): a factor used to create the first period working capital account on the basis of the initial cost of the fishing system (CPLAT + CFAPP + COBPRO) − SUBSID.

Liabilities Factor (COLIAB): a factor used to create the first period liabilities account on the basis of the initial cost of the fishing system (CPLAT + CFAPP + COBPRO) − SUBSID.

Tax Life (TAXLIF): the nominal tax life, for depreciation purposes, assigned to the fishing system.

Composite Market Life Expectancy of Fixed Assets (COMLIF): the estimated usable life of the fishing system.

Tax Basis for Corporation Normal Tax (TAXBAS): that value over which a corporation must pay a corporation surtax, Federal income tax.

Corporation Normal Tax Rate (TAXROB): the corporation's normal tax rate, Federal income tax.

Corporation Surtax (TAXREX): the corporation surtax rate, Federal income tax.

Rate of Interest of Borrowed Capital (RATINT): the basic annual interest rate paid by the operator for the use of borrowed funds used to purchase the fishing system.

Rate of Interest for Present Values Calculations (PVINT): the econometric rate of the value of money, usually higher than RATINT.

c. *Fishing System Biological and Landings Variables* (i) The Biological Model, Beverton–Holt. *Instantaneous Rate of Natural Mortality* (RNMORT): the Beverton–Holt term (M), instantaneous rate of natural mortality.

Instantaneous Rate of Fishing (FBH): the Beverton–Holt term (F), instantaneous rate of fishing.

Slope of the Walford Line (WALK): the Beverton–Holt term (k), slope of the Walford Line.

Age at Recruitment (AGEREC): the Beverton–Holt term (t_l), defined as the age in years at which a fish is recruited to the fishery and is vulnerable to catching.

Standard Minimum Age of Reference (AGEMIN): the Beverton–Holt term (t), defined as the minimum age at which the fish can be retained, captured, by the fishing gear in use.

Age at Zero Length (AGEZER): the Beverton–Holt term (t_0), a parameter of the Von Bertalanffy growth curve within the Beverton–Holt model.

Ideal Maximum Weight of the Fish (OMEGA): the Beverton–Holt term (W_∞).

Best, Maximum, U.S. Catch per Standard Day Experience to Date from the Fishery (TOPCAT): the best, to date, historical experience in catch per standard day for the fishery by U.S. fishermen.

Average Catch per Standard Day Experienced To Date from the

Fishery (AVGCAT): the average, to date, historical experience in catch per standard day for the fishery by U.S. fishermen.

Landed Fish This Year That Were Vulnerable at the Start of the Year (VULCAT): the weight of the total annual landings of fish that were vulnerable at the start of the year. Up to ten years of historical or predicted data may be used.

Rate of Exploitation (RATEX): the fraction of the fish in a population at a given time which is caught and killed by man during a specified time interval immediately following (usually a year). The term is applied to separate parts of the stock; in this case it applies to the landed fish vulnerable at the start of the year, VULCAT.

Yearly Total Catch for the Fishery (BIOCAT): the total catch of the specified fishery based on historical or predicted data, and is used in the computation of the surplus production function.

(ii) The Landings Model. *Quadratic Coefficient of the Empirical Landings per Standard Day Equation* (EMPA): the quadratic coefficient for use in the empirical landings model.

Linear Coefficient of the Empirical Landings per Standard Day Equation (EMPB): the slope of the empirical landings per day equation. For the haddock fishery this value is derived from the Woods Hole Biological Laboratory empirical data.

Minimum Standard Fishing Days Used in the Empirical Landings Model (EMPMIN): the lower boundary value of effort for the year, USAEFF, for use in the empirical landings model for the haddock fishery data as derived from Woods Hole Biological Laboratory studies.

Maximum Standard Fishing Days Used in the Empirical Landings Model (EMPMAX): the upper boundary value of effort for the year, USAEFF, for use in the empirical landings model for the haddock fishery data as derived from Woods Hole Biological Laboratory studies.

Seasonality of Landings Factor (SEAFAC): a period-specified factor that permits the adjustment of period landings value for seasonality.

U.S. Effort for the Year (USAEFF): used as the entry value for the landings model within the limits of EMPMIN and EMPMAX (see above), and is stated in terms of U.S. expended standard fishing days.

Landings per Standard Day This Period (GIVCAT): the use of this variable obviates the use of the empirical landings per standard fishing day equation. This variable may be specified for several periods, and thus introduce the effect of a declining or increasing catch as a function of time.

d. *Fishing System Market Variables.* There is in the computer program a (BETA) matrix. This matrix contains the fish product price computation coefficient for six fish product price equations, each with a maximum of 12 coefficients over four seasons or periods. The BETA matrix is not normally altered unless another econometric hypothesis is to be introduced.

There is also a (ZNOW) matrix that contains the values of the exogenous variables and intercept terms that determine the landed price of fish, such as cold storage, holdings, landings, price indices, etc. The initial values of each of the variables are stored in a third matrix (ZSTORE). A fourth matrix (CZ), specifies the time pattern of growth of the ZNOW matrix. The ZNOW matrix can be generated with linear, quadratic, or exponential functions of time and the initial values located in ZSTORE. The CZ matrix thus contains the coefficients which are to determine the exact growth pattern of the ZNOW values from the initial ZSTORE values. The price of landed fish is determined by specified operations on the current ZNOW, determined by time, CZ, and ZSTORE, and by the BETA matrix.

In order to operate on the three matrices, additional variable or indicator data is required: the number of variables (plus 1 for a constant term, if any) in the price equation (VARMAR); the number of fish species for which landed prices are to be computed (FTYPES); the number of variables, ZNOW, replaced by specific elements of the U.S. Catch (REPDEL); the species replacement indices of the variable values of the ZNOW matrix to be modified (SPCRPI); an indicator flag for the BETA matrix that indicates what function form is to be used in the ZNOW matrix (FLGJI); the variable, FLOW, used to adjust the values of the flow variables to conform to the BETA matrix (FLOW); the time period, in months, of flow variables required as input to the price equations (BB); the time period in months, of the flow variables located in the ZNOW matrix [Together with BB, FR transforms ZNOW so as to be compatible with the BETA matrix (FR)]; the factor required to convert prices generated in the price equations to common units of dollars per thousand pounds (BB2).

The resultant data from the biological model are operated on by the following conversion factors: (1) Landed weight to market weight (EQWT). This conversion factor adjusts the whole, round weight as taken from the water to the weight brought to the dock. (2) The number of pounds per unit weight of catch (PNO). This factor converts catch to 1000-pound units.

e. *Integrated Systems Program.* *Maximum Number of Years* (YRMAX): the specified number of years for which operating statement generator results are desired.

Number of Periods per Year (PERPYR): the specified number of financial "reporting" periods per year. This may be one, two, or four.

f. *Input Data Format.* No details of the type of input data forms required for the Integrated Systems Program are given here. Detailed program operating instructions are given in Appendix I of Vol. II of the final report.

3. *The Integrated Systems Program*

The Integrated Systems Program components are: a data bank section, a program control section, the landings and biological models, a market model, and an operating statement generator. The functioning of the data bank, the landings and biological models, and the market model are administered by the program control section. Input data and computed data are moved to the operating statement generator for compilation into financial results for each run.

a. *The Data Bank Section.* The primary function of this section is to hold and release data to the operational portions of the Integrated Systems Program. For a given set of fishing system configuration variables, the systems analyst can establish a data baseline in the data bank. He can then work with a single variable within the program, a CASES variable exercised over a range of values, and observe the effect induced on the systems performance measures when the numerical value of this variable is changed. The analyst may then establish a new data base, describing a new system configuration, and manipulate the variable as a new case and observe the effect on the measures of systems effectiveness. This procedure will thus allow the analyst to heuristically optimize a given system or to make trade-off comparisons between system configurations.

b. *The Program Control Section.* The function of this section of the program is to coordinate the movement of input data through the data bank to the landings and biological models, the market model, and the operating statement generator. The input control section checks for the correctness of the input parameter names, but not on their values.

c. *The Biological and Landings Models.* The effective exploitation of a natural resource, such as a fishery, requires that the operator or other regulative body have an understanding of the cause-and-effect relationships associated with the resource.

One of the most useful concepts developed in fisheries science to understand this relationship is that of equilibrium yield per recruit or recruitment. This concept is not only of theoretical interest but also provides the basic working method utilized by fisheries managers. Equilibrium yield calculations permit predictions regarding effects of fishing intensity as well as the effects of gear modification or changes. The historical background of equilibrium yield theory dates back to 1918, but the recent developments and significant advances were made during the past decade. The two currently utilized theoretical biological models of exploited fisheries are the dynamic pool model, which is due to Beverton and Holt, and the parabolic surplus production model, due to Schaefer.

The Beverton–Holt model has been developed for a demersal trawl fishery involving cod, plaice, and haddock. The Beverton–Holt model assumes constant recruitment and a constant natural mortality rate. The von Bertalanffy growth function is incorporated within this model. For the purposes of this study, the Beverton–Holt model importantly distinguishes between numbers of fish of an entering year class and the time they become vulnerable to a particular type of fishing gear. This permits the assessment of fishing technology as well as effect of various levels of fishing effort.

In the case of Georges Bank, reliable age composition data are available for the catch over a period of years as well as statistics of fishing effort in units proportional to the fishing mortality generated. In the haddock fishery with which we are concerned, age composition is expressed in units of catch per unit of effort, which is a valid index of abundance from which the total mortality coefficient may be estimated from a decrease in abundance of year-classes. The natural mortality coefficient was roughly estimated, and the parameters of the Von Bertalanffy growth function were empirically derived from size-at-age data.

In utilizing this model, it should be appreciated that although the von Bertalanffy growth function is justifiable on physiological grounds, it is only an approximation to the true growth function. Secondly the assumptions involved, including constant recruitment, constant natural mortality, and a growth rate independent of population size, are not valid over a large range of population sizes. However, for the purposes of the biological submodel, these simplifying assumptions do not defeat the usefulness of the model. The assumption of constant recruitment can be circumvented by expressing yields in terms of yield per recruit.

The Schaefer model differs from the Beverton–Holt model in that rate of recruitment, rate of growth, and mortality are combined into a single function of the exploited population. The assumptions implied in this model are: that the rate of natural increase of population responds immediately to changes in population density and that the rate of natural increase at a given weight of population is independent of any deviation of the age-composition of the population from the steady state age-composition of that population weight. Historically, the Schaefer model has been applied to tropical fisheries, where age determination is difficult, and to unexploited fisheries. The Integrated Systems Program has provision for two landings models and the theoretical Beverton–Holt model. The landings models have the following features:

Landings may be computed by the program utilizing the empirical data supplied by the Bureau of Commercial Fisheries, Woods Hole Biological Laboratory. The empirical relationship used in the model is

$$Q = a_0 - \frac{dQ}{dE} E,$$

where Q is the catch (landings) in metric tons; a_0 is the Q-intercept at $E = 0$, EMPC; dQ/dE is the rate of catch per unit of effort, EMPB; and E is the annual U.S. fishing effort in terms of standard fishing days, USAEFF. The value of E is an input variable, USAEFF, and may be established for each year, up to 10 years, on the basis of historical data or predicted data. Thus, landings per year are computed and adjusted for seasonality and system fishing effort.

Landings per standard day for each period, up to 40 periods, may be introduced into the program as an input profile. This data may be based on historical or predicted values. Landings for the period are then adjusted by the system fishing effort.

The biological model of Beverton and Holt has the following characteristics. Assuming a parabolic surplus function and having certain data available, the following population dynamics data may be computed: the maximum stock, the biologically-indicated fishing effort, the recommended change in fishing effort, the resultant change in catch, the number of fish at recruitment, and the Beverton–Holt yield in pounds per million fish at the standard minimum age of reference. The Beverton–Holt relationship can, with appropriate input data, be used to make comparisons between system-applied fishing effort and theoretical equilibrium yield effort. The landings models and the Beverton–Holt model are so structured that variable values relative to species other than the haddock can be inserted for purposes of studying other fisheries and fishing systems, viz., the herring fishery using purse-seining methods.

At this time, the lack of firm, researched, biological data regarding the seasonality of catch in the Georges Bank waters precludes the use of any rigorous analytical or accurate empirical statement of seasonality. However, as incorporated in the landings model, provision has been made for the comprehension of seasonality of catch and may be specified by the systems analyst as landings data input.

d. *The Market Model.* The net returns to the operator in the fishing industry are dependent on both revenues and costs. The market model deals with the first of these factors, that of the determination of gross revenues. To do so, the prices to be received by operators for particular fish products need to be calculated. Standard economic analysis classifies factors affecting demand as: tastes and preferences of consumers, prices of substitute commodities, and prices of complementary commodities. These factors influence the quantities persons are willing to buy at alternative prices. The relationship of quantity demanded and price is usually expressed as the demand or demand schedule. Classically it can be displayed as a curve plotting alternative prices and the respective quantities demanded, as in Fig. 4.

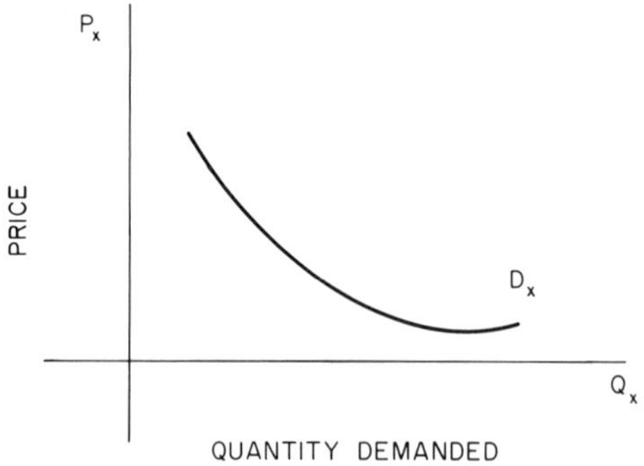

FIG. 4. A curve representing the form of the relationship between the quantity of fish demanded versus the price of fish, where D_X is the demand schedule for fish product X.

Equilibrium price is that at which neither potential sellers nor buyers have an incentive to offer prices different than the current market price. Thus, equilibrium is determined by the intersection of the demand and supply schedules, where the supply schedule indicates the quantities sellers are willing and able to sell at alternative prices. In Fig. 5, D_x is as in Fig. 4, and S_x is the supply schedule. The equilibrium price is then q_e. To predict the price which will result in a free market, we must determine the equations describing the demand and supply schedules and then solve these equations for p_{e_x} and q_{e_x}. There are limitations on how well this can be done. The complexity of the true relationships may involve random factors or a number of variables so great as to make the problem practically unmanageable. Under some circumstances it is, fortunately, possible to arrive at statistical approximations of the demand schedule. These approximations can be used so that various landings quantities, which represent the supply, can be translated into prices and these prices used to calculate revenues.

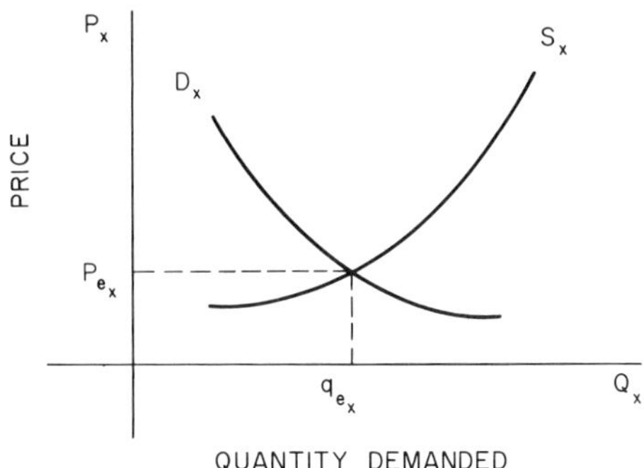

FIG. 5. The form of the relationship between the quantity of fish demanded versus price, where D_x is the demand schedule and S_x is the supply schedule (see text).

The market model has been designed to comprehend up to six fish products, with up to 11 variables in the supply and demand time function schedules. In the case of the haddock fishery all other fish—cod, hake, etc.—can be treated as a lumped market and the revenue derived. Also, in the case of the haddock fishery, the differentiation between adult and scrod may be accommodated with two sets of market function equations.

The market model is also capable of dealing with seasonality effects in demand. This has been accomplished by allowing the input of up to four matrices of seasonality coefficients.

To sum up, depending on the quality of input data, the market model can provide a reasonable conversion of units of specific fish products yielded by specified fishing systems during the time period under study in terms of dollars of gross revenue.

The modified market model used in this study was derived from research by J. F. Farrell and H. C. Lampe of the University of Rhode Island.

e. *The Operating Statement Generator.* The operating statement generator is the terminal section of the Integrated Systems Program. This section prepares an operating statement of the financial results generated by the operation of a given fishing system over time. This report is a combined operating results statement, a financial position statement, and a statement of the measures of system cost performance. Return on investment and present value of the system earnings flow have been selected as the two systems cost performance measures.

Biological and landings data and market data are also produced. The following elective arrangements are available to the analyst.

Financial results may be elected to be computed quarterly. These may be summarized annually or semi-annually. The number of periods over which the financial results are computed may be specified. Thus, if the analyst wishes to do a 10-year, 40-quarter operating analysis, he may do so. However, he must provide adequate input data for the number of periods to be studied. The availability of biological data and market-model data will be determining factors in the life of the analysis, as neither of these environments are static in the real world.

Distribution of labor and other operating costs may be specified as a shares option or as a paid-wages option. Thus, fishing systems that are not based on lay agreements can be treated, as might be the case in certain factory trawler systems.

4. *Integrated Systems Program Output Data. Output Listing Dictionary*

The following is a listing of the Integrated Systems Program outputs. The list is organized in the same order as the documents are produced by the computer system printer.

(1) Landings Model Output Data
System standard days by period of year.
Computed landings per standard day, in metric tons, by period or year.
Model period landings, in metric tons, by period or year.
Model annual landings by year.

(2) Market Model Output Data
Fish product weights in thousands of pounds by year or period.
Market model computed prices in dollars per hundred pounds by year or period.
Market model computed revenue by fish product by year or period.

(3) Operating Statement Generator Output Data
a. Beginning Year Balance Sheet
Fixed Assets (FASSTS).
Additions to working capital (ADDWC).
Working capital (WC).
Total assets (TASSETS).
Liabilities (ALIAB).
Net worth (ANETW).

b. Period per Annual Income Statement by Period with Annual Summary
Gross revenue for the period (PERREV).
Variable joint and trip-fixed costs for the period (VJTFC).[a]
Joint variable costs for the period (VJC).[a]
Net revenue for the period (REVNET).[a]
Labor share for the period (SHLABR).[a]
Labor costs for this period (ALABC).[b]
Straight-line depreciation for the period (SLDC).
Interest charge for the period (CHINT).
Fishing system insurance charges for the period, for hull and machinery, based on market value (CHINMV).
Fishing system insurance charges for the period, for hull and machinery based on replacement cost (CHINRC).
P & I insurance charges for the period (PIINS).
Repair and maintenance charges for the period (RM).
Administrative costs for the period (ADC).
Gross operating profit for the period (GOP).
Gross operating profit, after taxes, for the period (GOPAT).
Gross operating loss for the period (GOL).
Gross operating profit, after taxes, for the period (TPGPAT).
Working capital at the end of the current year (WCEOY).
Total assets at the end of the current year (TASEOY).
Net worth at the end of the current year (ANWEOY).
Accumulated depreciation at the end of the current year (ACDEOY).
Depreciated fixed assets at the end of the current year (DFAEOY).
Depreciated total assets at the end of the current year (DTAEOY).
Federal income taxes for the current year (FTFTY).

c. System Financial Performance Measures
Return on investment, before taxes for the current year (ROI).
Return on investment, after taxes, for the current year (ROIAT).
Return on investment for the current year based on depreciated total assets, DTAEOY (ROIDTA).
Return on investment for the current year, after taxes, based on depreciated total assets, DTAEOY (ROIDAT).
Year-end ratio of ALIAB to DFAEOY, i.e., liabilities to depreciated fixed assets (RATEOY).

[a] Indicates the lay system method of direct labor compensation.
[b] Indicates a nonlay, wages, method of direct labor compensation.

d. Surplus Production Function and Biological Model Yield Data
The vulnerable stock at the start of the current year (VULSTK).
The computed instantaneous fishing rate (F).
The maximum stock in metric tons (WINF).
The instantaneous rate of increase in stock, K (RATK).
The production function F value, RATK/2 (PF).
The equilibrium catch in metric tons (QMXCAT).
The indicated recommended change in fishing effort, in percent + or −, (F-PF/F)100, (PCTBIF).
The indicated change in catch, in percent (PCTQMC).
Number of fish at recruitment (RECNUM).
Beverton–Holt yield Y in pounds of fish per million fish at the standard minimum age of reference (YIELD).

e. Present Value Analysis

At the end of each operating statement generator run period, as specified by YRMAX, a subroutine within the Integrated Systems Program will perform a present value analysis on the earnings, profits or losses stream accrued by the fishing system operators.

Typical of the outputs of this subroutine are: the present value of the investment opportunity, the resale value of the system in the final year of analysis, YRMAX, and the highest attainable present value for specific year spans.

It is clear from the diversity of both input data and program outputs that a fantastic number of operating result analysis procedures and resource utilization planning decisions are possible. The Integrated Systems Program provides the user with a rapid and flexible methodology to evaluate alternative fishing system designs, management policies, and operational strategies.

C. Detailed Biological Model

Figures 6, 7, and 8 illustrate the detailed flow charts of the Biological Model used for this systems analysis study. The Biological Model is run or exercised once a year. The computation is broken down into two segments. The first segment computes various variables under the assumption of a parabolic surplus production function. The second segment computes the Beverton and Holt yield for a defined or calculated instantaneous rate of fishing.

Assuming a parabolic surplus production function, a rate of exploitation, a ratio of vulnerable to nonvulnerable fish at the start of the year, and a best catch per standard day to date, the following is computed in the first segment; a maximum stock, the biologically indicated fishing effort, the recommended change in fishing effort, and the resultant change in catch.

The second segment of the Biological Model computes: number of fish at recruitment, Beverton–Holt yield in pounds per million fish at the standard, and minimum age of reference.

The instantaneous rate of fishing employed may be calculated in the first segment of the Biological Model or it may be supplied exogenously.

The initial computation of the first segment of the Biological Model is the vulnerable stock present at the start of the year, VULSTK:

$$\text{VULSTK} = (\text{VULCAT})/(\text{RATEX}), \tag{1}$$

where VULCAT is the landed fish this year which were already vulnerable at the start of year, in metric tons, RATEX is the rate of exploitation which is determined from:

(1) the age composition of landings data;
(2) growth, weight, and length data.

The instantaneous rate of fishing, FRATE, is computed in one of two ways. If the yearly catch, CATVAL, is specified by the user as input data, BIOCAT, the instantaneous rate of fishing, FRATE, is computed as

$$\text{FRATE} = (\text{CATVAL})/(\text{VULSTK}). \tag{2a}$$

When CATVAL is BIOCAT, BIOCAT is 0, CATVAL is computed as

$$\text{CATVAL} = \text{USCATB} \times (\text{YRSFDY}/\text{USAEFF}). \tag{2b}$$

The instantaneous rate of fishing, FRATE, becomes a function of CATVAL and VULSTK:

$$\text{FRATE} = \left|(\text{USCATB}) \times \left(\frac{\text{YRSFDY}}{\text{USAEFF}}\right)\right| / \text{VULSTK}, \tag{3}$$

where USCATB is annual landings computed in the Landings Model, YRSFDY/USAEFF is percent of U.S. yearly effort that the computed annual landings represent. Maximum stock, WINF, if no fishing takes place, is a function of vulnerable stock present at the start of the year, VULSTK, best catch per standard day to date, TOPCAT, and average catch per standard day to date, AVGCAT:

$$\text{WINF} = \text{VULSTK} \times (\text{TOPCAT}/\text{AVECAT}). \tag{4}$$

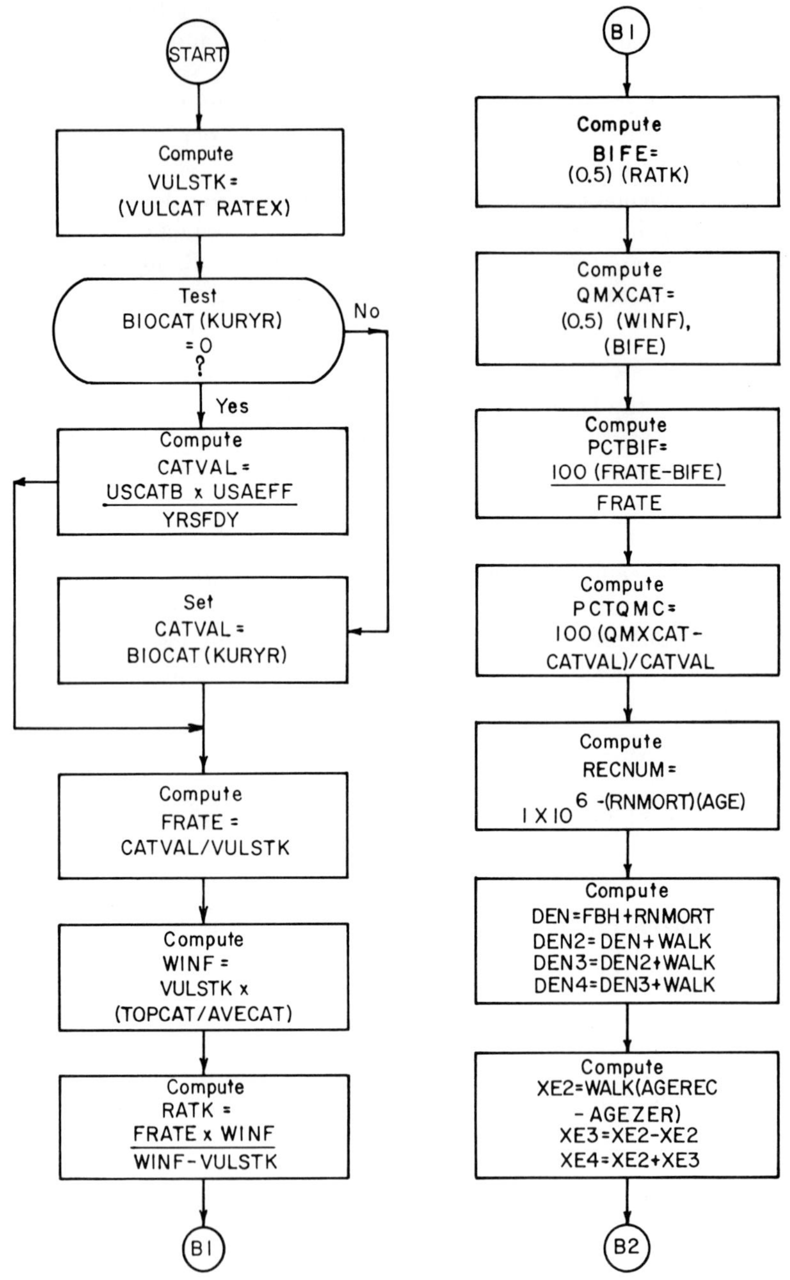

Biological Model flow chart:

FIG. 6. Part I. FIG. 7. Part II.

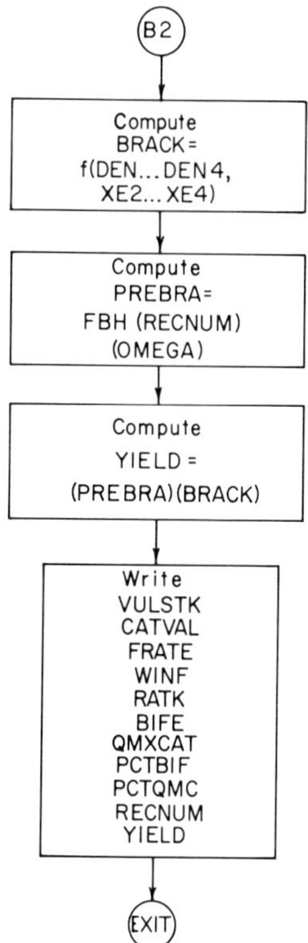

FIG. 8. Biological Model flow chart: Part III.

The instantaneous rate of increase of stock, RATK, is computed by the relationship

$$\text{RATK} = \frac{(\text{FRATE})(\text{WINF})}{\text{WINF} - \text{VULSTK}}, \tag{5}$$

where FRATE is the instantaneous rate of fishing [from Eq. (2) or (3)], WINF is maximum stock [from Eq. (4)], and VULSTK is vulnerable stock present at start of the year.

The biologically indicated fishing effort, BIFE, and the maximum equilibrium catch WMXCAT, are obtained by dividing the computed values of instantaneous rate of increase of the stock, RATK, and the maximum stock, WINF, by the geometric mean. As the parabola is symmetrical, the divisor has the value of 2:

$$\text{BIFE} = \text{RATK}/2, \tag{6}$$

$$\text{QMXCAT} = (\text{WINF})(\text{BIFE})/2. \tag{7}$$

The biologically indicated percent change in fishing effort, PCTOMC, is therefore

$$\text{RCTBIF} = 100(\text{FRATE} - \text{BIFE})/\text{FRATE}, \tag{8}$$

and biologically indicated percent change in catch PCTQMC is

$$\text{PCTQMC} = \frac{100(\text{QMXCAT} - \text{CATVAL})}{\text{CATVAL}}. \tag{9}$$

The initial computation of the second segment of the Biological Model is the number of fish at recruitment, RECNUM, which is computed by the relationship

$$\text{RECNUM} = 1 \times 10^6 e^{-(\text{RNMORT})(\text{AGE})}, \tag{10}$$

where RNMORT is the natural mortality rate (input by user), and AGE is the age at recruitment (input by user) minus the standard minimum age of reference (input by user). The estimation of equilibrium yield, YIELD, is obtained via the Beverton–Holt method incorporating the von Bertalanffy growth equation:

$$\begin{aligned} \text{YIELD} = (\text{FBH})(\text{RECNUM})(\text{OMEGA}) \Bigg[& \frac{1}{\text{FBH} + \text{RNMORT}} \\ & - \frac{3e^{-\text{WALK}(\text{AGEREC}-\text{AGERZER})}}{\text{FBH} + \text{RNMORT} + \text{WALK}} - \frac{3e^{-2\text{WALK}(\text{AGEREC}-\text{AGEZER})}}{\text{FBH} + \text{RNMORT} + 2\text{WALK}} \\ & - \frac{e^{-3\text{WALK}(\text{AGEREC}-\text{AGEZER})}}{\text{FBH} + \text{RNMORT} + 3\text{WALK}} \Bigg], \end{aligned} \tag{11}$$

where FBH is the Beverton–Holt instantaneous rate of fishing, OMEGA is the ideal maximum weight of the fish, WALK is the slope of the Walford line, and AGEZER is the age at which the fish would have zero length if it had grown according to the von Bertalanffy equation. The following is a condensation of Eq. (11) as used in the computer program:

$$\begin{aligned} \text{DEN} &= \text{FBH} + \text{RNMORT} \\ \text{DEN2} &= \text{FBH} + \text{RNMORT} + \text{WALK} \\ \text{DEN3} &= \text{FBH} + \text{RNMORT} + 2\text{WALK} \\ \text{DEN4} &= \text{FBH} + \text{RNMORT} + 3\text{WALK} \\ \text{XE2} &= (\text{WALK})(\text{AGEREC} - \text{AGEZER}) \\ \text{XE3} &= 2(\text{WALK})(\text{AGEREC} - \text{AGEZER}) \\ \text{XE4} &= 3(\text{WALK})(\text{AGEREC} - \text{AGEZER}) \\ \text{BRACK} &= [(1/\text{DEN}) - (3e^{-\text{XE2}}/\text{DEN2}) + (3e^{-\text{XE3}}/\text{DEN3})(-e^{-\text{XE4}}/\text{DEN4})]. \end{aligned}$$

The above is an illustration of the Fortran program for the biological segment of the total systems model. It is clear that this segment may be revised and improved without seriously affecting other segments of the total model. This is another advantage of the systems approach to fisheries science alternative strategy evaluation.

D. Some Results of the Analysis

The model of fishing systems based on the Georges Bank haddock fishery was demonstrated to be a valuable tool in planning and utilizing marine resources. The full potential of the model in evaluating various strategies involving one or more fishing platforms has yet to be realized since the model has been "exercised" in a limited manner to date. A sensitivity analysis was performed, and three of the most sensitive variables considered were: (1) the standard fishing day coefficient, a measure of boat efficiency; (2) the number of days a system is employed fishing in the course of a year; and (3) the level of total US effort for the year in the haddock fishery.

It also became clear due to the multidisciplinary nature of the entire study and the complexity of the program that a group of professional people from several disciplines is required to become familiar with the program and to structure a plan of investigation involving various operating strategies.

IV. Application of Bayes' Theorem for Decision Making in Fisheries Science

A. General

It has been suggested that the judgment of the experienced fisheries scientist or progressive fisherman can be integrated with the techniques of present day systems analysis in a formal manner. The integration of judgment with a specialized technique (systems analysis) poses some unique problems involving both personnel and techniques. The personnel problems remain open but Bayes' theorem appears to be a useful technique for decision making in fisheries science. Its application in other disciplines is fairly widespread. Use of Bayes' theorem requires an understanding of the dual meaning of "probability." This involves the concept of a priori (as opposed to a posteriori) probability, and the concept of conditional probability.

Probability, when considered in one restricted sense, refers to the relative frequency with which some event occurs when one experiment is repeated an indefinitely large number of times. Thus, when we say the probability of drawing the name of a given boat (say, the "Gail Ann") from a list of names of 40 boats is 1/40, we mean that if we put the names of all 40 boats on cards in a hat and then draw a card blindly from the hat an extremely large number of times, we will draw the card containing the name "Gail Ann" one time in 40, on the average. Probability, in another sense, denotes the measure of our belief in the truth of some assertion. For example, a sport fisherman may have the belief that he will catch a fish within the next hour, but he may not be completely certain of this. He may consider that the odds are 2 to 1 in his favor, or that he has a 67% chance of catching a fish or that the probability is 0.67 that he will catch a fish. All of the above are different ways of expressing the same measure of disbelief. The first interpretation of probability is important because it is the more rigorous interpretation used normally by mathematicians in the conceptual study of probability theory. The second interpretation is important because it is used in everyday life very frequently.

The concept of a priori and a posteriori probability can be illustrated fairly simply. Assume that the names of the 40 boats mentioned previously consist of 20 stern trawlers and 20 side trawlers, the Gail Ann being classed as a stern trawler. Suppose that a card from the bunch of 40 is drawn randomly and that we wish to assign a probability to the event that "the card drawn is the name ' Gail Ann'." With no hint as to which card is drawn, we say only that the a priori probability is 1/40. If, on the other hand, we obtain some special intelligence or knowledge that the card drawn was the card for a stern trawler, we can then eliminate the 20 side trawlers from consideration and say the a posteriori probability is 1/20 or 0.05 that the card contains the name of the vessel "Gail Ann."

The concept of conditional probability is inherent in the above-mentioned example. If we let "P" denote probability and "|" denote "given that," then we can make the following conditional probability statement: P(card drawn is the "Gail Ann" | card drawn is a stern trawler card) = 1/20. Bayes' theorem includes the concepts mentioned above.

B. Hypothetical Example

A fishing cooperative wishes to expand operations if it can be reasonably sure that the stocks of fish on new grounds will justify the effort. By "reasonably sure" is meant that the people concerned want to be 90% sure. The best judgment of the commercial fishermen conducting exploratory fishing is that they are only about 70% sure they should expand operations, based on their experience, a knowledge of operations, and conversations with other fishermen operating in the new area.

A systems analysis is made of the fishery and fishing grounds using all relevant information and analytical techniques applicable to determine the probability of adequate stocks on the new grounds. The systems analysis report qualifies the conclusions reached by using the concept of conditional probability as follows:

(1) Given that there are adequate stocks of fish on the new grounds the probability is 0.8 that the systems analysis would reach the correct conclusion, i.e., would predict significant new stocks on the grounds.

(2) Given that there are inadequate stocks of fish on the new grounds the probability is 0.1 that the systems analysis is in error, i.e., would predict adequate stocks.

We next combine the fishermen's judgment with the results of systems analysis as follows using Bayes' theorem.

A generalized formula for Bayes' theorem can be stated as

$$P(E_k \mid E_0) = \frac{P(E_k) \times P(E_0 \mid E_k)}{\sum_{j=1}^{n} P(E_j) \times P(E_0 \mid E_j)}. \tag{12}$$

This generalized formula may be reduced to the one shown below and used in the example which follows.

$$P(E_1 \mid E_0) = \frac{P(E_1) \times P(E_0 \mid E_1)}{P(E_1) \times P(E_0 \mid E_1) + P(E_2) \times P(E_0 \mid E_2)}. \tag{13}$$

The left side of Eq. (13) is the a posteriori probability of E_1, given E_0; on the right side appear the a priori probability of E_1 and E_2, together with the probability of E_0 given E_1 and E_2.

An explanation of the symbols, descriptions, and the measured values assigned to them are shown in Table I. Substituting values from this table into Eq. (13) we obtain

$$\begin{aligned} P(E_1 \mid E_0) &= \frac{P(E_1) \times P(E_0 \mid E_1)}{P(E_1) \times P(E_0 \mid E_1) + P(E_2) \times P(E_0 \mid E_2)} \\ &= \frac{0.7 \times 0.8}{0.7 \times 0.8 + 0.3 \times 0.1} = 0.95. \end{aligned} \tag{14}$$

In Eq. (14) the original a priori probability is used in conjunction with subsequent information (the systems analysis) to calculate a new a

TABLE I

LIST OF SYMBOLS, DESCRIPTIONS, AND NUMERICAL COEFFICIENTS FOR AN EXAMPLE USING BAYES' FORMULA

Symbol	Description	Value
E_1	The event: There are adequate stocks of fish on the new grounds.	
$P(E_1)$	The a priori probability that there are adequate stocks of fish on the new grounds. This is based on the fishermen's judgment.	0.7
E_0	In general, an observed event as a consequence of or in conjunction with the event E_1. In this case the systems analysis results in a prediction that adequate stocks will be found.	
E_2	The event: There are inadequate stocks in the new area.	
$P(E_2)$	The a priori probability that there are inadequate stocks in the new area. $[1 - P(E_1)]$.	0.3
$E_1 \mid E_0$	Conditional probability statement: There are adequate stocks in the new area given that the systems analysis results in a prediction that there are adequate stocks.	
$P(E_1 \mid E_0)$	The a posteriori probability that adequate stocks are present on the new grounds.	To be calculated
$E_0 \mid E_1$	The conditional probability statement: The systems analysis results in a prediction that adequate stocks are present, given that there are adequate stocks present.	
$P(E_0 \mid E_1)$	The probability of $E_0 \mid E_1$, obtained from the systems analysis.	0.8
$E_0 \mid E_2$	The conditional probability statement: The systems analysis results in a prediction that adequate stocks are present given that there are inadequate stocks present.	
$P(E_0 \mid E_2)$	The probability of $E_0 \mid E_2$, obtained from the systems analysis.	0.1

posteriori probability. Conclusion: On basis of a posteriori probability proceed with fleet expansion.

$$\{\text{Fisherman's judgment} + \text{systems analysis}\} + \text{Bayes' formula} = \text{new judgment} \rightarrow \text{decision}$$

From the above exercise it is suggested that the decision making of the future in fisheries science will embody many new concepts—not the least of which is a thorough systems analysis of the problem. It appears that in addition to the results of systems analysis it might be possible to employ Bayesian inference with personal probabilities to further refine our basis for decisions.

V. Other Systems Analysis Applications

A. General

Systems analysis programs of the complexity indicated in the previous section are as yet rare in fisheries science. However, a few models designed to investigate costs and benefits of various management strategies have already been described. For example, the salmon gear limitation policy program described by Royce *et al.* (1963) incorporated mixed species and economic factors into a model which was designed to investigate the economic and biological consequences of various schemes for restricting the entry of gear into the net fisheries for salmon in northern Washington State waters. In another salmon study, Paulik and Greenough (1966) described a model of the sockeye salmon fishery and the canning industry. Biological sectors of the model include a production function and spawner–recruit relations. Variable weather and other factors were separated by random error terms derived from analyses of past population fluctuations. The model output consisted primarily of seasonal operating margins under different sets of conditions. The above-mentioned reports demonstrate that many kinds of fisheries management strategies can be evaluated by designing a suitable model and "exercising" it on a computer. In this manner the consequences of various management techniques can be quickly evaluated on the computer, and the results of this work can be applied to "real world" situations as reasonably good approximations with predictive value. It seems safe to state that in spite of their cost, complexity, and multi-disciplinary nature realistic simulation models of fisheries resources will be one of the most effective tools of the resource decision maker in the near future.

Simulation models of a somewhat more restricted nature, at least from a management decision point of view, have been utilized to reveal certain features of the population biology of fishes. Saila and Shappy (1963) described a numerical probability model (Monte Carlo method) of the migration of salmonid fishes which was used in an effort to estimate the precision of mechanisms involved in the migration of salmon to the vicinity of their natal streams from distant feeding areas. Larkin and McDonald (1968) utilized a computer simulation model to describe the cyclic behavior of sockeye salmon stocks in the Skeena River. Other simulation models include a model of the population biology of Pacific salmon by Larkin and Hourston (1964) and a study of the cyclic behavior of salmon populations by Ward and Larkin (1967). Most simulation modeling which has been utilized by fisheries scientists has been primarily biological in nature. Paulik (1969) has summarized most of the available material of both published and unpublished nature which deals with computer simulation models in fisheries resources.

B. Lobster Simulation Study

An example of a simulation study which is primarily biological is given by Saila and Flowers (1966). In this model the effects of some regulations on the fishery for the American lobster, *Homarus americanus* are studied. The study may be of interest to biologists for two reasons: (1) it assesses certain regulations by means of a model study, and (2) it is a biological example of the application of sectionally continuous functions. The authors are unaware of many previous biological applications.

The basic assumptions applied to model construction were as follows: (1) uniform recruitment with a 50–50 sex ratio at the onset of maturity; (2) annual molts by male lobsters and molting every other year by females in alternation with the breeding condition; (3) the berried condition is assumed to last for 12 months.

Specifically, the exponential (logarithmic) model of the decrease of 1000 male and 1000 female lobsters was examined over a 10-year period for a range of total mortality coefficients from 30 to 80%. Survival of the male lobsters is given as

$$N_{t_L} = N_0 \exp(-Zt_L), \qquad j = 0, 1, 2, \ldots, \quad j + 1 \geqslant t_L \geqslant j, \qquad (15)$$

where $Z = F + M$, F the instantaneous fishing mortality rate and M the instantaneous natural mortality rate.

Protection of berried females was incorporated into the model in the following way: Relate the time intervals to size (t_L) by assuming a uniform length increase per molt. That is t_L may be defined as a length-dependent time scale, with a constant expressing the percentage increase in total dimension per molt. Since the females are nonberried in the first year and berried in the second without a change in size we have two sets of values for the females over t_L—one for the berried and one for nonberried specimens. This provides two sectionally continuous curves over the length-dependent time interval t_L which are related to survival. In the case of males one continuous curve (an exponential decrease) relates t_L to survival as shown in Eq. (15). It follows that by protecting females in the berried condition their rate of decline over t_L will be less than in the nonberried state.

In the nonberried state survival of females over t_L is given as

$$N_{t_L} = N_j \exp[-Z(t_L - j)], \qquad j = 0, 2, 4, \ldots, \quad j + 1 \geqslant t_L > j. \tag{16}$$

Survival of the berried females is given as

$$N_{t_L} = N_j \exp[-M(t_L - j)], \qquad j = 1, 3, 5, \ldots, \quad j + 1 \geqslant t_L > j. \tag{17}$$

Values of M utilized herein were 0.10 and 0.15.

At the outset we have two numerical coefficients for the females—N_0 for the nonberried and N_1 for the berried, where N_1 is the number of survivors from the first year in the nonberried state. The latter decays according to Eq. (17). Nonberried females decline according to Eq. (16).

Since in the case of females we are using survivors of the nonberried state, unchanged in size, as the new N_j, we are affecting a transformation of axis. The effect of this transformation is a shift of the curve for berried females a t_L-unit interval to the left so that we may write for the females over the interval $0 \leqslant t_L \leqslant j + 1$:

$$\begin{aligned}
N_{t_L} &= N_0 \exp(-Zt_L), && 1 \geqslant t_L \geqslant 0, \\
N_{t_L} &= N_1 \exp[-M(t_L - 1)], && 2 \geqslant t_L > 1, \\
&\vdots \\
N_{t_L} &= N_{(j-1)} \exp[-Z\{t_L - (j - 1)\}], && \geqslant t_L > j - 1, \\
N_{t_L} &= N_j \exp[-M(t_L - j)], && + 1 \geqslant t_L > j.
\end{aligned}$$

To find the average number of berried female survivors over a period [a, b] we must solve the equation

$$\bar{N} = \frac{N_0}{b - a} \int_a^b f(t)\, dt.$$

Since $f(t)$ is a sectionally continuous function it is necessary to find the average number of survivors over each period so that

$$\bar{N}_k = \frac{N_j}{1 - 0} \int_0^1 \exp(-Mt_L)\, dt_L\,, \qquad k = 1, 2, 3, 4, 5, \quad j = 1, 3, 5, \ldots. \tag{18}$$

The integrated form of Eq. (18) is simply $\bar{N} = Na/M$ for each of the segments where a is the annual mortality rate and $a = 1 - e^{-M}$. The average abundance during the year for each length is the area under the curve divided by the base (which is unity). The total fecundity of the lobster stock under various conditions of natural and fishing mortality rates can be found by multiplying the average abundance of each size class of berried females by its fecundity and summing for the entire population.

A Fortran program was written incorporating the above equations. The model results from this study included the following: (1) The theoretical effects of a double-gage regulation[1] were demonstrated and no justification for this regulation was found. (2) A consistent increase in total fecundity of model stocks which incorporated protection of berried females was demonstrated. (3) Stable sex ratios near unity were achieved with theoretical stocks which involved protection of berried females. (4) The advantage of model studies to make inferences concerning biological phenomena difficult to assess in natural stocks was shown.

REFERENCES

Baranov, T. I. (1918). *Nauch. Issledoviktiol Inst. Izv* **I**, 81.

Beverton, R. J. H., and Holt, S. J. (1957). *Min. Agr. Fish. Food (U.K.) Fish Investig.* Ser. II.

Cushing, D. H. (1968). "Fisheries Biology." Univ. of Wisconsin, Madison, Wisconsin.

Larkin, P. A., and Hourston, A. S. (1964). *J. Fish. Res. Bd. Can.* **21**, 1245.

Larkin, P. A., and McDonald, J. G. (1968). *J. Anim. Ecol.* **37**, 229.

Paulik, G. J. (1969). *Trans. Amer. Fish. Soc.* **98**, 551.

Paulik, G. J., and Greenough, J. W. (1966). *In*, K. E. F. Watt (Ed.) "Systems Analysis in Ecology." Academic Press, New York.

Ricker, W. E. (1958). *Fish. Res. Bd. Can. Bull.* **119**.

Royce, W. F. *et al.* (1963). *Univ. Wash. Publ. Fisheries, New Series* **2**.

Saila, S. B., and Shappy, R. A. (1963). *J. Cons. Cons. Perma. Int. Explor. Mer.* **28**, 153.

Saila, S. B., and Flowers, J. M. (1966). *Proc. Gulf Caribbean Fish. Inst.* **18**, 66.

Schaefer, M. B. (1965). *Trans. Amer. Fish. Soc.* **94**.

Ward, F., and Larkin, P. A. (1964). *Prog. Rep. Int. Pacific Salm. Fish Commun.* **11**.

Watt, K. E. F., ed. (1966). "Systems Analysis in Ecology." Academic Press, New York.

Watt, K. E. F. (1968). "Ecology and Resource Management: A Quantitative Approach." McGraw-Hill, New York.

[1] A double-gage regulation as used herein involves both an upper as well as a lower legal size limit

9

Digital Simulation Modeling in Resource Management and the Training of Applied Ecologists

G. J. PAULIK*

CENTER FOR QUANTITATIVE SCIENCE IN FORESTRY, FISHERIES AND WILDLIFE
UNIVERSITY OF WASHINGTON, SEATTLE, WASHINGTON

I. Introduction

The growth of simulation modeling using digital computers has been an explosion within the computer explosion. The last few years have been so productive that it is not possible to measure progress in terms of individual papers. Even tabulating the numbers of new periodicals launched, new books published, major symposia held, and new computer simulation languages written would be extremely difficult.

Because ecological systems are so complex and because the holistic approach is so important when considering the total ecological situation, ecology is a natural area for application of computer simulation models.

*Deceased.

More surprising than the limited use of simulation modeling in basic ecological studies is the small number of applications to current problems in environmental deterioration. Many such problems are caused by man's intervention in complex ecosystems whose existence is dependent on an intricate and delicate scheme of natural balances. Most of the sophisticated uses of simulation have been restricted to a few subject areas and to specific types of problems within these areas; the applications have not been large-scale operations.

Although we are currently witnessing a rapid advance in the use of quantitative techniques such as systems analysis and simulation modeling in a variety of branches of ecology, by far the majority of published work is still in nonecological areas, particularly in such business fields as manufacturing, marketing, corporate structure, finance, inventory control, and production. There have also been extensive applications of simulation modeling in transportation, communication, medicine, facility planning, artificial intelligence, and in a large variety of military and space fields. Educators, physiologists, physicists, nuclear engineers, chemists, geneticists, mathematicians, and statisticians have all made extensive use of simulation modeling. Fortunately for the ecologist, a number of excellent recent texts on computer simulation modeling are available. A list of selected references is given in Appendix A.

The analogies between systems problems in the various subdisciplines of economics and systems problems in ecology are not widely recognized. Five key titles from Appendix A will provide the interested ecologist with an overview as well as entry into the relevant literature:

(1) "Management Science" (Beer, 1968),
(2) "Systems Simulation for Regional Analysis" (Hamilton *et al.*, 1969),
(3) "Simulation in Business and Economics" (Meier *et al.*, 1969),
(4) "Simulation Models and Gaming" (IBM Scientific Computing Symposium, 1966), and
(5) "Simulation Programming Languages" (Buxton, 1968).

It is almost embarrassing today to continue missionary work to convince ecologists of the merits and advantages of such quantitative methods as simulation and systems analysis. In some instances there is reason to be suspicious that the acceptance lag-period for a new concept is over and the publicity may be starting to run ahead of performance and possibly even ahead of potential performance for some time to come. Yet in certain areas absurd arguments continue over the appropriateness of mathematical modeling for solving biological problems.

In my opinion, most of the authors whose work is reviewed in this chapter espouse either explicitly or implicitly a pragmatic philosophy. The quantitative methodology they employ is a computerized extension of logical thinking and the old-fashioned scientific method. The use of computer simulation models is part of a thoroughgoing attack on problems involving dynamic behavior of complex systems.

Levins (1966) describes three alternative strategies of model building in population biology:

(1) sacrifice generality to realism and precision;
(2) sacrifice realism to generality and precision;
(3) sacrifice precision to realism and generality.

Most computer-simulation modelers have chosen strategy (1) about which Levins says:

> "This is the approach of Holling, (e.g., 1959), of many fishery biologists, and of Watt (1956). These workers can reduce the parameters to those relevant to the short-term behavior of their organism, make fairly accurate measurements, solve numerically on the computer, and end with precise testable predictions applicable to these particular situations."

The circumstances surrounding operational management of a renewable resource such as a fish stock or a forest appear to be forcing the use of simulation modeling techniques. Managers of renewable resources are continually making decisions which are of considerable and immediate importance to a sizeable sector of our society. The applied ecologists who advise resource managers are attempting to circumvent obvious inadequacies of standard analytical models by resorting to computer models which can incorporate and hopefully mimic the dynamic behavior of a large number of interacting variables. The short-term accountability of managers to users of the resource is also forcing simulation modeling to assume an iterative character. New measurement capabilities resulting from advances in instrumentation and computerized data processing are providing the feedback from the real world to compare with earlier forecasts from models as to the status of certain variables. This type of feedback system causes rapid model evolution.

Although most forecasts being made today are simply statements of what could occur if a long list of assumptions are fulfilled, there is every indication that for certain renewable resources we are acquiring the ability to make unqualified short-term forecasts.

An interdisciplinary group meeting under the auspices of the Committee on Oceanography, National Academy of Sciences–National Research Council (NASCO) to discuss the use of mathematical models in biological oceanography noted that computer simulation models are not widely used in this field (Banse and Paulik, 1969). The NASCO-sponsored group identified a list of critical constraints to rapid expansion in the development and use of models in biological oceanography, most of which are also relevant to the entire field of ecology. It is of interest that this NASCO group was unable to reach agreement upon the single, most effective means of stimulating the use of modeling techniques in biological oceanography.

My purpose in this chapter is to review some current and recent work in applied ecology involving simulation modeling and also to examine the potential of simulation models as training devices. I hasten to emphasize that this review is not intended to be exhaustive, but rather concentrates attention upon a few representative examples. Many of the models described below are the subject of graduate theses and have not as yet been published in the open literature. There are two main reasons for lack of formal publication. The first is that simulation models are difficult to express in the usual publication format (Garfinkel, 1968a) and the second is that much of the work is so recent.

II. Solving Statistical Problems in Applied Ecology by Simulation

Schaefer (1968) has stated that one use of fisheries simulation is "... to arrive at estimates of the parameters in a model by varying them until an acceptable simulation of a series of observed data is achieved." A professional statistician, however, would probably view the task of determining values of parameters in a model to fit a given set of data as a problem in statistical estimation theory. Although there is no clear-cut separation between the domains of statistics and of simulation, the distinguishing characteristic of this use of simulation in applied ecology is the attempt to determine a set of parameters that will allow a dynamic model to generate an artificial sequence which mimics as faithfully as possible an observed time history involving a biological population. This type of problem is quite common in practical fisheries work. Often the only information available to a manager who must decide the size of the maximum sustainable yield that can be supported by a fish stock is the catch and effort history of the fishery.

A comprehensive treatment of this technique is given by Pella and Tomlinson (1969) who develop a generalized stock production model and a computer program for fitting the model to sequential catch and effort data. They describe the growth rate of an exploited population by

$$dP_t/dt = HP_t^m - KP_t - qf_tP_t , \tag{1}$$

where P_t is the biomass of the population at time t, f_t is fishing intensity at time t, and H, K, q, and m are constants. If the skewness coefficient m is equal to 2, Eq. (1) becomes the familiar Schaefer curve (Schaefer and Beverton, 1963) derived from the logistic equation of population growth. The triple product qf_tP_t in Eq. (1) is the instantaneous removal rate or "catch rate."

The harvesting intensity f_t as shown in Eq. (1) is time-dependent. If f_t is set equal to the constant f during the time interval $[0, t]$, it is possible to solve for the population size at time t:

$$P_t = \left[\frac{H}{K+qf} - \left(\frac{H}{K+qf} - P_0^{1-m}\right) e^{-(K+qf)(1-m)t}\right]^{1/(1-m)}, \tag{2}$$

where P_0 is initial population size.

The observations to be fitted to the process defined by these two equations consist of three vectors containing sequences of catches, efforts, and interval lengths

$$(C_1 , C_2 ,\ldots, C_n), \qquad (E_1 , E_2 ,\ldots, E_n), \qquad (\Delta t_1 , \Delta t_2 ,\ldots, \Delta t_n),$$

where C_i is catch during the ith interval, E_i is number of gear units operating during the ith interval, and Δt_i is length of the ith interval. Note that the fishing intensity during the ith interval is defined as

$$f_i = E_i/\Delta t_i .$$

By means of a numerical integration routine Pella and Tomlinson generate for given values of H, K, q, P_0, and m, sequences of estimated catches,

$$\hat{C}_1 , \hat{C}_2 ,\ldots, \hat{C}_n ,$$

produced by a fishery on a population described by Eqs. (1) and (2). They develop a computer routine that searches over the feasible space of parameter values to find the point that will minimize the sum of squares:

$$\sum_{i=1}^{n} (C_i - \hat{C}_i)^2.$$

An impressive application of the Pella–Tomlinson model is for data obtained from a *living* simulation of a fishery—an experimental population of guppies (*Lebistes reticulatus*) maintained by Silliman and Gutsell (1958) for the purpose of determining the reaction of an actual population to systematic removals at three-week intervals. Removal rates varied among different aquaria. Pella and Tomlinson represented the discontinuous removal process acting at discrete time points by assuming periods with open and closed fishing seasons. The open season lasted for only one one-hundredth of a period and the values that could be assumed by the parameter q were restricted. Figure 1 shows observed and predicted

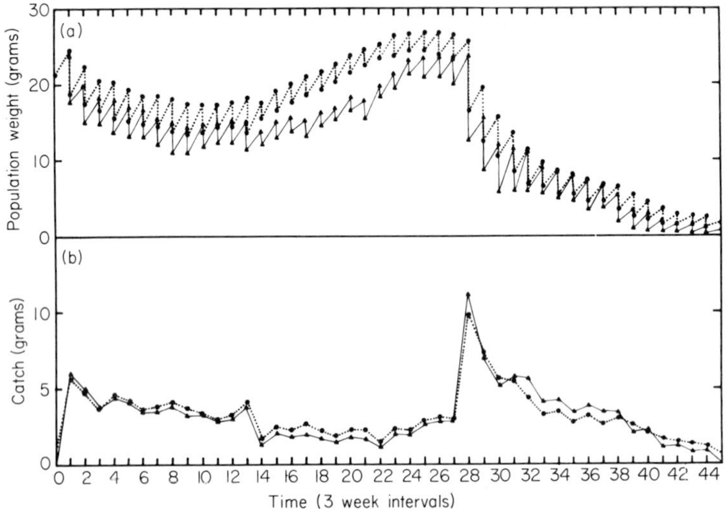

FIG. 1. Observed and predicted population size (a) and catch histories (b) for guppy population *A* of Silliman and Gutsell (1958). The catchability coefficient q was restricted in fitting the catch data. Solid line: observed. Dotted line: predicted. (After Pella and Tomlinson, 1969.)

population size and catch histories for one of the guppy populations. The only data required by the Pella–Tomlinson model are the catches, efforts, and interval lengths.

Other authors who have made important contributions to fisheries science by employing simulation methods and time-series data to estimate parameter values include Chapman *et al.* (1962), Murphy (1966), and Silliman (1966). Chapman *et al.* analyzed historical data from the fishery for Pacific halibut (*Hippoglossus stenolepis*); Murphy worked with the records of the fishery for Pacific sardine (*Sardinops caerula*); and Silliman with data on catches of Atlantic cod (*Gadus morhua*).

The use of simulation to solve a different but common type of statistical problem encountered in ecological work is exemplified by the computer experiments run by Braaten (1969). Braaten constructed artificial populations and exploitation schemes to generate the input data used in the DeLury estimator for initial population size when a population is subjected to cumulative depletion. For these artificial populations, he deliberately violated the assumptions upon which the estimation procedure is based to determine its sensitivity to such deviations. In addition to the usual DeLury estimator he also studied the robustness of three modifications (see Table I). Table I, which was taken from his paper, is typical of the sort of results Braaten obtained. An initial population size of 100,000 was used to generate the data given in Table I. From systematic variation of "environmental conditions" Braaten found that two of the modifications of the DeLury estimator had better overall performances than the original estimator.

Paloheimo (1961) provides a clear introduction to the use of simulation for studying procedures for estimating mortality rates from fishery data.

Francis (1966) and Pella (1969) have employed Monte Carlo simulation techniques to study the effectiveness of mark–recapture experiments for estimating population parameters and to design large-scale field experiments that satisfy program objectives and efficiently allocate available resources. The SIMTAG program developed by Pella for simulating tagging experiments on yellowfin and skipjack tuna in the equatorial Pacific is by far the most ambitious program of this type yet constructed. In the Monte Carlo approach, each animal in the simulated population is represented as a particle with attributes affecting its probabilities of migrating in different directions at different speeds and dying from various causes.

Like many other agencies concerned with management of oceanic fish stocks, the Inter-American Tropical Tuna Commission (IATTC) has carried on small-scale tagging experiments for a number of years. From the results of this tagging, the IATTC has learned the general pattern of movement for tuna of a given size in certain parts of the fishing grounds at different seasons and something about the magnitude of fishing and natural mortality rates. The IATTC now needs "hard" estimates of such important parameters as intraseasonal in-migration, the relative magnitudes of natural mortality and out-migration, and the fishing mortality rates generated on various components of the stock in different parts of the fishing grounds. To help design a tagging experiment to estimate these factors, Pella constructed a general simulation model (SIMTAG) that represents the IATTC's total understanding of the migration patterns and demographic characteristics of the tuna populations in the eastern tropical Pacific. SIMTAG serves not only as a memory extension of a sort but can be modified to study plausible, but unproven, biological hypotheses about tuna movements.

TABLE I

DeLury-Type Estimates of the Population Size Derived from 36 Combinations of Effort and Catchability in a Deterministic Model[a]

Catchability type	Effort type	DeLury	Generalized DeLury	Weighted Generalized DeLury	Effort-corrected DeLury
Constant	Constant	90,640	100,000	100,000	100,170
	Cyclic	90,170	102,410	100,160	100,200
	Gable	90,530	99,420	98,700	100,170
	Negative ramp	90,800	99,830	100,180	100,160
	Positive ramp	90,360	100,060	99,680	100,170
	Damped oscillation	91,100	98,550	100,660	100,170
Cyclic	Constant	80,880	90,750	88,380	90,960
	Cyclic	81,230	94,230	91,500	91,730
	Gable	81,110	90,460	88,240	91,260
	Negative ramp	81,390	90,900	88,730	91,470
	Positive ramp	80,230	90,460	87,950	90,410
	Damped oscillation	76,840	84,830	82,670	87,460
Downward jump	Constant	71,330	82,740	85,270	83,060
	Cyclic	70,570	85,740	85,530	83,070
	Gable	71,400	82,200	84,210	83,130
	Negative ramp	72,110	83,120	86,170	84,160
	Positive ramp	70,410	82,230	84,160	81,920
	Damped oscillation	70,670	79,730	84,720	82,000
Upward jump	Constant	152,210	159,590	183,940	159,650
	Cyclic	158,080	167,560	184,380	162,160
	Gable	149,980	157,010	179,940	157,770
	Negative ramp	148,580	155,690	171,860	154,940
	Positive ramp	156,090	163,740	201,380	164,910
	Damped oscillation	160,240	166,130	223,540	165,610
Negative ramp	Constant	73,310	84,050	81,370	84,320
	Cyclic	72,930	87,140	82,440	84,640
	Gable	73,660	83,810	80,820	84,660
	Negative ramp	74,050	84,400	82,450	85,260
	Positive ramp	72,440	83,560	80,100	83,330
	Damped oscillation	73,530	82,000	81,030	83,990
Positive ramp	Constant	128,080	136,120	181,990	136,200
	Cyclic	129,340	139,700	169,230	137,020
	Gable	125,600	133,260	176,940	134,290
	Negative ramp	126,320	134,070	170,950	133,700
	Positive ramp	129,840	138,170	198,070	138,930
	Damped oscillation	127,940	134,410	211,250	136,040

[a] After Braaten (1969).

SIMTAG is being used to generate particles representing tagged tuna released by research vessels on simulated cruise paths. The behavior of a "tagged" tuna is influenced by its age, the season, and various migratory cues incorporated in the model; the actual migration paths are stochastically determined by sampling from various types of probability distributions. Eight causes of "mortality" imitate tagging stresses, physical deterioration of tags, and natural mortalities; fishing mortality is controlled by a complicated function of the seasonal and spatial distribution of effort. If, during its simulated migration, a fish contacts land, different sorts of reflective behavior are incorporated in the model. Basic statistical measures, such as the mean and variance of death times by release category, are automatically computed by SIMTAG.

In my opinion, this type of model has great potential. It is able to simulate a proposed tagging experiment to determine if the experiment will yield the desired results under various possible configurations of fishing effort and regardless of the underlying migratory pattern.

Francis (1966) used a Monte Carlo simulation model to design a multiple mark–recapture experiment to estimate mortality rates for pink salmon during their estuarine and oceanic residence. Francis (1969) has completed a Ph. D. thesis in which he has coded an extension of SIMTAG to simulate tagging experiments for elucidating the migratory behavior of adult pink salmon moving from the open ocean into the complex network of inshore waters near the boundary between southeastern Alaska and British Columbia. Francis has employed this model to guide analytical derivations of statistical procedures for estimating critical parameters in population migration models.

Siniff and Jessen (1969) should be consulted for an ingenious application of Monte Carlo simulation techniques to aid in the interpretation and statistical analysis of animal movements as discerned from telemetry data for red fox (*Vulpes fulva*) and snowshoe hare (*Lepus americana*).

III. Management Simulation Models

Defining a "management category" for simulation models is almost as difficult as defining the functions of a resource management agency. Sound management policies must be based upon biological realities. However, the economic performance of a resource industry is often found to be intimately associated with the type of management policy adopted and economic considerations may exert either directly or indirectly a powerful influence upon policy decisions.

A strong natural relation exists between biological and economic studies in fisheries. I have attempted below to separate management models into those primarily concerned with unravelling the biological characteristics of fisheries resource systems and those primarily concerned with investigating benefits and costs of management policies. Admittedly, this separation is somewhat artificial since many models are constructed for the specific purpose of evaluating the economic consequences of management alternatives under different biological constraints.

A. Biological Management Models

The simulation model employed by Tillman (1968) to study the hake stocks exploited by a large Russian and much smaller American fishery off the coasts of Washington and Oregon in the months of May through November is typical of applications of computer models for integrated research and management studies. Until 1965 little detailed biological information was available for these stocks. In 1966 a Soviet high seas fleet entered the hake fishery at the same time American effort expanded. The Soviets harvested 291 million pounds of hake and the Americans 3.7 million pounds in 1966. Because of this sudden and vigorous exploitation, research studies on the Pacific hake were expanded and intensified. Tillman used data collected in this research program and what was known about the other five exploited stocks of hake in the world to piece together a mathematical model of the population biology of the Pacific hake.

Tillman's simulation model contained the 11-yr classes (4–14 yr of age) found in the fishery as well as detailed information on growth, fecundity, and mortality rates. Such factors as availability by age group, net selectivity, spatial distribution, and seasonal fluctuations in catchability were also incorporated into the model. The simulated population was self-generating through density-dependent spawner–recruit curves of the Beverton–Holt type as determined from field data (Beverton and Holt, 1957). Two alternative stock structures hypothesized by biologists studying these hake populations were included in the model. In one of these the entire recruitment to the exploited stock was produced by a Washington–Oregon population which undertook a bathymetric migration to reach offshore spawning grounds. In the other version the fishery operated on only a portion of a larger coastal hake population spawning in common grounds off California. In both versions fish were recruited to the Washington–Oregon fishery.

Simulation experiments were run on the model to evaluate the long-term performance of various management schemes involving mesh size regulations under different patterns of increasing effort. Tillman found that long-term behavior of the fishery was very sensitive to the stock structure (see Figs. 2 and 3) and made recommendations for studying both stock structure and the pattern of recruitment.

When Tillman employed standard yield-per-recruitment models of the Beverton–Holt type to study the population, he found increasing age at entry did not improve sustained yield. However, using the simulation model which included a density-dependent spawner–recruit relationship, he found sustained yield increased greatly when age at entry was raised from 4 to $5\frac{1}{2}$ or 6 yr; optimum age at entry was dependent upon the hypothesized stock structure as well as the form of the spawner–recruit relationship.

Figure 2 shows two of the yield isopleths Tillman obtained under simulated equilibrium conditions. The net-selectivity patterns in Fig. 2 define fractions of available four-year-old and five-year-old fish captured by the gear, a pattern of (0.0, 0.5) indicates no four-year-old fish are taken and the probability of capture for five-year-olds is one-half of that for older fish. For the California recruitment model, greater yields of hake are obtained only by increasing effort while for the Washington–Oregon pattern, the yield at a given level of effort can be changed by changing the selectivity pattern. The line BB′ in the Washington–Oregon isopleth defines the selectivity values that produce maximum equilibrium yields at fixed levels of effort.

In Fig. 3 the temporal behavior of the yield is shown for two recruitment possibilities and three net selectivity patterns when fishing effort increases according to a logistic-type curve. The difference between the yields obtained for the different recruitment patterns is especially striking.

Tillman's work provides an introduction to the problem of building a simulation model in the virtual absence of historical catch records and also provides an informative contrast between conclusions reached using a classic analytical approach and those reached using a computer simulation model.

Southward (1966) developed a large-scale simulation model to study three management techniques for determining annual catch quotas for halibut in the northeastern Pacific. The three schemes evaluated were: (1) an empirical approach based on trends in catch-per-unit effort and the proportion of the catch below a certain age, (2) the use of catch-and-effort statistics to estimate the maximum sustainable yield from an assumed logistic productivity curve, and (3) the use of the yields-per-recruitment analysis developed by Ricker (1958). The simulated management agency collected data from the simulated halibut populations in the biological sector of the computer model illustrated by the

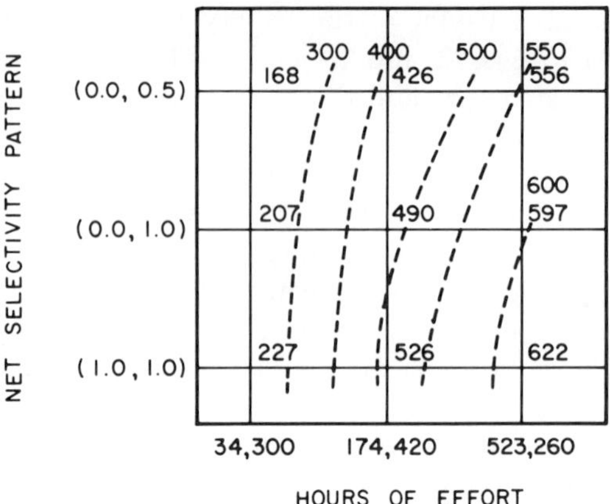

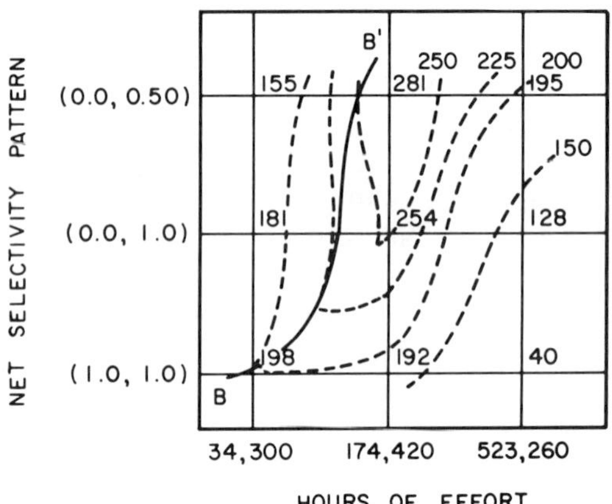

FIG. 2. Pacific hake yield isopleths for two possible recruitment patterns. Yield given in millions of pounds. The dotted lines show equal yield contours. (After Tillman, 1968.)

flow diagram shown in Fig. 4. Sampling plans for measuring basic life processes of growth, mortality, and natality were studied on stochastic and deterministic versions of the model. Measures of production level and stability were analyzed by analysis of variance techniques. Southward discusses the use of his model to develop an optimum regulatory policy in planning future research studies.

Larkin and McDonald (1968) used a computer simulation model to synthesize the main features of the population biology of sockeye salmon of the Skeena River and to generate simulated population time

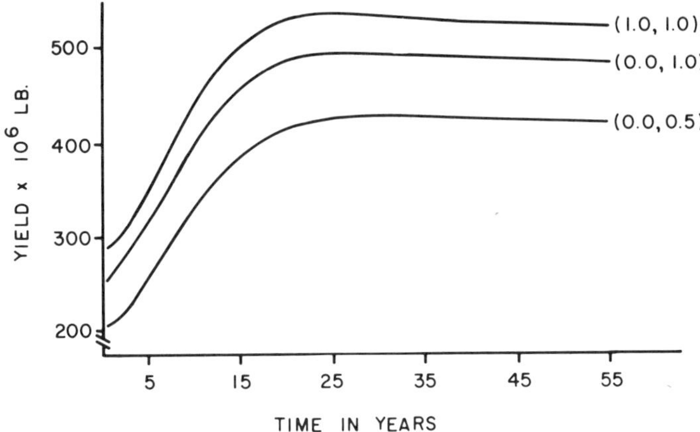

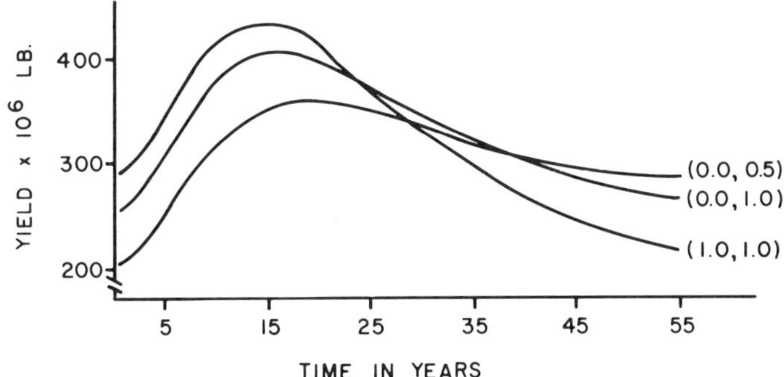

FIG. 3. Temporal behavior of the yield of Pacific hake for an increasing fishery for two recruitment patterns and three possible net selectivity patterns. The top figure was calculated assuming a common coastal population spawning off California and the bottom figure for a separate stock spawning off Oregon and Washington. (After Tillman, 1968.)

histories which could be compared with the observed time histories since 1908. The Skeena River run is made up of two main stocks, each of which is a composite and each of which contains a number of different year classes. Larkin and McDonald were especially interested in cyclic behavior of the stocks and the various mechanisms that affected yearly fluctuations in the relative proportions of various age groups. Their paper is an excellent introduction to the use of computer simulation models combining the interrelations of a large number of variables to generate new biological hypotheses and to narrow areas of ignorance

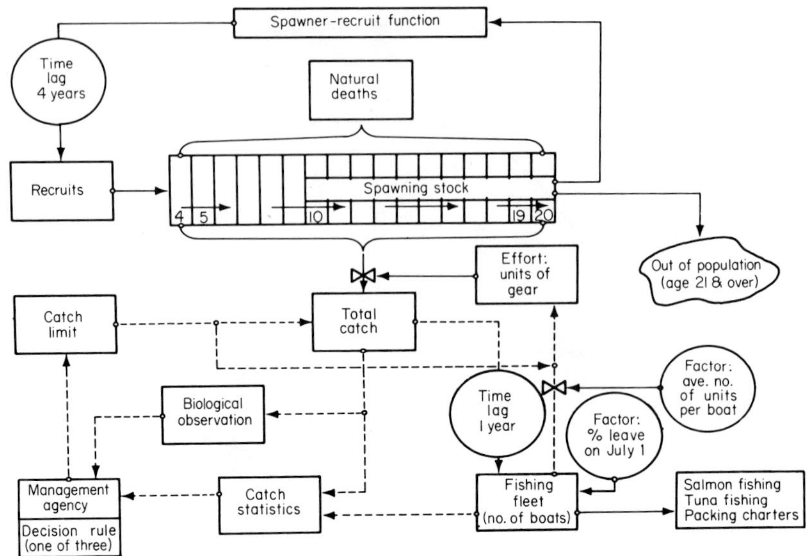

FIG. 4. Flow diagram of a simulation model of the Pacific halibut fishery. (After Southward, 1966.)

concerning population phenomena. Larkin and McDonald's model also serves as a dynamic archive of existing knowledge of the population biology of the Skeena River sockeye. Larkin and McDonald (1968) conclude:

> The simulation was kept virtually free from manipulation based on hindsight and may therefore be a fair reflection of present understanding of the population biology of Skeena sockeye salmon. It underlines a lack of understanding of mechanisms determining age of return and a lack of field studies on the possible existence of depensatory mortality from predation or other causes.

Further illustrations of this approach are given in two important papers: Larkin and Hourston (1964), and Ward and Larkin (1964).

Larkin and McDonald (1968) and Tillman (1968) share a particular world view in the way they use biological data concerning life history characteristics and the manner in which they employ computer simulation models to determine consequences of different hypotheses concerning population structure and population control mechanisms.

Larkin (1963) studied the consequences of exploiting either or both of a pair of competing populations governed by the Lotka–Volterra equations:

$$dN_1/dt = r_1N_1 - a_1N_1^2 - b_1N_1N_2 ,$$
$$dN_2/dt = r_2N_2 - a_2N_2^2 - b_2N_1N_2 ,$$

where r_i is the intrinsic rate of increase for population i, $i = 1, 2$; the a_i's represent intraspecific effects of population density, and the b_i's interspecific effects. Larkin studied the conditions under which a stable equilibrium could be maintained and how sustained yields were affected by different exploitation strategies. He also used a stochastic version of this model to study the transient behavior of small populations.

It should be noted that the degree of abstraction for this model was considerably higher than either model described earlier. No attempt was made to simulate realistically any given population; instead, the general aspects of a specified mode of competition were emphasized.

A large number of simulation models have been proposed which are based either in fact or in spirit upon a system of differential equations. The paper by Davidson and Clymer (1966) exhibits one such model of seasonal growth of phytoplankton and zooplankton and shows an analog computer solution of the system of differential equations. Riley (1963, 1965) has used this type of model to study relations between trophic levels in a food chain. Riley examined model steady-state solutions and applied these to field observations taken under quasi-steady-state conditions in different regions of the ocean to solve for model parameters. Although Riley's model involves a number of gross approximations, it seems to provide useful estimates of steady-state concentrations of phytoplankton, zooplankton, and phosphate. Steele's 1958, 1961, and 1965 papers as well as Dugdale and Goering (1967) should be consulted for further variations and extensions of this general approach.

Parker's (1968) modeling of the response behavior of biological productivity in Kootenay Lake to a reduction in the amount of inorganic phosphate pollution from a fertilizer plant located near the Kootenay River above the lake also involves a system of differential equations but is distinguished by the use of sophisticated digital computer techniques. Parker integrated numerically on a digital computer simultaneous differential equations representing rates of change in phosphate concentration, algae concentration, cladoceran density, and kokanee standing crop. Seasonal changes in river discharge and phosphate concentration, lake temperature and phosphate concentration, and photoperiod are included in his model.

Although Parker does not claim a high degree of realism for the results he obtained, the power of the methods he employed to study the transient behavior of the interacting elements in his model is evident. Of special interest is his use of a cathode ray display and a light pen to study system response to changes in parameters. His article provides an excellent example of the application of digital computer techniques for studying continuous systems.

Steelhead trout (*Salmo gairdneri*) growth and production in a minerally-deficient lake before and after fertilization have been studied by Olsen (1969) who developed a simulation model of the lake ecosystem as an accompaniment to an extensive field sampling program. His model, which was coded in the DYNAMO language (Forrester, 1961), includes a detailed representation of the feeding and metabolic energetics of fish. Some of his results are shown in Fig. 5.

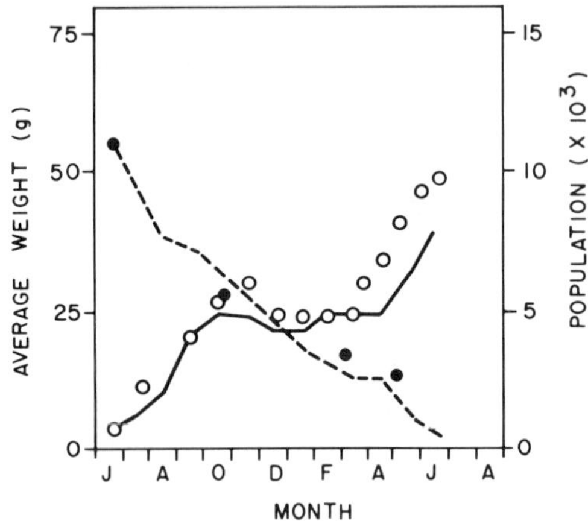

	Observed data		Simulated data	
	Age	Holdover	Age	Holdover
Total catch (4/24–6/18/67)	1438	529	1895	500
Weight catch (kg)	62	58	57	51
Residual population	330	90	400	110
Total production (kg)	232	18	119	12

FIG. 5. Simulated growth data for a population of steelhead trout in Fern Lake for the 1966–67 period compared to estimates obtained from field data, ○: observed weight, —: simulated weight, ●: observed population, – – –: simulated population. (After Olsen, 1969.)

King and Paulik (1967) list references to a number of the recent large-scale continuous simulators prepared for digital computers. Garfinkel's paper (1968b) on the development and use of a computer language for simulating complex chemical and biochemical systems should also be consulted. Silliman (1966) has pioneered the application of analog computer techniques to study fish population dynamics.

A good discussion of the interaction between modeling and laboratory research over an extended period is given by Yamamoto and Robb (1967). Although this discussion is concerned with physiological models, the article contains a number of suggestions which should be helpful to applied ecologists.

B. Bioeconomic Management Models

The model developed by Mathews (1967) is representative of an important type of fishery management application. He asked how much additional investment can be justified to increase the accuracy of forecasts of the size of the sockeye salmon runs to Bristol Bay, Alaska, and developed a model to determine the potential economic benefits to the canning industry of varying degrees of forecast reliability.

Forecasts should be especially valuable for the Bristol Bay sockeye salmon fishery since the area is isolated, the fishing season is extremely short, and the run size is highly variable. Two types of potential benefits were identified: (1) the canning industry could plan more efficiently and thus lower costs, and (2) the management agency could more accurately adjust spawning escapements to the most productive levels.

A comprehensive detailed model of the fishery and the canning industry was written in the DYNAMO language. Sockeye runs to eight major rivers in the Bristol Bay system were represented in the biological sector of the model. Figure 6 is a diagrammatic flow chart for Mathews' model. Production functions for seven of these rivers were estimated by fitting spawner–recruit relationships developed by Ricker (1958) to historical data. For the large and highly cyclical Kvichak system, a spawner–recruit relation which included a postulated interaction between year classes was employed. The simulated Kvichak stock cycled with a periodicity of four to five years which mimicked quite faithfully the actual behavior of the run. Variable weather and other factors were represented by random error terms derived from analyses of past population fluctuations. At least four major age classes representing various combinations of freshwater and ocean residence were employed for each river system. Close agreement was obtained between the frequency distributions of simulated run sizes and actual run sizes since 1940.

Catch data from the Japanese high-seas fishery were analyzed to develop an equation to simulate catching of mature and immature fish on the high seas. Similar analyses determined the relationship between ocean age composition and fishing mortality rates in the inshore fishery.

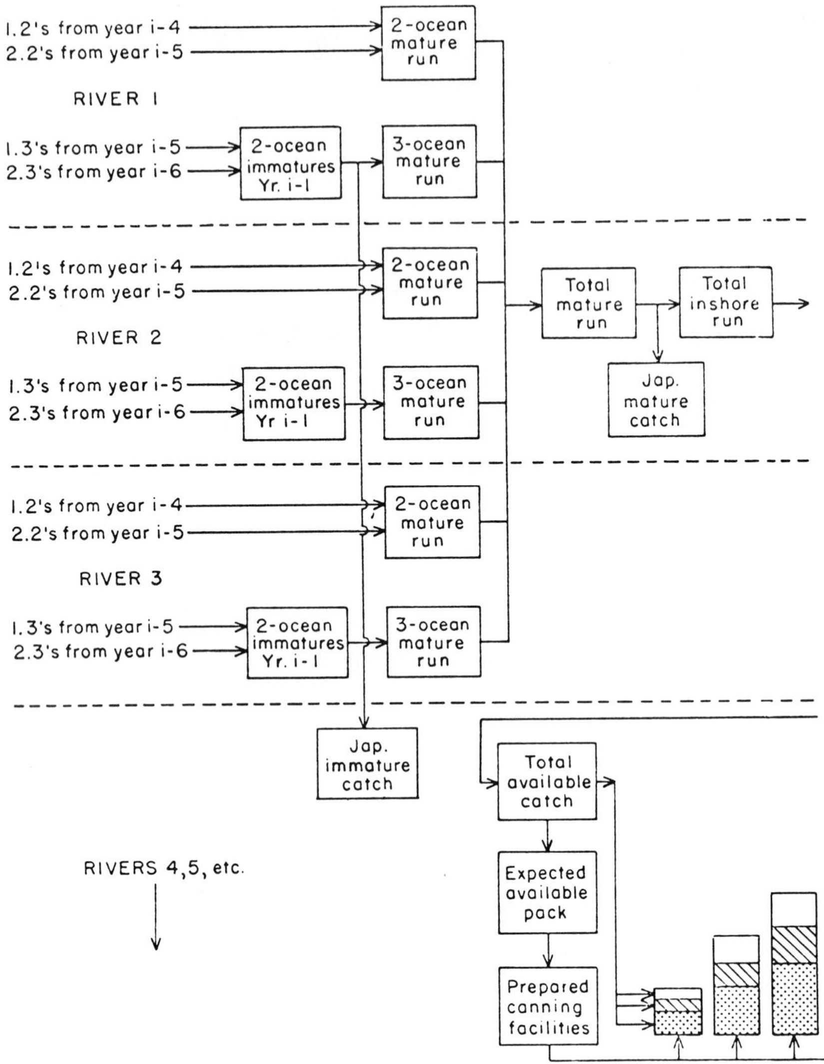

FIG. 6. Flow chart for simulation model of sockeye salmon

The financial structure, productive capacity, and organization of the canning industry were determined from an analysis of one of the major canning consolidations in the Bristol Bay area. The measure of economic performance employed in the model was the 100-yr average of the difference between receipts from canned fish sales and total operating costs.

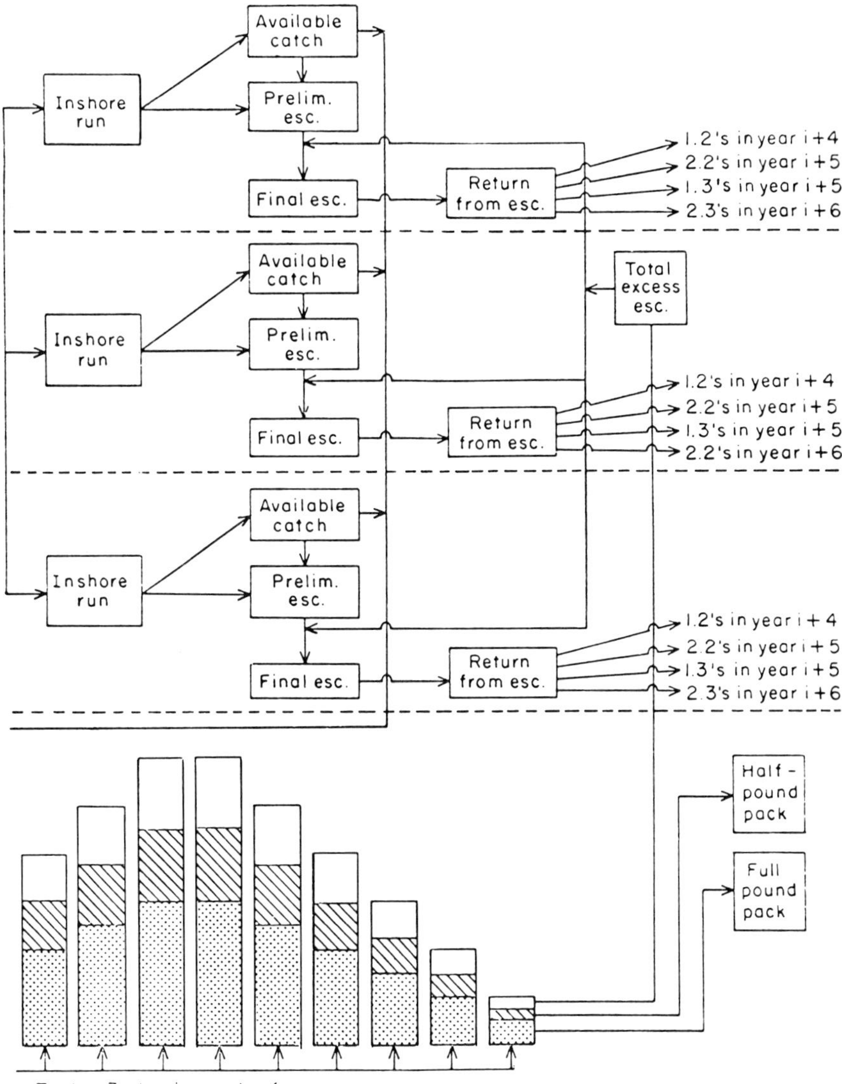

canning industry in Bristol Bay, Alaska. (After Mathews, 1967.)

The model was run to determine not only the average seasonal operating margins under different conditions but also the interrelations of factors such as forecast accuracy and the intraseasonal flexibility of the industry; e.g., if the run size were smaller than expected, the industry shifted more of the pack to half-pound tins which had a higher unit margin.

Users of this type of model are concerned with two "ifs." The first involves the large number of assumptions underlying the model. The second "if" is under the user's control. He uses the mathematician's logic to ask: *If* certain events occur, *then* what will follow? For example, *if* a reliable forecast is made and *if* the industry acts on the forecast, what will be the magnitude of the change in the operating margin? A few of Mathews' results illustrate the answers he obtained.

(1) Under a total run forecast of $\pm 50\%$ error, the operating margin increased about $574,000 over an optimum policy without a forecast. If the error were reduced to $\pm 10\%$, the operating margin increased to $697,000. When one considers costs likely to be involved with reducing the forecast error from 50 to 10%, one can see that there is a clear limit to the expenditure that can be justified for improving forecast precision.

(2) For individual river system forecasts with an error of $\pm 100\%$, the operating margin increased by $1,042,000. Individual river system forecasts of $\pm 50\%$ caused an increase of $2,102,000; when their precision was $\pm 10\%$, the increase rose to $2,409,000.

It should be mentioned that besides the results cited above, this model also provided information on the interrelations between such factors as price, output, and run-size variability.

Royce *et al.* (1963) developed a simulation model to investigate the economic and biological consequences of various schemes for restricting the entry of gear into the net fisheries for salmon in northern Washington State waters. Four species of salmon were represented in this model and their rates of travel and routes of migration were determined by an analysis of catch data as well as from tagging experiments. One of the basic features of the model was the management decision algorithm which simulated the regulation of the fishery by the International Pacific Salmon Fisheries Commission to control the rate of exploitation and also the division of the catch of sockeye and pink salmon between Canada and the United States. Costs functions for three types of gear—purse seines, gill nets, and reef nets—were developed and potential economic benefits of various amounts of gear reduction were estimated using the model.

Greenough (1967) applied the DYNAMO language to simulate a hypothetical intraseasonal gantlet fishery. According to Greenough,

> A "gantlet" fishery is characterized by the migration of exploited stocks through an area of intensive fishing so the fish must "run the gantlet" of the gear. Intraseasonal fisheries management can be contrasted to interseasonal management. Broadly speaking, interseasonal management deals with the establishment of catch and escapement goals that will allow full present and future use of the resource. The basic problem of intraseasonal management is to achieve the goals set by interseasonal

management within the framework of existing environmental, biological, economic, and political restraints. Historically, virtually all quantitative fisheries management models have been interseasonal models.

Greenough's model (Fig. 7) contains five interrelated sectors: (1) a fish migration sector, (2) a gear sector, (3) a mobile fishing gear decision-making sector, (4) a fishing sector, and (5) a diffusion species and target species migration sector. The purpose of this model is not to study any specific fishery, but rather to study how different types of management policies, particularly management decision rules, data collection subsystems, and information transmission subsystems affect measures of biological and economic performance of a fishery.

The papers by Newell and Newton (1968a, b) should be consulted for a number of references to the use of computer simulation models for studying management policies for forest resources.

IV. Simulation Models for Training

A. Population Modeling on Time-Sharing Terminals

The sheer convenience of having the computational power of a large digital computer as close as the nearest telephone makes an immense body of analytical techniques that were considered esoteric yesterday of interest to today's student of applied ecology. In the most popular and least expensive time-sharing mode, communication with the computer is accomplished via the keyboard of a teletype. A teletype can be carried about in a suitcase and by means of an acoustical coupler, any telephone can be employed to link the investigator with a digital computer containing his data files and his models. A single teletype can access any number of different computers and thus provides the user with a wide selection of services and capabilities.

A clear popular introduction to computer time sharing is given by Main (1967). Although the discussion below will be restricted to use of alpha-numeric input and output on a teletype terminal, subject-matter specialists can interact directly with large-scale digital computers via computer generated displays in the form of graphs or line drawings. The recent state of the art in the use of graphic display screens to communicate with computers is described in an issue of the *IBM Systems Journal* (Mehring, 1968) devoted entirely to interactive graphics in data processing.

The student's side of his conversation with the computer consists of a series of questions, answers, comments, and data input interspersed

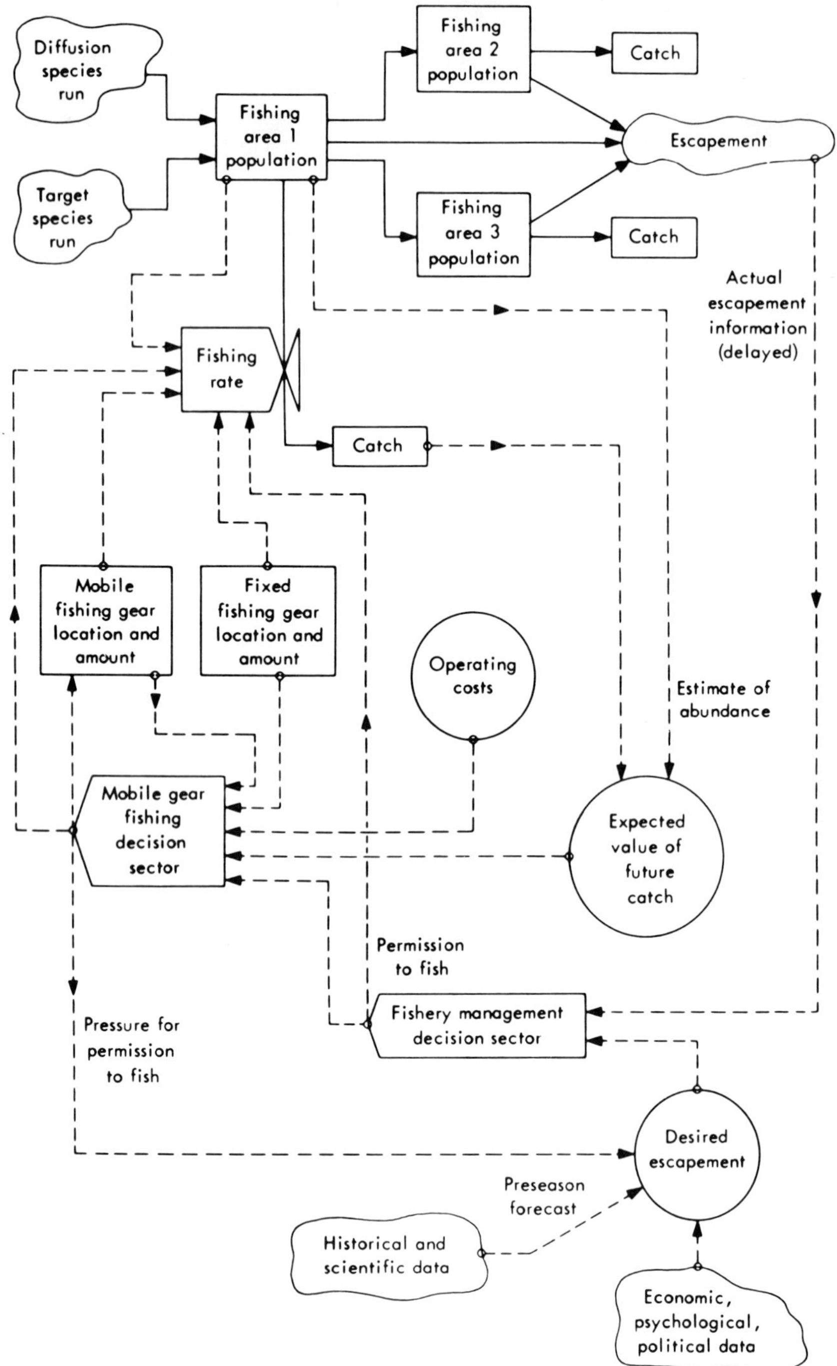

FIG. 7. Basic model of intraseasonal gantlet fishery system. By permission from Meier, R. C., Newell, W. T., and Pazer, H. L. (1969). "Simulation in Business and Economics." Prentice-Hall, Englewood Cliffs, New Jersey.

with computer replies, some of which may be more questions or output from analyses. Simplicity of this sort makes it possible for students to do back-of-the-envelope type calculations on a large-scale digital computer. Mistakes in typing and a high proportion of his errors in logic are called to the student's attention, and using various editing routines, he can immediately erase and correct these mistakes without losing the material already typed in. Storage of other programs, useful subroutines, and data bases are easy and inexpensive.

Two examples are given below demonstrating student use of a population projection routine in a time-sharing environment. This routine is employed in a standard graduate sequence in population dynamics in the College of Fisheries at the University of Washington. The reader should note that the students experience a degree of involvement similar to that experienced by an analog computer user. It is easy to experiment with the model by changing parameter values or its structure to get an instinctive "feel" for the behavior of the biological system being studied. The user can interrupt at any time to study the output or to change parameter values. He simply types in subroutines which are immediately incorporated in the model.

1. *Matrix Representation of Population Processes*

Matrix methods for simulating the growth of a population at discrete time intervals have not been widely used in applied ecology in spite of the fact that they are easier for most students to understand than the corresponding continuous methods and, to quote Skellam (1967), "... throw much light on a wide variety of fundamental problems: subspeciation by separation in time; the co-existence of species with similar biology; the avoidance of endemic parasitism; and the outbreak of pests." Skellam's remark concerns only a small subclass of population matrices.

Goodman (1967) attributes development of the deterministic theory of population growth in which the time-scale and the age-scale are continuous to Sharpe and Lotka (1911), Lotka (1939), and Rhodes (1940), and development of the deterministic theory of population growth in which time-scale and age-scale are discrete to Bernardelli (1941), Lewis (1942), and Leslie (1945, 1948a, b). Goodman himself provides not only a key entry into the literature but also greatly clarifies the relations between the discrete model and the continuous model of population growth.

The ability to compress time effortlessly by being able to project n generations in the future helps the student of population dynamics to acquire a meaningful understanding of both current population structure and assumptions regarding the operation and control of basic population processes. One reason that matrix methods have not found favor among ecology students and their professors is that the calculations are enormously laborious and, using either paper and pencil or desk calculator, devastatingly boring. Instant computer computations magnify greatly the armchair biologist's ability to speculate in a quantitative manner.

The matrix population routine actually used for teaching exists in two basic forms—one as a series of Algol programs (Appendix B shows one subroutine) on a General Electric Mark I time-sharing system, and the other as a Fortran IV program for a 7094–7040 direct-couple system. The Fortran program will handle much larger problems than the Algol program, but can only be operated in a batch-process mode. The user writes his own subroutines in either program to express the vital coefficients—age specific survivals and fecundities—as functions of the current size and composition of the population or as functions of characteristics of past populations.

Up to five interacting populations can be represented in the time-sharing system and up to 10 in the Fortran program. A wide variety of interactions between populations can be represented. For example, adults of population A may be predators on juveniles of population B while juveniles of population A may compete with juveniles of population C at the same time they are subject to predation from older age classes of population D. A clear introduction to the use of computer simulation models and modified Leslie matrices to study the growth of two interacting populations is given by Pennycuick *et al.* (1968).

The basic matrix model assumes a population with a constant sex ratio divided into $k + 1$ age groups. The population is represented at discrete points in time corresponding to the age intervals. Let $n_{i,t}$ be the number of females alive in the ith age group; i.e., in the age interval $(i, i + 1)$ at time t; $i = 0, 1, 2, \ldots, k$; s_i (the survival rate) be the proportion of females alive in the ith age interval who survive to be alive in the $(i + 1)$th interval at time $t + 1$; and f_i (the fecundity rate) be the number of daughters produced per female in the ith age group during the time interval $(t, t + 1)$ who will be alive in the 0th age group at time $t + 1$. Using the $(k + 1) \times (k + 1)$ survival–fecundity matrix M defined as

$$M = \begin{bmatrix} f_0 & f_1 & f_2 & \cdots & f_k \\ s_0 & 0 & 0 & \cdots & 0 \\ 0 & s_1 & 0 & \cdots & 0 \\ \vdots & \vdots & \vdots & \cdots & \vdots \\ 0 & 0 & \cdots & s_{k-1} & 0 \end{bmatrix},$$

and the vector of numbers in each age group at time t, defined as

$$\mathbf{n}_t = \begin{bmatrix} n_{0,t} \\ n_{1,t} \\ \vdots \\ n_{k,t} \end{bmatrix},$$

we can calculate by matrix multiplication the numbers-at-age vector at time $t + 1$, $\mathbf{n}_{t+1} = M\mathbf{n}_t$.

It is well known that when at least two f_i's are nonzero, the matrix M will have one dominant real root $\lambda_1 > 0$, and that $\ln(\lambda_1) = r$, the intrinsic rate of increase of the population. The eigenvector associated with λ_1 is proportional to the stable-age distribution of the population.

Extensions of this model to cases where the age-specific mortality and fertility rates may be complicated functions of population characteristics do not appear to be analytically tractable. However, it does seem highly desirable to retain the basic Leslie matrix structure when studying a population whose age-specific vital rates are affected by either population density or other biological or physical elements in an ecosystem. An easily understood model readily adaptable to matrix notation is that given by Edmondson (1968) which is described below.

a. *Intrinsic Rates of Increase for Rotifers.* Edmondson (1968) constructed a graphic model for studying effects of reproductive schedules on basic population parameters such as population growth rate, birth rate, death rate, age structure, and egg ratio. Edmondson says,

> The model is primarily developed to represent the activities of such rotifers as *Keratella chochlearis* that reproduce parthenogenetically, lay eggs one at a time, and carry the eggs during their development. A female hatches from the egg and spends some time in an immature condition. She then lays an egg which is carried attached to the outside of the body until it hatches. After an interval, which may be short, another egg is laid and carried.

Figure 8, which was taken from Edmondson's paper, is a complete, self-contained description of his model. In the figure the graphic model is illustrated for a rotifer with a four-day life span, a duration of immaturity of one day, and a one-day egg-development period.

For the matrix model which represents this system,

$$M = \begin{bmatrix} 0 & 1 & 1 & 1 \\ 1 & 0 & 0 & 0 \\ 0 & 1 & 0 & 0 \\ 0 & 0 & 1 & 0 \end{bmatrix} \qquad \text{and} \qquad \mathbf{n}_{1/2} = \begin{bmatrix} 1 \\ 0 \\ 0 \\ 0 \end{bmatrix}.$$

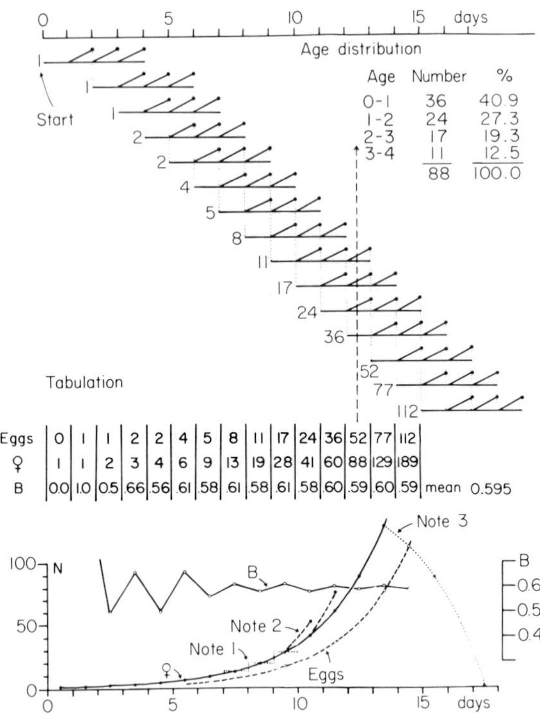

Fig. 8. A graphical model of the growth of a population starting with a single parthenogenetic female at time 0 ("start"). The female produces eggs at times 1.0, 2.0, and 3.0. The duration of development of the eggs is 1.0 day, as shown by the diagonal lines extending upward and to the right. Thus newly hatched females are added to the population at times 2.0, 3.0, and 4.0. The founding female dies at time 4.0. Each of the offspring shows exactly the same schedule, being immature for one day, continuously ovigerous for three days, and dying at the end of the fourth day after producing three eggs at daily intervals.

Every time an egg hatches, a new symbol is added to the diagram and marked with a number to show how many individuals it represents. For example, at time 4.0 two eggs hatch, and the symbol is marked "2." It is counted twice in the census. The model population can be counted at any time by looking vertically down and counting lines. In this example, the population is counted at noon each day (times 0.5, 1.5, 2.5, etc.), and the number of females and eggs recorded as in the "Tabulation." At time 12.5, there were 52 eggs carried by 88 females. The next midnight the 52 eggs hatched, and the symbol shows 52. The egg ratio (eggs per female) E is calculated, and from it, the reproductive rate B. In this case $B = E$ because the duration of development is 1.0.

The graph at the bottom shows values plotted on an arithmetic scale. The value of B oscillates widely at first when the numbers are small, but after the tenth day settles down to a value of 0.595, eventually 0.594. With the high degree of coordination in hatching, the population grows in steps (Note 1), but the noon values are well fitted by a smooth exponential curve, as are those of the eggs. The dashed lines show how the

Note that times are measured at the middle of each day, i.e., the initial age distribution vector exists at time $\frac{1}{2}$, the next at time $\frac{3}{2}$, etc. (see Fig. 8).

To calculate future generations of this rotifer population, the student only has to type on the teletype keybord the M matrix and the $\mathbf{n}_{1/2}$ vector. If he is not familiar with the program, he requests instructions on its use and receives from the teletype the listing shown in Fig. 9. By typing in statements 495, 970, 971, and 972 shown at the top of Fig. 10, he can produce in a few seconds any number of future generations of rotifers. Statement 495 is an instruction to the computer to print out the age vector which represents a census of the population at the middle of each day. This age vector is labeled AV in Fig. 10. Age vectors for 25 successive days are shown in this figure. These were generated from a single founding female and the total of each vector is in agreement with Edmondson's tabulation as shown in Fig. 8. The number of eggs present at time t can be read from the age vectors as the number of females in the 0th age group at time $t + 1$ (cf. Edmondson's calculations shown in Fig. 8). At the end of the 25 generations requested by the student, the status of the population is printed out (see bottom of Fig. 10). The student is then asked if he would like to continue calculations from this point on with the same population.

The results from any stage of iteration using the basic population-growth simulator can be immediately entered into another program also callable from the terminal. A description of how to use this second program, EGEN, is included in Fig. 9. Input and output listings for EGEN which computes the eigenvalue, the stable age vector, and the generation length are shown in Fig. 11 for the rotifer example. The eigenvalue of 1.466 corresponds to an intrinsic rate of increase of $r = 0.382$ (see Edmondson, 1968, p. 16). The generation length calculated by EGEN is defined by Goodman (1967) as, "... the average age of the mothers of all those who are in the 0th age-interval in the stable population, i.e., the 'average age at childbirth' in the stable population."

To change the M matrix to represent a rotifer with a two-day duration of immaturity, the student simply types in the statement 1250 as shown in Fig. 11. The eigenvalue, stable-age-distribution vector, and generation length for this maturity schedule are also shown in Fig. 11.

---

population would have grown had there been no deaths (Note 2). The finely dotted line shows how the population would have declined had egg hatching been abruptly cut off (Note 3). The age structure of the population can be determined by counting animals of different ages, as in the example for time 12.5. The final stable-age values are slightly different; 40.6, 27.7, 18.9, and 12.9%. (After Edmondson, 1968.)

```
PURPOSE--
GRSIM IS AN ALGOL PROGRAM FOR THE G.E. TIME-
SHARING SYSTYM WHICH SIMULATES THE GROWTH OF
UP TO 5 POPULATIONS OF LIVING ORGANISMS.
IT DOES THIS BY REPEATEDLY APPLYING GROWTH
MATRICES (G) TO AGE-DISTRIBUTION VECTORS (A).
THIS RESULTS IN SEQUENCES OF AGE-DIST. VECTORS
WHICH CHART THE GROWTH OF EACH POP. AS A
FUNCTION OF TIME.
    EACH N BY N MATRIX G (1<N<6)  IS COMPOSED
OF TWO VECTORS- 1) THE FECUNDITY VECTOR F =
F(1)...F(N)  WHICH FORMS THE TOP ROW OF G AND
2) THE SURVIVAL VECTOR S = S(1)...S(N-1) WHICH
FORMS ITS PRINCIPLE SUBDIAGONAL.
INPUT--
FOR EACH POP. THE USER INPUTS 3 VTRS.: A,
F, AND S.  EACH VTR. IS PRECEEDED BY A LINE
NO., EACH OF THE AGE VTRS. IS FOLLOWED BY THE
SYMBOL   9$9,    , AND THE LAST SURVIVAL VTR.
IS FOLLOWED BY THE SYMBOL   -9$9;     .
THE LINE NOS. START AT 970 AND ARE INCREMENTED
BY 1 FOR EACH ADDITIONAL VTR.
    E.G., THE STATEMENTS:

970 100,200,300,400,      9$9,
971 0,0,1,1,
972 .9,.8,.7,
973 100,100,100,       9$9,
974 0,1,2,
975 .5,.5,      -9$9;
RUN

WILL CAUSE GRSIM TO INPUT  2 POP. WITH
THE FOLLOWING  A,F, AND S VTRS.:
A[1]=100,  F[1]=0,  S[1]=.9
A[2]=200,  F[2]=0,  S[2]=.8
A[3]=300,  F[3]=1,  S[3]=.7
A[4]=400,  F[4]=1
FOR POP. 1, AND
A[1]=100,  F[1]=0,  S[1]=.5
A[2]=100,  F[2]=1,  S[2]=.5
A[3]=100,  F[3]=2
FOR POP. 2.

    GRSIM THEN PRINTS OUT THE INITIAL STATE
OF EACH POP. AND ASKS FOR FURTHER INPUT WITH
THE QUESTION:
      NO. OF GEN.,VTRS., AND NPR
      ?
THE USER REPLYS BY TYPING 3 NOS.  NIS,DCZ, AND
NPR.
NIS=NO. OF GENERATIONS SIMULATION IS TO RUN.  IF
NIS=0 THE PROGRAM ENDS.
DCZ=NO. OF OPTIONAL INPUT VECTORS GRSIM IS TO
ASK FOR.  IF DCZ=0 NO VECTORS ARE ASKED FOR.
GRSIM PRINTS OUT THE TOTAL POP. EVERY NPR
GENERATIONS.
E.G. IF THE USER TYPES 20,0,5,
THE SIMULATION WILL RUN FOR 20 GENERATIONS, NO
OPTIONAL VTRS. WILL BE ASKED FOR, AND THE
PROGRAM WILL PRINT OUT THE TOTAL POP. FOR EACH
GROUP EVERY 5 GENERATIONS.

    GRSIM CAN BE MODIFIED TO SIMULATE POP.
GROWTH IN THOSE CASES WHERE THE ELEMENTS OF
G ARE FUNCTIONS OF SEVERAL VARIABLES.  THIS IS
DONE BY WEAVING GRSIM TOGETHER WITH SOME SUB.
WHICH USURPS THE PLACE OF SUB. MYGR (LINE NOS.
460-550).  THIS IS DONE WITH THE STATEMENT:
         EDIT WEAVE GRSIM,(NEW MYGR)
WHERE -NEW MYGR- IS THE NAME OF A PREVIOUSLY
STORED SUB. WHICH OVERRIDES THE CURRENT MYGR.
```

FIG. 9. Listing of instructions on how to run the GRSIM and EGEN time-sharing programs on the teletype terminal.

```
THIS NEW SUB. MAY MAKE USE OF THE OPTIONAL
INPUT VTRS. MENTIONED ABOVE.
THE USER MAY GENERATE AND SAVE AS MANY NEW MYGR
TYPE SUB. AS HE DESIRES. IN THIS WAY HE MAY
TAILOR GRSIM SO AS TO BE ABLE TO INVESTIGATE
A VARIATY OF PROBLEMS IN POP. DYNAMICS, E.G.
CROWDING, CANNIBILISM, INTERSPECIFIC COMPET-
ITION, ETC.
     NOTE: THE LINE NOS. IN EACH NEW MYGR MUST
LIE WITHIN THE SAME LIMITS AS IN THE OLD MYGR
(460-550)

PURPOSE--
EGEN FINDS THE LARGEST EIGEN VALUE OF ANY N BY N
MATRIX (1<N<6) AND THE ASSOCIATED EIGEN VECTOR
FOR FECUNDITY-SURVIVAL TYPE MATRICES.
INPUT--
THE USER INPUTS EACH ROW OF THE INPUT MATRIX ON
A DIFFERENT LINE.  EACH ROW IS PRECEDED BY A
LINE NO., THE FIRST ROW IS FOLLOWED BY THE
SYMBOL    9$9,      , AND THE LAST ROW IS FOLLOWED
BY THE SYMBOL    0;     .  THE LINE NOS. START AT
1250 AND ARE INCREMENTED BY 1 FOR EACH ADDIT-
IONAL MATRIX ROW.   E.G.   THE STATEMENTS:

1250 1,1,2,0,1,      9$9
1251 1,0,0,0,0,
1252 0,1,0,0,0,
1253 0,0,1,0,0,
1254 0,0,0,1,0,     0;
RUN

WILL CAUSE EGEN TO FIND THE LARGEST EIGEN
VALUE AND ASSOCIATED EIGEN VECTOR OF THE
ABOVE 5 BY 5 MATRIX.
```

FIG. 9 (*continued*)

The same methods employed to study the rotifer populations can be extended without difficulty to more complicated population models of interest to resource managers. In the next section a brief description is given of a model used by students to design fishing regulations to obtain maximum productivity from a simulated chinook salmon population.

b. *Maturity Schedules and Economic Productivity of Chinook Salmon Stocks in the Columbia River.* The harvesting of an animal with a life cycle as complex as that of the chinook salmon (*Oncorhynchus tshawytsha*) can become quite involved. After spawning in the main stem or in one of the tributaries of the Columbia River, these salmon die. Their progeny migrate to the sea where they may spend less than two years or more than five before returning to their homestreams to spawn. Chinooks undertake extensive ocean migrations, both offshore and northward along the coast into Alaskan waters. Large numbers of these salmon are taken in commercial troll and sport fisheries in the ocean all along the western coast of North America. Substantial numbers are also taken in the lower Columbia River on their way to the spawning grounds. While the ocean fisheries take a mixture of mature and immature fish, the river fisheries operate almost exclusively on mature fish.

```
495 PRINT("AV=","",A[I,1],"",A[I,2],"",A[I,3],"",A[I,4]);
970 1,0,0,0,         9$9,
971 0,1,1,1,
972 1,1,1,      -9$9;
RUN


GRSIM       14:36    S2 FRI 04/25/69

******************************
 SET NO.=      0
 NO. OF ITER IN SET=  0
 TOTAL NO. OF ITER SO FAR=  0

----------
 STATUS OF POP.      1
 NO. OF AGES=    4
 TOTAL POP.=     1
 AGE DISTRIBUTION VECTOR-
 1      0      0      0

               AGE-POP. MATRIX
 0      1      1      1
 1      0      0      0
 0      1      0      0
 0      0      1      0
----------
******************************


NO. OF GEN.,VTRS.,AND NPR
? 25,0,0

AV= 1       0      0      0
AV= 0       1      0      0
AV= 1       0      1      0
AV= 1       1      0      1
AV= 2       1      1      0
AV= 2       2      1      1
AV= 4       2      2      1
AV= 5       4      2      2
AV= 8       5      4      2
AV= 11      8      5      4
AV= 17      11     8      5
AV= 24      17     11     8
AV= 36      24     17     11
AV= 52      36     24     17
AV= 77      52     36     24
AV= 112     77     52     36
AV= 165     112    77     52
AV= 241     165    112    77
AV= 354     241    165    112
AV= 518     354    241    165
AV= 760     518    354    241
AV= 1113       760    518    354
AV= 1632       1113      760    518
AV= 2391       1632      1113      760
AV= 3505       2391      1632      1113
******************************
 SET NO.=      1
 NO. OF ITER IN SET=  25
 TOTAL NO. OF ITER SO FAR=  25

----------
 STATUS OF POP.      1
 NO. OF AGES=    4
 TOTAL POP.=     12664
 AGE DISTRIBUTION VECTOR-
 5136      3505      2391      1632

               AGE-POP. MATRIX
 0      1      1      1
 1      0      0      0
 0      1      0      0
 0      0      1      0
----------
******************************


NO. OF GEN.,VTRS.,AND NPR
? 0,0,0
```

FIG. 10. Listing of input and output from time-sharing terminal using the GRSIM program for simulating population growth of Edmondson's rotifers.

```
1250  0,1,1,1,       959,
1251 1,0,0,0,
1252 0,1,0,0,
1253 0,0,1,0,       0;
RUN
WAIT.


EGEN          9:31    S2 MON 05/12/69


 ----------
 EIGEN VALUE=    1.46557
 INTRINSIC RATE OF INCREASE=    .382245
 EIGEN VECTOR-
 1       .682328       .465571       .317672
 GENERATION LENGTH=    2.75119
 PERCENT AGE COMPOSITION-
 40.5586       27.6742       18.8829       12.8843
----------


END OF DATA


USED       11.50 UNITS.
1250 0,0,1,1,        9$9,
RUN


FILE SIZE LIMIT


EGEN          9:33    S2 MON 05/12/69


 ----------
 EIGEN VALUE=    1.22074
 INTRINSIC RATE OF INCREASE=    .199461
 EIGEN VECTOR-
 1       .819172       .671044       .5497
 GENERATION LENGTH=    3.4503
 PERCENT AGE COMPOSITION-
 32.8956       26.9472       22.0744       18.0827
----------


END OF DATA
```

Fig. 11. Input and output listings for rotifer data using the EGEN program described in the text. To change the maturity schedule for the rotifers, the user has only to replace statement 1250. The bottom half of the figure shows the results of changing the length of the immature stage from one to two days.

The approximate timing and the sequencing of various events in the life history of a Columbia River chinook salmon have been described by Cleaver (1967). The extensive hatchery system for fall chinook on the lower Columbia River provides many possibilities for altering the genetic composition of a given stock to maximize its productivity as measured by the total value of its catch in various fisheries. As a class exercise, students try to develop hatchery breeding policies and to establish levels of fishing intensity and size selectivity characteristics of the gear used in the river and in the ocean to maximize the economic performance of the hatcheries. Chinooks of different ages taken in different fisheries vary greatly in economic value.

It is not difficult to define a population generation matrix for a typical chinook salmon life history. Most of the known properties of the classical survival–fecundity matrix will still apply to such a matrix. Using the symbol e_i for potential egg deposition per spawning unit for i-year-old spawners; and equal to (fraction female) $\times$ (eggs per female, age i), $e_1 = e_2 = 0$, $i = 1, 2, 3, 4, 5$; it is assumed the oldest spawners are 5 years old; N_1 is number of juveniles (yearlings) surviving from eggs spawned one year before; R_{i0} is number of i-year-old fish entering the ocean fishery; and equal to $N_i q_i \exp(-M_i)$; M_i is six-month instantaneous natural mortality rate in the ocean for i-year-old fish; p_i is fraction of mature survivors of age i. This fraction leaves the ocean to enter the river and $p_5 = 1$ where 5 is last age in fishery; $q_i = 1 - p_i$ is fraction immature, $q_1 = 1$; N_i is number of age i fish surviving the ocean fishery for $i > 1$; $N_{i+1} = R_{i0} \exp[-(c_i f + M_i)]$; c_i is coefficient of catchability for age i in ocean; f is fishing intensity in ocean between t and $t + 1$; S_i is number of spawners of age i, and equal to $N_i p_i \exp(-r_i \rho) = N_i k_i$; r_i is coefficient of catchability for age i in river; ρ is fishing intensity in river; k_i is proportion of N_i maturing and surviving river fishery to become spawners of age i for $i = 2, 3, 4, 5$; and equal to $p_i \exp(-r_i \rho)$; s_i is annual ocean survival fraction for age i, and equal to $\exp[-(2M_i + c_i f)]$; u_i is rate of ocean exploitation on age i,

$$u_i = \frac{c_i f_i}{(c_i f_i + M_i)} (1 - \exp[-(c_i f_i + M_i)]);$$

C_i is ocean catch in numbers of age i, and equal to $u_i R_{i0}$; W_i is ocean catch in weight (or dollars) of age i and w_i is average ocean weight (or value) of age i, and equal to $w_i C_i$; v_i is rate of river exploitation on age i, and equal to $1 - \exp(-r_i \rho)$; A_i is river catch in numbers of age i, and equal to $v_i N_i p_i$; V_i is river catch in weight (or dollars) of age i and y_i is average river weight (or value) of age i, and equal to $y_i A_i$; and

$$W_0 = \sum_{i=1}^{4} W_i , \qquad V_0 = \sum_{i=1}^{4} V_i , \qquad WV = W_0 + V_0 .$$

We can construct a relationship between the numbers of spawners of age i and of age $i - 1$:

$$S_i = S_{i-1} k_{i-1}^{-1} q_{i-1} s_{i-1} k_i \qquad \text{for} \quad i = 3, 4, 5.$$

The Beverton and Holt (1957) relationship can be applied to calculate from the numbers of chinook of different ages spawning together at time t the number of surviving yearlings produced at time $t + 1$:

$$N_{1,t+1} = \sum_{i=1}^{5} (e_i/D)\, S_{i,t} \qquad \text{for} \quad D = \alpha + \beta \left(\sum_{i=2}^{5} e_i S_{i,t} \right).$$

These relationships define the population generation matrix

$$M = \begin{bmatrix} 0 & 0 & e_3/D & e_4/D & e_5/D \\ s_1 k_2 & 0 & 0 & 0 & 0 \\ 0 & k_2^{-1} q_2 s_2 k_3 & 0 & 0 & 0 \\ 0 & 0 & k_3^{-1} q_3 s_3 k_4 & 0 & 0 \\ 0 & 0 & 0 & k_4^{-1} q_4 s_4 k_5 & 0 \end{bmatrix},$$

where $\mathbf{n}_t$ is a vector of spawners except for the first age group which contains the number of progeny surviving a year after spawning. Time is measured at the center of the spawning season each year. This is done for convenience; other time points and corresponding population generation matrices can easily be defined. The form of this population generation matrix is not unique.

A subroutine which generates the elements of the matrix M as defined above for the chinook salmon population for the time-sharing program GRSIM is shown in Fig. 12. This subroutine prints out catches in the various fisheries in addition to calculating future generations.

The student can easily manipulate maturity schedules and fishing rates and for each set of conditions run the program until the population converges to a stable age distribution. Because of the form of the reproduction function, the eigenvalue for the stable-age distribution will be one, i.e., the population numbers stabilize. The annual production of the stable population in pounds or dollars is used as an optimality measure.

In addition to determining optimum equilibrium production, the student enters various initial age distributions and population sizes to

```
PROCEDURE MYGR(I,K,N,NG,GR,GRI,A,D);
INTEGER I,K,NG;INTEGER ARRAY N;
ARRAY GR,GRI,A,D;
BEGIN ARRAY E,S,KI,P,R,M,C,Q,U,V,SS[0:5];
ARRAY NN,RO,CC,WW,W,AA,VV,Y[0:5];
REAL F,RHO,ALP,BT,D,WV,WO,VO;
IF K>1 THEN GO TO L8;
M[1]:=.5;M[2]:=.25;M[3]:=.2;M[4]:=.15;
C[1]:=.0001;C[2]:=.003;C[3]:=.003;C[4]:=.008;
R[1]:=.0001;R[2]:=.04;R[3]:=.08;R[4]:=.09;
P[1]:=.01;P[2]:=.30;P[3]:=.95;P[4]:=1;
E[1]:=0;E[2]:=0;E[3]:=2500;E[4]:=2900;E[5]:=3000;
W[1]:=4;W[2]:=10;W[3]:=21;W[4]:=26;
Y[1]:=5;Y[2]:=14;Y[3]:=23;Y[4]:=28;
F:=100;RHO:=20;ALP:=10;BT:=.00001;
FOR J:=1 STEP 1 UNTIL 5 DO BEGIN
S[J]:=EXP(-(2*M[J]+C[J]*F));
KI[J-1]:=P[J-1]*EXP(-R[J-1]*RHO);
Q[J]:=1-P[J];
U[J]:=((C[J]*F)/(C[J]*F+M[J]))*(1-EXP(-C[J]*F-M[J]));
V[J]:=1-EXP(-R[J]*RHO);
END;KI[0]:=Q[0]:=1;
FOR J:=1 STEP 1 UNTIL 4 DO BEGIN
GR[I,J+1,J]:=Q[J-1]*S[J]*(KI[J]/KI[J-1]);END;
L8: D:=0;FOR J:=1 STEP 1 UNTIL 5 DO BEGIN
SS[J]:=A[I,J];D:=D+E[J]*SS[J];END;
D:=ALP+BT*D;FOR J:=1 STEP 1 UNTIL 5 DO
GR[I,1,J]:=E[J]/D;
FOR J:=1 STEP 1 UNTIL 5 DO NN[J-1]:=SS[J]/KI[J-1];
FOR J:=1 STEP 1 UNTIL 5 DO BEGIN
RO[J]:=NN[J-1]*Q[J-1]*EXP(-M[J]);
CC[J]:=U[J]*RO[J];WW[J]:=W[J]*CC[J];
AA[J]:=V[J]*NN[J]*P[J];
VV[J]:=Y[J]*AA[J];END;
WO:=VO:=0;FOR J:=1 STEP 1 UNTIL 4 DO BEGIN
WO:=WO+WW[J];VO:=VO+VV[J];END;
WV:=WO+VO;
PRINT(CC[1],"",CC[2],"",CC[3],"",CC[4]);
PRINT(WW[1],"",WW[2],"",WW[3],"",WW[4]);
PRINT(AA[1],"",AA[2],"",AA[3],"",AA[4]);
PRINT(VV[1],"",VV[2],"",VV[3],"",VV[4]);
PRINT(WO,"",VO,"",WV);PRINT(" ");
L1: END MYGR;
```

FIG. 12. Procedure MYGR to be inserted in the Algol program GRSIM (see Appendix B) for simulating the population dynamics of Columbia River chinook salmon using the matrix method described in the text.

observe how the population behaves as it moves toward the equilibrium situation. The resulting sequences of well-known ecological parameters, such as the intrinsic rate of increase and the generation length, and also the sequences of economic measures generated under various discounting policies are most informative. He can also study breeding strategies under different assumptions about the speed with which the genetic characteristics of the stock can be altered.

Even though the program described here is primarily for teaching purposes, it begins to approach a degree of realism which is not far from the research frontiers of fisheries science. For example, Fig. 13 is a flow chart for a simulation program being used by the Washington State Department of Fisheries to study fish cultural policies in a hatchery producing both chinook and coho salmon (Bergman and Woelke, 1969).

B. Simulation Games for Teaching

Although simulation gaming has been used extensively in business and by the military to augment conventional lectures, laboratories, and seminars, there has been little interest to date in applying such games to teach the principles of resource management. At the present time I am aware of only two projects to develop computerized simulation resource management games. One is in the Fish and Wildlife Department at the University of Michigan (personal communication, James T. McFadden) and the other in the Center for Quantitative Science in Forestry, Fisheries, and Wildlife at the University of Washington. Both projects have similar goals, i.e., to construct a series of simulated resource management games that can serve as "Link trainers" to provide students with the type of learning experience that now requires several years in a responsible management position. Simulation models of this type permit students to test their analytical skills as well as their decision-making abilities in "realistic" management situations. The student-players have to analyze the management situation, formulate critical problems and apply their analytical tools to solve the problems and to develop short-term management tactics and long-range management policies.

In one version of this type of game, the simulated resource is imbedded in a complex and dynamic environment and a time history of several years is compressed into a period of a few days or months according to the student's training schedule. As the simulated system progresses through time, the student makes decisions on what factors to measure, how to collect the necessary data, and how to analyze and interpret the data to obtain numbers which can be converted into management action. The simulated resource will respond to his management decisions and provide him with the opportunity to evolve management policies. Particular emphasis is placed on involving the student in decision making under uncertainty.

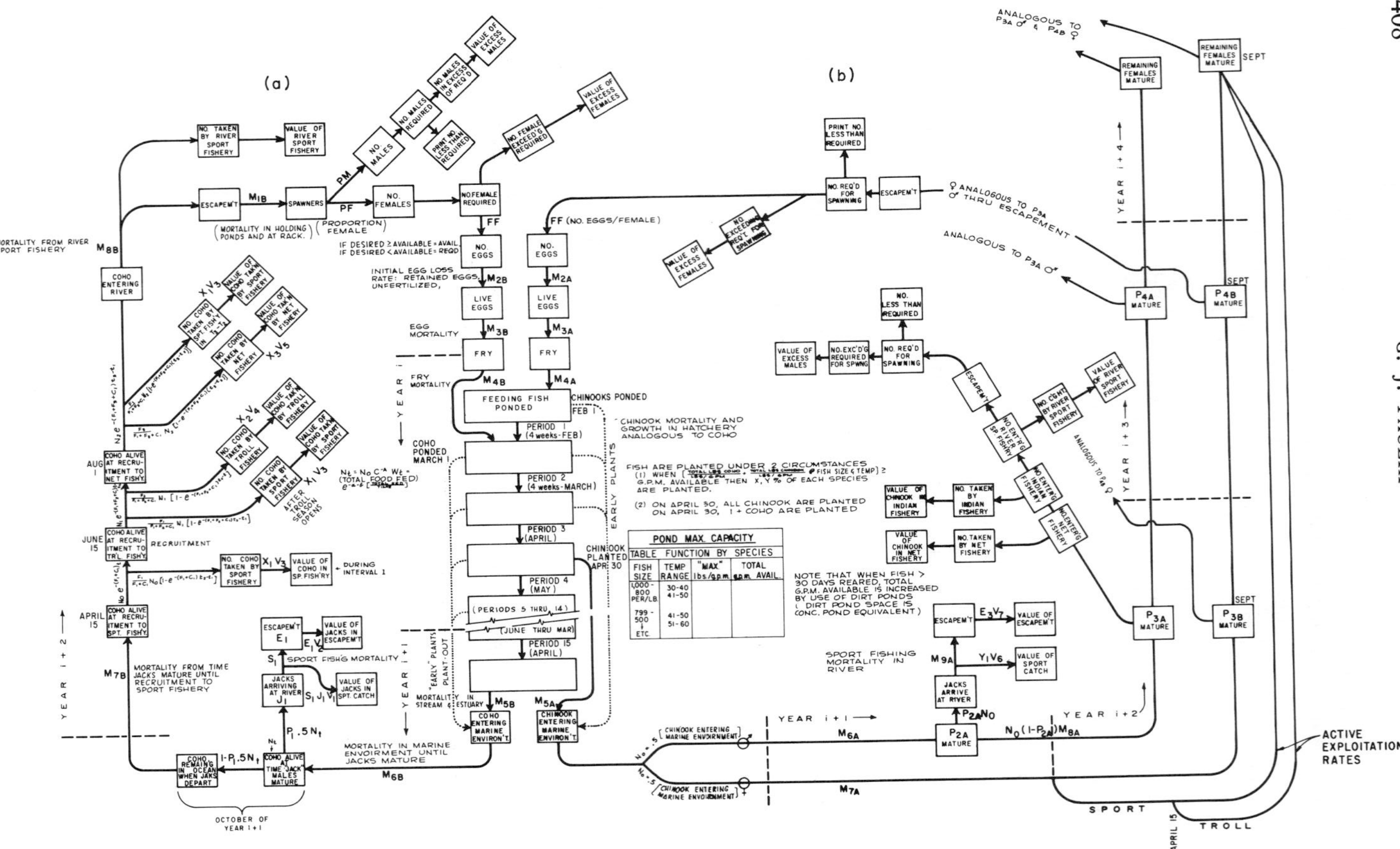
(a)
(b)
MORTALITY FROM RIVER SPORT FISHERY
M_{8B}
COHO ENTERING RIVER
NO. TAKEN BY RIVER SPORT FISHERY
VALUE OF RIVER SPORT FISHERY
ESCAPEM'T
M_{1B}
SPAWNERS
(MORTALITY IN HOLDING PONDS AND AT RACK.)
(PROPORTION FEMALE)
PM
PF
NO. MALES
NO. MALES REQUIRED
NO. MALES IN EXCESS OF REQ'D
VALUE OF EXCESS MALES
PRINT NO. LESS THAN REQUIRED
NO. FEMALES
NO FEMALE REQUIRED
NO. FEMALE EXCEED'G REQUIRED
VALUE OF EXCESS FEMALES
FF
IF DESIRED ≥ AVAILABLE = AVAIL.
IF DESIRED < AVAILABLE = REQD
NO. EGGS
INITIAL EGG LOSS RATE: RETAINED EGGS, UNFERTILIZED,
M_{2B}
LIVE EGGS
EGG MORTALITY
M_{3B}
FRY
FRY MORTALITY
M_{4B}
FEEDING FISH PONDED
COHO PONDED MARCH 1
YEAR i
PERIOD 1 (4 weeks-FEB)
PERIOD 2 (4 weeks-MARCH)
PERIOD 3 (APRIL)
PERIOD 4 (MAY)
(PERIODS 5 THRU 14) (JUNE THRU MAR)
PERIOD 15 (APRIL)
"EARLY" PLANTS
PLANT-OUT
MORTALITY IN STREAM & ESTUARY
YEAR i+1
M_{5B}
COHO ENTERING MARINE ENVIRON'T
MORTALITY IN MARINE ENVOIRMENT UNTIL JACKS MATURE
M_{6B}
COHO ALIVE TIME JACK MALES MATURE
COHO REMAINING IN OCEAN WHEN JAKS DEPART
OCTOBER OF YEAR i+1
$P_1 .5N_t$
$1-P_1 .5N_t$
JACKS ARRIVING AT RIVER J_1
S_1
$S_1 J_1 V_1$
VALUE OF JACKS IN SPT. CATCH
SPORT FISH'G MORTALITY
ESCAPEM'T E_1
$E_1 V_2$
VALUE OF JACKS IN ESCAPEM'T
MORTALITY FROM TIME JACKS MATURE UNTIL RECRUITMENT TO SPORT FISHERY
M_{7B}
YEAR i+2
APRIL 15
COHO ALIVE AT RECRUITMENT TO SPT. FISH'Y
NO. COHO TAKEN BY SPORT FISHERY
$X_1 V_3$
VALUE OF COHO IN SP. FISH'RY
DURING INTERVAL 1
JUNE 15
COHO ALIVE AT RECRUITMENT TO TR'L FISH'Y
RECRUITMENT
AUG 1
COHO ALIVE AT RECRUITMENT TO NET FISH'Y
NO. COHO TAKEN BY SPORT FISHERY
VALUE OF COHO TAK'N BY SPORT FISHERY
AFTER TROLL SEASON OPENS
NO. COHO TAKEN BY TROLL FISHERY
$X_2 V_4$
VALUE OF COHO TAK'N BY TROLL FISHERY
NO. COHO TAKEN BY NET FISHERY
$X_3 V_5$
VALUE OF COHO TAK'N BY NET FISHERY
NO. COHO TAKEN BY SPT. FISH'Y IN $T_3 - T_4$
VALUE OF COHO TAK'N BY SPORT FISHERY
$N_t = N_0 C^{-A} W_t$ = (TOTAL FOOD FED)
FF (NO. EGGS/FEMALE)
NO. EGGS
M_{2A}
LIVE EGGS
M_{3A}
FRY
M_{4A}
CHINOOKS PONDED FEB 1
CHINOOK MORTALITY AND GROWTH IN HATCHERY ANALOGOUS TO COHO
EARLY PLANTS
CHINOOK PLANTED APR 30
M_{5A}
CHINOOK ENTERING MARINE ENVIRON'T
FISH ARE PLANTED UNDER 2 CIRCUMSTANCES
(1) WHEN [ ] ≥ G.P.M. AVAILABLE THEN X, Y % OF EACH SPECIES ARE PLANTED.
(2) ON APRIL 30, ALL CHINOOK ARE PLANTED ON APRIL 30, 1+ COHO ARE PLANTED
POND MAX. CAPACITY
TABLE FUNCTION BY SPECIES
FISH SIZE | TEMP RANGE | "MAX" lbs/gpm | TOTAL gpm AVAIL.
1,000-800 PER/LB. | 30-40 41-50
799-500 ↓ ETC. | 41-50 51-60
NOTE THAT WHEN FISH > 30 DAYS REARED, TOTAL G.P.M. AVAILABLE IS INCREASED BY USE OF DIRT PONDS (DIRT POND SPACE IS CONC. POND EQUIVALENT)
$N_0 \times .5$ [CHINOOK ENTERING MARINE ENVIRONMENT] ♂
$N_0 \times .5$ [CHINOOK ENTERING MARINE ENVIRONMENT] ♀
YEAR i+1
M_{6A}
M_{7A}
P_{2A} MATURE
$P_{2A}N_0$
$N_0(1-P_{2A})M_{8A}$
JACKS ARRIVE AT RIVER
M_{9A}
$Y_1 V_6$
VALUE OF SPORT CATCH
ESCAPEM'T
$E_3 V_7$
VALUE OF ESCAPEM'T
SPORT FISHING MORTALITY IN RIVER
YEAR i+2
P_{3A} MATURE
P_{3B} MATURE
SEPT
NO. ENTER'G NET FISHERY
NO. TAKEN BY NET FISHERY
VALUE OF CHINOOK IN NET FISHERY
NO. ENTER'G INDIAN FISHERY
NO. TAKEN BY INDIAN FISHERY
VALUE OF CHINOOK IN INDIAN FISHERY
NO. ENT'R'G RIVER SP. FISH'RY
NO. C'GHT BY RIVER SPORT FISHERY
VALUE OF RIVER SPORT FISHERY
ESCAPEM'T
NO. REQ'D FOR SPAWNING
NO. LESS THAN REQUIRED
NO. EXC'D'G REQUIRED FOR SPWNG
VALUE OF EXCESS MALES
ANALOGOUS TO P_{4B} ♀
YEAR i+3
P_{4A} MATURE
P_{4B} MATURE
ANALOGOUS TO P_{3A} ♂
♀ ANALOGOUS TO P_{3A} ♂ THRU ESCAPEMENT
ESCAPEM'T
NO. REQ'D FOR SPAWNING
PRINT NO. LESS THAN REQUIRED
NO. EXCEED'G REQ'D FOR SPAWNING
VALUE OF EXCESS FEMALES
YEAR i+4
REMAINING FEMALES MATURE
ANALOGOUS TO P_{3A} ♂ & P_{4B} ♀
SPORT
TROLL
APRIL 15
ACTIVE EXPLOITATION RATES

Simulation games can be constructed at various degrees of complexity and resolution. Paulik (1968) describes a specific example of the simplest type of resource management game:

> . . . consider a computerized simulation model of a population of trout in a stream used to supply recreational fishing. As the human population in the locality of the stream grows, water quality deteriorates. The effects on the stream of increasing thermal pollution and industrial waste pollution are included in the model. A student or a small group of students is given the assignment of developing and implementing a complete research and management program for the stream during a simulated 50-year period. The student-player starts the game with a data base and biological and environmental information furnished by the instructor. More environmental data and information on population processes such as natality, mortality, and mobility are available at "prices" which vary according to the accuracy and precision of the information. The student's annual "purchasing power" is limited by the instructor. As the player uses this model he will have an opportunity to observe certain policies—for example, regulations on angling or planting strategies—in action and to recognize problems such as pollution or changes in public demands for recreational use of the stream as they arise. He will also have to respond to such things as fluctuations in his operating budget and random variability in weather and other exogenous factors affecting the productivity of the stream. At the end of the simulation exercise a number of measures of his performance will be made available to him; e.g., total production of harvestable biomass and percent utilization, degree of satisfaction of recreational demands, and quality of the catch produced. A more sophisticated and generalized version of the resource simulator could be employed to represent the economic and biological sectors of an expanding industry based upon a renewable natural resource. Such a general model could then be modified to simulate particular situations such as the development of the fish meal industry in Peru or the rise and fall of the California sardine fishery.

	Chinook	Coho
Adults return	Sept. 15–Dec. 15	Oct. 10–Jan. 20
Egg	Oct. 2–Dec. 15	Nov. 10–Feb. 1
Fry	Dec. 1–Feb. 1	Jan. 20–March 5
Ponded	Feb. 1	Feb. 26

FIG. 13. Flow chart and hatchery timing tabulation for Washington State Department of Fisheries salmon hatchery producing both (a) coho and (b) chinook salmon which are harvested in several types of fisheries. The variables used are: P_1 is proportion maturing as jacks; V_2 is value of jacks in escapement (per fish); V_1 is value of jacks in sport catch (per fish); C_1 is instantaneous natural mortality rate for marine coho; V_3 is value per fish during given interval for sport fish; V_4 is value per fish during given interval per troll fish; V_5 is value per fish during given interval for net fish; A is instantaneous mortality rate in hatchery pond; F_1 is inst. mortality rate of sport fishery (marine); F_2 is inst. mortality rate of troll fishery; F_3 is instantaneous mortality rate of net fishery; X_1 is number of coho taken in sport fishery during given time interval; X_2 is number of coho taken in troll fishery during given time interval; X_3 is number of coho taken in net fishery during given time interval; Y_1 is number of chinook taken in river sport fishery.

The idea of allowing students to deal with resource systems in which economic and political factors must be considered as well as biological factors is especially appealing. Newell and Newton (1968b), in their comprehensive treatment of the prospects for computer simulation gaming in the ecological and natural resources area, effectively summarize the arguments for large system models.

> The primary reason for large system models would be to present to the players some of the problems connected with decision making where many factors are involved. One of the major problems involving natural resources is the failure of the decision-making agencies to adopt a total system viewpoint dealing with resources. Games of this type could demonstrate the effects of suboptimization and accomplish a great deal toward demonstrating the value of a systems orientation in ecology. Other important benefits to be derived from modeling large systems include the understanding which inures to the benefit of the modelers, and the potential of using the model for additional research into ecosystem behavior.

Many of the proponents of simulation gaming claim that the emotional appeal and glamour of computerization increases the effectiveness of this technique as a teaching device. Simulation games add an element of immediate and direct involvement to the timc-honored case study method. Meier *et al.* (1969, p. 206) repeat a remark of Cyer's about the Carnegie Tech Management Game:

> Essentially, the game is a living case. The student is put in a situation with a variety of problems to be identified and solved. More important, the student must be prepared to live with his decisions. In this respect the game is unique. No other educational tool presents this opportunity and challenge.

a. *A Game to Illustrate How Management Regulations May Affect Economic Yields from a Renewable Natural Resource.* McKenzie and Paulik (1969) have designed an experimental game for illustrating to students how various types of management schemes influence sustainable biological yield and net economic yield from a renewable natural resource. This game is coded in BASIC (Kemeny and Kurtz, 1968) and is available on a remote terminal of the type described above. The game contains three interrelated sectors—biological, economic, and management.

The biological sector includes three salmon stocks harvested by a common fishery but spawning in separate watersheds. Watershed size can be varied by the faculty supervisor. He can also vary life histories and productivity characteristics of the stocks simulated. One current version simulates life histories for coho salmon stocks. Each coho stock has a three-year life span; thus in three consecutive years three genetically distinct lines (Ricker, 1962) use each watershed. Each stock is governed by a separate production curve whose parameters can be varied to allow interactions between the lines using a particular environment. The model contains neither genetic feedback mechanisms nor changes in environmental capacity with time. The system is closed to other stocks, the sex ratio is fixed, and during the season no mortalities other than fishing remove fish. All stocks migrate through the fishing area in 25 days.

The economic sector is concerned primarily with costs of fishing and the value to the fisherman of fish landed. Management agency costs are also included. The unit of gear represents a standard gill-net boat. The annual cost function for fishing is the sum of four factors: (1) the cost of minimal shore facilities, (2) fixed costs proportional to the number of fishing boats, (3) operating costs proportional to the number of boat days expended, and (4) catch-proportional costs. Value is directly proportional to weight of the catch. Management costs are in part determined by the player. He decides how much to spend to control the quantity and quality of the information he receives. The instructor fixes upper limit on management expenditures but the student does not have to spend the entire annual allocation.

Entry of new gear into the fishery each year is controlled by a weighted average over the past three years of the economic performance of existing gear. The function relating entry rate and either actual past earnings or "perceived" past earnings can be varied by the faculty supervisor.

The management sector provides the framework within which the player operates to diagnose and regulate the fishery. From his own analysis of data obtained by sampling the simulated resource, he deduces the biological characteristics of the salmon run and determines a "target" rate of exploitation; i.e., the common removal fraction that according to his analysis will produce the maximum sustainable yield from the mixture of stocks comprising the total run. The type of fishing regulations he can impose to achieve the exploitation rate desired will depend upon the type of management constraints built into the model by the instructor. The manager regulates the amount of gear and the amount of fishing time. Management constraints available along the gear axis range from no control by the manager over entry of new gear and no limits on the total number of units participating to complete administrative control over entry and fleet size. In the free entry version the manager controls the rate of exploitation generated by the fishery each year by varying the fraction of the potential 25-day season that boats are allowed to operate. In other versions, gear can be limited either by direct control of entry or by some indirect means such as imposing additional rents on fishermen. The power to limit the number of gear units challenges the player to achieve his target rate of exploitation by varying both the fleet size and the length of the season.

The player in the role of the management agency can lose a variable degree of ability to regulate the rate of exploitation by means of a "pressure" function which allows industry influences to modify his management recommendations. This pressure function is determined in part by the immediate past economic performance of the fishery. For example, if the player has permitted such a large surplus of gear to accumulate that he must dictate to prevent biological overexploitation so few fishing days that the average boat incurs a severe financial loss, the season he recommends may be arbitrarily lengthened by an amount dependent upon the projected loss and the current economic status of the fleet. Of course the player can try to counteract this sort of pressure by recommending a shorter season than he really wants. Sometimes such a strategy is successful but other times it leads to disasterous consequences. This is one of the most interesting feedback loops in the game.

The player can alter gear efficiency by allowing technological innovations or imposing inefficiency regulations at randomly selected times that occur with an expected frequency and variability determined by the faculty instructor.

The sequence of events during a game and the frequency of decision making is controlled to some extent by the player. He is given an upper limit on annual expenditures and then he sets regulations and the number of years for the simulation to run before he wants to consider changing the regulations. He allocates his resources between sample surveys on the various spawning grounds and in the fishery where both biological and economic information can be collected. He can reallocate available sampling funds and modify fishing regulations at the end of every year. A variety of analytical and plotting subroutines is available for the player's convenience.

When the player has completed one session of the game, his overall performance is graded in terms of the average gross value of the catch from the salmon resource under his management, a time history of annual net revenues per boat, the total net revenue produced, the funds spent by the management agency, and the condition of the stocks and the fishery at the end of the game. A current version of the program also gives him information about how his performance compares to the best performance observed to date.

Students quickly learn that even if they are clever enough to manage the fishery to obtain a high gross economic yield under certain management constraints, the entire value of the resource may be dissipated in excessive harvesting costs. For example, using currently realistic gill-netter costs and the free entry constraint, there is a buildup of a large fleet whose size and fishing power can be regulated only by making operations economically inefficient by drastically reducing the number of fishing days. Students always seem a bit surprised about the economic importance of their decisions which were made on sound biological bases.

Limited experience with teams of graduate students playing this game indicates that it is fulfilling its basic objective quite well. However, no attempt has been made to determine if the game is really superior to other devices for presenting the same concepts. The students have shown a tendency to regulate the fishery in such a way as to "beat" the programmer by determining the underlying structure of the game rather than trying to solve the biological and economic problems of resource management. It is extremely difficult to assess the true value of the skills they acquire.

V. The Future of Simulation Modeling

It would be foolhardy to attempt to predict the future impact of computers and simulation modeling on resource management science and teaching. The rapid development of time-sharing systems and the increase in numbers of new computer centers all over the country make it clear that within a few years a significant portion of the population will have computer terminals in their homes. With increased availability of terminals and time-sharing systems, use of interactive simulation modeling is obviously destined to increase enormously.

The management agency of the near future will maintain at least one, and more likely several, simulation models of natural resources in which they have interests. The models will be basic planning tools for studying total system response to both natural and man-made modifications of the environment. Decision makers will be able to try out various policies on the models to obtain better insights into possible long-range effects. Such models should be especially useful for determining potential costs and benefits of research alternatives and for identifying high leverage research programs. They will not only guide data collection but will be part of a feedback loop in which the models themselves are subject to continual modification and updating as knowledge increases about the structure and behavior of the natural system being modeled. Thus the management, the model, and understanding of the dynamics of a resource should improve simultaneously.

Microscopes and computers have often been compared in terms of their effect on knowledge. In their eminently readable book, Sterling and Pollack (1968) recall some predictions made shortly after microscopes were invented:

> One ambitious scheme for which the microscope was to be used was to count the number of generations that could be found in an ovum, in the naive belief that in this way the time remaining until Judgment Day could be assessed. It is a credit to man's imagination that not only were some investigators able to differentiate the yet unformed eggs into "male" and "female," but others were actually able to count the number of future generations contained in the confines of the ovum.

It seems safe to predict that many of the results of early simulation modeling will not be any more reliable than the early microscopists' estimates of the time remaining until Judgment Day. Old-fashioned flesh-and-blood ecologists are going to have to shoulder the responsibility of keeping the new electronic ecologists honest. Cormack's (1969) remark seems especially appropriate here, "Even the most general mathematical model is a plaything relative to the complexities of an animal population."

Appendix A. Selected General References on Computer Modeling and Simulation

American Statistical Association and Public Health Association of New York City (1967). *Proc. Simulation Business Public Health.* Amer. Statistical Assoc., Washington, D.C.

Beer, S. (1968). "Management Science. The Business Use of Operations Research." Doubleday, Garden City, New York.

Buxton, J. N., ed. (1968). "Simulation Programming Languages." North-Holland Publ., Amsterdam.

Chorafas, D. N. (1965). "Systems and Simulation." Academic Press, New York.

Desmonde, W. H. (1969). "A Conversational Graphic Data Processing System: The IBM 1130/2250." Prentice-Hall, Englewood Cliffs, New Jersey.

Evans, G. W., Wallace, G. F., and Sutherland, G. L. (1967) "Simulation Using Digital Computers." Prentice-Hall, Englewood Cliffs, New Jersey.

Forrester, J. W. (1961). "Industrial Dynamics." M.I.T. Press Cambridge, Massachusetts and Wiley, New York.

Forrester, J. W. (1968). *Mgm. Sci.* **14**, 398.

Garfinkel, D. A. (1966). A Simulation Study of the Effect on Simple Ecological Systems of Making Rate of Increase of Population Density-Dependent. Sci. Memo 54, N.I.H Information Exchange Group No. 3.

Green, P. E., Robinson, P. J., and Fitzray, P. T. (1967). "Experiments on the Value of Information in Simulated Market Environment." Allyn and Bacon, Boston, Massachusetts.

Hamilton, H. R., Goldstone, S. E., Milliman, J. W., Pugh III, A. L., Roberts, E. B., and Zellner, A. (1969). "Systems Simulation for Regional Analysis. An Application to River-Basin Planning." M.I.T. Press, Cambridge, Massachusetts.

Hollingdale, S. H., ed. (1967). "Digital Simulation in Operational Research." American Elsevier, New York.

Hufschimdt, M. M., and Fiering, M. B. (1966). "Simulation Techniques for Design of Water-Resource Systems." Harvard Univ. Press, Cambridge, Massachusetts.

IBM Scientific Computing Symposium on Simulation Models and Gaming (Proceedings). (1966). IBM Data Processing Division, White Plains, New York.

Klerer, M., and Reinfelds, J., eds. (1968). "Interactive Systems for Experimental Applied Mathematics." Academic Press, New York.

Krasnow, H. S., and Merikallio, R. A. (1964). *Mgm. Sci.* **11**, 236.

Ledley, R. S. (1965). "Use of Computers in Biology and Medicine." McGraw-Hill, New York.

Martin, F. F. (1968). "Computer Modeling and Simulation." Wiley, New York.

McKenney, J. L. (1967). "Simulation Gaming for Management Development." Graduate School of Business Administration, Harvard Univ., Boston, Massachusetts.

Meier, R. C., Newell, W. T., and Pazer, H. L. (1969). "Simulation in Business and Economics." Prentice-Hall, Englewood Cliffs, New Jersey.

Mize, J. H., and Cox, J. G. (1968). "Essentials of Simulation." Prentice-Hall, Englewood Cliffs, New Jersey.

Morse, P. M., ed. (1967). "Operations Research for Public Systems." M.I.T. Press, Cambridge, Massachusetts.

Mulvihill, D. F., ed. (1966). "Guide to the Quantitative Age. Readings from Fortune." Holt, New York.

Naylor, T. H., Balintfy, J. L., Burdick, D. S., and Chu, K. (1966). "Computer Simulation Techniques." Wiley, New York.

Newell, W. T., and Newton, J. (1968a). Annotated Bibliography on Simulation in Ecology and Natural Resources Management. Working Paper No. 1, Center for Quantitative Science in Forestry, Fisheries, and Wildlife, Univ. of Washington, Seattle, Washington.

Newell, W. T., and Newton, J. (1968b). Computer Simulation Game Models for Ecology and Natural Resources Management. Working Paper No. 2, Center for Quantitative Science in Forestry, Fisheries, and Wildlife, Univ. of Washington, Seattle, Washington.

Orr, W. D., ed. (1968). "Conversational Computers." Wiley, New York.

Paulik, G. J., and Greenough, Jr., J. W. (1966). *In* "Systems Analysis in Ecology" (K. E. F. Watt, ed.), pp. 215–252. Academic Press, New York.

Preston, L. E., and Collins, N. R. (1966). Studies in a Simulated Market. Institute of Business and Economic Research, Univ. of California, Berkeley.

Sterling, T. D., and Pollack, S. V. (1968). "Introduction to Statistical Data Processing." Prentice-Hall, Englewood Cliffs, New Jersey.

Van Court Hare, Jr. (1967). "Systems Analysis: A Diagnostic Approach." Harcourt, New York.

Van Dyne, G. M. (1966). Ecosystems, Systems Ecology, and Systems Ecologists. ORNL-3957, UC-48-Biology and Medicine. Oak Ridge National Laboratory, Oak Ridge, Tennessee.

Waggoner, P. E. (1968). *Biometeorology* **45**.

Watt, K. E. F., ed. (1966). "Systems Analysis in Ecology." Academic Press, New York.

Watt, K. E. F. (1968). "Ecology and Resource Management: A Quantitative Approach." McGraw-Hill, New York.

Appendix B. A Time-Sharing Algol Program (GRSIM) for Calculating Population Growth with a Generalized Version of the Leslie Matrix Model[a]

Figure 9 is a listing of instructions on how to use GRSIM. The test material in the section entitled *Matrix Representation of Population Processes* provides two examples of applications of GRSIM as well as a discussion of the matrix model.

Time-Sharing Algol is discribed in a General Electric reference manual (General Electric, 1967). The definitive reference to the Algol language itself is Rutishauser (1967). Rutishauser not only provides a clear introduction to Algol but he also gives a complete and comprehensive treatment of more advanced aspects of the language.

```
100 BEGIN
110 COMMENT
120 PROGRAM NAME--GRSIM
130 PURPOSE-- SIMULATES THE GROWTH OF UP TO 5 LIVING
140 POPULATIONS
150 ------------------------------------;
160 INTEGER I,J,K,M,NS,NACC,NORD,NG,NIS;
170 INTEGER TNP,NPR;REAL SS;
180 ARRAY A,F,S[1:5,1:5];ARRAY DC[1:25,1:5];
190 ARRAY GR,GRI[1:5,1:5,1:5];
200 INTEGER ARRAY N[1:5];INTEGER DCZ;
210 PROCEDURE SVM(NG,N,F,S,GR,GRI);
220 INTEGER NG;ARRAY F,S,GR,GRI;
230 INTEGER ARRAY N;
240 BEGIN INTEGER I,J,K,NI,NIM;
250 FOR I:=1 STEP 1 UNTIL M DO
260 FOR J:=1 STEP 1 UNTIL M DO
270 FOR K:=1 STEP 1 UNTIL M DO
280 GR[I,J,K]:=GRI[I,J,K]:=0;
290 FOR I:=1 STEP 1 UNTIL NG DO BEGIN
300 NI:=N[I];NIM:=NI-1;
310 FOR K:=1 STEP 1 UNTIL NI DO
320 GR[I,1,K]:=GRI[I,1,K]:=F[I,K];
330 FOR J:=1 STEP 1 UNTIL NIM DO
340 GR[I,J+1,J]:=GRI[I,J+1,J]:=S[I,J];END;
350 END SVM;
360 PROCEDURE MYAGE(N,I,A,G);
370 INTEGER I; INTEGER ARRAY N;
380 ARRAY A,G;
390 BEGIN ARRAY S[1:5];INTEGER J,K,NI;
400 NI:=N[I];
410 FOR J:=1 STEP 1 UNTIL NI DO BEGIN
420 S[J]:=0;FOR K:=1 STEP 1 UNTIL NI DO
430 S[J]:=S[J]+G[I,J,K]*A[I,K];END;
440 FOR J:=1 STEP 1 UNTIL NI DO A[I,J]:=S[J];
450 END MYAGE;
460 PROCEDURE MYGR(I,K,N,NG,GR,GRI,A,D);
470 INTEGER I,K,NG;INTEGER ARRAY N;
480 ARRAY GR, GRI,A,D;
490 BEGIN INTEGER J,L,NI;
500 IF K>1 THEN GO TO L1;
510 NI:=N[I];
520 FOR J:=1 STEP 1 UNTIL NI DO
530 FOR L:=1 STEP 1 UNTIL NI DO
540 GR[I,J,L]:=GRI[I,J,L];
550 L1: END MYGR;
560 PROCEDURE LABL(NS,NIS,NC);INTEGER NS,NIS,NC;
570 BEGIN NS:=NS+1;NC:=NC+NIS;
580 PRINT("*********************************");
590 PRINT(" SET NO.= ","",NS);
600 PRINT(" NO. OF ITER IN SET= ","",NIS);
610 PRINT(" TOTAL NO. OF ITER SO FAR= ","",NC);
620 END LABL;
630 PROCEDURE WROUT(NG,N,NS,A,GR);
640 INTEGER NG,NS;INTEGER ARRAY N;ARRAY A,GR;
650 BEGIN INTEGER I,J,K,NI,NIM;
660 REAL EMAX,S;ARRAY G,H[1:5,1:5];
670 FOR I:=1 STEP 1 UNTIL NG DO BEGIN
680 FOR J:=1 STEP 1 UNTIL M DO
690 FOR K:=1 STEP 1 UNTIL M DO BEGIN
700 H[J,K]:=G[J,K]:=0;END;
710 NI:=N[I];NIM:=N[I]-1;
720 FOR J:=1 STEP 1 UNTIL NI DO
730 FOR K:=1 STEP 1 UNTIL NI DO
740 H[J,K]:=G[J,K]:=GR[I,J,K];
750 S:=0;FOR J:=1 STEP 1 UNTIL NI DO
760 S:=S+A[I,J];
770 PRINT(" ");PRINT("----------");
780 PRINT(" STATUS OF POP. ","",I);
790 PRINT(" NO. OF AGES= ","",NI);
800 PRINT(" TOTAL POP.= ",S);
810 PRINT(" AGE DISTRIBUTION VECTOR-");
820 FOR J:=1 STEP 1 UNTIL NI DO
830 BEGIN PRINT("",A[I,J],"");END;PRINT(" ");
840 PRINT(" ");PRINT("          AGE-POP. MATRIX");
850 FOR K:=1 STEP 1 UNTIL NI DO BEGIN
860 FOR J:=1 STEP 1 UNTIL NI DO BEGIN
870 PRINT("",G[K,J],"");END;PRINT(" ");END;
880 PRINT("----------");
890 END;
900 PRINT("*********************************");
910 PRINT(" ");PRINT(" ");
920 END WROUT;
930 COMMENT START MAIN PROG. HERE;
940 L6: NS:=-1;NIS:=NACC:=0;
950 M:=5;
960 DATA POP:=
970 1;GO TO L20;
980 I:=0;
990 I:=0;
1000 L12: I:=I+1;IF I>M THEN GO TO L1;
1010 J:=0;
1020 L13: J:=J+1;IF J>M THEN
1030 BEGIN READATA(POP,SS);GO TO L15 END;
1040 READATA(POP,A[I,J]);
1050 IF A[I,J]<-8.99$9 THEN GO TO L1;
1060 IF A[I,J]<=8.99$9 THEN GO TO L13;
1070 L15: BEGIN N[I]:=J-1;
1080 FOR J:=1 STEP 1 UNTIL N[I] DO
1090 READATA(POP,F[I,J]);
1100 FOR J:=1 STEP 1 UNTIL N[I]-1 DO
1110 READATA(POP,S[I,J]);GO TO L12;
1120 END;
1130 L1: NG:=I-1;
1140 IF NG<=0 THEN GO TO L2;
1150 SVM(NG,N,F,S,GR,GRI);
1160 LABL(NS,NIS,NACC);
1170 WROUT(NG,N,NS,A,GR);
1180 L5: PRINT("NO. OF GEN.,VTRS.,AND NPR");
1190 READATA(TELETYPE,NIS,DCZ,NPR);
1200 IF NIS<=0 THEN GO TO L6;
1210 IF DCZ<=0 THEN GO TO L7;
1220 PRINT(" OPT. VTRS.---5 NOS.");
1230 FOR K:=1 STEP 1 UNTIL DCZ DO
1240 READATA(TELETYPE,DC[K,1],DC[K,2],DC[K,3],DC[K,4],DC[K,5]);
1250 L7: ;
1260 FOR K:=1 STEP 1 UNTIL NIS DO BEGIN
1270 TNP:=0;IF NPR>0 THEN BEGIN
1280 J:=ENTIER(K/NPR);IF J*NPR=K THEN
1290 TNP:=1;IF TNP>0 THEN PRINT(" ");END;
1300 FOR I:=1 STEP 1 UNTIL NG DO BEGIN
1310 MYGR(I,K,N,NG,GR,GRI,A,DC);
1320 MYAGE(N,I,A,GR);IF TNP>0 THEN
1330 BEGIN SS:=0;FOR J:=1 STEP 1 UNTIL N[I] DO
1340 SS:=SS+A[I,J];
1350 PRINT("ITER","",K,"",",POP.","",I,"","=","",SS);
1360 END;END;END;
1370 LABL(NS,NIS,NACC);
1380 WROUT(NG,N,NS,A,GR);GO TO L5;
1381 L20:PRINT("IF YOU WISH TO KNOW HOW TO RUN -GRSIM-");
1382 PRINT("THEN TYPE IN -1- OTHERWISE TYPE IN -0-");
1384 READATA(TELETYPE,NORD);IF NORD>0 THEN GO TO L30;
1386 PRINT("TYPE IN AGE, FECUNDITY, AND SURVIVAL");
1388 PRINT("VECTORS FOR EACH POPULATION STARTING AT LINE");
1390 PRINT("970.");GO TO L2;
1392 L30: PRINT("READ IN PROGRAM -GEXPLN- WHICH EXPLAINS");
1394 PRINT("HOW TO RUN -GRSIM-.");
1410 L2: END;
```

[a] Programmed by Lawrence E. Gales at the Center for Quantitative Science in Forestry, Fisheries, and Wildlife, University of Washington, Seattle, Washington.

REFERENCES[a]

Banse, K., and Paulik, G. J. (1969). *Science* **163**, 1362.

Bergman, P. K., and Woelke, W. E. (1969). Computer Simulation Techniques for Developing Management Policies for West Coast Salmon Resources (unpublished). Washington State Dept. Fisheries, Olympia, Washington.

Bernardelli, H. (1941). *J. Burma Res. Soc.* **31**, 1.

Beverton, R. J. H., and Holt, S. J. (1957). On the Dynamics of Exploited Fish Populations. *Min. Agr. Fish. Food (U.K.) Fish. Investig. Ser. II* **19**, 1.

Braaten, D. O. (1969). *J. Fish. Res. Bd. Can.* **26**, 339.

Chapman, D. G., Myhre, R. J., and Southward, G. M. (1962). *Int. Pac. Halibut Comm. Rep.* **31**, 1.

Cleaver, F. C. (1967). The Effects of Ocean Fishing upon the Columbia River Hatchery Stocks of Fall Chinook. Ph.D. thesis, College of Fisheries, Univ. of Washington, Seattle, Washington.

Cormack, R. M. (1969). *In* "Oceanography and Marine Biology Annual Review, 1968" (H. Barnes, ed.), pp. 455–506. Allen and Unwin, London.

Davidson, R. S., and Clymer, A. B. (1966). *Ann. N.Y. Acad. Sci.* **128**, 790.

Dugdale, R. C., and Goering, J. J. (1967). *Limnol. Oceanog.* **12**, 196.

Edmondson, W. T. (1968). *Oecologia (Berl.)* **1**.

Francis, R. C. (1966). Estimation of Survival Rates from a Multiple Tag-Recapture Experiment. M.Sc. thesis, Biomathematics Group, Univ. of Washington, Seattle, Washington.

Francis, R. C. (1969). Ph.D. thesis, Biomathematics Group, Univ. of Washington, Seattle, Washington.

Garfinkel, D. A. (1968a). *Computers Biomed. Res.* **2**, i.

Garfinkel, D. A. (1968b). *Computers Biomed. Res.* **2**, 31.

General Electric (1967). Time-Sharing ALGOL Reference Manual. General Electric Information Service Dept., Bethesda, Maryland.

Goodman, L. A. (1967). *J. Roy. Statist. Soc.* **A130**, 541.

Greenough, J. W. (1967). Simulation Model of a Hypothetical Intraseasonal Gantlet Fishery. M.Sc. thesis, College of Fisheries, Univ. of Washington, Seattle, Washington.

Holling, C. S. (1959). *Canadian Entomologist* **91**, 293–320.

Kemeny, J. G., and Kurtz, T. E. (1968). "Basic Programming." Wiley, New York.

King, C. E., and Paulik, G. J. (1967). *J. Theoret. Biol.* **16**, 251.

Larkin, P. A. (1963). *J. Fish. Res. Bd. Can.* **20**, 647.

Larkin, P. A., and Hourston, A. S. (1964). *J. Fish. Res. Bd. Can.* **21**, 1245.

Larkin, P. A., and McDonald, J. G. (1968). *J. Anim. Ecol.* **37**, 229.

Leslie, P. H. (1945). *Biometrika* **33**, 183.

Leslie, P. H. (1948a). *Biometrika* **35**, 213.

Leslie, P. H. (1948b). *J. Roy. Statist. Soc.* **A111**, 44.

Levins, R. (1966). *Amer. Sci.* **54**, 421.

Lewis, E. G. (1942). *Sankhya* **6**, 93.

Lotka, A. J. (1939). *Actualités Sci.* **780**, 1.

Main, J. (1967). *Fortune*, August, 88.

Mathews, S. B. (1967). The Economic Consequences of Forecasting Sockeye Salmon (*Oncorhynchus nerka*, Walbaum) Runs to Bristol Bay, Alaska: A Computer Simulation Study of the Potential Benefits to a Salmon Canning Industry from Accurate Forecasts of the Runs. Ph.D. thesis, College of Fisheries, Univ. of Washington, Seattle, Washington.

[a] Citations appearing in Appendix A are not repeated below.

McKenzie, D., and Paulik, G. J. (1969). The U.W. Salmon Fishery Management Game (unpublished m. s.). Center for Quantitative Science in Forestry, Fisheries, and Wildlife, Univ. of Washington, Seattle, Washington.

Mehring, H. E., ed. (1968). Interactive Graphics in Data Processing. *IBM Systems J.* **7**(3–4).

Murphy, G. I. (1966). *Proc. Calif. Acad. Sci.* **34**, 1.

Olsen, J. (1969). Steelhead Trout Growth and Production in Fern Lake, Washington, Determined by Sampling and a Simulation Model. Ph.D. thesis, College of Fisheries, Univ. of Washington, Seattle, Washington.

Paloheimo, J. E. (1961). *J. Fish. Res. Bd. Can.* **18**, 645.

Parker, R. A. (1968). *Biometrics* **24**, 803.

Paulik, G. J. (1968). *Publ. Fisheries, Univ. Washington, New Series* **4**, 295.

Pella, J. J. (1969). SIMTAG: A Computer Program for Simulating Mark-Recapture Experiments on Yellow-Fin and Skipjack Tuna (unpublished). Inter-American Trop. Tuna Comm., LaJolla, California.

Pella, J. J., and Tomlinson, P. K. (1969). *Inter-Amer. Trop. Tuna Comm. Bull.* (in press).

Pennycuick, C. J., Compton, R. M., and Beckingham, L. (1968). *J. Theoret. Biol.* **18**, 316.

Rhodes, E. C. (1940). *J. Roy. Statist. Soc.* **103**, 68–89, 218–245, 362–387.

Ricker, W. E. (1958). *Fish. Res. Bd. Can. Bull.* **119**, 1.

Ricker, W. E. (1962). Regulation of the Abundance of Pink Salmon Populations. *H. R. MacMillan Lectures in Fisheries, Symposium on Pink Salmon, 1960*, pp. 155–201. Univ. British Columbia Inst. Fisheries, Vancouver, Canada.

Riley, G. A. (1963). *In* "The Sea" (M. N. Hill, ed.), Vol. II, pp. 438–463. Wiley (Interscience), New York.

Riley, G. A. (1965). *Limnol. Oceanog. Suppl.* **10**, 202.

Royce, W. F., Bevan, D. E., Crutchfield, J. A., Paulik, G. J., and Fletcher, R. L. (1963). *Publ. Fisheries, Univ. Washington, New Series* **2**, 1.

Rutishauser, H. (1967). "Handbook for Automatic Computation. Description of ALGOL 60." Springer-Verlag, New York.

Schaefer, M. B. (1968). *Trans. Amer. Fish. Soc.* **97**, 231.

Schaefer, M. B., and Beverton, R. J. H. (1963). *In* "The Sea" (M. N. Hill, ed.), Vol. II, pp. 464–483. Wiley (Interscience), New York.

Sharpe, F. L., and Lotka, A. J. (1911). *Phil. Mag.* (Ser. 6), 435.

Silliman, R. P. (1966). *Fish. Bull. U.S. Fish Wildl. Serv.* **66**, 31.

Silliman, R. P., and Gutsell, J. S. (1958). *Fish. Bull. U.S. Fish Wildl. Serv.* **58**, 215.

Siniff, D. B., and Jessen, C. R. (1969). *Adv. Ecol. Res.* **6**, 185.

Skellam, J. G. (1967). *Proc. 5th Berkeley Symp. Math. Statist. Probability* **IV**, 179.

Southward, G. M. (1966). A Simulation Study of Management Regulatory Policies in the Pacific Halibut Fishery. Ph.D. thesis, College of Fisheries, Univ. of Washington, Seattle, Washington.

Steele, J. H. (1958). *Marine Res. Scot. Home Dept.* **7**, 1.

Steele, J. H. (1961). *In* "Oceanography" (M. Sears, ed.), pp. 519–538. AAAS Publ. 67, Washington, D.C.

Steele, J. H. (1965). *ICNAF Spec. Publ.* **6**, 463.

Tillman, M. F. (1968). Tentative Recommendations for Management of the Coastal Fishery for Pacific Hake, *Merluccius productus* (Ayres), Based on a Simulation Study of the Effects of Fishing upon a Virgin Population. M.Sc. thesis, College of Fisheries, Univ. of Washington, Seattle, Washington.

Ward, F., and Larkin, P. A. (1964). *Prog. Rep. Int. Pacific Salmon Fish. Comm.* **11**, 1.

Watt, K. E. F. (1956). *J. Fisheries Res. Bd. of Canada* **13**, 613–645.

Yamamoto, W. S., and Robb, W. F. (1967). *Computers Biomed. Res.* **1**, 65.

10

Systems Methods in Environmental Court Actions

ORIE L. LOUCKS
INSTITUTE FOR ENVIRONMENTAL STUDIES
UNIVERSITY OF WISCONSIN, MADISON, WISCONSIN

I. Introduction

Beginning in 1966 and continuing through the present, a growing number of complex court actions concerning environmental damage have provided a new application of systems methods. Complex scientific and medical testimony always has been a part of negligence litigation, but most cases have concerned only a few disciplines and have usually involved little time lag between the event and the damages alleged. Historically, negligence actions frequently involved expert medical testimony, evidence of physical injury, permanent disability, or reconstruction of an accident. All are relatively simple compared with, for example, the movement of radionuclides into and through an arctic ecosystem (lichen–reindeer–man) with resultant radiation hazard to remote eskimo populations.

The methods of presenting relevant scientific testimony in court actions involving damage to the environment and to the biological systems supported by the environment are very different from those of the more common damage suit. Important examples include the hearings held in 1968–1969 on the impact of DDT in Wisconsin waters (Loucks, 1971), the releases of radionuclides and thermal effluents to the atmosphere (Yannacone, 1970) or to water bodies supporting major fisheries, and the flooding of river basins such as the Oklawaha in Florida (Dasmann, 1970). In such litigations, the scientific testimony must include not only the deleterious effects (e.g., peregrine falcon mortality) but also the mechanisms by which these effects are brought about, the magnitude of the time lags and their basis, the response of the plant and animal populations being affected, and, in the case of DDT, the means by which physical transport of the material around the world takes place.

The steps listed here are only representative of the components in any environmental system, but testimony from several scientific experts representing the disciplines concerned in a complex system frequently does not tie together well enough to provide effective proof of the alleged damage. Environmental litigation, as any other litigation, requires that expert testimony be integrated so well that interdependencies and the functional integrity of the relationships as a whole are readily apparent. The overview provided by a systems approach is the only effective means of achieving orderly documentation of the integrity of the whole system and therefore of the upset it may have suffered. A systems analysis, even if only descriptive, frequently provides precise links between the environmental and biological subsystems involved.

The objective of this chapter, therefore, is to discuss the ways in which systems studies can be used in court actions. Many aspects of ecological systems studies already treated in these volumes are involved, but they are viewed here from a more elementary point of view. The common denominator of systems methods is as a vehicle for integrating data and processes. It is therefore also effective in presenting an integrated and holistic view of the environmental system before the courts.

This chapter will also examine the techniques by which scientific evidence in a systems format is developed for specific court cases. The methods will be illustrated through examination of several litigations in which systems studies have been used. The first and most comprehensive was the integration of evidence related to the movement and impact of DDT in a major regional ecosystem (Harrison *et al.*, 1970). The second was a part of the court action brought in Colorado to prevent a proposed release of tritium from Project Rulison. I am grateful for the opportunity of having worked with one of this country's best environmental lawyers, Mr. Victor Yannacone, Jr., in both of these cases. The other cases consider the long-range effects of proposed impoundments of streams in Wisconsin and the impact of the proposed nuclear reactor at the Shoreham site in Long Island, New York.

Many other examples of the application of systems methods to a wide range of environmental and social problems are now becoming available, but the approach and general methodology will be very similar in each new case as it develops. The following sections draw heavily on transcripts of hearings to illustrate the way in which the scientific record is written in court, and the concluding sections will generalize as to techniques and significance for ecological systems science.

II. Scope of Technical Evidence in Environmental Litigation

One of the best-documented examples of an environmental court action with too many scientific components for any direct comprehension was the hearing in 1968–1969 on whether DDT constituted a pollutant under Wisconsin statutes. The systems studies carried out as part of the AEC safety analysis before releasing radionuclides in Colorado were less complete, but the litigation raised new questions for systems ecology. Both can be used as examples of the way technical testimony in a complex case becomes more effective in a systems context.

A. Damages Alleged in the Wisconsin DDT Petition

Although opposition to the use and impact of DDT had been voiced strongly by conservationists throughout the 1950s, and most strongly by Rachel Carson in 1962, the outcry was partly emotional, and clearly based on scattered, piecemeal scientific evidence. The ingredients for a reexamination of the mechanisms and magnitude of DDT impact were coming together in the middle 1960s. These were: strong scientific evidence of its persistence, of the physiological mechanisms in organisms, and of its damage to populations of top carnivores and whole ecosystems. Also available were a substantial group of involved scientists, a small citizens group willing to raise the financial support necessary for a court battle with the agricultural and chemical industries, and a lawyer willing to immerse himself in the complex scientific and legal questions.

That the hearings on DDT took place in Wisconsin is primarily the result of some unique features of Wisconsin's administrative law. The most important of these features is that in cases of conflicting public interest, a citizens group of six may petition for a declaratory ruling. The water-quality legislation in Wisconsin gives authority to the Department of Natural Resources to control pollutants introduced in Wisconsin's lakes and streams. A pollutant is defined as any material that may be damaging to fish, game, or wildlife in state waters. The basis for the Wisconsin hearing on DDT, therefore, was a petition alleging that DDT was producing a deleterious effect on the fish, game, and wildlife of Wisconsin waters, and that DDT should be declared a pollutant under Wisconsin law.

Thus, the Citizens' Natural Resources Association of Wisconsin (and three days later, the Wisconsin Division, Isaak Walton League of America) filed a petition with the Wisconsin Department of Natural Resources alleging as follows:

(1) that DDT is an inherently dangerous, highly toxic, broad spectrum, persistent chemical biocide capable of cycling throughout the biosphere and increasing in ecological effect as a result of biological concentration;

(2) that the use of an inherently dangerous, persistent chemical biocide such as DDT will result in the direct mortality of many non-target organisms and reduction of biological diversity throughout the ecosystem within which it is applied;

(3) that the people of the State of Wisconsin are entitled to the full benefit, use, and enjoyment of the natural resource that is the Wisconsin Regional Ecosystem, without diminution from the application of the chemical biocide DDT, and that application of DDT will cause *serious*, *permanent*, and *irreparable* damage to this Wisconsin resource;

(4) that as a result of said DDT application for the treatment of Dutch elm disease and other uses, the waters of the State of Wisconsin have been polluted, contaminated, rendered unclean and impure, and further made injurious to public health, harmful for commercial or recreational use, and deleterious to fish, bird, animal, and plant life.

In the hearing that followed this petition, the Environmental Defense Fund, with Mr. Yannacone as chief counsel, had to be prepared to prove these allegations with substantial scientific evidence. Above all, the integrity of the system as a whole would have to be established, if the "serious, permanent, and irreparable damage" to the Wisconsin Regional Ecosystem was to be proven.

Furthermore, proof of the allegations would require evidence not only of the laboratory effects of DDT, but of the actual mechanisms by which DDT moves from the site of application to fish and game in remote areas. To provide evidence on transport of DDT, the strategy of the petitioners' case was to offer description of the entire environmental system to which DDT was being added; then after documentation of the chemical characteristics of DDT and its primary breakdown product DDE, the *response of the system as a whole* to this chemical would be analyzed. Section III.A describes the evidence introduced.

B. System Upstes Alleged from a Proposed Tritium Release

The proposal by the Atomic Energy Commission to release tritium after an underground atomic detonation in southwestern Colorado in late 1968 led to another court litigation (Yannacone, 1970) utilizing systems methods. Earlier, in 1968, the AEC, as part of its Plowshare Program, had detonated a small nuclear device to facilitate the collection of a diffuse natural gas pool in the deep bedrock strata. After a six-month period for decay of short-lived radionuclides, the pool was to be tapped, and the long-lived tritium formed by the blast was to be released and burned at the well-head so that the natural gas could be used. The complaint brought by three citizens' groups charged that the proposed release of tritium would produce an unwarranted health hazard to residents in Colorado. The AEC, on the other hand, had contracted for a systems analysis of potential tritium hazard in a representative Colorado ecosystem, and the results showed that there would be none. This part of the trial, therefore, came down to a difference of opinion in the quality and completeness of the systems study carried out. The discussion of evidence in Section III.B is therefore directed to comparative effectiveness of systems models.

III. Modeling and Systems Testimony in Two Judicial Actions

A systems approach is appropriate in many environmental court actions, but only if the complaint alleges damages over a long time period (past or future), and involves the processes of a large and complex system. Section II examined the conceptual problems of even stating the complex mechanisms by which a large environmental system suffers permanent and irreparable damage. The following sections are addressed to the actual mechanics of presenting testimony on systems modeling and analysis in a form suitable for a court. The question and answer format of the trial record has been retained to show how gradual is the development of each point and to illustrate the role of cross examination.

The systems ecologist may know very well the quality of interconnections that couple air, land, and water systems, and man's long-range impacts on them; but the court has to be satisfied that the testimony is both relevant to the case at issue and based on sufficient experience and familiarity with the case to warrant the witness being accepted as an expert. The systems scientist has to recognize that almost everything he would like to say could be challenged as irrelevant, immaterial, and incompetent on the grounds that he is not an expert in any of the fields the lawyers or the court have traditionally recognized. A good collaborating lawyer will prepare the systems analyst for these challenges (little different from the Ph.D. oral exam), ask sufficient questions to demonstrate the witness' qualifications, and guide the testimony through the myriad of objections. The following sections, therefore, include extensive excerpts of the court transcripts showing: (1) techniques used and representative responses needed to "qualify" the systems scientist as an expert witness, (2) the nature of testimony appropriate for complex environmental systems, and (3) the kinds of objections and cross-examination to be expected of an environmental systems model or analysis in the courtroom.

A. The Wisconsin Hearings on DDT Contamination

1. *Establishing the Integrity of a Regional Ecosystem*

The first step by Mr. Yannacone in the DDT hearings was to utilize a series of witnesses to describe the Wisconsin Regional Ecosystem, its internal relationships, and its interconnections with the surrounding states and the Great Lakes. In the case of the DDT hearings, many of the most important studies had been done in other states, and parallels between other ecosystems and that of Wisconsin had to be established if these data were to be accepted in the hearing as relevant. The following series of questions by Mr. Yannacone (State of Wisconsin DNR, 1970a) together with my answers, constitute an example of the system description that must be introduced if more specific evidence on variables and processes in the system is to be recognized as relevant later in the hearing. This excerpt followed introductory questions on qualifications, position, and experience:

Q Now, Doctor, in the course of your studies, and in the development of the field of ecology, do you study the mechanisms whereby nutrients and other materials are transferred from one area to another?

A Yes. One of my primary research areas at the present time is the investigation of water transport from its source, first of all as precipitation, its infiltration into the soil, and its movement back through vegetation eventually into the atmosphere again.

Q Would you describe some of these processes of water transport?

A Perhaps the best point to begin with is to note that everywhere we have vegetation there is a continual demand for water. Approximately a tenth of an inch every day moves from the soil, through the soil surface or through the vegetation into the atmosphere. As it moves into the atmosphere in one place, it must come down in another location. This is one of the mechanisms whereby impacts on the environment in one location can significantly influence the biota of another location.

Q And, Doctor, are there other mechanisms of transportation?

A There is transport of water along the surface of the ground and in streams, transport of water in the substrate, and there can be transports of plant and animal material in the atmosphere as well as in water.

Q Now, Doctor, as an ecologist, is part of your regular work the study of the interaction among independent dynamic systems of various areas?

A Yes.

Q Now, will you briefly explain for us the way in which the Wisconsin Regional Ecosystem interacts with those systems around it and the type of interaction that can be found?

A I think I can discuss this question best by going back to the movement of water that I mentioned a moment ago. I draw the Examiner's attention to the fact that during the summer months the primary atmospheric circulation over the central United States is a northward flow of air from the Gulf of Mexico all the way up the Mississippi basin, even into Wisconsin. This is interrupted periodically by fronts and air masses from the west and north; but as they move to the east, there is a replacement of the flow again from the south. Thus we have the major part of our precipitation coming in of southern origin. I might also comment on the movement of other materials in the atmosphere into the Wisconsin Regional Ecosystem, in particular nitrogen compounds released to the south of Wisconsin. Nitrogen from the atmosphere has been identified as one of the major sources of nutrient inputs in Wisconsin lakes.

Q Now, Doctor, does Lake Michigan play any part in the influence of, or is it influenced by, activity within the Wisconsin Regional Ecosystem?

A The major impact of Lake Michigan on the biota of Wisconsin is through the moderation of temperature along the Lake Michigan shore. This induces condensation of moisture in a shoreline area approximately twenty miles in width. Of course, there is more impact from Lake Superior on northern Wisconsin than there is from Lake Michigan. And this is particularly strong during the fall when Lake Superior is open and heavy snowfall occurs in that region from water picked up over the open lake.

Q Does the terrestrial ecosystem of Wisconsin itself have any influence upon Lake Michigan?

A Well, as I have already pointed out, the moisture moving from the soil and vegetation into the atmosphere at a tenth of an inch every day moves from south to north, and from west to east as we have pressure systems moving from west to east. Thus, much of the evaporation over Wisconsin moves across Lake Michigan, and, if condensation takes place there, materials carried with it go into the lake.

After brief cross-examination, this testimony was followed by that of Dr. Charles Wurster, biologist at the State University of New York at Stony Brook. Dr. Wurster was able to detail the physical and chemical characteristics of DDT and to show that these properties allow DDT to move in the environment by the same mechanisms that transport water as well as the processes in the food chain that lead to "biological concentration" of DDT into the higher trophic levels (Woodwell *et al.*, 1967; Wurster, 1969). During the weeks that followed, other witnesses testified on uptake and release of DDT, metabolism of DDT to the breakdown products DDE and DDD, physiological effects of DDT and DDE at the levels of molecular, enzyme, and nervous system processes, organismal effects on reproduction of fish and egg-shell thickness in birds, and population effects in the case of threatened species, such as the peregrine falcon. After six weeks of testimony, by both sides of the DDT issue, the hearing was being overwhelmed with scientific detail; and no testimony had yet been offered to couple quantitatively the DDT use, transport, and concentration processes with long-term deterioration of the regional ecosystem as a whole.

2. *Identification of Variables and System Description*

To provide a structured, quantitative base through which to draw the diverse scientific evidence on properties and effects of DDT in the natural ecosystem, an independent systems study of DDT was undertaken by a number of the hearing participants during the winter and spring of 1968. This work was based substantially on ongoing systems studies of nutrients in lake systems at the University of Wisconsin, and included contributions from Dr. Donald Watts, statistics; Drs. John Mitchell and Howard Harrison, mechanical engineering; Richard Tracy and William Reeder, zoology; Drs. Joseph Hickey and Dan Anderson, wildlife ecology; David Parkhurst and myself in botany, and the lawyer, Victor Yannacone.

The results of this work have now been revised and published (Harrison *et al.*, 1970), but the procedure by which it was introduced in the hearing record and what was achieved in doing so is presented in the following excerpts of transcript. They should be read not only from the point of view of an adaptation of systems science, but from the viewpoint of scientific proof in a court proceeding. The method of reaching "truths" is very different from scientific appraisal over years in scientific journals, but is nevertheless just as effective. Every participant in the systems study was ill-prepared to testify in some respects, and it was therefore with considerable reluctance that I agreed to serve as a witness on behalf on the entire study group. The first questions relate to establishing my qualifications and competence:

Q In the course of your regular professional activities, Professor Loucks, have you had occasion to investigate the waters of the State of Wisconsin?

A Yes. I hold an appointment in the Water Resources Center at the University of Wisconsin to carry out research on the pollution of Wisconsin waters.

Q What discipline is comprehended with this research?

A The research is specifically a program of systems modeling with a view toward systems analysis of the flows of energy and nutrients, particularly nitrogen and phosphorus, in Wisconsin water systems. And, as such, it is essentially independent of any discipline in the usual sense. Since it is a program of a team of investigators, the key discipline in this research is systems analysis itself.

Q And, Doctor, in the course of your regular professional activities, you have published, jointly with others of course, work in systems analysis, have you not?

A That's right.

After these general questions on experience and approach, the questioning proceeded to specifies of the DDT systems model, and identification of variables and transfers involved. Asterisks indicate the omission of interrupting comments, heated outbursts, or repetitive answers.

Q Now in the course of your regular professional activities, and with respect to evaluating the pollution of the waters of the State of Wisconsin, have you had occasion to review and evaluate the information available on the presence of DDT in various elements of the Wisconsin Regional Ecosystem?

* * *

A Yes, I have. In December my testimony was with respect to the transport mechanisms that exist in the Wisconsin Regional Ecosystem. Subsequent testimony has, I believe, shown that DDT follows this transport system. The program in eutrophication was initiated in September last year, and I joined it at the end of January. This program is a large-scale effort to apply the techniques of modern science to the analysis of pollution in water systems generally, but Wisconsin water systems in particular.

* * *

Q Now, Dr. Loucks, I show you a chart [Fig. 1, Exhibit 189] and I ask you if you can identify it?

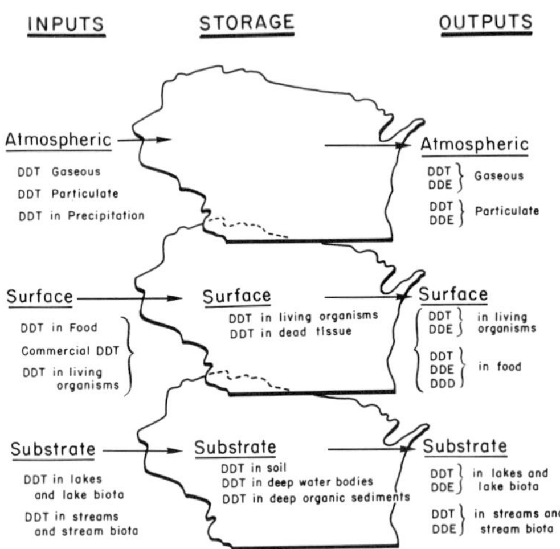

FIG. 1. A simple schematic introduced at the Wisconsin DDT hearing to show the multidimensional character of DDT exchange in the Wisconsin regional ecosystem. It served to draw attention to the diverse mechanisms involved in dispersal, to the differences between atmospheric, substrate, and surface exchange processes, and to the role of storage in the surface and substrate compartments. Testimony on this exhibit drew attention to the coupling of the Mississippi River system and Lake Michigan as integral parts of the regional ecosystem.

A Yes, I can.

Q All right. And this chart purports to be what?

* * *

A This chart is a pictorial summary of the inputs, outputs, and storage terms of DDT in the Wisconsin Regional Ecosystem as they can be best represented in such a form from our systems analysis of the pollutant materials in Wisconsin water systems.

* * *

Q Doctor, would you please explain in words just what Exhibit 189 purports to show?

A It purports to be a complete listing of the sources of DDT that come into Wisconsin and which are listed here as inputs. Perhaps a key one is the input, at the surface, of commercial applications of DDT. But there are other sources of DDT which are transported physically into the Wisconsin Regional Ecosystem at the surface—DDT in living organisms which may migrate into the state, the transport in the atmosphere, DDT in gaseous form, and DDT attached to particulate matter which has been carried by winds across the region, and DDT in precipitation. . . . These materials leave the Wisconsin Regional Ecosystem in similar transport mechanisms, except that at the surface there is no physical removal of a material comparable to the commercial application input of DDT. I show as "outputs" the breakdown products of DDT which can also be carried out of the system. I draw attention in particular, however, to the fact that there are certain storages of DDT which have been identified, which means that the total output leaving the Wisconsin Regional Ecosystem is less than the total input.

Following this testimony Mr. Yannacone attempted to have the pictorial description of regional inputs, outputs, and storages accepted into evidence:

[MR. YANNACONE] All right. I would now like to offer Exhibit 189, please.

[MR. STAFFORD] I renew my objection. I don't know enough about the qualifications of this witness who disclaimed any work with DDT when he appeared before. He testified now from his own knowledge and experience what the atmospheric surface, substrate, etc. inputs, outputs, and storage of DDT are. I don't think you have any proper foundation for him to give evidence on this subject.

[EXAMINER VAN SUSTEREN] 189 is merely a pictorial representation of how DDT comes into Wisconsin and how it leaves Wisconsin. That's about it, isn't that right?

* * *

[MR. STAFFORD] It doesn't state much more than that, no, but it says quite a little bit, I think, pictorially which I don't think this witness has any qualifications to testify to.

[EXAMINER VAN SUSTEREN] Well, Exhibit 189 is received.

With the general description of DDT exchanges in the region accepted, the testimony proceeded to a more detailed examination of systems processes at a local level. Variable identification began with water, both because of the Wisconsin group's experience in this area, and because it is the coupling mechanism between the introduction of DDT and its effect on aquatic life:

Q Now, Professor, I show you a chart entitled "Water Transfer Functions" [Fig. 2, Exhibit 190; Watts and Loucks, 1969] and ask you if you can identify it?

A Yes, I can.

* * *

Q Will you identify it, please?

A This is a chart listing the input water variables for a watershed that can be viewed as a subsample of the Wisconsin Regional Ecosystem. It lists the zones at which transfer functions operate on the input water, and therefore lists the multiple outlets for each of the eight inputs listed within the chart.

Q And you assisted in the preparation or supervised the preparation of that chart and its ultimate publication, did you not?

A Yes, I did.

Q Does it fairly and accurately represent the water transfer functions present in the waters of the Wisconsin Regional Ecosystem?

A Yes, it does.

[MR. YANNACONE] I offer it (Exhibit 190) at this time.

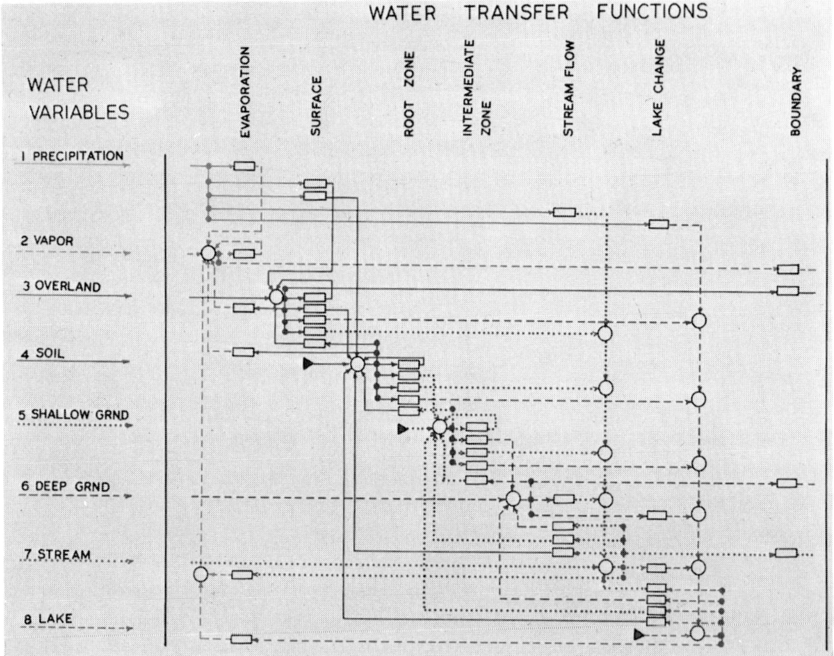

FIG. 2. A flow chart from the paper by Watts and Loucks (1969) illustrating in servo-system notation the redistributions of water in an ecosystem. It was introduced to show the flows of each form of water through the system, the processes involved, and the summing junctions integrating the sources of each form.

[MR. STAFFORD] I object to it. It hasn't been fully described, Your Honor, I don't know from the description that the witness has just given where the input or the output is, or what is all this. . .

[MR. YANNACONE] I will further describe it.

* * *

After much more questioning, Exhibits 190 and 191 (a related flow chart for energy), and then the entire paper (Watts and Loucks, 1969), were eventually accepted in evidence. The next step was to list and describe the variables (pools or compartments) which would specifically apply to the flux of DDT and DDE through the system.

Q Now, Professor Loucks, would you refer to the chart in Exhibit 193-A [Table I] and tell us what it is?

A This chart, 193-A, entitled "Transports and Transformation of DDT in Air and Water Exchanges, Wisconsin Regional Ecosystem" lists in detail all of the exchanges that our group in systems analysis has been able to identify for the material DDT as it is carried in the atmosphere and water transport systems.

* * *

Q Now, Doctor, would you summarize for us the method whereby the information in such a complex system is integrated and processed by means of systems analysis today?

* * *

A First of all, we might look at Exhibit 193-A. Simply by way of illustration of the kind of information that is listed here, we may take DDT that occurs as a gas in the atmosphere. We see that it can be redistributed as DDT in two forms, one that is DDT attached to any particulate matter in the atmosphere, or it may become DDT in precipitation. There is also potentially a transformation of DDT to DDE residues in the atmosphere. . . .

At this point, Mr. Willard Stafford, counsel for the Agro-Chemical Task Force, renewed his efforts to place in question my qualifications as an environmental systems expert by challenging the sources of my materials:

[MR. STAFFORD] I'm not being rude, I simply don't mean to be, but I would like to interrupt the witness, Mr. Examiner, and ask him at this point:
You refer now to what is designated item No. 1 in Exhibit 193-A; the name of that item is "Atmospheric DDT". Now you have said you have simply taken DDT that occurs as a gas in the atmosphere. Where did you get the information as to how DDT occurs in the atmosphere, and the amount of it and what form it's in? You know nothing about that yourself. I'm just wondering what the source of your information is.

[MR. YANNACONE] Mr. Examiner, it's already part of this record by their own witness.

[EXAMINER VAN SUSTEREN] It's in the record.

[MR. STAFFORD] Is that the source of your information, the material that's already in this record? That's all I'm trying to find out, what the source is.

[WITNESS] That's one of the sources. There are many sources whereby we arrive at the conclusion that's presented in this summary of the transports.

[EXAMINER VAN SUSTEREN] There's testimony in the record that DDT has a relatively low vapor pressure; and there's also testimony in the record that it has the power of adsorption to particulate matter; and the particulate matter that we may know as dust, spores, dirt and so on in the atmosphere may contain adsorbed particles of DDT. And also, the principle of codistillation was brought up in regards to the adherence of a molecule of DDT attached to a molecule of water.

[MR. YANNACONE] Could we go on, please.

[WITNESS] Yes. I don't want to go through all of these variables, but there are two or three, I think, that illustrate the mechanisms of systems description.

The DDT that may be in the atmosphere on particulate matter may have originated on the land surface and been carried into the atmosphere. In any case, DDT that is on particulate matter can be either accumulated as a storage of DDT on the particulate matter in the atmosphere; it may become DDT in precipitation; it may become DDT in the surface waters simply by the fall-out of the atmosphere as particulate matter with DDT on it; and by the same mechanism it may become DDT in streams or in lakes.

Each of these variables are followed through in essentially the same form. I draw your attention to "DDT in Streams" which again may result in a storage of DDT in the streams, or there may be codistillation of DDT to the atmosphere; it may move into interstitial soil water; it may become part of the surface groundwaters; it may become DDT in the aquatic vegetation; it may become DDT in the herbivores by moving through the skin of those organisms in the water systems; and so forth. We are simply listing all of the places that DDT can go from any particular source.

Similar testimony was used to describe the trophic structure of an ecosystem and the DDT variables needed to model the flow of DDT through the biological carrier system (Table II). Both Tables I and II are more conveniently shown in matrix form, but for the purposes of court action, the tabular form is more easily read. After Mr. Yannacone moved acceptance into evidence the following exchange occurred:

[EXAMINER VAN SUSTEREN] Any objection to exhibit sheet A and B?

[MR. STAFFORD] Yes. I don't think it's been properly qualified; no proper foundation. For example, speaking of under the column "Outputs" and now referring to 193-A, you referred to DDT in aquatic herbivores, first carnivores, second carnivores, third carnivores. What does that mean?

[WITNESS] We will be coming to an explanation of the dynamics of these figures in a moment.

TABLE I

Inputs, Transports, and Transformations of DDT in the Atmospheric and Water Carrier Systems, Wisconsin Regional Ecosystem

Inputs	Outputs	
	Transfer to	Transformation to
Atmospheric carriers		
Atmospheric DDT	DDT on particles DDT in precipitation	DDE, DDD residues
DDT on atmospheric particulates	Storage of DDT DDT in precipitation DDT in surface waters DDT in streams DDT in lakes	DDE, DDD residues
Water carriers		
DDT in precipitation	DDT in surface waters DDT in streams DDT in lakes	DDE, DDD residues
DDT in surface waters	Storage of DDT Gaseous DDT DDT in streams DDT in lakes DDT in aquatic vegetation	DDE, DDD residues
DDT in interstitial soil water	Storage of DDT DDT in groundwater DDT in streams DDT in lakes DDT in terrestrial vegetation	DDE, DDD residues
DDT in streams	Storage of DDT Gaseous DDT DDT in interstitial soil water DDT in groundwater DDT in aquatic vegetation DDT in aquatic herbivores DDT in aquatic first carnivores DDT in aquatic second carnivores DDT in aquatic third carnivores DDT in aquatic scavengers	DDE, DDD residues

Table continued

TABLE I (*continued*)

Inputs	Outputs: Transfer to	Outputs: Transformation to
DDT in lakes	Storage of DDT	DDE, DDD residues
	Gaseous DDT to atmosphere	
	DDT in interstitial soil water	
	DDT in streams	
	DDT in lakes	
	DDT in aquatic vegetation	
	DDT in aquatic herbivores	
	DDT in aquatic first carnivores	
	DDT in aquatic second carnivores	
	DDT in aquatic third carnivores	
	DDT in aquatic scavengers	

TABLE II

TRANSPORTS AND TRANSFORMATION OF DDT IN THE BIOMASS WISCONSIN REGIONAL ECOSYSTEM

Inputs	Outputs: Transfer to DDT in	Outputs: Transformation to
Biomass carriers		
DDT in		
Aquatic vegetation	Aquatic vegetation	
	Dead aquatic vegetation	
	Bottom detritus	
	Aquatic herbivores	
	Aquatic decomposers	
Dead aquatic vegetation	Bottom detritus	
	Aquatic decomposers	
	Aquatic scavengers	
Bottom detritus	Bottom detritus	
	Aquatic decomposers	
	Aquatic scavengers	
Aquatic herbivores	Aquatic herbivores	DDE, DDD residues
	Bottom detritus	
	Aquatic first carnivores	
	Aquatic decomposers	
	Aquatic scavengers	
	Interstitial water	
	Streams	
	Lakes	

Table continued

TABLE II (*continued*)

Inputs	Outputs: Transfer to DDT in	Outputs: Transformation to
Aquatic first carnivores	Aquatic first carnivores	DDE, DDD residues
	Bottom detritus	
	Aquatic second carnivores	
	Aquatic third carnivores	
	Aquatic decomposers	
	Aquatic scavengers	
	Interstitial waters	
	Streams	
	Lakes	
Aquatic second carnivores	Aquatic second carnivores	DDE, DDD residues
	Bottom detritus	
	Aquatic third carnivores	
	Aquatic decomposers	
	Aquatic scavengers	
	Interstitial water	
	Streams	
	Lakes	
Aquatic third carnivores	Aquatic third carnivores	DDE, DDD residues
	Bottom detritus	
	Aquatic top carnivores	
	Aquatic decomposers	
	Aquatic scavengers	
	Interstitial water	
	Streams	
	Lakes	
Aquatic top carnivores	Aquatic top carnivores	DDE, DDD residues
	Bottom detritus	
	Aquatic decomposers	
	Aquatic scavengers	
	Interstitial water	
	Streams	
	Lakes	
Aquatic decomposers	Aquatic decomposers	DDE, DDD residues
	Bottom detritus	
	Surface waters	
	Interstitial waters	
	Streams	
	Lakes	
Aquatic scavengers	Aquatic scavengers	DDE, DDD residues
	Aquatic first carnivores	
	Bottom detritus	
	Aquatic decomposers	
	Interstitial water	
	Streams	
	Lakes	

[MR. STAFFORD] I would ask that the examiner then reserve his ruling upon the admissibility of 193-A and B until we have had information to do with dynamics.

[MR. YANNACONE] I will withold the offer.

3. *Mathematical Analyses of the DDT Model*

The testimony that followed the initial characterization of the ecosystem (above) was concerned with an explanation of the trophic structure of an ecosystem, using frequent reference to Exhibit 65, the study of DDT levels reported by Woodwell *et al.* (1967) for the Carmans River marsh on Long Island. Reproduction, predation, and mortality, by trophic level, were described so as to provide a basis for a mathematical statement of rates of transfer for both biomass and the DDT it carries. The differential solubility of DDT in lipids compared with water, which provides the basis for selective retention of DDT in living organisms, was implicit from the extensive earlier testimony of Dr. Wurster and others. My testimony on the trophic structure dynamics concluded as follows:

A The trophic structure that I have just described gives us a mechanism for following the buildup in concentration of DDT in each of the trophic levels in the system. The aquatic vegetation (the primary producers) will have DDT in their tissues as a function of the concentration of DDT in the surrounding aquatic environment. Similarly, the aquatic herbivores will have DDT in their tissues as a function of the period of feeding, the concentration of DDT in their food (the phytoplankton), and direct uptake of DDT into epidermal lipids from the water. As a result, the concentration of DDT in the herbivore level will be greater than that in the phytoplankton as a function of the total food consumed over the period of its life.

I should explain that each of these organisms must consume daily in the order of twice as much food as they retain. The additional food is expended as burned energy in maintaining the living system. Thus, as long as the organisms in any trophic level are living and expending energy, there is continual intake of food containing DDT in a concentration characteristic of the trophic level being fed on. Expended energy does not breakdown any quantity of DDT. There may be partial breakdown of DDT to DDE and other residues in organisms, but most of the DDT is retained in the body of the organism in the higher trophic level.

At this point, the lawyer for the Agro-Chemical Task Force, Mr. William Stafford, objected strenuously on the grounds that I had no qualifications in the area of DDT chemistry. This is an objection that will face every systems analyst in court, and the only response is to continue to demonstrate the integrative role of systems testimony, and to get the court to recognize that portions of all the technical testimony must appear somewhere in the systems presentation.

[MR. STAFFORD] This man has no qualifications on his own to come to these conclusions.

[MR. YANNACONE] You can test it on cross-examination.

[MR. STAFFORD] It's argumentative. I ask it be stricken from the record.

[EXAMINER VAN SUSTEREN] It's a recap of the testimony and evidence that we have received so far in regard to accumulation and residues.

[MR. STAFFORD] I don't think it's that at all. It doesn't purport to be. It's based on Exhibit 65, as I understand his testimony.

* * *

[EXAMINER VAN SUSTEREN] All Dr. Loucks is doing is apparently giving a recap as to exactly how, through the various ecological systems, there is a final accumulation of residues.

[MR. STAFFORD] That is a function of an advocate, not a scientist.

* * *

[EXAMINER VAN SUSTEREN] The objection is overruled.

The direct examination continued with the introduction of a draft manuscript presenting equations summarizing the diffusion of DDT through ecosystems:

Q Dr. Loucks, would you please describe for us, using the methods of modern systems analysis, how the equilibrium concentration of DDT in any given trophic level is determined?

A I have portions of a manuscript discussing the mathematical models for DDT accumulation through the trophic levels of an ecosystem.

Q Doctor, would you describe that paper for us?

A This section is entitled "Ecosystem Analysis: Diffusion of DDT".[†] It follows the same principles that we have been discussing: the inputs of DDT at a certain concentration; taken up at a certain rate; destruction of DDT in the

[†] See the Appendix.

organisms in a particular trophic level; a certain retention and accumulation of it at that trophic level; and certain outputs. The general equation can be described as the net sum of inputs, destruction by metabolism, accumulation, and outputs, formulated on the basis of concentration in the trophic level below that for which the concentration is being predicted, on the rate of ingestion, on the death rate of the organism, on the amount destroyed by metabolism, and lastly, on the amount lost by excretion.

Q Now, Dr. Loucks, in the terms that you just described in that equation you did include, did you not, the breakdown of DDT at any given level into its metabolites?

A Yes, that is the term, "amount destroyed by metabolism" included as one of the components.

* * *

Q And in order to chart the behavior of the system, Doctor, you do not need to know either the exact mechanism of this breakdown or the exact proportion of this breakdown in order to predict the ultimate behavior of the system, do you?

A That is true for the conclusions that can be derived from this analysis.

Q In other words, all you have to do to model the system is to recognize the fact that the process exists; is that correct?

A That's what has been done in this analysis.

After testimony on representative levels of DDT and DDE concentration in the various trophic levels, and more strenuous objections, the examination proceeded to analysis of impact on populations and on the system as a whole.

Q Now, Dr. Loucks, without belaboring us with the actual equations, have you in the course of your professional activities considered the actual mathematical relations involving the transfer functions across various elements of the Wisconsin Regional Ecosystem, in particular as it relates to water, animals, plants, and wildlife that are related to the water?

A Yes, I have.

Q Can you with a reasonable degree of certainty form an opinion as to the effect of DDT on the Wisconsin Regional Ecosystem, in particular the waters thereof, the plants, the fish and wildlife therein and related thereto?

A Yes, I can.

* * *

Q Doctor, will you give us that opinion?

[MR. STAFFORD] I object to this again, because there's no proper foundation laid in this witness' qualifications for him to give an opinion on this subject. Furthermore, it, of course, invades the province of the examiner. Nothing but a conclusion and speculation on this gentleman's part in view of his admitted lack of experience with DDT in the ecosystems of Wisconsin.

* * *

[EXAMINER VAN SUSTEREN] And if it's in relation to the systems analysis, the objection is overruled.

Q Will you give us your opinion with respect to the ecological factors and the systems analysis thereof?

A We have considered a whole sequence of evidence, beginning with the transport mechanisms that we discussed, and the cumulative concentration of DDT toward higher and higher levels in the ecosystem. I conclude from this analysis, both from the predictions that can be made from the mathematical models, and from the evidence with respect to DDT and its effect on populations, that DDT and its toxic breakdown products found in Wisconsin waters form a dynamic process in which the concentration levels at the higher trophic levels in the ecosystem are still building up.

The only true sink of DDT in the system can be shown to be the lipid biomass of plant and animal tissue. The DDT concentrations in this tissue at each trophic level can be shown to be building up now, and will likely be building up for some time in the future in certain species. . . .

Again, the integrative nature of conclusions drawn from systems studies drew objections from the opposing lawyer, Mr. Stafford, and heated exchanges with Mr. Yannacone.

[MR. STAFFORD] Now, your Honor, this last statement about building up in the lipid solutions by this witness is clearly incompetent. It's the grossest sort of speculation.

* * *

[EXAMINER VAN SUSTEREN] Mr. Stafford, the entire record is replete with testimony and evidence from both sides of the table on the concentration of DDT in lipid tissue. It just is.

[MR. STAFFORD] Well, of course it is, and we acknowledge it. But his opinion that it's building up and continues to build up—he has no basis whatsoever for that statement.

[MR. YANNACONE] Test it on cross-examination.

[EXAMINER VAN SUSTEREN] I have already ruled that he can proceed on this only in regards to his own systems analyses that he has worked up, in regards to his equations, and the relationship between mass and so on.

The testimony on the DDT systems model and the preliminary analysis of it concluded with the following description of potential impacts on populations of wild species, both herbivores and carnivores.

A The analysis by differential equations of the diffusion of DDT through the trophic structure of the ecosystem indicates that the equilibrium concentration will be a function of the lifetime of the organisms in the system. Thus, to the extent that there are some species whose lifetime is longer than the period which we have used DDT, we conclude that there can be further buildup in the concentration in some of these species.

There is no evidence from the systems analysis that we have done that the breakdown of DDT is yet equal to the additional introduction of DDT in the system. It is therefore most probable, in my opinion, that there will be substantially more degradation of the stability characteristics of the ecosystem if the input of DDT is not stopped.

* * *

I form the opinion, on the basis of our analysis of population fluctuations in each trophic level, which can be shown to be nonlinear differential equations, and from which we can expect stable but oscillating systems, that in such a system the top carnivores can be shown to play an important regulating mechanism on the numbers of prey at a lower trophic level. We conclude that the removal or decrease in numbers of important predator species in this system can lead to exploding populations of the prey species, a response that is acknowledged as a serious upset and degradation of the ecosystem. We submit that DDT is capable of, and is bringing about, this upset in the Wisconsin Regional Ecosystem.

[MR. YANNACONE] Thank you very much. Doctor.

Cross-examination of this testimony was minimal. It is always difficult to cross-examine mathematical relations, so questions focused on the competency of the witness. Mr. Stafford concluded as follows:

[MR. STAFFORD] I move that the testimony and the conclusions of this witness which pertain to the buildup of DDT in animals and men, and in the environment, be stricken because they are not based upon any research whatsoever of this witness or any competency on his part, and they are simply gratuitous conclusions.

[EXAMINER VAN SUSTEREN] The objection is overruled. They are based obviously on the work of others, and his opinion and/or conclusion is based on his analysis of the systems analyses.

4. *Outcome of the DDT Hearings*

The outcome of this systems testimony can be recognized in two forms. First, it was considered by the DDT hearing examiner along with the testimony of many other witnesses in reaching his decision on the evidence. Secondly, the original charts, equations, and interpretations, drawn together for the purpose of synthesizing diverse technical testimony, were later published as a paper in the open literature (Harrison *et al.*, 1970). Both results have since stimulated interest in applying these methods to other environmental contaminants such as mercury and polychlorinated biphenyls (PCBs).

In summarizing the evidence in the DDT hearing, the examiner, Maurice Van Susteren, concluded as follows:

> DDT, because of its chemical/physical properties and once applied to crops, or placed in the atmosphere, moves throughout the environment in water, air, soil and food (Exhibits 73, 74, 189, 198–204; T-209–214, 665–670, 723, 728, 831–837, 869, 2204, 2205, 2209, 2212, 2215). A minor transport mechanism is in organisms such as birds and fish (T-239). It has been found in filtered air, untreated forest soil, and the fish in the untreated watershed had in some instances 2.4 ppm DDE (T-243–245). DDT and metabolites have been found in oceanic food chains from zooplankton to gulls, osprey, cormorant, petrels, pelicans, and peregrine falcons with a corresponding biological concentration at each level of the chain (Exhibit 65; T-251–256, 661, 662, 715).

Then, in stating an opinion as to the conclusion that could be drawn from the evidence, the examiner said:

> DDT is ubiquitous. It is found in the atmosphere, soil, water, and in food in what might be considered minute amounts. The chemical property of being soluble in lipid of fat tissue results in storage primarily in the body fat and nervous systems of all organisms in all levels of food chains. It is therefore impossible to establish levels, tolerances or concentrations at which DDT is toxic or harmful to human, animal or aquatic life.

Finally, the examiner wrote a ruling, under Wisconsin law, to the effect:

> DDT, including one or more of its metabolites in any concentration or in any combination with other chemicals at any level, within any tolerances, or in any amounts, is harmful to humans and found to be of public health significance. No concentrations, levels, tolerances or amounts can be established. Chemical properties and characteristics of DDT enable it to be stored or accumulated in the human body and in each trophic level of various food chains, particularly the aquatic, which provides food for human consumption. Its ingestion and dosage therefore cannot be controlled and consequently its storage is uncontrolled. Minute amounts of the chemical, while not producing observable clinical effects, do have biochemical, pharmacological, and neurophysiological effects of public health significance.
>
> DDT and its analogs are therefore environmental pollutants within the definitions of Sections 144.01(11) and 144.30(9), Wisconsin Statutes, by contaminating and rendering unclean and impure the air, land and waters of the state and making the same injurious to public health and deleterious to fish, bird and animal life.

B. Court Action to Limit Release of Radionuclides in Colorado

1. *Drawing of the Complaint*

During the summer of 1969 a court action began in Colorado against the Atomic Energy Commission and two natural gas companies to prevent the carrying out of Project Rulison (see Yannacone, 1970, for Plaintiffs' Brief). This project was a part of the U.S. AEC Plowshare Program of peaceful uses of atomic energy. It was designed to use a small nuclear device set off deep in the bedrock of southwestern Colorado to create small cracks in the deep rock strata where diffuse natural gas could collect in a marketable-sized pool. The project was opposed first on the grounds of a danger during detonation because of earthquake and venting hazards, and later to prevent opening of the pool because of the presence of large quantities of long-lived radionuclides—tritium and krypton-85.

The Colorado Open Space Coordinating Council, with Mr. Yannacone again serving as counsel, filed a complaint against Austral Oil Company, the Geonuclear Corporation, and the Atomic Energy Commission for a Declaratory Judgment (Yannacone, 1970):

> seeking declaration of rights and legal relations of the parties to the matter in controversy, specifically:
>
> (a) The rights of the people of the State of Colorado to the protection of their health and safety and the health and safety of

those generations yet unborn from the hazards of ionizing radiation resulting from the distribution of radioactive materials through the Colorado Regional Ecosystem as a result of the actions of the defendants herein complained of.

(b) The rights of the people of the State of Colorado to the full benefit, use and enjoyment of the national natural resource treasures of the State of Colorado without degradation resulting from contamination with radioactive material released by the actions of the defendants herein complained of.

* * *

(d) That the release of any radioactive material which might enter the permanent biogeochemical cycles of the Biosphere will violate the rights of the people guaranteed under the Ninth Amendment of the Constitution of the United States.

The complaint against the detonation of the device alleged among other points:

That it appears from the public statements of representatives of the Defendant CER Geonuclear Corporation made at Aspen, Colorado during July, 1969, that the radioactive gases, krypton-85 and tritium, will be ultimately released to the atmosphere, at a time considerably less than one half-life of either radioactive gas.

That the defendants have not demonstrated, at any time prior to the commencement of this action, any evidence indicating that the total effects of Project Rulison upon the State of Colorado and the permanent biological, chemical and geological cycles of the Biosphere have been studied or evaluated, and in any event no such information has been made available to the Plaintiffs, or the general environmental science community through publication in recognized scientific journals of general circulation.

Finally, the plaintiffs' complaint, drawn by Mr. Yannacone, alleged irreparable damage to the regional ecosystem:

The release of tritium or krypton-85 into the atmosphere or their introduction into the permanent biogeochemical cycles of the Biosphere as a result of Project Rulison will cause serious, permanent and irreparable damage to the national natural resource treasures of the State of Colorado and will present a direct threat to the health and safety of the people of Colorado not only of this generation but of those generations yet unborn.

That there is no evidence now available to the environmental science community indicating that the radionuclides trapped in the blast underground, including strontium-90, cesium-137 and iodine-131 will not enter the permanent biogeochemical cycles of the Colorado Regional Ecosystem, in particular the underground water supply.

The plaintiffs therefore asked judgment declaring:

The rights of the people of the State of Colorado to the protection of their personal health and safety and the health and safety of those generations yet unborn from the hazards of ionizing radiation resulting from the distribution of radioactive materials through the Colorado Regional Ecosystem as a result of Project Rulison, and restraining:

the defendants from proceeding with the detonation of any nuclear bomb in the State of Colorado, until such time as the defendants have shown good cause supported by substantial evidence that such detonation of a nuclear bomb will not cause contamination of the permanent biogeochemical cycles of the Biosphere with radioactive materials, and that such detonation of a nuclear bomb will not release any ionizing radiation into the environment.

As in the DDT hearings, the key elements of the complaint were (1) that the case was being brought as a class action in which damage was threatened for all citizens residing in the regional ecosystem; and (2) the agents potentially capable of causing the class damage, the radionuclide releases, followed complex pathways through the ecosystem and would require systems methods as a part of any court proceeding.

After some pretrial hearings, and the taking of depositions by AEC witnesses, the action to prevent detonation of the nuclear device was denied. A second series of complaints by three different Colorado groups was then brought to prevent release, or venting, of the natural gas pool because of the same hazards cited above. The cases were combined, and trial was set to begin 12 January 1970, in Federal Court in Denver before Judge A. A. Arraj.

2. *A Systems Model for Regional Impact Analysis*

One of the first steps in bringing a litigation to prevent the proposed release of long-lived radionuclides from Project Rulison was to place into the court record a description of the interacting physical and biological processes and transport systems referred to in the complaint as the Colorado Regional Ecosystem. In this case, however, the AEC had already carried out a systems analysis for its safety evaluation, so the central question would turn on the adequacy and competence of the systems model and its analysis. Mr. Yannacone served as counsel for the plaintiff, asked the questions to put my testimony in evidence, and raised objections to inappropriate questions during cross-examination.

After preliminary questions on qualifications, the early questioning sought to establish the principles that govern completeness and competence of a regional systems model (U.S. D.C., District of Colorado, 1970). The questions are by Mr. Yannacone, and the answers are mine.

Q So that we understand what we are talking about, what do you mean when you use the phrase "systems ecology?"

A Systems ecology to me is the investigation of the system that is acting on biological materials in the natural environment, and it has three major components: These are the atmospheric transport system as it influences biological materials; the water transport system, the redistribution of water from the atmosphere to the surface, to the vegetation, and to the groundwater; and thirdly, the biological transport system itself, where we have movement of many materials by grazing and predation activities.

Q Doctor, would you tell us what you mean by the phrase "transport system"?

A By a transport system, I mean the system in the environment that involves a movement, transfer, or exchange of material from one point to another or from one form to another, as in the transformation of carbon dioxide by photosynthesis.

Q All of these transport systems are functions of time and some of them are functions of distance, aren't they?

A That's right, and functions of other properties of environmental systems.

Q Now, Doctor, tritium is a biologically active material, is it not?

A It can be in certain systems, yes.

Q Would you please tell us the basic elements of the atmospheric transport system of a biologically active material?

A The basic elements of this transport system include the circulation of the atmosphere, particularly the circulation of the lower atmosphere, and, in the case of tritium, we are primarily concerned with the water components in the atmosphere. This transport system includes such features as the lateral flow of air, including the flow of water vapor over a landscape where it may encounter areas of high topography which can result in cooling of the air and a resultant condensation of the water vapor, where it enters the water transfer system.

Q All right, Doctor, would you summarize briefly the elements of the water transport system as they influence a biologically active material such as tritium.

A The water transport system is much more fully understood than the atmospheric transport system, and we do have a computer simulation capability for predicting the movement of water through the land system from the moment that precipitation strikes the surface. The water is then redistributed to a number of variables within the model. I am describing the Stanford watershed model developed at Stanford University over the past ten years.

At this point, we attempted to introduce published papers into the evidence, and, after objections and questioning by opposing lawyers, Mr. Yannacone asked the following questions elaborating the extent of my own research, and its relevance to radionuclides. The goal was to demonstrate that systems analysis is an integrative science requiring direct participation in many disciplines.

Q Now, with respect to that paper (Watts and Loucks, 1969) and the water transport system we have under consideration, are there elements of the work done in that study that relate to the transport of biologically active materials such as tritium in a system such as the Rulison regional transport system?

A The continuing objective of the systems studies at Wisconsin is to investigate the transport of nutrients, nitrogen and phosphorus, from various sources on the landscape to lakes and streams. It is the consensus of our group and of other groups across the country that a simulation capability of the carrier material, water, is the best means of achieving good prediction of a transported material, such as nitrogen or phosphorus.

We have also applied this technique to investigation of the transport of DDT. Any other material that enters water can be modeled and simulated by the techniques that Professor Dale Huff published in his paper.

Further questioning by opposing lawyers was directed to trying to demonstrate incompetence and immateriality of my testimony because of insufficient work with radioactive materials:

Q [BY MR. EARDLY] Doctor Loucks, what courses did you take in college or since college which led you to an understanding of the major functions of these radioactive nuclides?

A I am presently engaged in work that is primarily concerned with systems analysis of materials moving through the environment. Most of my preparation has led up to this activity. In developing this experience we have relied very heavily on the systems studies carried out at Oak Ridge National Laboratory, which I have followed closely for the past fifteen years. It has been one of the leading centers in the development of systems analysis of biological systems. And it is from that association that I have some experience with the movement of radioactive materials in systems. As a systems analyst, I cannot appear as a specialist in research in radioactive materials themselves, per se, but on their movement in complex systems.

Q Now, what practical personal experience have you had in tracing the movements of radioactive substances in the ecology?

A My personal experience in that area has been in the utilization of the published literature from Oak Ridge National Laboratory on the movement of these materials in biological systems, and which we are presently adapting through the systems analysis group at the University of Wisconsin.

The questions then returned to more detailed statements of the characteristics of the systems involved in Colorado, with the questioning again by Mr. Yannacone.

Q Doctor, will you discuss the basic elements of the water transport system in two aspects, the first being that portion which is variant, and a function of the transport mechanism, and that which is dependent upon the chemical and physical properties of the material to be transported?

A Yes, the primary transport system is the movement of the water through the system itself. I listed seven or eight variables into which incoming water precipitation can be partitioned. The processes whereby it goes through this partitioning are such processes as evaporation, infiltration, gravitational flow, absorption by plants, and related transformations. This is essentially the system for material that is the carrier system. In addition, we have the transporting material, a material that may be in the water, for example, nitrogen, DDT, or tritium, but each of these materials will go through the processes at some rate that is somewhat different than that of the water.

Q In other words, then, Doctor, once you have adequately described the water transport function in the water transport, and then not only the physical and chemical properties and the biological activity of the material, you can utilize your systems model and come up with some predictable statements about the distribution of the biologically active material, is that correct?

A If the chemical and physical properties are well-enough known, one can make the adjustment in the system to achieve a prediction of the flow of the carried material. In many cases the properties are not well-enough known to make this prediction, however.

This characterization of the physical transport systems was followed immediately by characterization of the biological transport system:

Q Doctor, would you describe for us the biological transport system as an element of a systems analysis as it affects the transport of a biologically active material such as tritium?

A The biological transport system consists of plant roots which absorb water from the surface layers of the soil; water, which enters the conducting system of the plant and in the leaves, and may be either transpired into the atmosphere or may at that point react with carbon dioxide and through photosynthesis become implaced in sugar molecules in the leaf which in turn can become implaced in starch or cellulose molecules of the plant tissue. The transport system continues then through what is referred to as another trophic level, another partitioning of the movement of materials upward through the food web when grazing animals, and this may include mice as well as cattle, feed on that herbage and utilize primarily the carbon and water that has been mixed in sugars and cellulose. But to the extent that tritiated water that may have been a part of the water incorporated in sugar molecules, it can continue to be a component of cellulose entering the digestive tracts of the animals feeding in the next trophic level.

Considerable testimony was introduced on the photosynthesis process in primary producers, the role of water in the equation for photosynthesis, and the potential for tritium to be used equally in the formation of sugars. The questioning then proceeded to characterization of other trophic levels in the system.

Q Now, Doctor, when the green material, green plants, are grazed by the next trophic level above, what happens?

A Of course, the contents of the green plant are ingested and any water in the plant, which might include tritiated water not found in cellulose, will move into the water circulation of the grazing animal. On the other hand, the sugars, starches, and cellulose can be broken down by the grazing animal and utilized as building materials in the tissue of that animal.

Q In other words, then, Doctor, there are two separate processes and mechanisms involved within this grazing animal, one for the water which is not bound in the green plant and is moved on, and the other which is bound in the elements by the chemical elements of the green plant?

A That's right.

Q And, Doctor, how much of this material that is ingested is retained by that carnivore or grazing animal?

A It would depend on the age of the grazing animals. What we are concerned with here is that in general about fifty percent of the energy intake, that is, the energy contained in the bonds of the sugar or cellulose molecules, will be utilized in respiration. This release of energy is to maintain body warmth and the activities of the grazing animal.

Of the remaining fifty percent, a portion will be excreted and a portion will be utilized in the building of tissues. We then have in the grazing animal the reduction of approximately fifty percent of the sugars or cellulose, to its components, carbon dioxide and water, and these are returned to the atmosphere.

Q Some of it remains with the grazing animal?

A A portion of the remainder with the grazing animal—that which is not excreted.

Q When the grazing animal is preyed upon, eaten or otherwise consumed, does the same process repeat?

A Yes, the same process repeats as we move through each predator–prey level, referred to as a new trophic level. We have the same utilization of approximately fifty percent in simply burning off the intake, and the storage of a major part of the remaining intake.

3. *Examining the Completeness of System Description Models*

The above testimony concluded the development of a relatively detailed outlining of the physical and biological components of a regional ecosystem model. The questioning then proceeded to a comparative evaluation of the adequacy of the AEC systems studies relating to the proposed release of tritium into the central Colorado regional ecosystem.

Q [BY MR. YANNACONE] Now, Doctor, is it possible to describe the regional transport systems in order to predict the transfer and distribution of a biologically active toxic material such as tritium? Yes or no?

A It's possible to make a description, yes.

* * *

Q Doctor, at this time can you tell with a reasonable degree of ecological certainty, based on the data contained in Exhibit designated as the O-40, Revision No. 2, the technical discussions of over-the-site safety programs for underground nuclear detonations published, and in Exhibit NVO 61, entitled "Project Rulison Postshot Plans and Evaluations," is it possible, based upon those two documents and the material contained therein, to adequately predict the transfer and distribution of the material tritium throughout the Rulison regional transport system?

A I do not think so.

Q Now, Doctor, would you elaborate on your answer to why they are not?

A I would like to contrast the completeness of the system description in these two documents with one in a paper by Stephen Kaye,[†] if I might.

Q Doctor, let's lay a proper foundation for that paper.

* * *

Q Now, without quoting therefrom, would you indicate briefly the subject matter of that Kaye paper?

A This paper is concerned with the feasibility and safety, particularly the safety, of the proposed sea-level canal in Panama, and it offers a systems model that they use to answer some questions with respect to the redistribution of radionuclides that may be expected in the tropical environment if and when the blast for the sea-level canal is set off.

Q And is that a systems model that was prepared by, through, or under the ægis of the Atomic Energy Commission?

A Yes, it is.

Q Have you examined the model as purely a systems model?

A Yes, I have.

Q Is the substance of that paper fairly representative of the basic elements of compartmentalized systems models as now being performed under the ægis of the Atomic Energy Commission?

A Yes, it is.

Q Now, would you indicate where, if anywhere, in Exhibit F2, Exhibit N, being the postshot Rulison memorandum, and Exhibit F2 being the preshot memorandum, where there appears any reference to the systems modeling or systems considerations for the purpose of predicting the ecological effects?

[†] Kaye and Ball (1967).

A In the Exhibit F2 there is a Chapter 15, "Environmental Safety," by R. G. Fuller, ecologist for Battelle Memorial Institute. In Chapter 15 there is a system model, Figure 15.2, "Generalized Materials Transfer Program," which has some similarity to a figure in the paper by Mr. Kaye, Figure 1, entitled "Preliminary Diagram of Environmental Pathways for Transfer of Radionuclides to Man in a Tropical Environment."

* * *

Q Doctor, in the course of your regular professional activities have you had occasion to take and review the systems model set forth in Exhibit F2?

A Yes, I have.

Q Doctor, can you with any reasonable degree of ecological certainty evaluate that model, first of all with respect to its capability as a fair and adequate description of the Rulison regional transport system, based on other data in the same documents?

A This system model represents a relatively advanced description of the system at Rulison, but it is deficient in several major respects. The information is simply not yet available to provide a fully satisfactory description of the regional transport system around Rulison; the model provided here represents the state of the art as of two or three years ago.

* * *

If you examine the model in Figure 15.2A in Exhibit F2 you will see that it is primarily a model of the biological transport system and the redistribution of materials in that biological system to the environment. It does not provide an adequate model of the uptake of materials in the environment into the biological materials.

At this point Judge Arraj took over questioning to assure that he fully understood the specifics of the criticism directed at the AEC preshot systems model.

[THE COURT] Aren't the dotted lines merely to show that part which results from the fallout and then it gets into the system, and then the solid line gives the pathway or transfer into the system? Is that correct?

[THE WITNESS] The solid line represents the transfer between variables within the system.

[THE COURT] Yes, but on the fallout, it has to start someplace. It falls out as shown by the dotted line, as I understand this. I'm not arguing about it.

[THE WITNESS] I say that the fallout and uptake by the plant represents a series of processes such as infiltration, absorption, and uptake through the leaves, evaporation, both from the surface and vicinity of the plants, and then from the leaves themselves through transformation. These are all processes involved in that uptake, and the complete system and the complete description of the transport system from the point source represented in the plans for the postshot evaluation ought to be incorporated in the model.

* * *

Q [BY MR. YANNACONE] What elements in your study designated Exhibit 33 do not appear in the Defendants' Exhibit 15.2?

A There are no elements showing relationship to the various water variables and the water transfers appearing in Figure 15.2 of Exhibit F2.

The discussion of biological aspects of a regional systems model was centered largely around the systems study offered by the AEC as part of their safety evaluation program. The criticisms were really ones of scale or precision, not error. With the examination of the AEC systems model complete, therefore, Mr. Yannacone began asking about the regional environmental system that controls the biological system, carries waste materials such as radionuclides, and which had not been treated in the AEC safety evaluation model.

Q Doctor, would you outline briefly for us what are the elements of an adequate description of the Rulison atmospheric transport system, and would you refer to the NVO-61 and indicate what if any differences there are?

A I would like to draw a diagram in support of this answer. May I? I would like to discuss first of all the induction of precipitation by orographic effects over a plane that is followed downwind by some local elevation, perhaps 1000 feet [Fig. 3]. We may have horizontal flow of air carrying a volume of water, but as it moves over this topography the air naturally is forced upward. As it is forced upward it is cooled because of the adiabatic lapse rate of temperature, in the order of three degrees Fahrenheit for 1000 feet.

This cooling by upward motion frequently results in the induction of cumulus clouds at some point near the top of the hill, and if the atmospheric system is unstable, with air at that point having a dew point near the ambient air on the plane, the cloud will build in sufficient size so that we get rain. Orographic rainfall of this type is what occurs all summer long in the mountain systems, and is what accounts for the differences in the forest composition that are described in Appendix B, "Ecological Considerations," of Exhibit N.

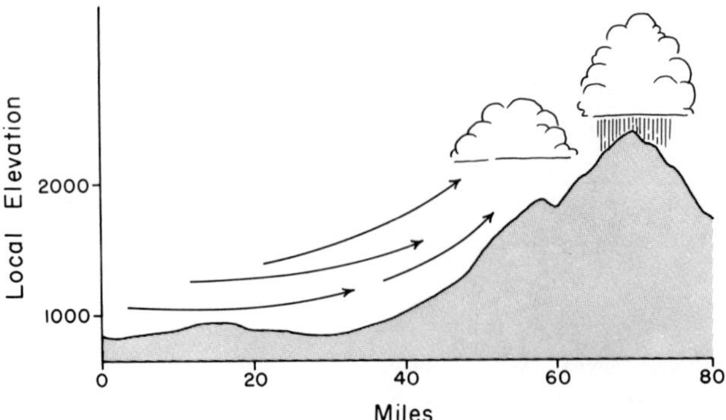

FIG. 3. A chart drawn free-hand during testimony in federal court in Denver as part of the description of tritium dispersal mechanisms in the Colorado regional ecosystem. The chart served to draw attention to water vapor transport and condensation patterns in regions of strong topography, and the importance of identifying critical points within the landscape where tritium fallout might be expected.

The differences, of course, in forest composition that I am talking about, are the presence of alpine fir and Englemann spruce, both species with relatively high demands for water. They occur on the upland in the White River National Forest, the so-called Battlement Mesa, south and east from the ground zero site, whereas at lower elevations in the Battlement Creek area you have species such as piñon pine that are tolerant to droughts and will survive with very little water.

So, if I may draw a specific cross section of the Battlement Creek and adjacent topography, we have the high topography in the White River National Forest at approximately 10,000 feet in elevation, and we have the Battlement Creek Valley with a point at which flaring will be done somewhere in the vicinity of 6500 feet. Thus we have a difference in elevation of 3500 feet, which under normal adiabatic lapse conditions would give a temperature difference of ten degrees, a magnitude which clearly is sufficient to bring about considerable cooling and, therefore, considerable condensation of water vapor as air masses move from the west to east over the Battlement Mesa plateau. The differences in composition which are recorded in Appendix B indicate a major difference in precipitation, and this difference is predictable as a function of the topography, the temperature differences, and the regional flow conditions.

Now, since this precipitation is induced locally, over a distance of approximately two and a half miles, we can expect that tritiated water released into the atmosphere at the flare point will be precipitated in the immediate vicinity when showers occur.

Now, as long as there is stability in the air mass, and there is no shower occurring, the tritiated water of course will be dispersed over a considerable distance, but the primary time for testing for contamination in this area must be when you are getting local precipitation induced as a result of the orographic effect.

I might point out that the report also shows that there will be a considerable release of heat from the flaring, and the heat itself will initiate updrafts that will reinforce the buildup of cumulus clouds and shower activity on this upland.

Now, since the shower activity will not be initiated until close to the top of the mountain, the continuation of that shower into the next valley is really the site at which most of the contamination would be expected to occur. This is in the Plateau Creek Valley, and I would point out that although the postshot plans and the evaluation documents show the location of residences in the Battlement Creek Valley system, it does not take into consideration the distribution of residences in the Plateau Creek area, the area where a system model of the regional atmosphere transport system predicts much or most of the contamination would take place.

The questioning continued to develop ramifications of a safety evaluation that had been based in part on an incompletely described system model. The central issue was whether the AEC studies did in fact represent the "state of the art" in environmental systems research at the time they were done, and if they did not, was the omission likely to have underestimated the potential health hazard to any of the people living in the vicinity of the proposed release. Objections by AEC defense counsel were frequent, but several major points were established. The first dealt with the feasibility of a more complete predictive model, and the second with the independence of system characteristics from quantities of tritium transported.

Q Now, Doctor, can you with a reasonable degree of scientific certainty indicate what, if any, studies will be needed before the actual transfer transport, and distribution of tritium as released during the flaring process of Project Rulison can be accurately predicted in a quantitative sense?

A It is my opinion from analysis of these two documents, and my understanding of ecological systems, that we would require a major program of study relating specifically to tritium and its activity, and its differences from water in movement through the atmospheric, water, and biological transport systems.

The model I envisage would be approximately twice as complex as Figure 15.2 (Exhibit F2). This isn't impossible. There are groups at several locations across the country that are dealing with models that are this complex, but these are people that are primarily concerned with water and nutrient transport, and the Atomic Energy Commission probably has not had access to these particular kinds of studies.

* * *

They allow us to examine the extent of infiltration of the water coming down on Battlement Mesa, its infiltration and subsequent reappearance in the stream water in the valley of Plateau Creek, and the potential contamination of those reservoir systems.

It seems to me that this is the kind of program which if carried out could give us the assurance that the proposed postshot plans and evaluation could be carried out safely, and I am very much struck by how far short of an adequate program the materials in Exhibit N are.

* * *

Q Doctor, does the actual amount of tritium to be released go to the quantitative description of the system or only the qualitative predictability of the system?

A No, the characteristics of the system and the characteristics of the material moving through the system will determine the essential properties of where that material will turn up at other points within the system and this is independent of the total load entering the system.

Q In other words, then, Doctor, the water transport system, once it is described for the Rulison regional transport system, will still be the Rulison regional transport system for water, in spite of the fact that you inject tritium, cesium-137, or I-131 into the water system?

A Yes, and it will still be the same system if you double the quantity of material or change the levels in any way.

Q Do I understand your testimony today that the missing element in the Atomic Energy Commission statements here is an adequate description of the systems transport functions so that quantitative predictions can be made once we know how much radionuclides are coming out of the flaring?

A Yes, I have examined the proposed meteorological monitoring and the proposed systems models that will be used in carrying out the postshot evaluation, and I find them grossly deficient in certain important segments of the transport system in this area, so much so as to be misleading to the public and to the scientific community.

4. *Cross-Examination Strategy*

Cross-examination of this testimony by lawyers for the AEC and the natural gas industry continued for almost two hours. It followed the pattern described previously of trying to demonstrate lack of experience with tritium and a lack of familiarity with the Project Rulison preparations. They also raised the interesting question of whether the public is protected better by monitoring programs than by predictions of responses based on systems studies:

Q [BY MR. SEARLS] Don't you consider the actual monitoring to be more accurate than predictions made prior to the reentry?

A No, I think—

Q Answer yes or no.

[MR. YANNACONE] I am going to object. If the question can be answered yes or no—

A No.

Q In other words, you think predictions made prior to reentry can be more certain than monitoring that might take place after the venting to the air?

A I didn't say they would be more certain. I say that we must have predictions that we can have some confidence in as part of the assurance that monitoring will be satisfactory.

Q Can monitoring be made more certain than predictions made prior to the reentry?

A Certainly, but it is after the fact.

At another point in the cross-examination, counsel for the AEC, Mr. Eardley, tried again to raise doubts as to the utility of a quantitative predictive model for examining movement of tritium in western Colorado environment. In this instance he was suggesting that the total proposed release would have to be known in order to develop a model:

Q [BY MR. EARDLEY] One thing that is difficult for me to understand since this model business is brand-new to me—I have never heard of it until today—is, if you don't have any idea what's going to come into the atmosphere, how can you predict by any means what the ecological effect of that substance is about whose possible nature and quantity you have no idea?

A We have learned through the last twenty years of use of DDT that we simply cannot introduce these kinds of materials into the environment if we don't have the basis for predictions. We can get it by appropriate studies. We can develop this predictive capability without having contaminated the environment.

[THE COURT] He asked how do you do that?

Q When you have an unknown, how do you arrive at a known?

A By investigating the physical and chemical and biological characteristics of the material moving through the system under laboratory conditions so that you get individual rate functions or process equations that can be utilized in the system as a whole.

Q Now, if we wanted to spend four million dollars to have you go out to Rulison, what would be your assumption in making a model as to how much tritium was going to come up? You would have to know that, wouldn't you?

A Not while you develop the model. Once you have a functioning model with the physical and chemical and biological characteristics of tritium operating in the transfer process, then, simply in the computer you can say we will put in 500 curies and see where it comes out. Then, we would put in 10,000 curies and see where it ends up in the environment. It's got to go somewhere and we would know where it's going to be.

[MR. EARDLEY] No further questions.

5. *The Court Decision*

Many issues were at stake in the Rulison court case: a number of issues of law, particularly the standing of the plaintiffs to sue the U.S. Atomic Energy Commission and ask for relief on behalf of the people of Colorado; secondly there were issues of fact, relating to whether the plans for the waste releases made "reasonably adequate provision for the protection" of human and other forms of life; and thirdly, are these plans within the protection standards of the AEC, and are the radiation protection standards themselves "adequate to protect life, health, and property." Testimony had been taken on all of them, and all are somewhat interrelated.

U.S. District Judge A. A. Arraj issued his decision in a Memorandum Opinion and Order on 16 March 1970, supporting the plaintiffs on most points of law, but finding in favor of the AEC in most points of fact. The systems testimony was one exception, but it was counterbalanced by the relatively small release involved. On these questions Judge Arraj wrote:

> The plaintiffs' challenge to the defendants' claim that the planned release of radionuclides will not present a threat to health is on two levels. At the one level, they challenge the assertion that the plans themselves provide adequate protection for health and safety. At the other level, they claim that although the plans may be adequate in terms of the AEC standards and other accepted standards, the standards themselves do not provide adequate protection for health and safety.
>
> The only significant evidence introduced by the plaintiffs in challenging the adequacy of the plans was through the witness, Dr. Orie Loucks. Dr. Loucks is a Professor of Botany and Forestry at the University of Wisconsin who has been working as a systems analyst in environmental problems. His opinion is that the AEC has made an inadequate ecological study, that distribution and resultant concentration of the radionuclides cannot be predicted, and that therefore the potential threat from the release is not accurately predicted in the plans. He thinks that a major study is necessary of tritium, its activity and movement through the atmosphere, water, and the biological transport systems. Such a study would cost $4 million and would take about four years.
>
> Defendants countered by offering the opinion of Dr. Vincent Schultz, formerly of the Division of Biology and Medicine of the AEC and currently a Professor of Zoology at Washington State University. His opinion is that the release of tritium from the Rulison flaring is of such an insignificant amount that no detectable ecological effect will result. This opinion is in agreement with the results of the AEC study found in Exhibit N, Appendix B.
>
> The Court is not in a position to evaluate a scientific controversy of great sophistication, and this controversy as to methodology is certainly more sophisticated than the conventional problems with which we are faced. However, we fortunately need not make such an evaluation to decide the issues presented in this case. The question that we must resolve here is whether or not the evidence establishes that the plans for the release and flaring of the gas are inadequate to provide a reasonably certain and rational basis for predicting that no danger to health and safety will result therefrom. The controversy as to the necessity of a complete ecological analysis of tritium distribution need not here be resolved if in fact an accurate prediction can be made from the information provided by the defendants' studies.

Judge Arraj concluded his decision by stating that:

... plaintiffs have failed to show the probability of irreparable damage if the flaring is not enjoined, and have failed to establish a right to the specific injunctive relief sought.

However, he then qualified the decision by emphasizing:

... This opinion, our findings, conclusions and ruling apply only to the specific factual situation presented by this litigation. We approve only of the flaring of the gas from the one well in the Rulison unit in which a nuclear device was detonated on September 10, 1969. We are not here and now approving continued detonations and flaring operations in the Rulison field. Such determination must be made in context of a specific factual situation, in light of contemporary knowledge of science and medicine of the dangers of radioactivity, at the time such projects are conceived and executed.

Further, although we have found that the plans for the flaring do provide reasonably for the health and safety of the public and that the specific plans for surveillance are reasonable, we determine that the Court should retain jurisdiction in order to insure that the plans we today approve as reasonable are in fact reasonably and safely executed.

The final action of the decision was to order that the complaint be dismissed, and that:

defendant Glenn Seaborg or his responsible agent comply fully with the information and data dissemination plan outlined in Appendix A to this opinion, insuring the distribution of such data to the Rulison Open File as indicated, the Colorado State Public Health Department, and also to this Court, when they first become available.

The Appendix referred to in the decision required more extensive "off site" sampling for radiological safety data than had been proposed initially in the Rulison Post Shot Plans, organized environmentally to follow atmospheric, hydrological, and biological transport.

IV. Brief Statements of Systems Characteristics

A more common type of participation by systems scientists in environmental litigation can be described simply as a professional statement as to the primary response characteristics of the system in question. Such statements are particularly appropriate in public hearings before a hearing examiner or examining board. Although there may be intensive cross-examination, there are rarely many interruptions to the testimony or challenges to the systems scientist's qualifications.

As in the more complex court cases just cited, a statement of primary response characteristics involves a certain amount of preparation, primarily in the form of adapting well-known system description and definition procedures to the system being considered in the litigation. Usually, the preparation will involve the development of flow charts and the general form of the equations for a mathematical model. Solutions are not usually possible, of course, but the model and equation forms provide a much more structured means of looking at the systems as a whole, and of addressing the question of probable impact of a proposed perturbation in it.

A. The Cross-Florida Barge Canal

An interesting example is the preliminary systems model of phosphorus for the Rodman reservoir in Florida. This analysis was undertaken as part of the preparation for the case against the U.S. Army Corps of Engineers to prevent construction of the Cross-Florida Barge Canal along the Oklawaha River in Florida. The materials brought together by Dasmann *et al.* (1970) summarized what was known about the physical and biological processes of the Oklawaha Regional Ecosystem. One of the last sections of this report is a brief paper by Ariel Lugo of the University of Florida. In it he states:

> Systems analysis is a tool for the analysis of ecosystems, and its techniques can be applied to situations created by the Cross-Florida Barge Canal Project. Decisions regarding the future benefit/cost relationships of the Rodman Reservoir, particularly as related to recreation and maintenance costs, are in part affected by the phosphorus budget and by its exotic aquatic weeds, yet systems analysis techniques or other holistic studies of the Rodman Reservoir apparently have never been undertaken. Lack of the predictive capabilities of systems analysis makes current estimates and future decisions ill founded.

As an example of the type of systems study needed for analysis of the Cross-Florida Barge Canal, Lugo proposed a model, Fig. 4, for the phosphorus budget of the Rodman Reservoir. He stated that phosphorus and nitrogen are the two limiting nutrients in plant growth, and selected phosphorus for discussion since data on its abundance in the reservoir and its effects on plant growth were available. Dr. Lugo outlines three steps necessary for analysis of phosphorus.

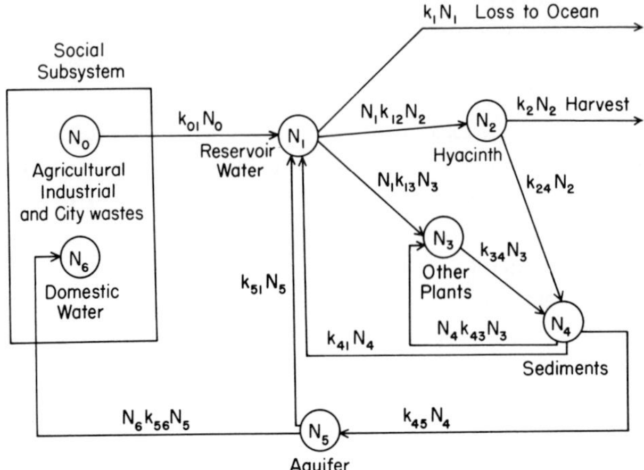

FIG. 4. A flow chart for phosphorus exchange in the Rodman Pool of the proposed Cross-Florida Barge Canal drawn from one developed by Dr. Ariel Lugo as part of the proposed systems testimony on the environmental impact of the Cross-Florida canal.

The first step in applying systems analysis to a problem is the identification of the compartments involved. In [Fig. 4] seven compartments are identified:

1. N_0, representing the phosphorus (P) concentration in the input to the reservoir, originating from agricultural runoff, and the industrial and city wastes.
2. N_1, representing the amount of P in the reservoir's waters.
3. N_2, representing the amount of P stored in water hyacinth, the expected dominant aquatic plant in the reservoir.
4. N_3, representing the stored P in other aquatic plants.
5. N_4, representing the amount of P stored in the sediments.
6. N_5, representing the amount of P stored in the aquifer.
7. N_6, representing the amount of P in the water man uses for drinking and other consumptive domestic requirements.

The second step, after all compartments have been identified, is to trace the movement of the material (P) through the various compartments. In [Fig. 4] the pathways are represented by solid lines connecting the compartments. Arrows indicate the direction of flow. The third step, following the description of pathways, is to write equations describing the behavior of each compartment.

B. Time Lags in the Eutrophication of Over-Developed Artificial Reservoirs

Another example of a qualitative systems analysis was developed for a hearing on a proposal to build a dam and recreation area on Hulbert Creek, a trout stream in central Wisconsin. It did not involve the preparation of a flow chart, or mathematical statement, but the equation forms and their time constants were presented in a statement. The hearing was held 10–23 March 1970 and involved a proposed reservoir of 300 acres suitable for the sale and development of some 1500 cottage lots. Because of the steep topography, shallow soils, and proximity to the town of Wisconsin Dells, the developer proposed municipal sewer and water services.

After a series of introductory questions and answers, I was allowed to make the following statement:

> I'd like to discuss three mechanisms by which we get delayed response in the buildup of pollutant characteristics in lakes. And I'm concerned about these because the public interest with respect to Lake Dells requires that we consider the characteristics of this lake during the later years of ownership by the cottagers who will come to this lake, and acquire lots. We can expect them to hold the lot probably at least twenty or twenty-five years. I'd like to discuss the parameters that allow us to make some estimation of the characteristics of this lake at that time rather than simply over the next two or three years.
>
> The first process is the accumulation of nutrient in the lake because of incomplete substitution of new water in the lake body each year. We have talked in the hearing about turnover time. And, of course, turnover time by definition is simply the time required for the complete substitution of water in the lake by new inflowing water. But we have to recognize that the water that flows out of the lake is partly new water that flowed in during that year, and therefore a portion of the lake body is annually made up of water that came in the previous year and which now has a higher concentration of nutrients. What this says is that for an equilibrium level of nutrients to be achieved in the lake, we have to have a higher concentration of nutrients in the outlet water than in the input waters.
>
> The length of time required for this equilibrium to develop is some exponential of the turnover time. I can give you an equation that describes this process as a function of the number of years that you must accumulate inputs. But perhaps we can simply say that this is an exponential that probably takes in the order of three to six years to reach equilibrium for this size of lake and with the turnover time that has been entered in the testimony.
>
> The second process concerning buildup in nutrients is essentially the process of urbanization around this lake body. The testimony of the developer is that we can expect perhaps fifty percent development,

as I understand it, in eight or ten years, and, if we accept the sigmoid curve in initiation of construction and addition of new construction, 100 percent development is twenty to twenty-five years away. This is supported by the rate of development at Redstone Lake, which is four or five years under way now and has approximately fifteen percent development. We then have a time lag in development which is going to result in a corresponding time lag in the nutrient loading associated with the high runoff and high content of particulate material that will take place when the lake is fully developed.

* * *

Now I want to point out these two mathematical relations operate in tandem and that since the second one is the sigmoid curve of nutrient input loading, and the first is an exponential of nutrient concentrations in the lake, we don't have an equilibrium of nutrient concentrations in this lake until some three to six years after we have reached the peak of the curve for nutrient buildup associated with development. This brings us to somewhere in the order of thirty years before we reach peak levels of nutrient loads.

The third consideration that operates to bring this nutrient system into equilibrium is the development of aquatic weed beds. We must recognize that aquatic weeds are the most effective means of tying up nutrients in a lake body, and that we will not have the aquatic plant growth developed strongly enough in this lake to be an effective means of reducing the nutrient load until there has been a certain amount of erosion along the banks and a slight terracing below water level on which the aquatic weeds can be rooted. We will not get the full development of the aquatic weeds in these shallow areas for ten or fifteen years. This will follow a curve with only a little development during the first few years, and then a period of relatively rapid expansion of aquatic weeds, and finally a tapering off. Again the expansion of the aquatic weed growth to its peak level will have a time lag after the peak in nutrient loads. Thus, we can expect substantial development of aquatic weeds within ten years, but the peak levels would not be expected until somewhere in the range of thirty to fifty years.

All three of these time lags have to be taken into consideration in judging when the water quality of this lake system will be at its best and when it will be at its worst. This whole time spectrum needs to be taken into consideration in approving a permit to build this lake.

At another point in the hearing I briefly addressed the question of what type of system response is acceptable in recreational developments.

I firmly believe that it is not appropriate to simply oppose all kinds of development. I think we do have to accept a kind of development. The only development that in my opinion would be stable in this area would be one in which the cottage units were constructed in the wooded areas on the slopes at a considerable distance from the trout stream, and in which the trout stream itself could be established as communal property to be used for fishing, hiking, and other scenic values by residents in the valley system.

The hearing had involved considerable testimony on other issues in the creation of impoundments for recreational purposes, particularly that of the loss of natural biological and scenic values. On 1 December 1970, the hearing examiner issued a Finding of Fact, Conclusions of Law, and Order. He listed and summarized the key facts relating to the dam and other physical developments, the economic growth, and aesthetic considerations, and concluded with the following three:

20. The Department finds that it is in the public interest to protect, foster and preserve, if possible, the archaeological, geological, botanical, aquatic and other ecological values now existing in a 500-acre area of the Hulbert Creek environmental corridor that would be destroyed by the proposed impoundment.
21. The Department finds that the proposed impoundment will adversely affect the water quality of Hulbert Creek, a cold soft water stream, and the impoundment will itself be adversely affected by high nutrient input.
22. The Department finds that the proposed dam is not in the public interest.

C. Systems Testimony in a Long Island Nuclear Reactor Hearing

A third example of a brief systems analysis relating to an anticipated environmental perturbation is the statement which I filed as part of the intervention by the Lloyd Harbor Study Group in the Shoreham nuclear reactor hearings on Long Island. The hearings were on the application by the Long Island Light Company before an AEC Licensing Board for a license to construct a boiling-water nuclear reactor power station at Shoreham, on Long Island. They began in May 1970 and continued intermittently for a full year. Testimony offered by the intervenors (many other groups had filed interventions in the hearings) ranged widely, from reactor design to environmental effects, although all statements on thermal effects were ruled inadmissible for consideration by the AEC Licensing Board.

The role of my statement was to outline a systems approach to the coupling mechanisms that tie together the physical and biological resources of eastern Long Island and Long Island Sound in an integrated statement. After introductory paragraphs on qualifications and experience with the proposed site and design specifications, the statement continued as follows:

> ... I want to cover two major areas in which there are long time lags before the response to the proposed modification in the environment can be observed. The two attributable to the location of this nuclear plant at the Shoreham site are: (1) the responses attributable to the introduction of radioactive wastes into the air and water in the vicinity of the site; and (2) the response in the marine biology of Long Island Sound to the introduction of the thermal pollutant.
>
> * * *
>
> One would like to think that with our modern technology, we would not introduce any more environmental contaminants and perturbations in natural ecosystems at levels that ultimately are very harmful for man. However, the record over the last two years is clear: We know so little about the mechanisms by which materials such as PCB or mercury compounds are *first* stored in the substrate of aquatic systems, and *secondly*, over very long periods of time, converted to high concentrations in the living biota of the system, at levels harmful to man, that the present response by regulatory agencies can be of only one form—to set standards prohibiting public use of the fish or game or agricultural products that become contaminated by these materials.
>
> I submit, however, that such regulatory mechanisms are the result of society depending on *monitoring after the perturbation*, a defensive process that is in the long run not in the public interest. Instead, we must begin to rely on prediction of these responses in advance, using a full and open hearing such as we have now for this reactor.
>
> * * *
>
> We must also be assured that no materials are being released for which the response properties in this air, land and marine system are any less than completely predictable. It is not a question of what is to be viewed arbitrarily as a safe concentration, but a question of

knowing beyond any reasonable doubt that there is no possibility that the low concentration will somehow build up to significant and potentially toxic levels. My analysis, taking into consideration what we know and do not know about the release of radionuclides into a marine system (via the atmosphere and hydrologic cycle) shows that our understanding is far too incomplete to warrant the long-range risks involved.

My statement proceeded to a more detailed review of recent papers on cycling and concentration of radionuclides in native ecosystems. In another part of the statement I commented on the need for a systems approach to thermal effects:

We can understand the long-range effect of thermal releases if we view the impact of a temperature change on the system in a context similar to the way we viewed the effect of biological concentration of contaminants as they move through the structure of an ecosystem. The increase in temperature of perhaps one degree has many potential effects, as others have testified. The ones of greatest consequence from a systems point of view are those which affect breeding, reproduction, and the life cycle at the base of the food pyramid. Anything that affects the population levels of species at the base of the food-web is going to have a very great secondary effect on the species in the next higher trophic level which are dependent on the food base. Computer analysis of this kind of perturbation has shown that there will be an even greater effect on the next higher trophic level. The greatest influence will be in the top carnivore species that for the most part provide the commercial and recreational resource in Long Island Sound.

In other words, we can say nothing about the long-range stability of this resource system when we determine that the game fish itself is not significantly affected by the small temperature change. The question that must be answered by this hearing, if we are to fully evaluate all aspects of the public interest, is whether there will be any significant upset to any part of the food-web that supports the commercial fish species in Long Island Sound, either now or at any time during the lifespan of the Shoreham plant. To do this, I would want clear evidence that phytoplankton and benthic organisms in the lower part of the trophic structure of the Long Island Sound ecosystem will not be affected as to species composition, reproduction, or as to rates of feeding—all of which we know to be very sensitive to temperature.

Finally, my concluding remarks focused on the use of systems analysis of the environmental effects of large power stations to assure either a design or standards about which there will be no question, or assurance that the natural resources will not be degraded in any significant way:

> They (the Company) could, by themselves, or in conjunction with other utilities and state or federal agencies, undertake a complete analysis of the regional land and water transfer systems to determine whether any long-range environmental deterioration will take place. We know how to do computer simulation of perturbations in these systems, provided adequate data are obtained, and we must begin using them to look ahead at the long-range effects of actions we proposed to take in 1970. Such studies cost significant sums of money, but the costs are small in comparison to the potential loss of natural resources.

At the time of writing, no decision has been reached on licensing the Shoreham power plant. If the license for construction is approved, the decision of the Licensing Board could be appealed in court and the testimony reevaluated there. The process is slow, but this in itself is no reason for not having the best scientific understanding available as part of the record from the beginning.

V. Discussion and Outlook for Systems Science

The systems related testimony of the DDT hearings, the Rulison case, and the other brief statements have been presented in some detail to illustrate precisely the level of scientific or technological preparation necessary in environmental cases. It is not as sophisticated as most of the material in these two volumes, because it represents a frankly applied point of view. As such, the level of the analysis is only as good as the least fully understood component of the environmental system. Unfortunately, this weakest link is still often at the level of system definition and description. Cases stemming from actual or threatened environmental upsets are being encountered more and more frequently, however, and effective prosecution is going to require sophisticated scientific preparation, including the use of systems methods.

From a practical point of view, what are the impediments, if any, to a more widespread application of our technical understanding of environmental systems in court litigations? These developments already are causing a considerable stir among scientists, but most seem to wait in the wings expecting the outraged citizens' groups or environmental lawyers to come to them. This is a long-standing general practice in the scientific community, and systems scientists are only being true to historical precedent and the conventions of our society. Times are beginning to change, however, and it is time that they should. The major technological advances of the late 1940s and 1950s produced both the affluence and the environmental degradation and public deprivation of rights that were recognized late in the 1960s. The next few years in the early 1970s will see the inevitable response as larger numbers of citizens seek redress in courts.

Environmental lawyers have pointed out for years that new precedents in the courts will precede by some years the new legislation to protect environmental systems that many scientists are asking for. Civil rights litigation led civil rights legislation by years, and the scientific input for litigation to control technological impact on the environment will similarly precede responsible legislation in this area. Scientists are accustomed to contributing through their disciplines to the drafting of new legislation in many areas, including the protection and regulation of natural ecosystems. To my knowledge, ecological systems science is not yet contributing as effectively as it should to the drafting and passing of environmental legislation.

There are some very plausible reasons for the absence of ecological systems studies as part of new legislation on environmental protection, but they generally parallel the relative lack of systems testimony in environmental litigation. The latter is being overcome, and we should be looking ahead to a much broader role for ecological systems work in state and federal agencies, in social organization, and in the setting of priorities by our society. It has been only a decade since the time lags inherent in the use of technology in the environment caught up with the scientists' ability to consider them through direct observation. The result then was that environmental scientists began adapting systems methods and mathematics to evaluation and to quantitative linking of data on a changing response pattern. The public will be drawn to use a similar capability within the next ten years.

For the present, some of the most interesting applications of ecological systems studies are in court litigations. Consider the dozens of factors that led to the concentration of mercury in fish of aquatic ecosystems. Consider the time lags before the response is recognized as an environmental hazard. A considerable part of the mercury pollution in three hundred miles of the Wisconsin River came either from a plant that terminated a contaminating process twelve years earlier, or from atmospheric transport out of modern plants.

Interesting environmental litigation is beginning to build from these past mistakes. More interesting to me, however, are the lawsuits that will center around the new mistakes that are threatened now, in the early 1970s—mistakes, whose disastrous effects we in systems science can predict over periods as long as decades, despite the interaction of ten to thirty factors. Some of these effects are in the form of an environmental upset with delayed responses of up to twenty-five years. Thus the aggrieved who will eventually seek redress in the courts are the children and grandchildren of the systems ecologists using this text. I submit that in the long run there are no more important applications of ecological systems science.

Appendix. Ecosystem Analysis: The Diffusion of DDT

The following equations were developed jointly with H. Harrison, J. Mitchell, D. Watts, D. Parkhurst, R. Tracy, and V. J. Yannacone, Jr., and were introduced into the record as a five page manuscript. This material was later revised and published (Harrison *et al.*, 1970).

Consider that for any level (i) represented in the food pyramid, the inputs and outputs of DDT mass (m) can be shown schematically as

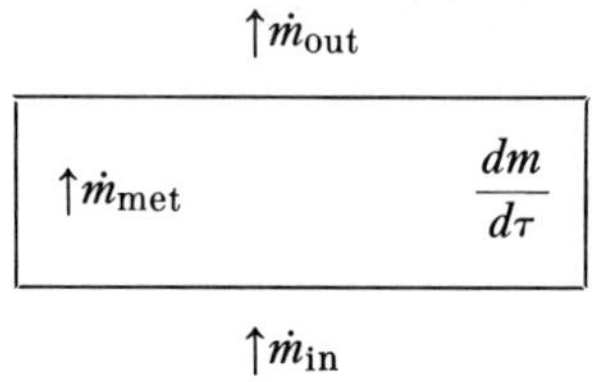

which can be summarized in the equations

$$\dot{m}_{\text{in}} = \dot{m}_{\text{out}} + \dot{m}_{\text{metabolized}} + dm/d\tau,$$

$$\dot{m}_{\text{in}} \propto c_{i-1}\dot{m}_{i-1}\,,$$

$$\dot{m}_{\text{in}} = \sum_{j=1}^{i-1} c_j\dot{m}_j\,,$$

and

$$\dot{m}_{\text{out}} = c_i\dot{m}_{\text{D}} + \dot{m}_{\text{ex}}$$

where

c_{i-1} = concentration of level $i - 1$,

$\dot{m}_{i-1}$ = rate of ingestion to account for input from the previous level,

j = trophic levels up to $i - 1$,

$$\dot{m}_{\mathrm{D}} = \text{death rate (mortality plus predation)},$$

$$\dot{m}_{\mathrm{met}} = \text{rate of metabolic destruction},$$

$$\dot{m}_{\mathrm{ex}} = \text{rate of excretory loss}.$$

Now,

$$\frac{dm}{d\tau} = m_i \frac{dc_i}{d\tau},$$

where

$$m_i = \text{mass of pesticide in the } i\text{th trophic level},$$

and

$$\frac{dc_i}{d\tau}\left(\frac{\dot{m}_{\mathrm{D}}}{m_i}\right) + c_i = \frac{\sum_{j=1}^{i-1} c_j \dot{m}_j}{m_i} - \frac{(\dot{m}_{\mathrm{met}} + \dot{m}_{\mathrm{ex}})}{m_i}.$$

Assume a time constant can be defined

$$\simeq \frac{m_i}{\dot{m}_{\mathrm{D}}} = \frac{\text{mass of } i\text{th level living at any time}}{\text{mass of the mortality rate losses}},$$

for

$$m_i = NM,$$

where

$$N = \text{number of individuals},$$

$$M = \text{mass of individuals, and}$$

$$\dot{m}_{\mathrm{D}} = MN_{\mathrm{D}},$$

where

$$N_{\mathrm{D}} = \text{number of individuals dying per unit time}.$$

Now, if the animals live a time τ before they die, and there are N animals in a population,

$$N_{\mathrm{D}} = (1/T)N,$$

and

$$T = \text{average life time}.$$

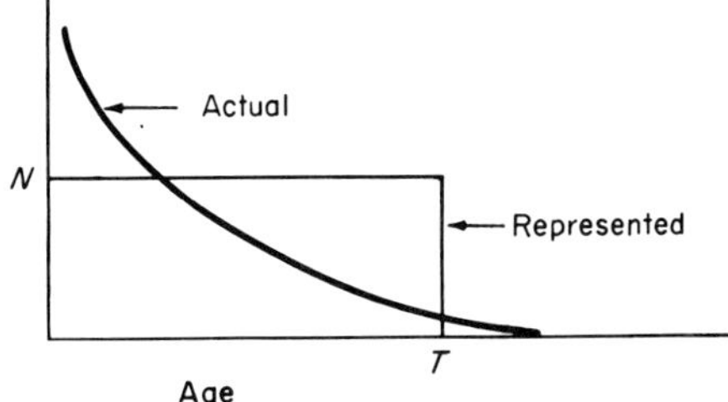

Thus:

$$\frac{m_i}{\dot{m}_D} = \frac{NM}{MN_D} = \frac{N}{N_D} = \left(\frac{N}{N/T}\right) = T,$$

and

$$\frac{dc_i}{d\tau} + \left(\frac{1}{T}\right) c_i = \frac{\sum_{j=1}^{i-1} c_j m_j}{m_i} - \frac{(\dot{m}_{\text{met}} + \dot{m}_{\text{ex}})}{m_i}.$$

This holds for $\tau \leqslant T$; the relation can only hold until the animal dies.

The system also has equilibrium levels. These can be determined from consideration at steady state:

$$dc_i/d\tau = 0 \qquad \text{for no input.}$$

Then,

$$c_i = \frac{T}{m_i}\left[\sum_{j=1}^{i-1} c_j \dot{m}_j - (\dot{m}_{\text{met}} + \dot{m}_{\text{ex}})\right].$$

Thus, equilibrium concentration is:

(a) Directly proportional to lifespan,
(b) Directly proportional to net retention,
(c) Inversely proportional to total mass in the trophic level.

REFERENCES

Dasmann, R. R., ed. (1970). Environmental Impact of the Cross-Florida Barge Canal, with Special Emphasis on the Oklawaha Regional Ecosystem. Florida Defenders of the Environment, Gainesville, Florida.

Harrison, H. L., Loucks, O. L., Mitchell, J. W., Parkhurst, D. F., Tracy, C. R., Watts, D. G., and Yannacone, V. J., Jr. (1970). *Science* **170**, 503.

Kaye, S. V., and Ball, S. J. (1967). *In* "Symposium on Radioecology, Proceedings of the Second National Symposium." Ann Arbor, Michigan, May 15–17, 731.

Loucks, O. L. (1971). *In* "The Patient Earth" (J. Harte and R. H. Socolow, eds.), Chapter 7, pp. 88–107. Holt, Rinehart and Winston, New York.

State of Wisconsin, Dept. of Natural Resources (1970a). Transcript of Hearings on Petition of Citizens Natural Resources Association, Inc. and Wisconsin Division, Izaak Walton League of America, Inc. for a Declaratory Ruling on Use of Dichloro-Diphenyl-Trichloro-Ethane, Commonly Known as DDT, in the State of Wisconsin. Docket No. 3-DR-1.

State of Wisconsin, Department of Natural Resources (1970b). Transcript of Hearings on an Application of N. E. Isaacson & Associates, Inc. for a Permit to Construct, Operate and Maintain a Dam in Hulbert Creek in the Town of Delton, Sauk County, for Recreation and Subdivision Purposes. Docket No. 3-WR-699.

U.S. District Court, District of Colorado (1970). Court Record of Combined Civil Actions C-1702, C-1712, and C-1722, Crowther versus Seaborg, Colorado Open Space Coordinating Council versus Seaborg, and Dumont versus Seaborg. U.S.D.C., Denver, Colorado.

Watts, D. G., and Loucks, O. L. (1969). Models for Describing Exchanges within Ecosystems. Inst. for Environmental Studies, Univ. of Wisconsin, Madison, Wisconsin.

Woodwell, G. M., Wurster, C. F., and Isaacson, F. A. (1967). *Science* **156**, 821.

Wurster, C. F. (1969). *Biol. Conservat.* **1**, 123.

Yannacone, V. J., Jr. (1970). *Cornell Law Rev.* **55**, 761.

11

Systems Ecology and the Future of Human Society

THEODORE C. FOIN, Jr.

ENVIRONMENTAL SYSTEMS GROUP, INSTITUTE OF ECOLOGY, UNIVERSITY OF CALIFORNIA, DAVIS, CALIFORNIA*

* *Present address:* Division of Environmental Studies, University of California, Davis, California.

I. The State of the Human Ecosystem

A. Human Society as an Ecological Problem

Can the global ecosystem survive the presence of man? This is a question that is just beginning to receive some long overdue attention. No matter what else we do, we are covering the Earth with our populations; this is our greatest problem. In the United States, concern for the environment has increased dramatically within the past two years, but mostly for the human environment, not for the global ecosystem. We worry about our health, our resources, and our environment, and in so doing we reveal our basic concern for our own species—not for others or for the way the whole system is integrated. The critical question is whether or not our normal modes of economic development are viable over the long term. This is a simple extension of the same arguments associated with conservationists for decades, but increased public awareness of ecological issues has forced a new and more serious examination of the issues.

The situation is much the same in most the rest of the developed countries, such as the Netherlands and Israel (Bentham, 1968), Australia (McMichael, 1968), and Canada (Cragg, 1968). Some have better conservation programs. Still, the history of conservation in the U.S. is better than in most of the underdeveloped countries, such as in South America (Eichler, 1968) and the Caribbean (Carlozzi and Carlozzi, 1968), and even in some relatively developed countries like New Zealand (Salmon, 1969) and the Soviet Union (Gerasimov, 1969). Each nation has the sovereignty to manage its own internal affairs, including the rights of uncontrolled population growth and unchecked economic development. Because national problems generally become world problems, any nation that claims these rights is like a company that claims the profits from manufacturing goods but passes the costs of waste disposal to the surrounding environment. The deleterious effects of excessive population and destructive development are passed on to the rest of the world as social, ecological, economic, and military problems.

For most of human history little attention has been given to the ecological consequences of large populations and unchecked economic development. Populations grew and regions developed with the expectation that the system would continue to function satisfactorily no matter what we did. There is no question, however, that men are subject to the same influences as other animal populations, and that the complexity of his brain in no way invalidates his vulnerability to biological and thermodynamic principles. Thus, the use of fossil fuels to drive the distribution system, to moderate weather and climate, and to support his populations is a credit to the ingenuity of man, but none of these things invalidate the natural ecological laws. Rather, man has built and continues to build an artificial system that could not exist were it not for the constant input of energy used to maintain it. If these sources of energy were to be removed, the artificial system would collapse because it could not function on solar energy alone. In this example, the answer is not to abandon civilization because our energy resources shall sooner or later be exhausted. It is simply a cautionary note that our planning should be based on the recognition of finite supply.

Ecologists are in a difficult but useful position. We know that man's survival is linked to that of numerous other species because we are all part of the global ecosystem. If we are to contribute to planning for survival and quality existence under the pressure of economic demands and population growth we must find ways for handling the complexity of the system. Two of the tools that have the potential for handling this complexity are systems analysis and simulation. In California the Environmental Systems Group has begun to investigate the use of simulation modeling as a tool for evaluating alternative courses of development and their implications. This chapter reports what we have done, our evaluation of its value, and the outlook for simulation and systems analysis for human socioecological problems.

B. Social Complexity and Problems of Human Society

Many of our social problems result from the complex webs of feedback loops that characterize all processes. A resource manager or a politician rarely has an adequate information base for making a decision, much less an appreciation for all the interaction effects that can appear within entirely unexpected areas. For example, one of the first requirements for life is adequate nutrition. With increasing numbers of people to feed, greater productivity of food is required. The use of pesticides was one step necessary for higher productivity. Unfortunately, the ecologically unsophisticated attitude behind the use of pesticides is

$$\text{pesticide} \rightarrow \text{pest}$$

with the destruction of other vulnerable organisms termed "side effects" (Moore, 1967). Moore also showed that so-called side effects are ecologically as important as destruction of the pest because the history of pesticide use reveals a positive feedback loop between pest numbers and pesticide. Each increased dosage or more deadly pesticide has yielded progressively more pests. For example, there is overwhelming evidence of the effects of DDT on predatory birds such as the Western grebe (Herman *et al.*, 1969) and peregrine falcon (Hickey, 1969), on the productivity of marine algae (Wurster, 1968), and even on human health (Radomski *et al.*, 1968). DDT is being abandoned as a pesticide largely because pest resistence has made it ineffective. The evidence clearly shows that the simple action of pesticide application engenders complex reactions in unexpected places.

A second example is the question of land use. Some planners and futurists foresee great megalopolises in the United States, covering entire states in some cases (Kahn and Wiener, 1967) and vast regions in others (Doxiadis, 1968). Such speculation is subject to the assumption that we could adequately feed the denizens of these cities. This is not a certainty, especially when different regions are all simultaneously urbanizing on the assumption that food will be produced elsewhere. There are a number of issues which affect the minimum quantity of land necessary to produce adequate food besides population size and pesticide use. These include human attitudes toward the use of land; the ownership and desired size of residential and recreational land; acreage taxation policies; regional planning, zonation, and the policy of granting variances; the impact of future technology; the availability of fuel and power inputs; and the variability of weather and climate systems affecting productivity. It should be evident that a single, simple, answer will not be sufficient, and that the only satisfactory solution is to assess and account for the observed complexity. Babcock (1966) provides a discussion of the complexity in zoning policies and the variances granted from them that illustrates this point.

C. Population Growth and Planning for the Future

In the United States, planning has always been the prerogative of the local community and of the family. Because of implications of socialism, state and regional planning have always been strongly resisted. Consequently, the little planning for economic development that has resulted means that what planning we do is mainly accommodative, a reaction to trends that have already been established. A simple illustration is the water problem in southern California. In our democracy, no one has the authority to limit the size of a city, no matter how unsuited the region may be for urbanization. Los Angeles has jobs and an attractive climate, so it has continued to grow rapidly. Having tapped all other cheaply available sources, southern California turned to the Sacramento Valley for water, and the California Water Plan was authorized in 1951 to supply it. When the plan is finally completed, there may be disasterous consequences for San Francisco Bay and the Sacramento–San Joaquin River Delta because of greatly reduced water volumes, more pollution, and salinity intrusion. Hence, northern California would have to pay the costs for water sent to southern California, in the form of a deteriorating Bay–Delta system—the price of accommodating to the growth of Los Angeles and irrigated agriculture in the southern part of the state.

Accomodation is a common feature of all three examples presented above. In land use, in water distribution, and in pesticide application we find ourselves doing things we felt it was imperative to do because the need was already established. Little or no look-ahead planning was attempted to prevent the problem instead of solving it. Apparently little attention has been paid to desalinization of water in Los Angeles after it was discovered that it would be cheaper (for the consumer) to bring water down from the north. But perhaps it would have been better if the California Water Plan had not been extended to the Los Angeles metropolitan region, for then the higher cost of water may have helped slow the city's growth. The essential difference is that accommodative planning tends to initiate and perpetuate positive feedback loops, where negative feedback is necessary.

The process of accommodation continues because once it is started it must continue. The root cause is our failure to appreciate the deleterious effects of population growth compared to the benefits. In the first place, there are strong biological, psychological, and instinctive drives for reproduction, reinforced over much of the world by social and theological conventions. Second, a growing population supplies national manpower for labor and the military. It is the growing population, however, that also provides the impetus for accommodative planning and the need for progressively more technological advancement. It is also the factor that reduces personal freedom (Cook, 1969), and that increases the necessity of broad planning above the local level. It now appears that global population stabilization is unlikely in the near future. A recent Gallup Poll (1969) for the National Wildlife Federation showed that of the 89% of U.S. citizens interviewed who thought our environment was deteriorating, only 48% thought population control was necessary to stop it. It is quite clear from our history that it is our nature to emphasize immediate problems and ignore future ones because we often require first-hand evidence of system malfunction before we act. We are beginning to see many of the problems developing because we have already overpopulated the world—wars for land of another nation, tideland oil spills, DDT residues in Antarctic penguins—and as long as we continue to increase without rigorous planning the future does not promise to be any better. Local planning is permissible as long as society is simple and highly personal. It becomes increasingly less tolerable as complexity grows because regionally interdependent social and economic processes begin to break down. In the United States the need for integrated regional planning is only now being recognized and the process started (Michael, 1968). Our first attempts at planning in an ecological context (McHarg, 1968) reflect the basic accommodation to population growth we make. In order to find better ways of developing, our society not only requires broad planning, but, even more, needs new methods of assessing planning decisions and potential courses of development. This will undoubtedly be difficult because each delay increases the social complexity, which in turn constrains the available options in planning policy.

II. Systems Analysis and Simulation of the Human Ecosystem

A. The Potential of Systems Ecology

The potential of systems ecology as a method for attacking problems of human society reflects the similarity of its methods and the nature of the global ecosystem. Feedback loops occurring between different processes are commonplace, and it is unreasonable to assume that an action will not produce reactions in other areas. These reactions in turn produce other reactions, so that the total course of events resembles a vast system of perturbations and damping mechanisms, of which those observed in natural ecological systems and outlined in the pesticide example above are only a small part. Systems ecology is an appropriate means to study the feedback loops and behavior of the whole system, and particularly the interrelationships between classical disciplines like community medicine, economics, law, ecology, and engineering. Systems ecology is particularly appropriate for human ecosystems because it is a means for synthesizing data from discrete disciplines into a single conceptual framework; it can supply this framework, and it can become part of a feedback loop between systems analysis and data collection.

These properties are not those of ecology but of systems methods; ecology is only one of several integrative disciplines. Hence, it is quite likely that systems engineers and others will be studying the human ecosystem using similar methods. The unique advantage of the systems ecologist is that sooner or later man must become part of a viable ecological system, and the systems ecologist is the one who can best guide development along these lines. The highest potential of systems methods in human problems will only be realized, however, if efforts are truly interdisciplinary, involving scientists from all relevant disciplines.

B. The Application of Systems Ecology

1. *Modeling*

The methods of systems ecology center upon the construction of a simulation model. Our approach is to construct a very general conceptual model, then to break it up into progressively smaller and more detailed modules. When the flowcharts are sufficiently detailed, data can be sought to quantify the static and dynamic variables, functions selected, and algorithms written for the functions. Particular care is exercised to include all the feedback loops possible within any developed submodel, and constant checks are made to insure all possible integration between modules.

The central tool of model building is a sufficient computer to handle the simulation model. Any model that accurately simulates the natural complexity of human activity tends to become very large rapidly; even with fairly large machines (such as the Davis IBM 7044:32K) submodels can easily use all available internal storage and require programming tricks and high efficiency to run.* The computer is also an appropriate tool for the systems analysis that goes into selection of mathematical functions.

2. *Interaction Effects*

One of the major benefits one wishes to obtain from building simulation models is to expose effects resulting from the interaction of two or more processes. Many of these may be unanticipated by any other method, but the principal difference is in emphasis. Our concept of systems ecology includes a very broad interdisciplinary viewpoint which may sometimes emphasize scope over detail. In contrast, academic and governmental research tends to concentrate within a range of subjects characteristic of a discipline or to be very mission-oriented, and detail is usually emphasized over scope. In either case, the result tends to be a narrow view of problems and solutions, with a consequent lack of interest or expertise in the important interface areas.

In the case of the United States Federal Government, the narrow view has led to a bewildering array of agencies which split a problem into discrete, but unnatural, pieces. Water pollution is the responsibility of the Federal Water Quality Administration, United States Department of Interior, while air pollution is under the jurisdiction of Health, Education, and Welfare, National Air Pollution Control Administration.

* We have since converted to the Burroughs B6700.

Thermal pollution is the responsibility of HEW, environmental radioactivity that of the Atomic Energy Commission.* One of the largest scale polluters of air and water in the country is the Department of Defense, and this situation persists because environmental quality is not one of their primary missions. In each of these examples potential interactions are easily discovered, such as air pollutants that are soluble in water and the location of nuclear power plants. In recognition of these gaps, the President has recently established the Environmental Quality Council and directed all Federal angencies to do systems-type planning. It is ironic that Congress has simultaneously directed all agencies granting research funds to insure that any research supported be directly relevant to the missions of that agency (Anon., 1970).

3. *Prediction, Gaming, and Planning*

There are three principal motives for building a simulation model. One of these is for prediction of the future of human society, such as an econometric model would do for short spans into the future. An example of one of these is the Bay Area Simulation Study (BASS) developed in the Center for Real Estate and Urban Economics at Berkeley (California State Water Quality Control Board, 1968). This model predicts population parameters, volume of employment by industry, and land-use patterns for the San Francisco Bay Area. In many respects the model is empirical, and there are some serious errors which weaken the projections, but still it is a flexible model that represents a significant advance over nonmodel projection techniques. Other models in which prediction was a central objective include the model of urban dynamics developed by Forrester (1969) and the growth of the Detroit urban region (Doxiadis, 1966). The latter is similar to the BASS model; both differ from the Forrester model in that they depend on mapping variables to grid locations, while the Forrester model is process-oriented. All three models are similar in that they assume minimal feedback effects arising from limited resources at the growing edges of the urban region. Because of the explosive population growth now being experienced, and the decline in resources reported in Cloud's book (1969), this assumption may be unwarranted.

The second motive for building a simulation model is to use it for gaming and as a teaching tool. The principal use of such a simulation model is to test assumptions, to validate a model, and to explore alternative strategies. There are a number of biological models available for

* All these functions were invested in the Environmental Protection Agency (EPA) in 1970.

these purposes, such as the Harvard Forest Simulation of Gould and O'Regan (1965). This model simulates the growth of a forest under management, and it specifically includes options for varying harvesting rates. A basic strategy is to cut an even annual harvest at sustained yield, but this program, though scientifically sound, does not respond to the economic system very well. This suggests a number of different cutting strategies which may or may not approximate sustained yield, but which can be evaluated using the simulation model as the basis of the game. The forest simulation is even more complex then this, with options for fluctuations in weather, the incidence of catastrophies, long-term strategies using past profits and losses, loans, expenses, and tax strategies. This model is also highly suited as a teaching tool because it teaches an appreciation for whole-system; i.e., forest, behavior under various influences.

The third motive for building a simulation model is for use as a planning tool. The Gould and O'Regan model may also be used to illustrate this purpose. Suppose that one wishes to consider the effect of four different cutting strategies, all taking an annual sustainable yield but differing in the distribution of cutting rates of the year. One could use the simulation model to evaluate the economic and biological effects of the four different strategies under a series of other assumptions, e.g., high, moderate, and low probability of fire, including the use of edaphic climaxes; weather, including human effects due to air pollution; and random fluctuations in growth parameters of the trees. Our own experience shows great interest at the state and federal levels of government in using models of human society for evaluating legislative bills, and even for suggesting appropriate ones.

It should be obvious that the three uses of a simulation model are not clearly separable, but are largely indivisible. The forest model can be predictive if realistic parameter values are used for each term in the model. By altering parameters and examining the effects, the model becomes a game and a teaching device; by implementing the results of gaming into guidelines, rules, laws, and proposals, the model is useful in planning. Of course, the user can theoretically opt for any or all of these uses, for none are mutually exclusive; the constraints are derived from the quality of the simulation model itself. This topic is discussed in further detail in Section II.C.

4. *Political Input and the Decision-Making Process*

The development of a systems model represents a potential boon to the politician and planner (in general) of human activities because it is a new means of obtaining further insight into a process and a means of making better decisions.

The political machinery can also aid the systems modeler. One of the ways is to furnish or to influence governmental research support; the other is to control those governmental agencies collecting data. A realistic appraisal of systems ecology would mean that the majority of one's time should be spent on systems analysis rather than simulation. One would expect to find areas of weak or missing data, most of which would not be amenable to statistical manipulation. The only ways to correct these faults would be to modify collection procedures or to present strong arguments as to why particular data are urgently needed. Since the government collects more data pertinent to human ecology than anyone else, feedback loops between politicians and systems scientists would be potentially highly productive. In our experience, however, more aid is given to the politicians than to the scientists. We have made numerous statements that were intended to influence the political system, but we have been particularly grateful for the help we have received from the California State Department of Public Health, which has made statistics available to us that were otherwise unobtainable. We anticipate that we and other researchers interested in human ecology will benefit from more such loops.

C. Limitations of Systems Ecology

1. *Data*

No simulation model can be better than the data used to construct it. A model built with imprecise or inaccurate data can still be a valuable educational device, but it is of little value for planning or prediction of future events. We have recognized a number of types of data faults which emphasize the improbability of having complete sets of adequate data. The two major categories are data bases which are incomplete or inadequate, and good data that are unavailable when needed. An example of the former is the available migration data for the State of California (California State Department of Finance, 1964). A demography model is fundamental to the construction of a model system, and one of the basic components of a demographic analysis is change in population size or characteristics due to migration. We have found evidence (Section III.D.7) that local migration patterns can be highly significant; yet the only available data are for multicounty State Economic areas. These data are suggestive but not nearly sufficient. Similar remarks apply to an assessment of the economic impact and future of forests in California. In some of the northern counties, only one company has logging operations. To protect their sources, the State Department of

Employment combines these figures with those of other counties. This is a serious matter when other submodels are based on the county as the fundamental unit.

Public health statistics have been an example of time-lag defects. Mortality records are reasonably complete and accurate, but if they are published only after a number of years, their accuracy may be wasted. Consider the recent trends of lung cancer and emphysema mortality in California (Fig. 1). Both have been rising rather spectacularly, and may

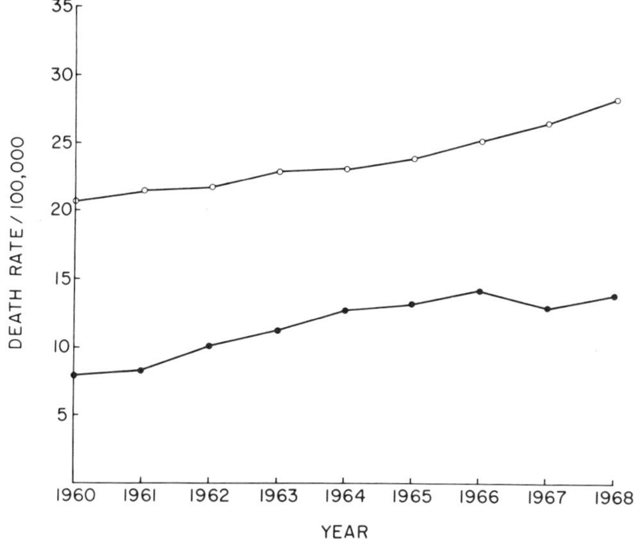

FIG. 1. Mortality trends for respiratory cancer (○) (lung, bronchus, trachea) and emphysema (●) in California, 1960–1968.

continue to do so. If this is the case, the public health officer in charge might have a threshold mortality rate at which he would begin to pay more attention to the problem or would demand drastic action. If he cannot obtain the data he needs, he would have to make an educated guess and take action on this basis, or he would have to delay his decision. In either case the potential consequences of an erroneous decision are serious. Assume, for example, that he had to decide what to do about emphysema in 1967, when the mortality rate fell slightly. Should he assume that the long-term trend was downward, and thereby risk no action when it was necessary, or should he react strongly on the assumption that the rate was only temporarily depressed? If he could expect the datum for 1968 in 1969, it might make a considerable difference than if he had to wait until 1971.

2. *The Timing of Innovation*

Assuming that the data base is adequate, the systems scientist still has the general problem of stochastic elements with which to contend. There are a number of textbooks available dealing with stochastic elements in biology, and these methods may be used with random-number generators and Monte Carlo methods in the computer to simulate stochastic processes adequately. The impact of technological innovation is a particularly difficult case to deal with; even if a probability value can be assigned to the rate of advance of technology in solving a particular type of problem, it is a different matter to predict what the advance will be and what its impact will be on the system.

Let us examine the issue of petroleum sufficiency as an example of how important technological innovation can be. M. King Hubbert (1962, 1967, 1969) has argued that under certain assumptions the United States will use 80% of its oil reserves of all forms (from natural gas to the tar shales) in less than 70 more years. In his 1969 paper, Hubbert assessed energy production from other sources in the future. Of these, the most important is the availability of power from the nuclear reactor, which will divert demand away from fossil fuels. Cloud (1968) has assessed the nuclear reactor program, and his conclusions are quite pessimistic: Unless we slow the use rate of ^{235}U and achieve functional breeder reactors or controlled fusion soon, nuclear power sources will not be sufficient to replace the fossil fuels as an energy source. Ryan (1966) has presented an entirely different argument. He believes that raising prices for petroleum would stimulate the search for oil, and that because we have looked in so few areas, there probably is an adequate supply remaining to be discovered.

The controversial issue is that we do not know how much petroleum or uranium can ultimately be found, and how much of them can be recovered. Demand can also influence supply by driving up prices and stimulating exploration (Ryan, 1966; Landsberg *et al.*, 1963), and each technological advance has resulted in upward adjustments of ultimate supply (Hubbert, 1962). Likewise, we do not know how much of the ultimate supply will be recoverable, since this is also a function of technological innovation (Landsberg *et al.*, 1963). Thus, who will ultimately be proved correct depends on an unknown factor, technology, and simulation models cannot be expected to add further insight by themselves.

How long the United States domestic petroleum reserves last is very much a function of how rapidly petroleum can be replaced as a stored energy source. If we accept Cloud's argument (1968), we can expect a shorter time to exhaustion; if controlled fusion becomes practical soon, we can expect a much longer lifespan for our fossil-fuel resources. We can only conclude that in the face of such uncertainty, a simulation model will always be vulnerable to sudden and dramatic changes in technological input. This certainly weakens the use of simulation for prediction and detailed planning unless innovation is unlikely. Nevertheless, for gaming and exclusion of alternatives in planning, the model that is an acceptable simulator will still be useful.

3. *The Information-Value System*

The human ecosystem is a complex system of positive and negative feedback loops, of interactions between processes, and of a set of institutions which are supposed to mediate both of these, collectively called the "decision-making process." The decision process is the most complex of the three issues, involving the collection and use of data and the discovery and implementation of technological advances. The single largest problem is not that the probability of a certain kind of decision cannot be assessed, but that the quantification of human attitudes and human values has not progressed sufficiently for use in simulation. The only way we have to circumvent this problem is to use branching processes in programs as operator options, to specify an option subject to the value system. Unfortunately, any simulation model of realistic size would rapidly become intractable with a great number of operator options. Our solution to this problem has been the only real option left: to assume a set of prevailing attitudes, and build the model with explicit options having attitude and value changes implicit within them.

The controversy over the proposed but unfunded Boeing Supersonic Transport (SST) is a good example of the influence of human values. The SST was a technological innovation which would have made very high air speeds available to the general public. It is a design that had some operating advantages over more conventional aircraft (Howard, 1968), could have helped correct the balance-of-payments deficit, and might have bolstered American international prestige. On the other hand, there were potential noise and energy problems (Kryter, 1969; Shurcliff, 1970) resulting from SST operation, and perhaps financial disaster if the plane did not win broad acceptance. These effects lead in turn to other effects, such as the economics of and unemployment in the aircraft industry and popularity for local politicians. Which arguments are accepted depends on the value systems of the people. Is national prestige worth more than lowered rates of fuel consumption, or is high air speed worth possibly greater congestion at both terminals?

In principle, we could build a model of the socioeconomic effects of the SST. What we cannot do is to say whether or not the SST will every fly, or when, or how many of them. In the decision not to fund the project, the SST did not appear to be a rational investment for reasons of noise and efficiency, but to assume that it will never fly is to ignore the possibility that some Congress in the future might weight these factors less heavily than national prestige or the economic impact on the aircraft industry. The SST example is simply a microcosm of the way society functions when data are imperfect or when a decision is not perfectly rational. There may be several possible options, each more or less rational to different decision-makers. While simulation may be valuable for ordering the available information and for providing insight into the nature of the problem, it cannot generally predict the resolution of any problem subject to the value system and the decision process.

III. The CALSIM Model

In June of 1968, the Ford Foundation provided funds to K.E.F. Watt and the Environmental Systems Group to begin construction of a mathematical model of California society, as a microcosm of a simulation model that might ultimately be built for the United States. The project was given the acronym CALSIM, for California simulation. The following sections describe the CALSIM model and what we have learned from it during the two years of the study.

A. Objectives of the CALSIM Model

The primary objective of the CALSIM project was to determine how feasible it was to attack human social, economic, and ecological problems through simulation models. The primary plan was to assess the quantity and quality of available data, familiarize ourselves with the literature, select the most urgent or most tractable of the problems, and begin construction of appropriate submodels. In the interest of precision the emphasis was on analysis. After enough submodels had been built, we were to undertake the integration of submodels into a functional system.

In the process of completing these tasks, the CALSIM model that emerged was to attain five goals:

(1) The model should account for all important processes accurately. The model should also include all areas of the state, preferably down to the city and county level, and even to finer levels where possible.

(2) The model should yield precise predictions, given the assumptions made, for any desired aspect.

(3) The model should be highly integrated, so that all the consequences that develop in discrete areas from any particular series of actions or decisions are revealed.

(4) Particular submodels should reflect the findings of any disciplines they represent, and wherever possible should contribute to those disciplines.

(5) The process should train individuals in the solution of complex, urgent, and interdisciplinary problems.

B. General Model Structure

As systems ecologists we conceived of the California population as part of a feedback loop with the physical, chemical, and biological environment. This is basically an ecosystem, with the major addition being the strong role of the human decision process in mediating flow rates in the system. The underlying principle is that no life can exist independent of its physical, biological, or chemical environment. One of the unique features of Man is his ability to modify his environment so it suits him; another is his technological achievements, which have temporarily freed him of dependence on solar energy and natural productivity; a third is the power of his communication networks (Platt, 1969). These three achievements have permitted the growth of the human population to continue explosively (von Foerster *et al.*, 1960). Population growth is a fundamental variable in human affairs because it increases demand and simultaneously makes it more difficult to maintain a specific life-style. In addition, population growth increases the number as well as the kinds of alterations of the total environment, and reduces the possible options available for future types and rates of environmental manipulation.

Recognition of the impact of human population dynamics on the California ecosystem is built into the basic conceptual model (Fig. 2). The basic loop is the California population-environment interaction; everything else forms part of the control system. Any activity of the population, such as cutting trees, building structures, and just existing, causes fundamental changes in the system that are detectable as effects. After quantification, the costs and/or benefits are subject to the decision process to moderate the actions of the population. Consider the cutting of trees. If too many are cut, the cost of excessive harvesting rates will be reflected in lower lumber prices, loss of forest productivity, and ultimately loss of forestry jobs. If cutting rates are too low, prices may be too high and sales too low, housing consequently more expensive, and forest disease incidence increased. Thus, a decision through pathway (1) is appropriate to change the interaction (cutting rates) to minimize the costs and maximize the benefits of forest harvesting.

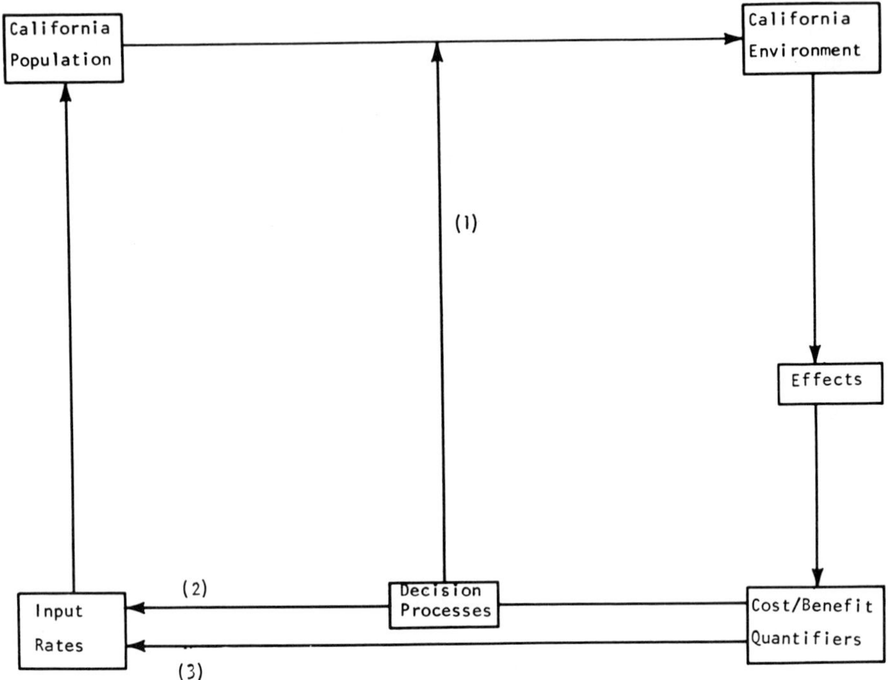

FIG. 2. The basic conceptual framework for the CALSIM project.

A second decision-mediated pathway concerns population change rates. Since the state of the population can only be influenced by changing birth, death, and migration rates, pathway (2) represents awareness of population effects on the ecosystem, and conscious effort to minimize these effects primarily through reduction of the birth rate, e.g., abortion, birth control, and perhaps through differential migration. Increased death rates are not usually voluntary, so most of these occur through pathway (3), which is not directly decision-mediated. An example is respiratory mortality due to air pollution.

C. MODEL DEVELOPMENT PROCEDURE

Model development consists of identifying and interrelating components within any one of the operational units of the flowchart, then repeating the process as systems analysis proceeds, until one has a series of flowcharts sufficiently detailed to permit direct conversion into Fortran statements. Figure 3 illustrates the initial expansion of the basic

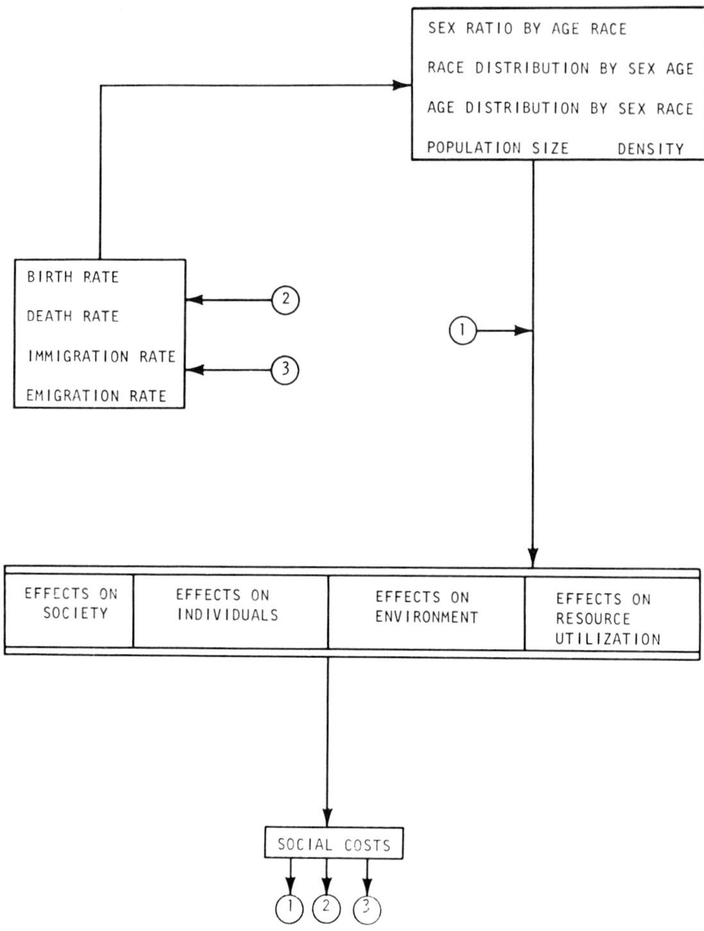

FIG. 3. The first expansion of the basic conceptual framework of CALSIM (equivalent to basic methodological flowchart).

model. Population input is broken into its component parts: births, deaths, immigration, and emigration. Similarly, the population may be separated into its characteristics, such as distributions by race, age, occupation, religion, sex ratio, population density, and degree of aggregation.

In this expansion we can eliminate the state "California Environment," because it is solely conceptual, and the "Decision Process" because it appears ubiquitously. We have recognized four major categories of effects:

(1) effects on social structure and institutions,
(2) psychological and physiological effects on humans,
(3) effects on environmental quality, and
(4) effects on resource allocation procedures.

We have not succeeded in finding a uniform system for quantifying benefits and costs. Clearly not all of these can be expressed in dollars; some may be expressed in two or more units, like agricultural energy input (calories or dollars). Others are not conveniently expressed in either unit, such as population attitudes toward use of leisure time. Thus, quantification of particular costs and benefits is likely to remain arbitrary, and variable.

Further expansion of the basic model requires development of modules from the four major effects. Each module requires its own set of cost–benefit quantifiers and a specific set of population parameters, so the detail that must be obtained from a demography model is specified by the requirements of all the submodels. For example, a particular module may require the distribution of a social characteristic by sex and age, while another may require one by age, race, and income. One submodel may need age-specific birth rates for all ages from 13 to 49, while another needs age-specific rates only from 18 to 30, and a third requires only a gross birth rate for the entire population. It is obvious that a complex series of submodels requires a highly sophisticated and flexible demography model.

Figure 4 traces the development of one of the modules. The submodel is the hypothesized air pollution–respiratory disease relationship, developed from effect (2)—psychological and physiological effects on humans. The flowchart is not yet detailed enough to convert to a program, for several of the modules contain groups of other submodels. Any one of a number of available air pollution models can be adapted for use in the respiratory disease model, e.g., that of Miller and Holzworth (1967), and others reviewed by Moses (1969). Each of these may contain other models, such as the distribution of automobiles as a function of socioeconomic status.

The use of disease rates as a measure of social cost illustrates that neither dollars nor energy need be used. The loss of individuals from the population can be translated, however, into the dollar cost of lost productivity, the cost of hospitalization prior to death, or any number

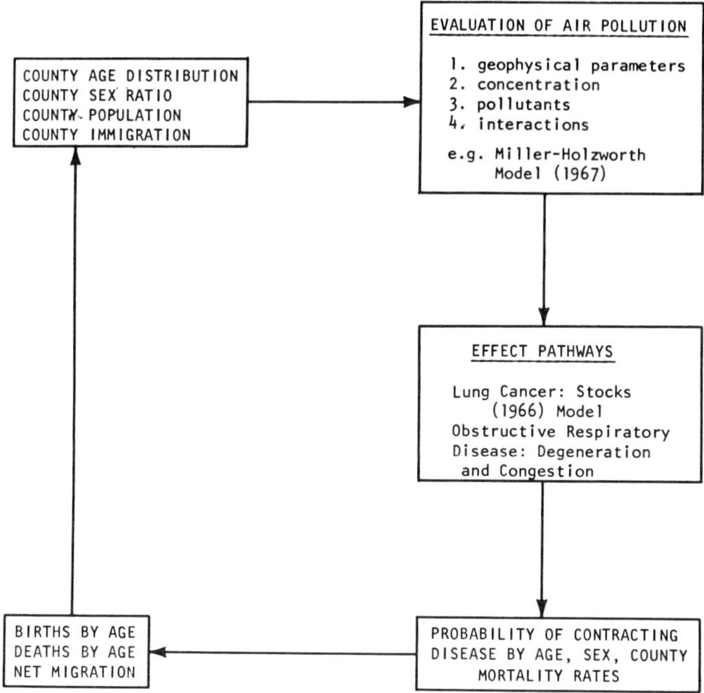

FIG. 4. The second expansion of the CALSIM concept, for the demographic-public health-air pollution effects-pathway.

of costs which ramify throughout the system. Since the cutoff point is purely arbitrary, a rate of death, safety on the streets, or high birth rates are as defensible as social costs as dollars or calories. This is particularly true when attitudes that have previously defied quantification are involved.

D. AREAS OF INVESTIGATION

In the course of our study we investigated a number of nominally discrete problems and made efforts to link some of them together. In fact, there are a number of overlaps because the definition of problem areas is often quite arbitrary, and because output from one submodel may be input for one or more others.

The size of the task we accepted with a small staff meant that we could not possibly build a functional model for California, and the unconnected submodels we worked on reflect this limitation. Nevertheless, we discovered a number of things by following different research strategies in the various areas. Some submodels were attempted using a broad viewpoint and sacrificing some detail for greater scope in the model, while others featured concentration on a smaller subject area but in relatively greater detail.

1. *Demography*

The demography model is the basic model of the CALSIM system because it supplies input directly or indirectly to the rest of the submodels. All models would require accurate predictions of age, sex, race, and socioeconomic distributions, and many of these would supply input via negative feedback to the demography model. Two general approaches were available to us. One of these was the empirical approach of the BASS group (California State Water Quality Control Board, 1968), in which the demography model is inseparable from an econometric model, and the other was the matrix approach of Rogers (1968).

We obtained copies of the BASS model from the authors, and the parts we tested appeared to be compatible with our computer. The BASS model consists of four major modules. The population model projects the population and attractiveness of an area; the "shift" model is a reconciliation routine for adjusting California trends to those of the United States; the structural model predicts the economic structure of a region and the shifts that would be expected with changes in population; and the allocation routine (time–distance matrix) distributes the people over the region. Unfortunately, the allocation routine was not made available to us, and because it was impossible for us to duplicate it, we decided not to use an expanded BASS routine for a demography model.

Rogers' method (1968) uses matrices for the standing population and its parameters, i.e., birth, death, and migration rates, so that the resulting demography model becomes nothing more than a set of matrix operations. Rogers' method is straightforward and we have programmed his method into a demography model, which is flowcharted in Fig. 5. Progress in this area has been halted, however, by inadequate data on the volume and characteristics of inter- and intracounty migrants. In 1960, 54.3% of the white inhabitants of California were migrants into the state (U.S. Bureau of the Census, 1960), far exceeding any other region. Yet, the only migration data available are those of the 1960 Census, summarized in "California Migration, 1955–1960" (California State Department of Finance, 1964). The chief weakness of the data is that information is not available at levels finer than the State Economic Area (SEA), which is frequently a multicounty unit.

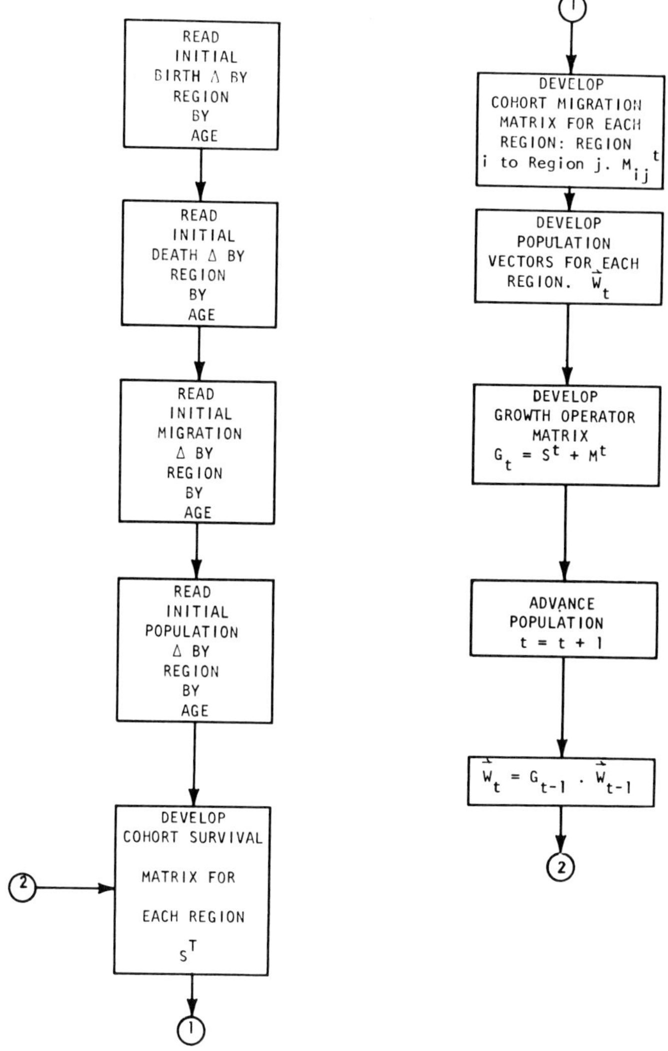

FIG. 5. Flowchart for a regional demography model, adapted from Rogers (1968).

There are a number of migration models available (see Haggett, 1965, for review) and several detailed studies which are valuable in elucidating causal pathways in the desire to migrate (Lee, 1964; Masnick, 1968; Miller, 1966, 1967). The basic problem is that what we know about migration patterns and the causation of migration is not suitable input for a detailed simulation model. There is nothing to prevent us from building simplified demography models or from making a series of assumptions in a more detailed one; in fact, these have been done for other models requiring a demographic input. Unfortunately, a sufficient demographic model cannot be built with the present data base.

2. *Education Taxation*

The problem of education taxation illustrates one of the social costs of population growth and demonstrates the necessity of a demographic model. When a population is growing, the age distribution shifts to higher and higher proportions of younger individuals. As a result, the ratio (education tax producers per education tax consumer) would be expected to decrease. In order to test this hypothesis we built a simulation model for a hypothetical population having the initial age distribution, birth, and death rates of the California population in 1965; to simplify the model, we assumed no net migration. In this elementary demography model the vectors for births and death were applied once per year for a period of a century, and the final age distribution was used to determine proportions of the population in school, the ratio of tax producers to tax consumers, the proportions in elementary school, high school, and college, and the proportion of taxes available at a given growth rate compared to that in a stable population.

Figure 6 shows the ratio of producers to consumers as a function of

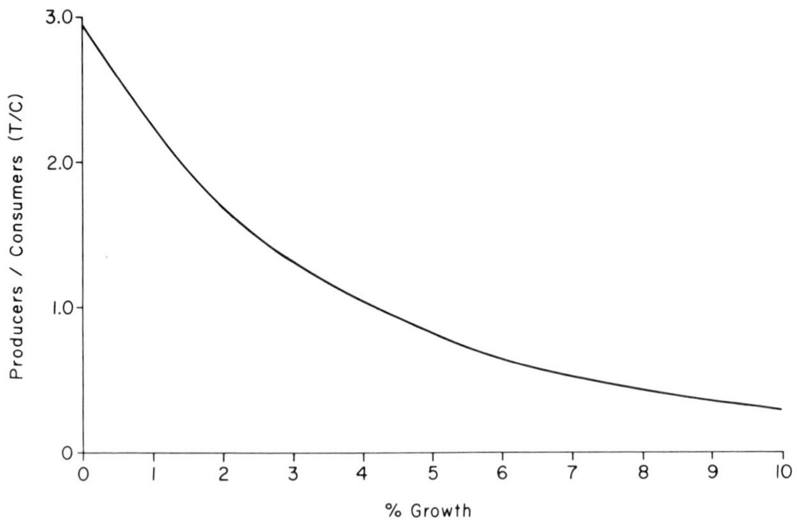

FIG. 6. The ratio of educational tax producers to consumers as a function of population growth.

population growth rate. This confirms that the producer per consumer ratio declines. The decrease is nonlinear, the rate of change being greatest at the lower, more realistic, population growth rates. The declining producer per consumer ratio and the rising proportions of students at all educational levels (for further detail, see Glass *et al.*, 1971) are reflected in the relationship between population growth and the proportion of taxes produced of those available in a stable population (Fig. 7). Essentially the same type of relationship results, but with the

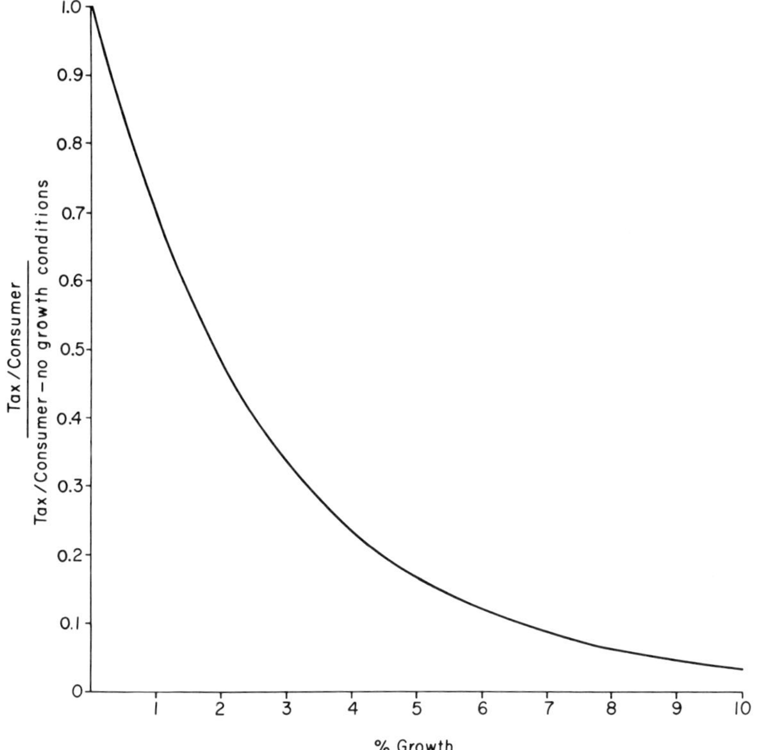

FIG. 7. The proportion of education taxes available in a growing population compared to those available in a stable population: an illustration of one social cost of population expansion.

social cost of population growth expressible in dollars. One percent population growth leads to approximately 25 % loss of taxes per consumer capita, 2 % to a halving of available educational funds.

These serious losses of revenue implicate population growth as one of the major contributors to deficits in school taxes, even without the influence of differential migration. A particularly fascinating positive feedback loop is the effort many communities make to increase the tax base by attracting industry. If in the process more workers are brought into the community, workers with young families, the effect will be a net fund loss if the additional education needs exceed the new tax revenues. Another positive feedback loop is the spiral between educational quality and the tax-producing population. If deficits are incurred, the educational system must raise more money, become more efficient, or accept declining quality in education. When the taxpaying public refuses to pay, for whatever reason, a circular causal system of declining educational quality and shrinking tax base is established. The city of Richmond, California, has been currently faced with such a situation. In Los Angeles, teachers went on strike not for higher wages, but to protest the impoverished situation of the schools—inadequate books, too few janitors, no heating in the winter. The strike was reportedly settled only by promises for an improvement of these conditions.

3. *Population Growth and Major Crimes*

This simulation model is very similar to the preceding one, but is concerned with a very different subject. The basic procedure is identical: a simple demography model coupled to vectors of characteristics of the population. In this case the vectors are the age-specific arrest rates for two general types of crime (personal crimes: murder, forcible rape, and aggravated assault; property crimes: burglary, robbery, auto theft, grand theft) drawn from the FBI data, 1960–1967. This model also demonstrates the influence of population growth clearly. For example, Fig. 8 shows the percentage increase in arrests over those in a stable population as a function of population growth rate where the age-specific per capita rates were held constant. In a population growing at as little as 1%, one could expect 70% more arrests for personal crimes and 153% more for property offenses. At 2% population growth, these estimates are 398% and 537%, respectively. For property crime arrests

$$Y = 0.7234e^{0.9563X}, \tag{1}$$

and for personal offense arrests

$$Y = 0.6157e^{0.8849X}, \tag{2}$$

where Y is arrests per arrests in a stable population, and X is population growth rate. These functions not only estimate the effect of population growth, they also indicate the magnitude of changes within the distribution of age-specific crime rates necessary to account for any remaining discrepancies.

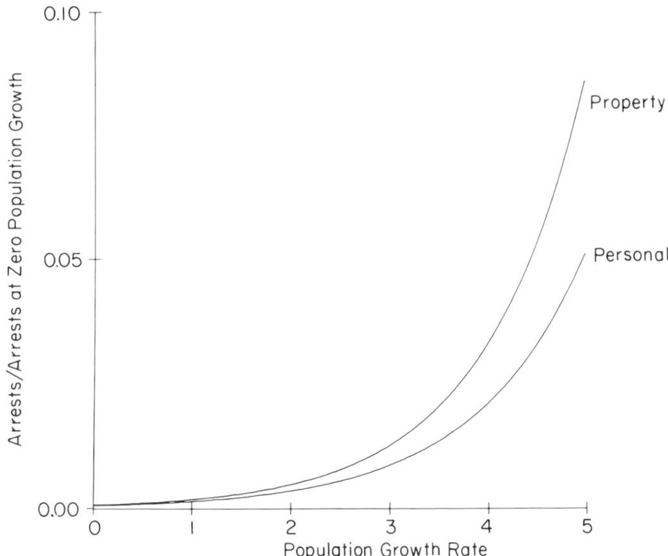

FIG. 8. Increase in two categories of criminal arrests as a function of population growth given constant age-specific arrest rates.

The curves of arrest rate per capita as a function of population growth (Fig. 9) reflect the sensitivity of the system to the shape of the age-specific arrest distribution. The per capita arrest rate for property crimes rises with population growth because these sorts of crimes are largely a product of the young (73% of arrests in 1960 were of persons aged 24 or less). This also accounts for the decline in rates at greater than 6% population growth, because in these cases a substantial proportion of the population is concentrated in persons too young to commit crime. Similar remarks also apply to the per capita arrest rates for personal crimes. The distribution of age-specific arrests for personal crimes approximates that of a normal curve, with only 45% of arrests of persons aged 24 or younger. This explains why the per capita personal crime arrest rate is maximal in a stable population. One could argue that one cost of population stabilization would be increased incidence of personal crime, but this is compensated for by decreasing property crime arrest rates; if population stabilization had been attained in 1960, the net decrease in criminal arrests would have been 0.55 arrests per thousand, all other factors being constant.

Perhaps the most important result of this simulation model is prediction of expected shifts in the age-specific proportions of arrests (Federal Bureau of Investigation data, 1960–1967). When the model was set up

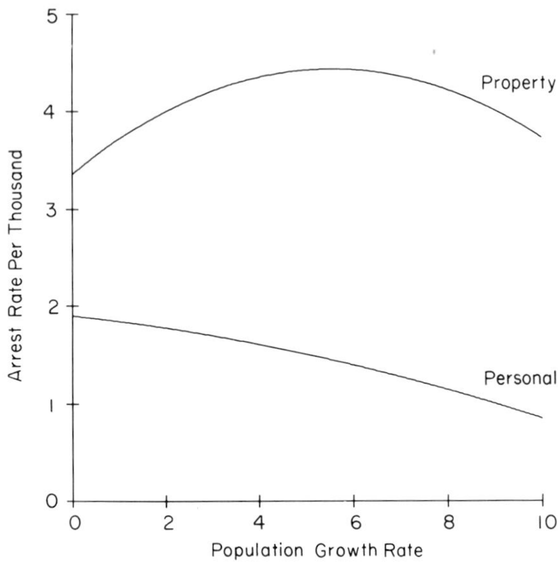

FIG. 9. Curves of arrest rate per thousand as a function of population growth.

to mimic the behavior of the system for the years 1960–1967 inclusive, all shifts in age-specific proportions were accounted for by the model or were statistical artifacts, except for increases in the number of arrests of juveniles (less than 15 years old) for personal crimes. With this exception, the model showed that recorded increases in arrests between 1960 and 1967 were not concentrated in the youth as is widely believed, but represented proportional increases at all ages or predictable shifts due to population growth. Unless our law enforcement agencies are doing a remarkable job increasing their efficiency proportionately to population growth and former arrest distributions, the nature of increased arrests has some alarming implications for social behavior in present American society. If we are showing increased propensity for crime and violence, it is occurring in all of us.

4. *The Spatial Distribution of Property Crime in the Los Angeles Metropolitan Area*

We have partially completed a mathematical model relating the four property-oriented crimes to some socioeconomic characteristics in 52 cities and 14 police districts in the Los Angeles metropolitan region. A number of socioeconomic conditions, reported as significantly correlated to the incidence of crime in the literature (Schmid, 1960a, b;

Boggs, 1965), were plotted against the California reports of offenses to police of 1960. From these, the six most precise relationships were chosen and described by empirical expressions derived from nonlinear least squares (Conway *et al.*, 1970) and polynomial fitting techniques. The six factors were percent unemployed, percent nonwhite, median value of owner-occupied housing, percent college graduates, median income, and percent homes built before 1950, recorded in the 1960 U.S. Census (U.S. Bureau of the Census, 1962).

In order to avoid intercorrelation between independent variables, the six estimated values ($\hat{y}$) were used to generate a mean crime rate ($\hat{\hat{y}}$) and its standard error. This approach was generally adequate except for the central districts of Los Angeles. To treat this problem, the region was divided into five arbitrary parts: Orange County; six police districts of the central business district; the western districts and suburbs, from the Western San Fernando Valley to Rolling Hills Estates; the eastern suburbs, from the San Gabriel Valley north into the San Fernando Valley and south through Downey; and in the southern area, from Long Beach through Compton. The data suggested that central district property crimes were being partially imported, and that an appropriate correction for underestimation in the Central Business District (CBD) should be incorporated. This was done by fitting a curve to crime rates as a function of distance from the city zone. Table I summarizes the model output with this correction.

TABLE I

MEAN REPORTED OFFENSES AND THEIR STANDARD ERRORS FOR *N* CITIES AND POLICE DISTRICTS IN FIVE SUBREGIONS OF THE LOS ANGELES METROPOLITAN REGION, WITH A COMPARISON OF MODEL OUTPUT

		Mean Offense Rate		Standard Error	
Region	*N*	Reported	Model	Reported	Model
Los Angeles	6	52.8	58.7	4.4	5.2
South area	7	22.7	22.9	3.0	2.2
Eastern area	29	17.4	20.4	1.8	1.7
West area	24	6.8	6.2	1.8	1.7
Orange County	10	10.5	15.1	1.5	0.9

There are a number of discrepancies apparent in these results, such as overprediction of the CBD and Orange County. This can be explained by a number of causes, like bias in reported rates, incorrect assumptions, missing data and variables, and human attitudes. At its present stage of development, the spatial model is not appropriate for prediction, but it is useful for certain other purposes:

(1) The model confirms that property crimes are largely a function of prevailing economic and sociological conditions.

(2) The model provides a systematic quantitative framework for examining whole regions and comparing subunits of the area. This is possible in principle without the aid of a computer model, but is facilitated by one.

(3) Perhaps the most important attribute of this model is that it is useful for focusing attention on areas where there is a large discrepancy between reported crime rates and model predictions. The model is a comparative analysis embodying correction for social and economic conditions. For example, Orange County communities were overpredicted and overdispersed. One problem is the extremely high report rate of Newport Beach, but as a whole, Orange County is homogenous economically, certainly more so than its crime rates. When compared to other regions of Los Angeles having similar social and economic parameters, Orange County appears to display a different set of attitudes, whose net effect is a suppression of offenses reported to the police.

The Los Angeles model, while still incomplete, may represent an excellent way to channel research efforts, and it will probably be useful as a means for expressing some of the current sociological theories about the causation of crime. Unlike the preceding two models, this one was intended in expanded form as an independent module for CALSIM. It does not now appear likely that this objective can be met because human attitudes we presently know so little about are too important to be ignored (e.g., Orange County). Data bearing on human attitudes are obtainable in principle, but it can only be done with detailed studies in the community, such as are being presently conducted by members of the Law School at Davis. Far from invalidating the systems model approach, this finding reemphasizes that focusing attention on areas like Orange County can be a useful function for the systems analyst.

5. *A Model of Physician Supply and Demand*

Adequate medical service availability depends strongly upon the number of physicians. As noted in the press and analyzed by Fein (1967), physician shortages are developing in many areas of the country where there was no problem recently. Since health services are important components of society, we undertook development of an interactive model of physician supply and demand for California. The model was designed for easy extention into related health services (dentists, nurses, medical technologists, and others), and it deals with a number of abstract, previously unquantified, social values.

The basic algorithm of this model is quite elementary. Physician utilization in average visits for year i (TU_i) is a function of age- and sex-specific visit rates available from the National Center of Health Statistics (1965, 1968) and from the U.S. National Health Survey (1960). Thus,

$$TU_i = \sum_i \sum_j \sum_k P_{ijk} U_{ijk} , \tag{3}$$

where P_{ijk} is population category by age j and sex k, and U_{ijk} is the utilization rate in visits per year. This measure was taken as an index of physician demand. It was the best one found for a previously unquantified variable. If we define D_i as the demand for physicians in year i, and W_i as the mean productivity per physician,

$$D_i = \frac{TU_i}{W_i}. \tag{4}$$

For supply the situation is somewhat more complex because doctors have the freedom to move and change professions. Let G_i be physician input into California for year i,

$$G_i = I_i + E_{i-5} , \tag{5}$$

where I_i is immigrating doctors, and E_{i-5} is doctors educated within the state and remaining but whose education started five years previously. The corresponding yearly loss L_i is given by

$$L_i = DC_i + RL_i + TEM_i + EM_i + AD_i + GPE_i , \tag{6}$$

where L_i is simply the sum of all the components of physician loss. In this expression DC_i is the number of physicians dying in year i, RL_i is the number of doctors whose licenses to practice were revoked, TEM_i is physicians temporarily out of state, EM_i is the number of permanent emigrants, AD_i is the number of doctors in full time administration or teaching, and GPE_i is the number of medical school graduates who never set up practice within the state, but emigrate upon graduation. The physician supply component is then

$$S_i = S_{i-1} + G_i - L_i , \tag{7}$$

and assuming no time lag in supply–demand adjustments,

$$R_i = S_i - D_i \,, \tag{8}$$

where R_i is the direct system measure of physician supply adequacy.

In compiling the data for this model, a number of interesting trends were discovered. From the standpoint of competition for doctors with other states, California has an advantage. In 1959, 69% of California physicians were immigrants, while only 25–30% of the state medical graduates emigrated (Stewart and Pennell, 1961), although there also appears to be a trend for increase in emigration. In recent times (1950–1965), about 250 graduates entered practice each year from California medical schools. This number largely just replaces the doctors who die each year; it is grossly insufficient for keeping up with the population growth of the state, so California's reliance on immigrating doctors is total. At the present time, efforts are being made to increase the number being trained in the state, but California will have to continue to rely on physician immigration. The consequences of decreased immigration for any reason are alarming, and it is not pure conjecture to think that could occur. As California becomes even more heavily populated, its attraction will surely continue to decrease. It is not in the interest of other states to pay for the education of California physicians, and one may expect restrictions to be placed on emigration of physicians. Finally, to rely on continued immigration is foolish if all states decide to follow this course.

Future California physician supply and demand to the year 2000 based on present trends is shown in Fig. 10. Note that R_i becomes negative after about 1976; this is mostly the result of greatly increasing demand. We also investigated the consequences of four other situations:

(1) no physicians immigrating after 1970,
(2) work load per doctor up 50% after 1970,
(3) work load per doctor down 50% after 1970,
(4) number of California medical graduates leaving state at 50% after 1970.

Only in case (2) would R_i remain positive; even in this case R_i declines toward 0 after 1985. In the other three cases, $R_i < 0$ in 1971, 1975, and 1976, respectively. All these things imply that California will have a difficult time maintaining its patient per doctor ratios without increasing the number of trainees, restricting emigration, shifting duties to lesser trained individuals, or increasing doctor load. It is of more than passing interest that the latest bond issue, specifically for new medical training facilities, was rejected by the voters (June, 1970, election).

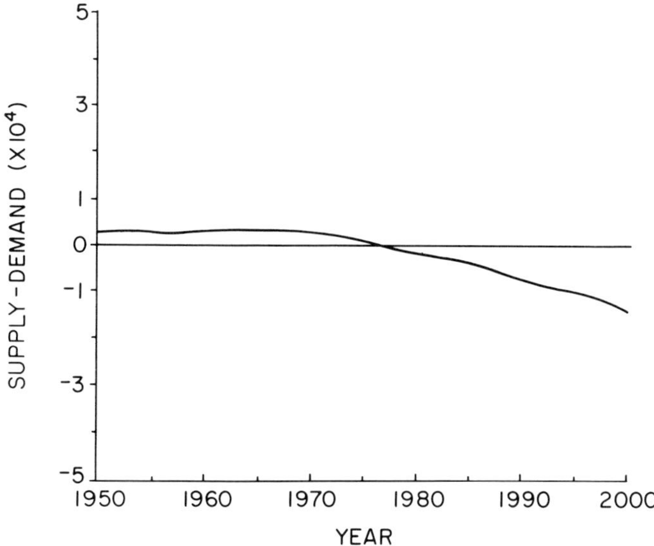

FIG. 10. The curve of physician supply and demand in California: extrapolation based on current trends in the health care industry.

6. *Transportation Demand, Air Pollution, and Lung Cancer*

In California the bulk of air pollutants are ultimately derived from the automobile; it is not accidental that one of the centers of oxidant concentration over Los Angeles occurs just east of the axis of the freeway system (Hamming *et al.*, 1967). With this background, we attempted to link several models in the literature into a system of submodels that would cover the pathway from automobile to lung cancer. The three main models used were that of Miller and Holzworth (1967), one of a class of air pollution diffusion models based on the normal curve, that of Stocks (1966), who constructed a probability model for the sequence of events from initiation to termination of lung cancer, and that of Slade (1967) for the rate of destruction of various air pollutants.

The incorporation of these models into a flowchart (Fig. 11) constitutes the third expansion of the basic model and is a direct development from Fig. 4. This flowchart is only a single expansion from the one used to write the program. This model was not only to develop one social cost of the operation of our transportation system, it was our first attempt to build a model centering on interfaces between sciences. Specifically, it was to bridge the gap between the atmospheric sciences and public health and between air pollution and the initiation of lung cancer. With

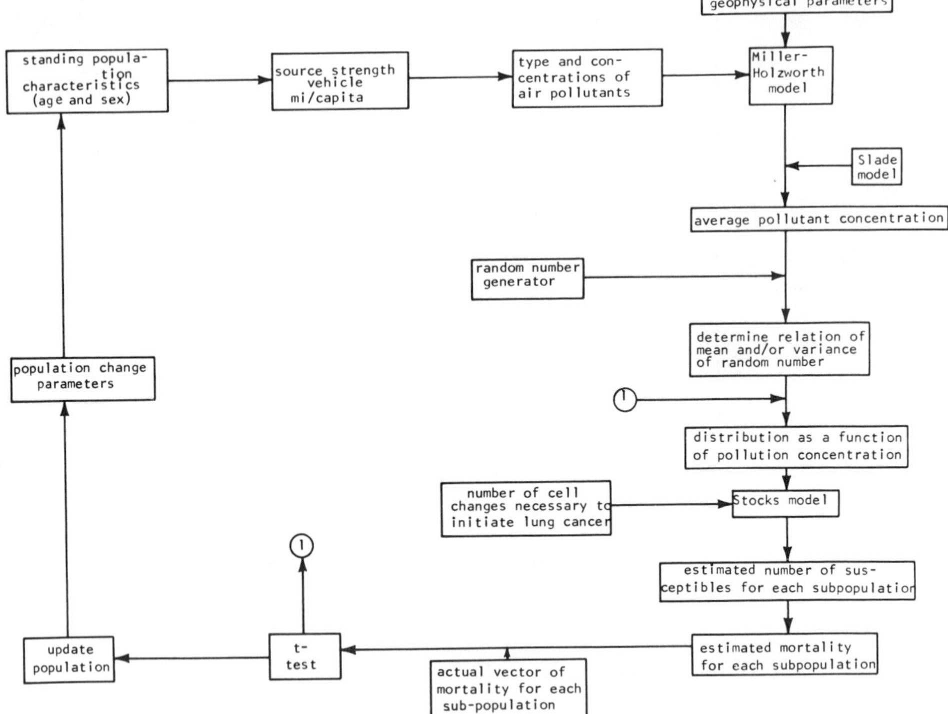

FIG. 11. The third expansion of CALSIM, for the air pollution–public health submodel.

our present state of knowledge, the model is theoretical and only represents a method for the isolation of some deterministic elements in a system composed of numerous stochastic variables. The major problems are an inadequate data base and incomplete understanding of mechanisms involved. One of the features of a flexible model is that switches and branches can be built in and options left for the operator or for specification as data, but as a model becomes more complex and as greater predictability is desired, more of these switches must be replaced with data. Even when we realized that the data requirements of this model could not be met, we continued to construct the model to see exactly what data would be needed and if we could stimulate any agencies to obtain it for us.

This model was not a success, because it could not be applied or validated in the real world. In addition, there are structural weaknesses that would invalidate the model. The Miller–Holzworth model, or for that matter any model presently available, is not accurate for describing pollutant concentrations over the large areas we were using it for, and no amount of iteration on parameter values could have overcome this problem. In fact, the only accurate model we are aware of is that being built by the Rand Corporation, one that has been applied only to a very small area (one cubic meter). Additional problems include the enormous demand of the model for computational space and its long computer time relative to real time, reportedly 8 to 1 (Greenfield, 1970). Despite these problems, the model is a method for organizing information that is most meaningful across interfaces, such as the role of weather in air pollution or the effect of air pollution on respiratory systems. The failure of the transportation–air pollution–lung cancer submodel is only that of the specific model; it does not invalidate the approach.

7. *Air Pollution and Respiratory Disease*

A thorough analysis of air pollution effects on respiratory health was carried out in connection with the preceding model. Data published by the State Department of Public Health (1960–1968) demonstrate highly significant increases of respiratory cancer (neoplasms of lung, bronchus, and trachea) and emphysema deaths (Fig. 1). The slopes of the linear regressions on these data are 0.762 and 0.843, respectively, both of which are highly significant ($p < 0.01$). There is some evidence of the public health hazard of air pollution (Winkelstein *et al.*, 1967; Hickey *et al.*, 1967), and the hypothesis of air pollution mortality is often accepted explicitly, as in the transportation–pollution–lung cancer model above. In truth, there are a number of conflicting causal pathways which obscure any real relationships between air pollution and public health, and for chronic effects the relationship is complex and unproved (Hatch, 1962; Wolkonsky, 1969). Part of the problem arises in differences in classification (Anderson, 1963, 1968; Mitchell *et al.*, 1968; Deane, 1965), another from differences in individual tolerance (Stocks, 1966; for lead by Rieke, 1969), and a third part in competing causes of death, such as smoking (U.S. Public Health Service, 1964).

There does not seem to be any real problem of misclassifying respiratory cancer, but the literature strongly suggests that emphysema has been and continues to be confused with other diseases of similar clinical manifestations, such as bronchitis, bronchiectasis, asthma, and even pneumonia in older patients. These diseases are not difficult to separate with detailed analyses, but in practice such determinations are often made on no more than gross evidence and often by medically untrained persons (Mitchell *et al.*, 1968). Therefore, we combined the five diseases into a class we termed "obstructive respiratory disease with pneumonia" because each features congestion and obstruction of one or more parts of the respiratory system. The regression on the 1960–1968 trends for the obstructive respiratory diseases declined nonsignificantly (regression coefficient $= -0.269$, $p > 0.95$), a reversal of the regression on emphysema deaths alone. When pneumonia was omitted, the coefficient was significantly positive but reduced (0.481, $p < 0.05$).

Since we had no capacity to gather data, it was impossible to deal with individual variation in susceptibility to air pollution, except for the general increase in susceptibility which is age- and sex-related. To correct for these factors, we used stepwise regression procedures (Draper and Smith, 1966) on county sex- and age-specific mortality records plotted against several variables of age, sex, and migration. We then examined residual patterns for evidence of air pollution effects. The dependent variable was the mean county mortality rate for 1960 through 1968 for the 27 counties that exceeded 100,000 in total population.

The models obtained from the stepwise regression were

$$Y = 40.913 - 0.363X_1 - 1.373X_5 \tag{9}$$

for respiratory cancer and

$$Y = 58.060 - 19.162X_4 - 1.361X_5 \tag{10}$$

for obstructive respiratory disease (without pneumonia) where Y is death rate per 10^5 capita; X_1 is percent migration into the county, 1955–1960; X_4 is county population sex ratio, 1960; and X_5 is county population age ratio ($<65/\geqslant 65$) in 1960. Model (9) accounted for 68.3% of the variance, and (10) for 84.2%. Thus, age and sex differences in the 27 counties were sufficient to account for the majority of place-to-place differences.

The patterns of residuals from these two models are presented in Table II. We were not able to find any correlations that supported air pollution effects; large residuals were scattered, and there were more in the Central Valley than in urban areas. In some urban areas, many counties such as Marin, Santa Clara, and San Mateo counties had negative residuals, indicating that there was less mortality than could be accounted for on the basis of age and sex. These macropatterns do not support the hypothesis of significant air pollution mortality, although they do not deny that air pollution effects may be contained in the residuals; nor do they forecast future trends for air pollution and respiratory mortality. They do suggest, however, that other factors may be more important, and that these factors also deserve attention from medical research. They also emphasize the complex nature of a highly interrelated system characterized by numerous feedback loops and interactive effects.

TABLE II

REGRESSION RESIDUALS FROM THE STEPWISE EQUATIONS FOR TWO DISEASES[a]

County	Respiratory Cancer		Obstructive Disease	
	$(y - \hat{y})$	$\left[\frac{100(y - \hat{y})}{y}\right]$	$(y - \hat{y})$	$\left[\frac{100(y - \hat{y})}{y}\right]$
Alameda	−2.97	−11.93	1.60	8.01
Butte	3.79	11.35	0.54	1.78
Contra Costa	3.86	18.12	11.75	12.07
Fresno	−2.42	−11.10	−0.32	−1.45
Humboldt	−1.32	−6.38	−1.44	−7.7
Kern	4.24	18.20	3.44	15.64
Los Angeles	−1.26	−5.06	0.40	2.33
Marin	−2.52	−12.60	−2.07	−18.00
Merced	0.00	0.00	−1.29	−6.45
Monterey	1.84	10.05	1.65	10.12
Orange	−3.06	−17.79	0.34	2.58
Riverside	−0.68	−2.91	3.64	12.17
Sacramento	2.31	10.18	−2.09	−12.37
San Bernardino	0.32	1.47	0.36	1.60
San Diego	2.36	10.93	1.05	6.08
San Francisco	7.76	19.80	0.91	3.47
San Joaquin	−1.04	−3.97	2.26	7.56
San Luis Obispo	−2.78	−11.73	−4.83	−20.82
San Mateo	−0.57	−2.66	−0.36	−2.57
Santa Barbara	−2.44	−11.67	−2.45	−14.94
Santa Clara	−3.62	−22.77	−0.64	−4.74
Santa Cruz	0.91	2.83	−1.61	−5.61
Solano	2.15	10.19	−0.79	−5.34
Sonoma	2.33	8.01	−0.66	−2.40
Stanislaus	−2.09	−8.10	6.72	20.61
Tulare	−4.02	−18.11	−4.47	−21.70
Ventura	−1.05	−5.68	−1.64	−10.65

[a] Here y is mortality rates per 10^5, and $\hat{y}$ is estimated y.

We are confident that the incidence of obstructive respiratory disease mortality has been relatively stable, but that respiratory cancer has not. At present we have no explanation for the steep rise in cancer incidence, although we know that environmental influences can change the internal milieu, which in turn affects the behavior of DNA and RNA. Braun

(1970) has recently reviewed the process of cancer cell initiation including the external environment; one influence mentioned is the effect of ionizing radiation on the environment of the nucleus through fluxes in ion balances.

In Table II the regression residuals do not show an expected urban–rural contrast, nor are they similar to the patterns for obstructive respiratory disease. Nine of the 27 counties had residuals of the opposite sign, and Marin, Santa Clara, and Orange counties are urban examples that all had negative residuals. Only one spatial pattern was discerned for these two groups of diseases. For obstructive respiratory disease, year to year variation in rates becomes more pronounced proceeding northward in the San Joaquin Valley, except for Fresno County. For the respiratory cancer complex there is no such pattern. This observation suggested a link to the dynamics of *Coccidioides immitis*, a pathological fungus that is the causative agent of Valley Fever (coccidioidomycosis) (Ajello, 1965). The growth of *C. immitis* is quite variable because of erratic rainfall patterns and biological competition. With too little rain, *C. immitis* cannot grow frequently enough to be a significant health hazard. We do not know why *C. immitis* does not do well in wet areas, but this fact is well known (Smith *et al.*, 1961). On the other hand, there are a number of studies which suggest adaptation of *C. immitis* to semiarid and arid climates, because of its salt tolerance and resulting competitive advantage over other micro-organisms (Egeberg *et al.*, 1964; Elconin *et al.*, 1964). In California, these conditions are found in the San Joaquin Valley and to a lesser extent in the Mojave Desert. On the eastern side of the Valley, coccidioidomycosis is only found north to Fresno, but is considerably more prevalent on the western side, spreading over the Coast Range and up the Central Valley as far as Contra Costa and Yolo counties. The smaller standard error in Fresno County may reflect the differential incidence from east to west (Table III).

We have been investigating the coccidioidomycosis–respiratory disease problem under the hypothesis that a previous case can predispose a person for more serious respiratory disease, including the obstructive ones. The literature generally supports this hypothesis. Pappagianis (1970) has reported that many coccidioidomycosis cases are relatively mild and go undetected, so many more people must be susceptible than are indicated by cases recorded by the Department of Public Health (1960–1968). *C. immitis* spores are inhaled, so the primary infection is usually in the lungs; about 40% of these require medical care, and many suffer permanent pulmonary damage (Winn, 1965). A few (0.1–10%, Salkin, 1965) get the disseminated form (coccidioidal granuloma), that spreads to other systems and is often fatal.

TABLE III

STANDARD ERRORS OF COUNTY DEATH RATES, 1960–1968, FOR TWO DISEASES IN THE SAN JOAQUIN VALLEY

	Standard Error	
County	Obstructive Respiratory Disease	Respiratory Neoplasms
Kern	2.00	0.93
Tulare	2.33	1.23
Fresno	1.07	0.99
Merced	2.66	2.35
Stanislaus	2.95	1.17
San Joaquin	3.94	1.23

We have not been able to establish a correlation between coccidioidomycosis and obstructive respiratory disease mortality using either environmental indicators for *C. immitis*, e.g., rainfall, farming activity, and soil alkalinity, or reported cases. The data generally supported a relationship but not a statistically significant one. We do not think this reflects lack of a causal relation as much as technical problems, although we may yet be proved wrong. More clearly than any other, this work illustrates the importance of data-gathering capacity, preferably a feedback loop between systems teams and collecting agencies. There are several means for correlating the incidence of coccidioidomycosis and obstructive respiratory disease that would be conclusive if the data were available: the intracounty distribution of obstructive respiratory disease mortality in the San Joaquin Valley; coccidin skin tests for patients suffering from obstructive disease, contrasted to a control population; and checks to see whether obstructive respiratory mortality rates are correlated to changes in incidence of coccidioidomycosis with a suitable time lag.

The coccidioidomycosis analysis also illustrates the complexity of any system or subsystem, and the increased data base demanded to attack the problems of such systems. There are undoubtedly a number of other etiological agents involved in respiratory pathology, of which *C. immitis* is only one. Even for such a restricted universe as the San Joaquin Valley, there is the impact of urban centers in counties which are very much rural otherwise, there is the influence of the Delta in the northern part of the valley, and there is mobility and the transient problem. Many people catch coccidioidomycosis in the Valley and display the symptoms elsewhere (Cheu and Waldmann, 1965). In the public health data, every one of the 27 counties except Humboldt recorded the disease during the years 1960–1968, even though many lie outside the endemic area for *C. immitis*.

A link between coccidioidomycosis and obstructive respiratory disease mortality is a promising hypothesis to account for the relatively high mortality of the latter in the lower Central Valley, but in the absence of further data we shall still have to guess about how significant Valley Fever really is.

8. *Agricultural Energy Efficiency and Land-Use Policy*

There were several motives for building this submodel. Energy is the foremost alternative measure of system performance to currency, especially when dealing with problems of resource sufficiency (Section III.D.10), when money may be a worthless substitute. One of the most interesting questions is not so much whether we can feed ourselves, but can we do so at a reasonable energy cost. For example, if massive inputs of fossil fuel or chemical energy were required, they might have to be supplied at the expense of another, perhaps equally vital, human activity. Second, for the model system chosen, the future magnitude of population and energy requirements could be determined as indications of the future for California, because Yolo County will be a microcosm for problems of urbanization in the near future (a conclusion of the BASS model). Third, the Yolo model was a means for testing several alternative urbanization policies and their effects on agricultural productivity and energy efficiency.

The Yolo County model is still in process of development. As presently constituted, the submodel includes routines for graphic display of the results for purposes of parameter sensitivity testing, an output routine, a simple population simulator (which again demonstrates the need for a demography model), a productivity routine, energy-calculating functions, a land-use simulator, and a consumption model.

The basic population projection technique assumes the existence of a simple negative feedback loop relating population increase per unit time to present densities. Thus,

$$Pop_{j,t} = Pop_{j,t-1}\{1 + (k_1 - k_2(Dens_{j,t-1}))\}, \tag{11}$$

where $Pop_{j,t}$ is population and $Dens_{j,t}$ is density at time t and for area j. The constants k_1 and k_2 are the parameters for rate of increase and inflection point; they are the variables that were used to determine the rate of increase. The algorithm is subject to a maximum density estimate

$$Maxdens_j = c_1 + c_2(Pop_{j,t}), \tag{12}$$

where c_1 and c_2 are the constants selected to attain the observed relationship between density and time. Hence, relation (12) is the estimator for k_1 and k_2. The additional urban land needed follows as a simple function of maximum density and population size,

$$UL_{j,t} = Pop_{j,t}/Dens_{j,t}, \tag{13}$$

where $UL_{j,t}$ is land.

Three land-use strategies were simulated:

(1) The city was assumed to grow continuously at uniform rates at its perimeter; remaining farmland is given by

$$A_{j,t} = A_{j,t-1} - (UL_{j,t} + TR), \tag{14}$$

where $A_{j,t}$ is agricultural acreage and TR is the land allowed to lie fallow between the perimeter of the urbanized area and the cultivated land.

(2) The city was assumed to grow discontinuously, with new urban growth occurring within limits at some distance from the city core and at a minimum distance from other such suburbs. The remaining agricultural land is evaluated by the same algorithm but depends more heavily on a functional grid system, and an additional subscript for city within a subregion of the county is necessary.

(3) The city was assumed to grow discontinuously in much the same manner as (2), except that the probability of an area becoming urbanized is also influenced by the transportation system. An area close to a major highway has a higher probability of urbanization, as does an area close to a highway compared to one which is not, at any given distance from the city core. The agricultural algorithm is substantially the same as in the preceding routine.

An additional constraint was placed on all three strategies. We recognized four categories of potential agricultural land and compartmented these in the county grid. Urbanization was assumed to proceed only on to the best land available; poor quality land was not used until all better land was urbanized.

Agricultural output for the county is evaluated after the quantity of land remaining after each cycle of urbanization is known. First, the proportion of acres under each crop is determined as a function of land of different classes presently being used for it. Second, yields per acre are evaluated as a function of water, fertilizer, and power inputs. For this purpose the following relationships were found in the literature or assumed:

$$Y = Y_{\min} + aX_1, \quad y \leqslant y_{\max}, \tag{15}$$

$$Y = Y_{\max}(1 - e^{aX_2}), \tag{16}$$

$$Y = (P - P_{\min})^a, \tag{17}$$

where Y is yield per acre, X_1 is quantity of irrigation water, X_2 is fertilizer added, X_3 is power input, and a, b are constants. Total agricultural output is easily evaluated using these relationships and the grid system.

Consumption of petroleum, electricity, natural gas, and various foodstuffs per capita population are straightforward estimates utilizing present levels or slowly rising demand. Consumption of power, fertilizer, and water by agriculture is output by the productivity submodel. The consumption output is then converted to BTU's in the energy subroutine for printed output.

The Yolo County model is not yet sufficiently complex to produce unexpected results. It was anticipated that the third land-use strategy would mimic reality most closely, and it did. This demonstrates that when urban regional planning is effective, it is chiefly effective by positioning the highway network. The model was found to be most sensitive to the coefficients governing population size, density, and agricultural energy input. For the output discussed below, these coefficients were chosen so that the population of Yolo County would increase 10 times, density would increase to 7 persons per acre, and agricultural power would double by the year 2020.

The total urban acreage in Yolo County was forecast to rise from 17,300 in 1965 to 108,011 acres in 2020, a 524% increase. This resulted in a decline from 200,000 to 109,289 prime agricultural acres, a loss of 45.4% of the total acreage in 1965. Since one assumption was that only prime land would be used so long as it was available, none of the other classification totals were affected. An immediate implication of these changes is for agricultural productivity. Because of our assumption about power input, the total output rose because of rising productivity per acre despite the loss of land (Fig. 12), but the increase became asymptotic to zero by 2000 because of declining land area, despite rising power input. Even with rising inputs of water, fertilizer, and power energy, the magnitude of solar energy is so great that its constancy was imparted to total energy input to agriculture (total input, 50×10^9 BTU; solar energy, 49.8×10^9 BTU; fertilizer, water, power, 0.2×10^9 BTU).

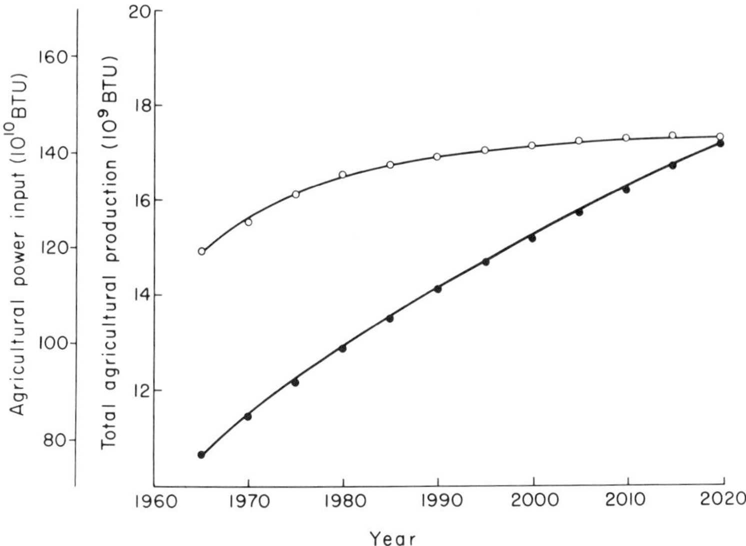

FIG. 12. Trends in agricultural power input and total agricultural production in Yolo County projected to the year 2020 under the third land-use scheme (see text for fuller detail). (○, production; ●, power.)

Because of the increased human input, despite its relatively minor magnitude, higher agricultural output meant greater efficiency both for agriculture and for the county. Finally, Yolo County would drop from 65 to only 7 times agricultural self-sufficiency.

Each of the component modules of the land-use–energy efficiency–agricultural productivity model is at this stage of development fairly crude and can be markedly improved. In addition, there are at least three other submodels which should be incorporated. The most important of these is an econometric routine incorporating cash flow for the farmer and how this affects what and when he plants, governmental subsidy systems, and parity ratios (the benefit–cost ratio of the farmer). In the present model the problem of supply and demand for energy input, water, fertilizer, and agricultural produce was not treated in detail. A second requirement is an appropriate weather and climate submodel, because plant productivity is also strongly linked to weather systems. In the present model, these factors were ignored. A third submodel needed is the most difficult one, for it must take into account the human attitudes influencing population growth and allocation in any area, including migration factors, community structure, and reactions to density. The present model made no attempt to include these factors, other than whatever attitudes are implied in the algorithms [Eqs. (11), (12)] used.

In summary, this model represents an initial effort to build one of the fundamental parts of the CALSIM project, which is essentially a spatial, grid-oriented model. As an initial effort the model was quite successful. With certain modifications, it is being used as an integral part of our continuing efforts (Section IV, below).

9. *Fossil Fuels and Energy Sufficiency in the United States*

One of the assumptions of the foregoing model was that the energy input necessary to keep agricultural productivity rising would be available through the year 2020. This may be a fallacious assumption, for there is considerable controversy about the question of energy resources in the United States. Errors have not been restricted to either the pessimists or the optimists. As late as 1960, Schurr *et al.* (1960), writing for Resources for the Future (RfF), could not foresee any shortage before 2000, while back in 1920 the chief geologist of the United States Geological Survey forecasted exhaustion by 1934 (Landsberg *et al.*, 1963). More recently, RfF has revised their opinion (Landsberg *et al.*, 1963), and a great deal of effort has gone into forecasting oil reserves. We have attempted to assess this problem.

Any attempt to determine remaining reserves of fossil fuels must deal with a number of uncertainties, like the prediction of ultimate supply, future trends in demand, potential and probable impact of future technology for extracting fossil fuels, and new sources of energy. Landsberg *et al.* (1963) have examined some of these questions. Among other things, they discovered that changes in demand in America historically were connected with changes in energy sources, like wood to coal, coal to oil, and presumably oil to nuclear power. Because of the uncertainty associated with supply and future technology, they concluded that prediction of future trends in fossil fuel supplies had to depend on historical trends and extreme caution. Hartley (1969) has used United Nations statistics to examine the world energy situation. He did not deal with questions of changes in the future at all. Restricting his attention to probable occurrences and present energy stores, Hartley calculated that world energy reserves were adequate for about 30 more years.

Zapp (1962) provided methods for petroleum estimation based on the assumption that cumulative discoveries and the rate of discovery per foot of exploratory drilling are a function of the cumulative footage of exploratory drilling up to one well for every two square miles. Using his methods, McKelvey and Duncan (1965) estimated a total oil reserve of

650 billion barrels for the United States, and Hendricks (1965) estimated a total of 400 billion barrels for the same area. Hubbert (1967) has shown that Zapp's basic hypothesis is untenable because the average rate of discovery had risen from 89 barrels/ft (1859–1966) to 276 barrels/ft by 1930, only to fall to 35 barrels/ft since then. In addition, the discrepancy of estimates using Zapp's method illustrate the sensitivity of the dQ/dh vs h curve to parameter values (Q is cumulative discoveries, h is cumulative length of hole in exploratory drilling). This casts doubt on the value of the estimates.

Hubbert's (1962, 1967, 1969) methods depend on two critical points:

(1) that the rate of production $P = dQ_p/dt$ follows the rate of cumulative discoveries dQ_d/dt by a time lag of 10 to 12 years. Hubbert claims that this has been the case for the last 40 years;

(2) that the complete cycle of resource exploitation will be logistic,

$$Q = \int_0^\infty P\,dt = \frac{P}{1 - ae^{bt}}, \tag{18}$$

and that the maximum rate of discovery dQ_d/dt has already been passed.

Fitting the function to the curve of cumulative proved discoveries for a number of fossil fuels, Hubbert (1969) concluded that the time required to exhaust the middle 80% of the world's petroleum resources—crude oil, natural gas, natural gas liquids, tar sands, and oil shales—will be about a century, which means that the time to exhaustion will be sometime around 2100. His estimates are based on an ultimate recoverable supply of between 1350 and 2100 billion barrels. Under present conditions of usage, the corresponding estimates for coal show that these are sufficient for 3 to 4 centuries.

Moore (1965) used Hubbert's approach of fitting an appropriate function to historical data, but instead of the logistic curve used the Gompertz equation, $Q = ab^{c^t}$. The Gompertz curve is similar to the logistic, but the inflection point is not fixed. Moore obtained an estimate of ultimate U.S. production (excluding Alska) of about 300×10^9 barrels using the Gompertz function, compared to Hubbert's estimate of 165×10^9 barrels for the same area. The discrepancy, using the same data, is notable.

Using the history of exploitation of such natural resources as the bison and whale (Watt, 1968), Watt (unpublished) has extended the statistical methods of Hubbert and Moore with a third function. Noticing that when whales became scarcer exploitation pressure was intensified, he reasoned that demand for petroleum should be similar. Using Fig. 8.2 from Hubbert (1969), Watt assumed that exponentiating production rates reflected exponential demand, and that this situation would continue until exhaustion of the resource. Thus, the historical trend in production of petroleum could be expressed as $Y = ae^{bX}$; linearizing and solving by simultaneous linear equations yielded $a = 0.033$ and $b = 0.0682$. Hence, supply = demand = 6.832% per annum.

Using 1880 as $t = 0$, the complete cycle of crude oil production may be expressed as

$$Q = \int_{t_1}^{t_\infty} ae^{bt}\,dt, \tag{19}$$

where Q is cumulative production of crude oil.

Integrating,

$$Q = \frac{a}{b}\left[e^{bt_\infty} - e^{bt_1}\right]. \tag{20}$$

Setting $t_1 = 1965$ and solving for t_∞ using Hubbert's estimates for Q (high = 2100 billion barrels; low = 1350 billion barrels), he obtained brackets in real time of 1996 and 2003. Watt's estimates indicate 100% exhaustion time between 26 and 33 years from the present (1970). Watt's estimates depend on a constant rate of increase in demand at nearly 7% per annum. Unless the rate of growth of petroleum production shifts to even higher rates, Watt's estimates represent the most pessimistic case.

Ryan (1966) has criticized the basic concept of statistical extrapolation. He challenged a number of features of the Hubbert method in particular, and the Moore method by implication:

(1) The production curve can be used as easily and probably more reliably than the proved discoveries. Fitting a logistic function to this curve leads to an estimate of ultimate recoverable capacity of 100 billion barrels, lower than the 170 billion barrels Hubbert used and even lower than the quantity already known or produced.

(2) The logistic curve cannot be defended as a fundamental consequence of economic, physical, or geological laws, but is purely empirical. How well it holds into the future is open to question. In partial answer, Ryan draws attention to Moore's (1965) projections made using the Gompertz curve. This curve is also empirical but is not any less defensible than the logistic function. Moore's estimate was 300 billion barrels, a considerably inflated figure compared to Hubbert's 170 billion barrels. Using different means to estimate the parameters, Ryan obtained even larger figures, on the order of 900 billion barrels.

The basic problem that Ryan discusses is one familiar to ecologists. If projections are to be made for planning policy, such methods must be well grounded in theory and must have precise data, for otherwise the theoretical confidence limits about the projection will be too wide to make the projection useful. This is the same problem as extrapolation from a regression line; on scientific grounds one would be justified in rejecting the estimates of Hubbert and Moore. Similar remarks may be applied to Watt's thesis even more easily. The most obvious point is that the production curve probably cannot sustain a 7% growth rate indefinitely, especially not as competition for crude oil intensifies, oil becomes harder to find, and prices begin to climb.

Individuals connected with the oil industry tend to emphasize the economic aspect of petroleum depletion. They see the time lag between production and discovery as a product of time lags due to gearing for production, drilling limitations, and other factors, and changes in the discovery curve as the consequences of economic shifts. An example is the shift that occurred in the changeover from coal to petroleum, when costs per barrel began to fall below corresponding costs for coal. Similar economic constraints also lie on related resources and the technology for exploiting them. For power generation in the breeder reactor or by controlled fusion the technology still does not exist; for the extraction of oil shales it does, but costs are still too high to be competitive with liquid petroleum. Strangely enough, it is not the oil industry that is unaware of the energy reserves problem: Ryan centers his criticisms on the dimensions of the reserves forcasts as guidelines for future planning, not the concept of shortage. Rather, it is in energy-dependent industries and a public, conditioned to the idea of infinite economic growth, that an awareness is lacking. Hubbert (1969) discusses this problem at length.

The value of the forecasts and procedures discussed for precise predictions is debatable. Emphasis should really be placed on the perspective derived from assessment of the energy problem, and estimates to an order of magnitude. While the precise predictions are of doubtful value, such estimates as Hubbert's or Watt's emphasize that petroleum energy cannot last forever, especially under constantly increasing demand. Such estimates also work to disperse the euphoria that seems to pervade human affairs, and may lead to better planning (Hubbert, 1966). If Watt is more correct than Hubbert, these results emphasize the importance of accelerated development of breeder reactors, controlled fusion, and stabilization of demand.

E. Conclusions from CALSIM

Conclusions drawn from each submodel have been discussed above. The six major conclusions that follow constitute an integration of all the submodels plus experience gained and not reported. What we have learned may contain valuable lessons for other teams interested in interdisciplinary approaches to human ecology, particularly systems modelers.

It should be obvious from a number of the submodels discussed that there are real problems with data quantity and quality, including some of the most important submodels, e.g., demography, or most urgent problems, e.g., air pollution, energy depletion. The CALSIM project demonstrated the strong reliance on good data that useful models have, and the fact that we do not have very much good data. There is undoubtedly a multiple causal system responsible for this state of affairs, but the part we should be concerned with is neglect due to lack of planning and a failure to treat the human ecosystem as an opportunity for experimental design.

The air pollution–respiratory disease analysis represents one example where a mutually beneficial relationship was established between the data collection agency (The Department of Public Health) and the systems modeler (The Environmental Systems Group). The extra effort entailed in breaking out the data in new ways was rewarded by a further analysis of it. This loop promises to be fruitful in the future, and also shows signs of further extension. Other agencies interested in systems analysis and simulation have also approached us for an exchange of services with offers of data, or financial support, or both. We have concluded that the systems analysis–data collection loop is a fruitful activity for all concerned, for it must lead to further integration of agencies and advancing knowledge of optimum means for guiding human affairs. This is a major benefit we intend to continue to encourage.

Modeling also stresses interdisciplinary integration because it is a teaching tool of great potential. The CALSIM project showed what sorts of problems could arise—data gaps, shifting human attitudes, uncertain technological inputs, how problems associated with one discipline could interact with problems of others, and what the potential of systems analysis for these problems could be. CALSIM was one example of the broad viewpoint systems methods impress on the student. We believe, as does Platt (1969), that this is a necessary ingredient for scientists who wish to attack and solve large-scale problems of the human ecosystem.

Our basic objective was to evaluate the possibility of systems methods for human problems. There is no real question that systems analysis is a fruitful activity no matter how one defines it. On the basis of our experience we also feel that systems simulation is feasible as long as its limitations are realized. The main problem is the data problem; this primarily affects the process of prediction. Planning is to some extent still a valid activity, and gaming is even moreso when the benefits of model system–data loops are considered. There are at least three reasons why systems simulation should be considered feasible even with acknowledged poor data. Data quality can be improved through operation of the data–model loop, which requires a series of progressively improved systems models; systems simulation is a natural extension of systems analysis; and for broad interdisciplinary problems there is no method that surpasses its potential. Systems methods possess no magic means for making better predictions than any other analytical tool, and for the present our achievements will probably continue to be modest, but we must continue to build systems models in order to get better ones.

The most important model emerging from CALSIM was the land-use energy flow routines discussed in III.D.8. There are several reasons for this. First, initial attempts at modeling in this area revealed that most key processes could be simulated without insurmountable difficulties. Second, the land-use energy flow model is spatially oriented, incorporates a demography model, and is basic to the operation of a model system. Third, energy dynamics is a key system measure that needs to be explored in much greater detail. Fourth, many important processes, such as health care, criminal violence, transportation dynamics, and cities' services can be developed as submodels of a land-use routine. Consideration of all these factors has led to the development of a plan for expanding efforts in modeling human processes (Section IV).

When CALSIM was organized and the Environmental Systems Group founded, our philosophy was that the human problem was an ecological one that could be handled by systems ecologists with suitable help from various consultants. This research strategy was excellent for training generalists and for developing broad horizons in thought processes, but we found that one could easily become a very superficial generalist, with ideas about many topics but sufficient depth in few. We therefore recommend that systems teams be broadly interdisciplinary in nature, although we do not yet know exactly to what extent. Steinhart and Cherniack (1969) have examined many institutes and other interdisciplinary teams and found that most are interdisciplinary only in name. To avoid repeating this situation the necessary quality is the ability and most importantly the intellectual commitment for truly cooperative research on the part of individuals in such teams. Whether or not such commitment can be obtained in an academic structure is still open to question.

IV. An Expanded Model

A. Structural and Philosophical Additions

The Environmental Systems Group is continuing to model some elements of human society. In recognition of some of the problems with CALSIM structure, several changes have been instituted for the second stage of systems modeling. The most important of these changes is the diversification of the group. Environmental Systems is now just one element of a much larger team, including systems engineers, civil engineers, lawyers, policial scientists, planners, agronomists, meteorologists, and economists. Most of these men have experience in systems modeling, and are thus in a position to apply their specific competence directly. Each member of the team was chosen for particular purposes, as is evident in Fig. 13.

The major conceptual change is that we now intend to develop a hierarchy of global, regional, and subregional models with much less emphasis on spatial locations and gridding techniques. Each systems model developed can always be considered as a subsystem of something higher, although above the global level the few interactions left with higher levels can be handled easily. The CALSIM model was arbitrarily limited to California, and we never really had to deal with import–export problems. This viewpoint is myopic for processes which are global or regional in character, such as atmospheric turbidity and global climate (Mitchell, 1970; Bryson and Wendland, 1970; Bryson, 1968). Suppose one were building a model for Nevada: California dynamics would have to be considered for her exports of air pollution, money, and people. A crude global model (Ratzlaff, 1969) already has been built and awaits refinement. A hierarchy of models from the global level downward is much less arbitrary about which interactions are to be considered. In addition, deemphasis of spatial locations avoids some of the problems encountered in the CALSIM model. The data requirements are reduced, emphasis can be placed on especially interesting or important problems, and internal computer storage is minimized. Our own experience has taught us that even relatively small models can have extremely large storage demands.

A second conceptual change is that it is impractical at the present time to attempt a model of the complete spectrum of human activities on limited budget and with limited staff. We have decided to concentrate our activities on a land-use, energy flow, agricultural productivity urban dynamics model for the reasons outlined in Section III.E.

B. Proposed Model Structure

We intend to build a tri-level systems model consisting of a global model, several regional models, and a number of subregional models. The global model will deal with gross features such as atmospheric gas balance, temperatures, and solar energy partitioning. The regional models will be concerned with the use and development of areas such as Lake Tahoe and the San Francico Bay Area, and regional problems such as land-use zoning. The subregions have largely not been chosen yet, but these will be defined by the specific problems they illustrate.

A very generalized flowchart of the proposed model structure is shown in Fig. 13. While centered about land-use problems, the model

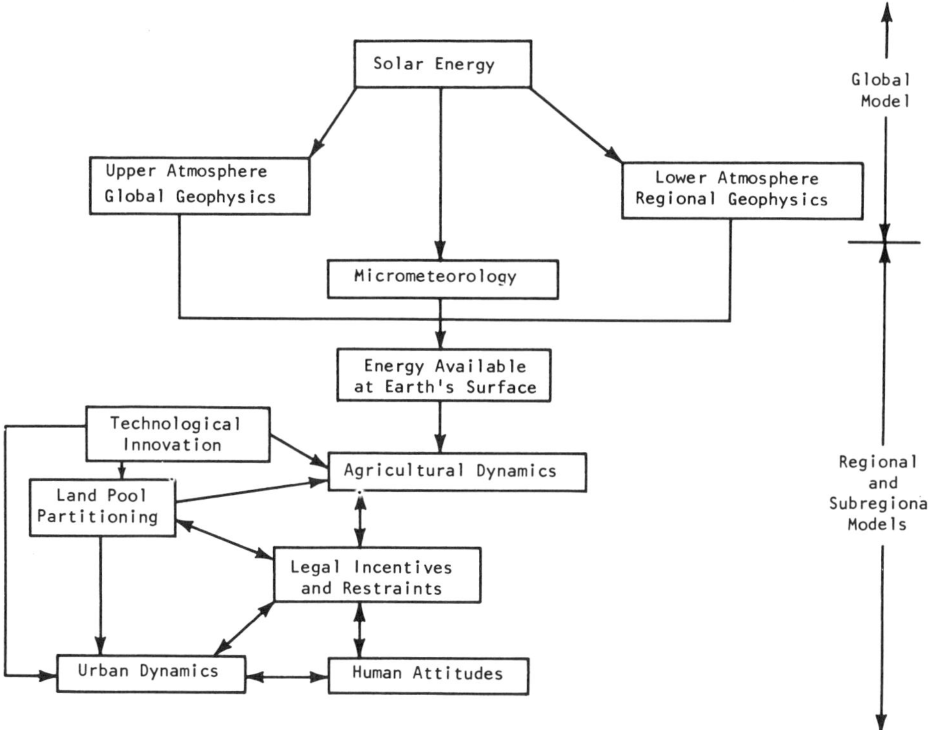

FIG. 13. Basic structure of an expanded version of the land-use model.

is actually considerably broader and can be expanded without difficulty. The group that has been assembled has the expertise to construct such a model, and in particular those difficult areas which would be most meaningful in a model: human attitudes, legal incentives, and technological innovation. At the present time the investigators involved in this model are integrating their efforts by specifying the input required by each submodel so that the necessary output can be generated.

Our flowchart representation clearly recognizes the interactive nature of the human ecosystem, especially the central roles human attitudes and legal decisions occupy. It is quite possible to build a model assuming a particular set of attitudes, but such an exercise could be quite sterile without attempts to estimate their influence. The flowchart reflects the way we propose to stimulate interaction between representatives of various disciplines: most of the social and professional sciences are instrumental in quantifying "human attitudes" and "urban dynamics," the legal profession and political science in "legal incentives and restraints," the engineers with "technological innovation," and so forth. It remains the primary responsibility of the Environmental Systems Group to stimulate and maintain interdisciplinary integration.*

V. Prospects for Human Ecosystems Simulation

Ecological systems analysis is relatively new, and ecological systems analysis applied to human problems is even more recent. Some general conclusions are appropriate for the development of other groups working on human problems using systems analysis to help them avoid some of the problems that were obvious, but best seen in retrospect. The strengths and weaknesses of systems methods have been previously discussed, and we now need to apply our conclusions to three major questions:

(1) What are the data requirements of a realistic model?

(2) What will the expense be?

(3) What are the unique features of a human ecosystem model?

Our work has shown clearly that the data requirements of systems models are enormous, becoming progressively more so as the models become more complex and more realistic. In our laboratory, Conway (1969) built a realistic reproduction model for insects. As narrow in scope as his model was, it still required an extensive literature search, which eventually yielded 74 papers containing significant data. This was the result of a high demand for realism. Many of the CALSIM submodels had data requirements that were much less exacting, but at

*Work on the expanded project has been in progress since 1970. Interested readers should contact Dr. Kenneth E. F. Watt, Dept. of Zoology, University of California, Davis, California, 95616, for further information.

the cost of realism. This trade-off is sometimes necessary, but the ultimate cost is always in precision of prediction from such models.

The Conway model required about two years labor and \$22,000 in salary and computational time, plus peripheral costs for library search. The education taxation model required about \$7,000 over only three months. The cost of systems modeling is high, although it is not certain that it is higher than other types of scientific investigation. The bulk of the costs are in salaries and computation time; the costs can be substantially reduced by efficient programming and clever use of available machinery, such as mixing batch processing and time sharing to suit the type of job. Another part of the high cost can be attributed to inefficiency due to the experimentation with methods that occurs in any new endeavor, and the lags in communication that occur among loosely coordinated individuals. The simplest justification for the extra cost is in the unique features of the human ecosystem model.

Systems simulation has one unique advantage over other scientific methods. Model building is a formal means for forcing attention on the logical structure of the system, which in turn leads to a multitude of beneficial effects. Causal pathways can be defined or erroneous ones rejected. Data faults can be detected or critical experiments formulated. System interactions can be found that may have been undetected otherwise. Some of these can be found by good inductive thinking and by the performance of critical experiments, but these methods lack the continuity of purpose and execution which are the real advantage and justification of systems techniques. Systems models are not to be regarded as a unique means for finding new discoveries; they are simply extensions of the scientific method.

We can now evaluate the present and future prospects of human ecosystem analysis. There are undoubtedly great problems facing anyone who wishes to work on human social dynamics: data problems; prediction of future technology; random, unpredictable elements; irrational decision-making; the danger of self-fulfilling prophesies; nonquantifiable human attitudes. These are formidable difficulties for anyone working on human dynamics, but they are worse for the systems modeler whose results depend on a strong data base and thorough understanding of underlying causal mechanisms. Our present understanding of human dynamics is minimal, well below that necessary to build a whole systems model that is predictive for any time span into the future or which can be used for detailed planning. There are simply too many unknowns to allow it. The question of whether it can be done at all is quite relevant; it will take a number of generations of increasingly sophisticated models before we can even begin to evaluate this question, and for the foreseeable future the answer probably is no.

Given such a dismal prospect for human systems modeling, the obvious conclusion is that we should abandon human systems modeling for the present. I do not believe that this is the case. Attempts to attack problems of the human ecosystem through simulation must be expanded if we are to take full advantage of the data collection–systems model feedback loop. Since progress depends on the data base, it is logical to build such a base that is neither too large (and therefore needlessly costly) nor too small (inadequate for good decision-making). Systems modeling is one way to improve the data base, whether or not policy is ultimately determined through use of a simulation model. In the process of collecting better data, better models should also result, which will help reveal what the potential of model systems can be. What we need to find are ways to operate the data–model loop optimally, to coordinate scientists in the most harmonious and productive manner, and to reduce the cost of modeling so it can be used more widely.

Small scale systems models dealing with particular problems can also be useful as tests of hypotheses and assumptions. If a future is constructed that assumes increasing conversion to nuclear power by 1975, a model could be built incorporating all known information about nuclear reactor power generation and uranium supplies to test this. If a complex plan is proposed for a region, a simulation model can be built to see if any of the elements of the plan are mutually exclusive, and therefore if the plan is faulty. In general, where there is enough understanding of mechanisms and sufficient data, simulation models can be used to reject hypotheses and assumptions, including those generated by other models. Much like the test of the null hypothesis, a model which fails to reject a premise may not be used in support of it.

An analysis of the success or failure of each of our submodels reveals that better models can be built for some problem areas than for others, and that usually a faulty data base is the root cause. This may lead some to deny the potential of systems modeling, and to claim that it is not a worthwhile activity. This ignores the fact that for certain discrete problems, such as health care services or the role of population growth in education taxes and crime, sufficient models can be built, and it ignores the potential for stimulating logical thought and better data collection. On the other hand, some modelers may claim more for their products than can reasonably be justified by their assumptions, understanding of underlying causal mechanisms, or data. It is the responsibility of the systems scientist to keep his models in perspective and to remember that while we may not be able to build very good large-scale human simulations now, systems methods have an integrative potential that must be exploited fully.

VI. Summary and Conclusions

There is increasing awareness of the problems of human existence: explosive human population growth, the artificiality of exploitative economic systems, the limits of technological solutions, the ecological limitations on human activity. The most basic of these problems is the process of uncontrolled human population growth. The resulting complexity and exponentiating problems due to their interactions demand new ways to cope with them. Systems analysis and systems simulation are techniques that are problem oriented, interdisciplinary, and broad-scale, and that can potentially provide predictions, a gaming device, and ultimately planning guidelines for decision makers.

This chapter reports on a two-year feasibility study funded by the Ford Foundation to assess the potential of systems methods in human ecological studies. The model system chosen was California; our expressed purpose was to construct a whole systems model for the state in order to reveal alternative courses of development and the social costs of some human activities, particularly population growth.

The mode of development was to start with a very general model, followed by progressively more detailed breakdowns of each component into modules that could easily be translated into Fortran statements. This was done for a number of discrete submodels, nine of which were discussed in detail. Six conclusions were drawn from this work:

(1) Data quality and understanding of causal pathways in human ecology are generally weak.

(2) Systems analysts and data collectors can form a mutually beneficial feedback loop from which the decision maker draws the maximum benefits.

(3) Systems training is valuable for stressing a broad, interdisciplinary, problem oriented, research philosophy.

(4) Systems models can only be improved by building them and striving to correct their weaknesses.

(5) The land-use energy flow model was the most promising.

(6) Systems teams must be broadly interdisciplinary.

These conclusions were incorporated into a plan for a hierarchy of models on land-use problems. Work on this model developed from CALSIM.

The prospects for systems models of human affairs in the present and near future were discussed, and the following conclusions drawn:

(1) Systems models demand a great quantity of high quality data, and they can be very expensive. In return for this investment, one hopes

to obtain a powerful tool for prediction and planning, a set of procedures for formally applying logical processes, and input for the data collector.

(2) For a number of reasons it seems unlikely that systems models will provide detailed plans for future development of any region, but systems methods have a strong impact on planning by forcing operation of the feedback loop on data collection and as a unique method for testing hypotheses.

(3) It would be unwise to abandon systems modeling because of its difficulties, and it would be foolish to ascribe properties to models that cannot be justified by the data input and logical structure. The application of systems techniques should continue because these methods are largely new and have untested potential. At the same time, we need to find ways to lower the cost and increase efficiency.

Acknowledgments

The work reported in this chapter represents the cumulative labors of many people who contributed ideas, algorithms, literature, and criticism to our efforts. I could not begin to name them all. Much credit is due to K. E. F. Watt, the principal investigator of the Environmental Systems Group, for the many ideas he has contributed and for editing the manuscript. Many of the ideas presented have been the product of group discussions and group efforts. To name just a few, John Wilson and Richard Bradley were mainly responsible for the idea and construction of the land-use model. The former collaborated with Miss Dana Thode to construct the air pollution–lung cancer model. Håkan Norelius and Miss Thode built the physician supply–demand model. N. R. Glass and the author built the education taxation simulation. The air pollution analyses were done in collaboration with Miss Thode, and the crime models with Miss Lynn McFarlane.

The entire project was carried out under a grant given to the Environmental Systems Group by the Ford Foundation. The work reported here would not have been possible without the support of the Foundation.

REFERENCES

Ajello, L. (ed.) (1965). "Coccidioidomycosis." Univ. Arizona Press, Tucson, Arizona.

Anderson, D. O. (1963). *Can. Med. Ass. J.* **89**, 709.

Anderson, D. O. (1968). *Can. Med. Ass. J.* **98**, 231.

Anon. (1970). Editorial, *Nature (London)* **226**, 297.

Babcock, R. F. (1966). "The Zoning Game." Univ. Wisconsin Press, Madison, Wisconsin.

Bentham, R. J. (1968). *Biol. Conserv.* **1**, 11.

Boggs, S. L. (1965). *Amer. Sociol. Rev.* **30**, 899.

Braun, A. C. (1970). *Amer. Sci.* **58**, 307.

Bryson, R. A. (1968). *Weatherwise* **21**, 56.

Bryson, R. A., and Wendland, W. M. (1970). Climatic effects of atmospheric pollution. *In* "Global Effects of Environmental Pollution" (S. F. Singer, ed.), pp. 130–138. Reidel, Dordrecht, Holland.

California State Department of Finance (1964). "California Migration 1955–1960." Dept. Finance, Sacramento, California.

California State Department of Public Health (1960–1964). "California Public Health Statistical Report, Part I. Vital Statistics." Dept. Public Health, Berkeley, California.

California State Department of Public Health (1965–1967). "Vital Statistics of California." Dept. Public Health, Berkeley, California.

California State Department of Public Health (1968). "Deaths from Selected Causes, California Counties by Area." Dept. Public Health, Berkeley, California.

California State Water Quality Control Board (1968). "San Francisco Bay–Delta Water Quality Control Programs Task II-3. Employment, Population and Land–Use Forecasts for the 13 Northern California Counties of the San Francisco Bay–Delta Area, 1965–2020." State Water Quality Control Board, Sacramento, California.

Carlozzi, C. A., and Carlozzi, A. A. (1968). "Conservation and Caribbean Regional Progress." Antioch Press, Yellow Springs, Ohio.

Cheu, S. H., and Waldmann, W. J. (1965). Unusual complications of ventriculo-atriostomy for communicating hydrocephalus in coccidioidal meningitis' *In* "Coccidioidomycosis" (L. Ajello, ed.), pp. 25–30. Univ. Arizona Press, Tucson, Arizona.

Cloud, P. E. (1968). *Texas Quart.* **11**, 103.

Cloud, P. E. (ed.) (1969). "Resources and Man." Publ. 1703, Committee on Resources and Man, National Academy of Sciences, National Research Council. Freeman, San Francisco, California.

Conway, G. R. (1969). "A Basic Model of Insect Reproduction and its Implications for Pest Control." Ph.D. dissertation, Univ. California, Davis, California.

Conway, G. R., Glass, N. R., and Wilcox, J. C. (1970). *Ecology* **51**, 503.

Cook, H. H. (1969). *Arch. Environ. Health* **19**, 560.

Cragg, J. B. (1968). The Present. *Biol. Conserv.* **1**, 13.

Deane, M. (1965). *Medicina Thoracalis* **22**, 24.

Doxiadis, C. A. (1966). "The Developing Urban Detroit Area. Volume I. Analysis." Detroit Edison, Detroit, Michigan.

Doxiadis, C. A. (1968). Remarks reported by G. H. Favre, *The Christian Science Monitor*. Monday, September 9, 1968.

Draper, N. R., and Smith, H. (1966). "Applied Regression Analysis." Wiley, New York.

Egeberg, R. O., Elconin, A. F., and Egeberg, M. C. (1964). *J. Bacteriol.* **88**, 473.

Elconin, A. F., Egeberg, R. O., and Egeberg, M. C. (1964). *J. Bacteriol.* **87**, 500.

Eichler, A. (1968). *Biol. Conserv.* **1**, 95.

Federal Bureau of Investigation (1960–1967). "Uniform Crime Reports." Federal Bureau of Investigation, Washington, D. C.

Fein, R. (1967). "The Doctor Shortage, an Economic Diagnosis." The Brookings Institution, Washington, D. C.

von Foerster, H., Mora, P. M., and Amiot, L. W. (1960). *Science* **132**, 1291.

Forrester, J. W. (1969). "Urban Dynamics." MIT Press, Cambridge, Mass.

Gallup, G. (1969). "The Public Considers its Environment." Poll taken for the National Wildlife Federation, Washington, D. C., February, 1969.

Gerasimov, I. P. (1969). *Nat. His.* **15**, 24.

Glass, N. R., Watt, K. E. F., and Foin, T. C. (1971). Human ecology and educational crises: one aspect of the social cost of an expanding population." *In: Is There An Optimum Level of Population?,* (S.F. Singer, ed.). McGraw-Hill, New York.

Gould, E. M., and O'Regan, W. G. (1965). "Harvard Forest Papers: Simulation." Harvard Forest, Petersham, Massachusetts.

Greenfield, S. (1970). Personal communication to J. L. Wilson, Environmental Systems Group, at meeting at RAND, January, 1970.

Haggett, P. (1965). "Locational Analysis in Human Geography." Edward Arnold, London.

Hamming, W. J., MacBeth, W. G., and Chass, R. L. (1967). *Arch. Environ. Health* **14**, 137.

Hatch, T. F. (1962). Changing objectives in occupational health. *Amer. Ind. Hyg. Ass. J.* **23**, 1.

Hartley, H. (1969). *New Sci. (suppl.)* Energy for the world's technology, *In New Sci.* **44**, 3.

Hendricks, T. A. (1965). *U. S. Geol. Surv. Circ.* **522**.

Herman, S. G., Garrett, R. L., and Rudd, R. L. (1969). Pesticides and the Western Grebe. *In* "Chemical Fallout" (M. W. Miller and G. G. Berg, eds.), Chap. 2, pp. 24–53. Thomas, Springfield, Illinois.

Hickey, J. J. (ed.) (1969). "Peregrine Falcon Populations." Univ. Wisconsin Press, Madison, Wisconsin.

Hickey, R. J., Schoff, E. P., and Clelland, R. C. (1967). *Arch. Environ. Health* **15**, 728.

Howard, L. R. (1968). A cost-effectiveness example: a hypothetical supersonic transport. *In* "Cost-Effectiveness" (J. M. English, ed.), pp. 166–213. Wiley, New York.

Hubbert, M. K. (1962). "Energy resources." Publ. 1000-D, National Academy of Science, National Research Council, Washington, D. C.

Hubbert, M. K. (1966). *J. Petrol. Technol.* **18**, 284.

Hubbert, M. K. (1967). *Amer. Ass. Petrol. Geol. Bull.* **51**, 2207.

Hubbert, M. K. (1969). Energy resources. *In* "Resources and Man" (P. E. Cloud, ed.), Publ. 1703, pp. 157–242. NAS-NRC, Washington, D. C.

Kahn, H., and Wiener, A. J. (1967). "The Year 2000: a Framework for Speculation." MacMillan, New York.

Kryter, K. D. (1969). *Science* **163**, 359.

Landsberg, H. H., Fischman, L. L., and Fisher, J. L. (1963). "Resources in America's Future." Johns Hopkins Press, Baltimore, Maryland.

Lee, E. S. (1964). *Demography* **1**, 56.

McHarg, I. (1968). "Design with Nature." Natural History Press, Garden City, New York.

McKelvey, V. E., and Duncan, D. C. (1965). United States and world resources of energy, *In* "Symposium on fuel and energy economics, joint with division on chemical, marketing, and economics. *Natl. Mtg. 149th Amer. Chem. Soc., Div. Fuel Chem.* **9**, 1.

McMichael, D. F. (1968). *Biol. Conserv.* **1**, 83.

Masnik, G. (1968). *Demography* **5**, 79.

Michael, D. N. (1968). "The Unprepared Society." Basic Books, New York.

Miller, A. (1966). *Demography* **3**, 58.

Miller, A. (1967). *J. Amer. Stat. Ass.* **62**, 1418.

Miller, M. E., and Holzworth, G. C. (1967). *J. Air Pollut. Control Ass.* **17**, 46.

Mitchell, J. M. (1970). A preliminary evaluation of atmospheric pollution as a cause of the global temperature fluctuation of the past century. *In* "Global Effects of Environmental Pollution" (S. F. Singer, ed.), pp. 139–155. Reidel, Dordrecht, Holland.

Mitchell, R. S., Walker, S. H., Silvers, G. W., Dart, G., and Maisel, J. C. (1968). *Amer. Rev. Resp. Disease* **98**, 601.

Moore, C. L. (1965). Analysis and projection of the historic pattern of supply of exhaustible resources. *Natl. Mtg. 27th Operations Res. Soc. Amer. Boston, May*, 1965.

Moore, N. W. (1967). A synopsis of the pesticide problem. *Adv. Ecological Res.* **4**, 79.

Moses, H. (1969). "Mathematical Urban Air Pollution Models." Doc. ANL/ES-RPY-001, Argonne National Laboratory, Argonne, Illinois.

National Center of Health Statistics (1965). "Volume of physician visits, U. S., July, 1963–June, 1964." Series 10, No. 18.

National Center of Health Statistics (1968). "Volume of physician visits, U. S., July, 1966–June, 1967." Series 10, No. 49.

Pappagianis, D. (1970). Remarks on coccidioidomycosis quoted by T. Fourkas *Sacramento (Calif.) Bee*, Tuesday, June 2, 1970, page A10.

Platt, J. (1969). *Science* **166**, 1115.

Radomski, J. L., Deichmann, W. B., and Clizer, E. E. (1968). *Fd. Cosmet. Toxicol.* **6**, 209.

Ratzlaff, E. D. (1969). "Application of Engineering Systems Analysis to the Human Social–Ecological System." M.S. thesis, Univ. California Davis, California.

Rieke, F. E. (1969). *Arch. Environ. Health* **19**, 521.

Rogers, A. (1968). "Matrix Analysis of Interregional Population Growth and Distribution." Univ. California Press, Berkeley and Los Angeles, California.

Ryan, J. M. (1966). *J. Petrol. Technol.* **18**, 281.

Salkin, D. (1965). "Clinical examples of reinfection in coccidioidomycosis." *In* "Coccidioidomycosis" (L. Ajello, ed.), pp. 11–18. Univ. Arizona Press, Tucson, Arizona.

Salmon, J. T. (1969). *Biol. Conserv.* **1**, 255.

Schmid, C. (1960a). *Amer. Soc. Rev.* **25**, 527.

Schmid, C. (1960b). *Amer. Soc. Rev.* **25**, 655.

Schurr, S. H., Netschert, B. C., Eliasberg, V. F., Lerner, J., and Landsberg, H. H. (1960). "Energy in the American Economy, 1850–1975." Johns Hopkins Press, Baltimore, Maryland.

Shurcliff, W. A. (1970). "S/S/T and Sonic Boom Handbook." Ballantine Books, New York.

Slade, D. H. (1967). *Science* **157**, 1304.

Smith, C. E., Pappagianis, D., Levine, H. B., and Saito, M. (1961). *Bacteriol. Rev.* **25**, 310.

Steinhart, J. S., and Cherniack, S. (1969). "The Universities and Environmental Quality." Rept. to the President's Environmental Quality Council, Office of Science and Technology, Washington, D. C.

Stewart, W. H., and Pennell, M. Y. (1961). "Health Manpower Source Book—Eleven Medical School Alumni." U. S. Public Health Service, Washington, D. C.

Stocks, P. (1966). *Brit. J. Cancer* **20**, 595.

U. S. Bureau of the Census (1960). "State of birth." Tables 7 and 12. Washington, D. C.

U. S. Bureau of the Census (1962). "County and City Data Book." Washington, D. C.

U. S. National Health Survey (1960). "Health Statistics—Volume of Physician Visits, U. S., July, 1957–June, 1959." Washington, D. C.

U. S. Public Health Service (1964). "Smoking and Health." Report of the Advisory Committee to the Surgeon General of the Public Health Service, Publ. 1103, Washington, D. C.

Watt, K. E. F. (1968). "Ecology and Resource Management." McGraw-Hill, New York.

Winkelstein, W., Kantor, S., Davis, E. W., Maneri, C. S., and Mosher, W. E. (1967). *Arch. Environ. Health* **14**, 162.

Winn, W. A. (1965). A working classification of coccidioidomycosis and its application to therapy. *In* "Coccidioidomycosis" (L. Ajello, ed.), pp. 3–10. Univ. Arizona Press, Tucson, Arizona.

Wolkonsky, P. M. (1969). *Arch. Environ. Health* **19**, 586.

Wurster, C. F. (1968). *Science* **159**, 1474.

Zapp, A. D. (1962). Future petroleum producing capacity of the United States. *Geogr. Surv. Bull.* 1142-H.

12

Next-Generation Models in Ecology*

A. BEN CLYMER
OHIO DEPARTMENT OF HEALTH
COLUMBUS, OHIO

I. Introduction

This chapter looks forward in time to trends that may be expected to develop in mathematical modeling of ecological systems, particularly the public problems aspects of such systems. Personal biases of the author will become evident in the various sections, but the viewpoint is purposely intended to be provocative to stimulate the reader to pursue and develop his own ideas, if only to counter those presented here.

Although the chapter title and its inclusion in this book appear to

* Based on a presentation at the working seminar on Mathematical Models of Public Systems, San Diego, California, 24–26 September 1970, co-sponsored by Simulation Councils, Inc. and the Department of Electrical Engineering, University of Southern California, and adapted with the permission of Simulation Councils, Inc.

restrict the treatment to "ecological" areas, narrowly interpreted, in actuality a broad spectrum of public problems—human ecological problems—is addressed. Ecology is adopted as a perspective on all public systems, and at the same time public problems are regarded as a major emphasis for ecology, one that will provide a prime motivation for future work in ecological systems analysis and simulation. Man, as a semirational, semidisciplined exploiter, manager, and manipulator of the ecosystems he occupies, and yet also as an utterly dependent species of animal with characteristics reaching into the top trophic level, cannot afford to exclude himself from the "ecomodels" that he constructs to represent nature.

II. Current Status of Ecomodeling

Developments to the present serve to indicate both directions and rates of change in the ecomodeling art. The references cited in this section should not be regarded as complete, but only as examples of past emphasis and methods. More comprehensive bibliographies are provided by Davidson and Clymer (1966), IBM (1966), Schwartzman (1970), Watt (1966, 1968), and Kadlec (1971). The subject of radioisotope movements in ecosystems has by itself generated an enormous literature, entries into which can be made with the help of AEC contractors, e.g., the Ecological Information and Analysis Center, Battelle Memorial Institute, Columbus, Ohio.

Table I indicates a variety of ecomodeling applications up to the present time, and Table II summarizes the different kinds of mathematical models that have been used. "Compartment modeling" has been most popular, involving descriptions of the rate of change of some time-dependent variable by ordinary differential equations, obtained by subtracting compartmental outputs from inputs. Ecosystems are partitioned into compartments in a variety of ways, e.g., portions of a forest (roots, boles, branches, leaves, litter, soil, etc.), sets of species, age classes, reaches of a river, depth strata in a water column or in soil, and geographic regions, etc. This approach is so straightforward and well adapted to the interests and methodology of current ecological research that few ecologists have seen fit to "graduate" to other types of models, models with more demanding mathematics.

Some impression of the historical growth of ecological modeling may be obtained by recording the years in which papers with at least one mathematical model were published. Based on about 60 references, a rate of the order of one paper per year was seen in 1957. By 1964, this rate

TABLE I

Ecosystem or phenomenon modeled	References
Hydrology	Clymer and Bledsoe (1969), Cooper (1969), Cooper and Jolly (1969), Hufschmidt and Fiering (1966), Van Dyne (1969b), Woo *et al.* (1966)
Predator–prey phenomena	Holling (1966a, b)
Aquatic ecosystems	Beverton and Holt (1957), Cushing (1968), Davidson and Clymer (1966), Patten (1965), Silliman (1967), Schaefer and Beverton (1963), Watt (1956)
Radionuclide fluxes and concentrations	Cowser *et al.* (1967), Olson (1963a, b, 1965), Turner and Jennrich (1967), Witherspoon *et al.* (1964)
Forests	Olson (1963a, b, 1964)
Population dynamics	Brennan *et al.* (1970), Pennycuik *et al.* (1968), Pielou (1969), Smith (1963), Taylor (1967)
Pesticide fluxes and concentrations	Robinson (1967)
Photosynthesis	Duncan *et al.* (1967)
Reproductive behaviors	Bartholomew and Hoel (1953)
Grasslands	Bledsoe and Jameson (1969), Goodall (1969), Schwartzman (1970), Van Dyne (1969a)

TABLE II

Mathematical type of model	References
Partial differential equations	Clymer and Bledsoe (1969), Section V herein
Ordinary differential equations	Atkins (1969), Bledsoe and Jameson (1969), Cowser *et al.* (1967), Garfinkel (1967), Garfinkel and Sack (1964), Olson (1963a, b 1965), Patten (1971), Van Dyne (1969a)
Stochastic and probabilistic	Bailey (1967), Bartlett (1960), Garfinkel *et al.* (1964), Neyman and Scott (1959)
Theory of sets	Robinson (1967)
Input–output analysis	Ayres (1964)
Functions of several variables	Holdridge (1947)
Linear programming	Barea (1963)

had increased to seven per year, a level more-or-less maintained until 1969. The papers of this period were characteristically concerned with some simple model as a basis for enthusiastic predictions about the potential benefits of ecomodeling. A new phase of growth ensued in 1969, with grants and contracts becoming available to support more substantial modeling efforts. Now, many universities are being funded to train ecomodelers, particularly as a result of the International Biological Program (IBP). It can thus be anticipated that the second period of growth may reach a higher rate level within five years, and the time constant for doubling will be much shorter, perhaps two years.

Most early ecomodels were stimulated by the studies of Lotka and Volterra in the 1920s on the dynamics of two interacting animal populations. Nuclear weapons testing motivated modeling to predict radionuclide fluxes through ecosystems and into man. More recently, interest in ecological modeling has been stimulated by problems like human population growth, future food resources, pollution, peaceful uses of atomic energy, resource management, and a whole plethora of complex public problems involving ecosystems.

To date, relatively little productive work has been done toward developing a more powerful methodology for ecological modeling. Present approaches are discussed in Brennan *et al.* (1970), Holling (1966a, b), Levins (1966), Schwartzman (1970), and Watt (1968). There is ferment within the U.S. IBP's Analysis of Ecosystems project, as evidenced in the proposals and progress reports of the Grasslands Program, mainly, but by and large the literature appears to be in a methodological rut.

III. Trends in Ecomodel Evolution

Ecological models are evolving profusely in many directions. In organizing the following discussion, some trends already noted by Van Dyne (1969a, p. 353, Fig. 11) have been drawn upon.

Ecomodels are getting bigger rapidly, growing in both number of equations per model and number of terms per equation. This trend has several important consequences. First, in larger models the variables are more richly interconnected, there are more feedback loops, etc. with the result that a wider variety of system responses is generated. The feedbacks tend to lend stability. Secondly, "problem bandwidth" (the ratio of largest to smallest inverse time constants represented in individual equations) is increasing, and with it the demand for more computer bandwidth. This problem bandwidth trend may create demand for computer bandwidths now available only in hybrid computers.

Nonlinearities are entering ecomodels with greater frequency and more variety. Early models contained only products of variables as sources of nonlinearity; now there are thresholds, limits, discontinuities, general nonlinear functions of one or more variables, and conditional logic.

Empirical equations are giving way to more meaningful representations of mechanisms. As realism increases, so do predictive power, scope of applicability, and access to parameters of control.

Probability is increasingly important, past models having been mainly deterministic. Now there may be time-varying coefficients based on deterministic seasonal changes over an annual cycle, but with a stochastic term added. Or, discrete events may be introduced at randomly distributed times. Some models may be driven by forcings that have a probabilistic component. Others have stochastic effects adapted to Monte Carlo simulation—selecting parameters from distributions and then statistically analyzing results of a set of simulation trials.

The scope of mathematics being employed in ecological modeling is already broadening (Table II). Dimensional analysis is an example of something for which greater use may be anticipated in the future, although the only reference to date, apparently, is Van Dyne (1969b, p. 252). Wider use of differential equations, both ordinary and partial, is certainly to be advocated. Partial differential equations have received surprisingly little application, with the exception of a few studies in fishery dynamics (see Section V and reference citations therein). Hydrology is another area of potential application, e.g., in connection with processes of runoff and infiltration on different terrains (Clymer and Bledsoe, 1969). Vertical diurnal movements of fish and plankton are also appropriate for description by partial differential equations, with depth and time as independent variables. Shrub invasions in semiarid grasslands is another. More generally, partial differential equations could be used to model spatial dynamics of any given species in a nonuniform environment. "Assumed mode methods" for solving systems of partial differential equations (Braun and Clymer, 1963), which should be especially effective, have apparently never been employed in ecology.

Modeling efforts of longer duration are beginning to materialize. The earliest papers were developed along the lines of: construct model, find data in the literature, program model, do a few computer runs, adjust some parameters, do a few more runs, and write the paper. Now, studies ranging over several years are beginning to emerge, involving both modeling and measurement and their progressive reconciliation. This is an advance, but it still falls short of the hierarchical approach which will be discussed below in later sections.

Ecological models are becoming more detailed. Population models, for example, now attempt to account in some degree for the mechanisms of physiology, behavior, and growth of individual organisms. It is no longer sufficient to finesse such details, e.g., in the case of representing crowding effects by a negative second-degree term in the logistic equation. As a consequence of attention to mechanisms, more disciplines are being drawn in to contribute specific submodels for use within ecomodels, and the latter are tending to become more systematically organized and hierarchically structured. This trend is facilitated further by the nesting of "macros" in simulation languages (Brennan *et al.*, 1970), and by increasing use of modern simulation languages such as CSMP (e.g., Chap. 1 of Vol. I). McLeod and Defares (1962) have successfully employed this same approach with analog computers in biomedical applications. A single-echelon model is too narrow to be effective in present-day terms.

Contemporary ecomodels are helping to fill gaps left by earlier efforts. New ecosystems are being modeled, on longer and shorter time scales, as are multispecies interactions and the dynamics of energy and material movement over different paths and cycles in ecosystems. Transfers between ecosystems have already been modeled (Cowser *et al.*, 1967), and in general higher echelons of scope are being approached, including the global level which is already being modeled in various aspects by a number of different groups (see, e.g., the description of world simulation projects in Anon. (1970).

The increasing development of models which treat similar aspects of ecosystems has produced a trend toward generalization of such processes as grazing and predation (Holling, 1966a,b). The availability of generalized models will facilitate the treatment of special cases by specialization of the models. Also, generalized models often represent alternative mechanisms of a process, making it possible to assess by methods of parameter identification which mechanisms singly or in combination are actually operative in a given experimental situation.

There is clearly a trend away from academically motivated models to models that concern the manipulation or management of ecosystems for man's use. This is evident in an increasing literature on management of such natural resources as forests, fisheries, watersheds, and ranges (e.g., Van Dyne, 1969a,b; Watt, 1966, 1968). Natural resource models must differ from descriptive models of the more academic nature in providing convenient access to parameters and inputs that are likely to be manipulated in management procedures.

The trend in modeling public systems is similar. The integrated management of man–nature interactions is so complex in its multifarious aspects that adequate planning is hardly possible without some kind of modeling. Problems of pollution, urbanization, and industrialization are mind-boggling in their social, economic, and ecological aspects when considered as different parts of the same whole. In problems of this magnitude the funding tends to follow fright, and a more sensible means of ongoing support needs to be developed. The demands of viewing public systems in their total ecosystem context have impressed upon ecologists the need for better understanding of the physical environment. This need is being readily met, since physical system modeling already has a long and successful history. The carryover to ecosystems includes models of relevant aspects of hydrology, meteorology, heat transfer, radiation geometry, etc.

The growing importance of problems in private and public management has underscored the need to develop value functions (objective functions, performance functions, etc.) which are to be optimized by proper management procedures. A value function quantitatively represents all considerations of importance to a problem in mathematical formulations that are consistent with the role of each factor. The design of value functions for ecosystems is in its infancy as an art, despite the fact that the variables to be included should be of major importance in management planning. As the art develops and value functions begin to appear more consistently in ecomodels, there should be a trend toward semiautomatic computer optimizations under human supervision. Meanwhile, the standard procedure will be trial-and-error, with error correction done offline.

The representation of man in ecological models has almost never been adequate in terms of realism. Man deserves to be included as a species in every ecosystem that is modeled as an applied problem, and all of his interactions with other components of the ecosystem should at least be considered before being ruled out of the model. In the past, man has made an appearance in ecomodels as an exogenous entity, i.e., as an off-stage forcing function presumed in advance to be known. It is much preferred to include man within the model so that his responses can be generated by the model rather than a priori. Human ecology at the present time cannot provide the needed information to make this possible, and neither can sociology, economics, or political science. A new synthesis of existing information is necessary in order to integrate concepts of human ecology with the variety of social, economic, and political models now available. The development of this synthesis needs to be accelerated because it is not occurring rapidly enough left to itself. A recent UNESCO report provides some useful insights in this area (UNESCO, 1969).

As models of ecological systems grow to the point of becoming unwieldy, a type of model which is likely to become popular is one with a modular organization. That is, some of the parts of the model will be replaceable by other versions, in more or less detail, to make it possible to represent a shifting focus of concern. Also, hierarchies of smaller models will eventually become preferable to a single large model as a more efficient means of representing a variety of man–ecosystem interactions (Clymer, 1969a,b, 1970; Clymer and Bledsoe, 1969). At the present time there is little indication in the ecological literature of progress in the direction of developing model hierarchies, although the notion is implicit in Fig. 1.1 of Levins (1968). In the following section some general considerations for hierarchical modeling are presented, followed by a number of examples of ecomodels and hierarchies of ecomodels described and discussed in some detail.

IV. Hierarchical Modeling

The concept of hierarchies in modeling is consistent with the hierarchical nature of natural systems. In the area of public problems there exists a hierarchy of management, either public or private, which should be performing a hierarchy of planning activities within the framework of a hierarchy of models. The output of such activity should be a continuous stream of mutually reconciled plans, organized hierarchically, and subject to change as conditions change, especially on the lower echelons. Such planning should be pursued at every level of government and industry. The hierarchies of planning efforts should extend outward from local and regional considerations to form a global long-range planning function which is constantly adjusted to moving planning horizons. Right now, a long-term transient must be designed and implemented to contain the present explosions of population and technology and gradually bring them into balance in the form of wisely governed moderate rates of change and equilibria. The hierarchy of planning efforts should be disciplined by a corresponding hierarchy of value functions, few of which now exist in any form.

As an example, consider the problem of optimizing a ranching operation, which is a business directly connected to an ecosystem. Figure 1 illustrates a hierarchy of some of the more obvious value functions that are relevant. The rancher might elect to have a different employee responsible for each of the functions, in which case the diagram would also serve as an organization chart and a chart of the planning process. If the ranch were represented by a model, the model would have a hierarchical internal structure. It would permit quantitative evaluation of numerous activities which might be incorporated into a hierarchical ongoing plan for the ranch. Examples of activities to be considered are stocking, mowing, seeding, planting, shrub control,

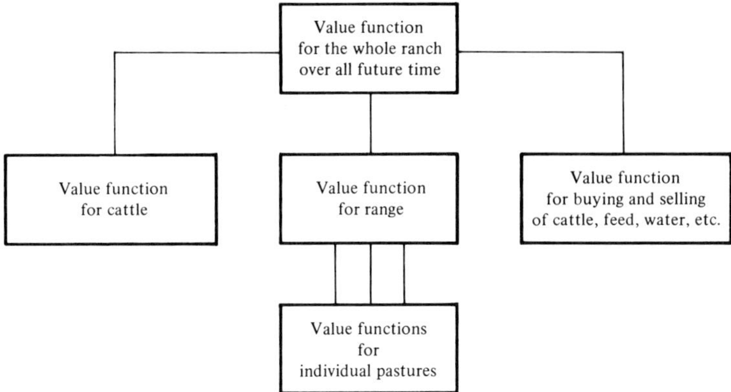

FIG. 1. Hierarchy of value functions for a ranch.

supplementary feeding, cattle rotation in pastures, control of insect and mammalian pests, refencing to redefine pastures, changing the herd size and the mix, etc. Any given activity might be favorable in terms of one value function but unfavorable in terms of another on the same echelon. Reconcilation of such conflicts would occur on the next higher echelon which contains both. The submodels representing each particular plan would have to be simulated to find the best resolution. The value functions should be nonlinear to account for the fact that some considerations may be unimportant until a danger point is neared, e.g., the length of grass blades approaching the point of becoming overgrazed. The rancher's responsibility would be to develop a suitable hierarchy of value functions, models, and a planning procedure, at least verbally if not mathematically, in order to arrive at good management decisions.

In this example the rancher's problem is not too different from that of any company or govermental unit that has responsibility for a resource. In order for effective planning to be developed and implemented, the modeler's responsibility is to produce appropriate hierarchies of ecomodels. But the concept of each hierarchy has first to be elaborated by individuals with expert knowledge of the problems.

One approach to conceiving a model hierarchy is to draw a map of some possible models. Although the procedure will be illustrated in more detail later (Section VI), the general way to go about it is to prescribe two dimensions of the ecosystem which together best specify the contents and behaviors that are known to be characteristic. For example, a preliminary design for a hierarchy of models concerning an aquatic ecosystem might include a sketch of a rectangular space defined by two selected variables such as trophic level and time scale. The trophic level scale might be partitioned into segments denoting phytoplankton, zooplankton, larvae, fish, and man. The time scale, preferably logarithmic, could range from an hour or less to a century or more. This scale might be divided into segments covering an hour to a day, a day to a month, a month to a year, a year to a decade, a decade to a century, etc. Any region of the plane can be thought of as suggesting some type of ecomodel. If there were five trophic levels and five time segments, then the 25 resultant regions would each be suggestive of a model having some interest to somebody. Two or more adjacent regions might be considered as a single model. The upper right-hand corner (man, long-time scale) indicates a long-term aquatic resource management model. The plankton blocks in the center time scale suggests a model of plankton population dynamics such as that developed by Davidson and Clymer (1966). By writing words into each block appropriate to the location of the region in relation to each coordinate axis, capsule descriptions of 25 ecomodels are obtained which can then be further developed through block diagrams and equations. A useful exercise at this point would be for the reader to draw up such a table of models for an area of interest to him. From such an exercise it will become obvious that no single model is adequate to represent all aspects of an ecosystem that may be of concern. Time scales range from long-term management down through succession of species; annual cycles of growth and productivity; diurnal oscillations of photosynthesis, respiration, and transpiration; and vertical gradients of water vapor, oxygen, and carbon dioxide. The trophic levels range downward from man to animals, birds, insects, plants, and microorganisms.

Models may be variously detailed in terms of the number of trophic levels included and also in how the processes between or within organisms are represented. Numerous types of dependent variables can serve as outputs of models, such as biomass, growth rates, concentrations, and fluxes of various ions, species populations, allowable harvests of resources by man, vertical stratification in terrestrial and aquatic communities, spatial distributions of organisms, temperature or gas concentration vertical profiles, reproduction rates, litter composition profiles, etc. Thus, a number of distinct but related models would be appropriate. Such a set of models might be thought of as a hierarchy with the vertical axis representing some measure of model scope. For example, a total ecosystem model might be at the top, with various echelons of subsystems ranging downward toward the bottom. Possibly a word like "pyramid" would be more appropriate than hierarchy, but a pyramid is a twofold hierarchy and so the term "hierarchy" will be used loosely when it is not clear what the structure of a well-designed set of models should be.

One example of a practical problem involving ecosystems in which a single model would have been completely inadequate or intractably complicated was the assessment of safety in constructing an isthmian sea-level canal with the use of subterranean nuclear detonations. The problem was complicated by the fact that many different radionuclides released by the explosions had to be considered in terms of their passage through the different ecosystems of the region, and particularly as they entered the complex food web leading to humans living in various localities with different life styles and diets. To make this problem tractable it was necessary to reduce it to a number of smaller problems, each of itself posing an ecomodel design task. Time scales varied due to the broad range of half-lives of the radionuclides, and this dictated different ecological processes which become of importance on different time scales. The strategy finally evolved was to screen the different radionuclides for importance, beginning with very simple and conservative models, and gradually developing more detail and accuracy for the most important radionuclides and food chains. Human ecology had to be considered. Although no clear hierarchy of models had emerged by the end of the project, had such an approach been applied at the outset it would have provided a powerful method of attack. But this is hindsight.

Another public problem area which requires a hierarchical modeling approach is the urban system. The city is a subsystem of the total human ecosystem. Within it are hierarchies of other subsystems such as the health care services discussed in Sections VIII and IX. Although models treating such special aspects of urban systems have been developed, models which consider the entire system have been few (e.g., Forrester, 1969). Notably missing are adequate models of the psychological and social processes responsible for the traumatic behavioral phenomena observed in cities in recent years. Economic aspects seem to be most easily modeled. Models which employ ordinary differential equations appear to offer the best promise in representing most features of urban systems, as, for example, the model of a community total health care system presented in Section IX. Forrester (1969) also employs ordinary differential equations to characterize housing construction, housing occupancy, population, employment, and other socioeconomic dependent variables. A recent review of the techniques is provided by Axelband *et al.* (1969).

The prescription that urban models should be developed as a hierarchy of ordinary differential equations models results from consideration of human ecology. Other approaches include applications of dynamic analysis and control theory, as, for example, the work of Krendel (1970) on the impulse response of a city government to citizen complaints. A major bottleneck in achieving more cogent and realistic urban models is the present inability to characterize the emotions, behaviors, communications, and other interactions of people. The model concept outlined in Section VII may stimulate others to develop capabilities for modeling the emotional phenomena in cities which underlie many social problems such as alcohol, drugs, crime, riots, divorce, ghetto environment, etc.

The most inclusive hierarchy of all, of course, is one which would attempt to represent man–environment interactions on a global scale. Some ideas on world modeling have already been presented elsewhere (Clymer, 1970), and the reader interested in the modeling of world-wide problems of all kinds should consider participating in the Simulation Councils', Inc., World Simulation Organization.*

The following sections will outline some types of models which may be foreseen as "next-generation" ecomodels.

V. A Fish Population Ecomodel

The literature on fish population dynamics is already substantial (e.g., Beverton and Holt, 1957; Cushing, 1968; Ricker, 1954; Schaefer and Beverton, 1963; Silliman, 1967; Watt, 1956). This section outlines the derivation of a fish population ecomodel to illustrate a possible application of partial differential equations to ecological modeling.

Consider a finite body of water assumed to be well mixed. This assumption precludes the details of predators searching for prey, which are important when prey population densities are small, and in effect implies an abundance of food. Different species are not distinguished.

The dependent variables will be taken to be population and biomass as functions of size or age and time, If it is assumed that size and mass of an individual fish are known functions of age, then there is really only one dependent variable, the mass of fish per unit age class as a function of time and age. Thus, age and time are the two independent variables. The population size per unit age class can be derived from the biomass per unit age class if the mass of an individual fish as a function of age is known.

Let $m(t, A)$ denote fish biomass per unit age class as a function of time t and of age A. Let $p(t, A)$ be the population size per unit age class. Let $\rho(A) = m/p$ denote the biomass of an individual fish of age A.

* Contact John McLeod, Executive Director, World Simulation Organization, Simulation Councils, Inc., P.O. Box 2228, La Jolla, California, 92037.

Then, the derivation task is to represent m by an equation depicting death due to natural causes, death by predation, deaths by catches and kills by man, growth by eating while aging, and any other effects that seem appropriate.

A common mathematical approach would be to group age classes at the outset, then derive an ordinary differential equation for each age class for the net rate of entry into the class, writing an entry or exit term for each effect considered. An alternative approach is to derive a single partial differential equation instead of a set of ordinary differential equations, applying the same kind of reasoning. Given the partial differential equation, it can either be reduced by finite-difference techniques to a set of ordinary differential equations (not necessarily the same as those that would have been obtained by lumping), or, alternatively, modal methods (Braun and Clymer, 1963) could be used to derive ordinary differential equations for modal amplitudes as dependent variables. Or, the partial differential equation could be programmed for machine solution by any of a number of available methods.

Consider the entries and exits from an age class of width ΔA centered at age A. The net rate at which biomass enters the class, which is the left-hand side of the desired partial differential equation, is $\partial(m\,\Delta A)/\partial t$ or $\Delta A\;\partial m/\partial t$. On the right-hand side is a negative term representing natural mortality, which may be assumed to be a known age-dependent death rate function $f_{\mathrm{d}}(A)$ multiplied by the class biomass $m\,\Delta A$. An additional term is needed on the right-hand side to represent net entry into the class by aging. Gross entry by aging is $m(A)\;\partial A/\partial t$, or just $m(A)$, since age advances exactly with time. Gross exit by aging is $m(A + \Delta A)$, or $m(A) + \Delta A\;\partial m/\partial A$, for small ΔA. The net entry rate is $-\Delta A\;\partial m/\partial A$.

Predation may be formulated by assuming that a fish of age A eats a mass $M(A)$ per unit time. Then, all fish in a unit class of age A consume prey at a rate pM, where p is the population size of the age class. Since not all of the biomass consumed is converted into increased body mass, the function $M(A)$ must exceed at all ages the slope of the function $\rho(A)$. The food supply pM is always available by virtue of the initial assumptions.

The mass flow of foods pM may be assumed to come from smaller fish. The size of a predator's mouth and digestive tract limits from above the size of fish prey eaten. This size is also limited from below because it takes more effort to catch a given mass of smaller fish, which usually reside in regions not desirable for larger fish to enter and which are probably less satisfactory in terms of food per unit mass. Therefore, a fish of age A will eat prey fish having a sharply limited range of lower ages. For simplicity, it can be assumed that the actual distribution of food fish ages is a pulse at the age such that $\lambda(A)\,\rho(A)$ is the mass of one fish eaten by a fish of age A. Thus $\lambda(A)$ is the ratio of prey mass to predator mass. Even if the range of food fish ages is finite, it could be represented by the mean age.

An expression for the biomass lost to predation by the age class of width ΔA centered at A can now be derived. Let A' denote the age of predators feeding on fish of age A. A' is determined by the relation $\lambda(A')\,\rho(A') = \rho(A)$, which follows from the following analysis:

$$\begin{aligned}\rho(A) &= \lambda(A')\,\rho(A')\\ &= \lambda(A+\delta A)\,\rho(A+\delta A)\\ &= \lambda(A)\,\rho(A) + \left(\lambda\frac{d\rho}{dA} + \rho\frac{d\lambda}{dA}\right)\delta A + \cdots,\end{aligned}$$

whence

$$\delta A \doteq \frac{[1-\lambda(A)]\,\rho(A)}{d(\lambda\rho)/dA},$$

and

$$A' = A + \frac{(1-\lambda)\rho}{d(\lambda\rho)/dA}.$$

The required term on the right-hand side of the partial differential equation for biomass lost to predation is

$$\begin{aligned}-p(A')\,M(A')\,\Delta A &= -\left\{p(A)\,M(A) + M(A)\left[\frac{(1-\lambda)\rho}{d(\lambda\rho)/dA}\right]\frac{\partial p}{\partial A}\right.\\ &\qquad \left. + p(A)\left[\frac{(1-\lambda)\rho}{d(\lambda\rho)/dA}\right]\frac{dM}{dA}\right\}\Delta A\\ &= -\left\{pM + \left[\frac{(1-\lambda)\rho}{d(\lambda\rho)/dA}\right]\frac{\partial}{\partial A}(pM)\right\}\Delta A.\end{aligned}$$

Catches and kills by man are usually age dependent. Catches by means of a net would be representable as a distribution corresponding to a random sample of the population of sufficient age to be caught by a specified mesh size. Moreover, since both catches and kills can be regulated according to some schedule and policy, these effects could be time dependent. There might also be a random component to account for the stochastic nature of both fishing and other population-influencing factors such as pollution. Without prescribing a particular function, an arbitrary function to represent catches and kills can be denoted

$F(A, t)\,\Delta A$. The entire partial differential equation may then be written, substituting $p = m/\rho$ to eliminate p as a dependent variable in favor of m,

$$\frac{\partial m}{\partial t} = -\frac{\partial m}{\partial A} - f_{\mathrm{d}}m - \frac{mM}{\rho} - \left[\frac{(1-\lambda)\rho}{d(\lambda\rho)/dA}\right]\frac{\partial}{\partial A}\left(\frac{mM}{\rho}\right) - F(A, t).$$

This equation is highly nonlinear and for many purposes it may be desirable to linearize it either partially or completely. General solutions in closed form would appear to be infeasible, but certain special cases may yield to analytical approaches. For example, if λ is assumed constant (independent of age), ρ and M are taken as proportional to age, and deaths due to natural causes and catches and kills by man are considered insignificant compared to predatory deaths, then the steady-state population distribution is such that biomass m per unit age class plotted log-log against age is a line with slope $\lambda/(1 - \lambda)$. In other instances it might be preferable to solve the partial differential equation by simulation. This would require setting boundary conditions at the ends of the age scale. At the low age end it might be assumed that the fish creation rate was simply related to the population capable of reproduction. Or, alternatively, one might set $m = 0$ beyond some assumed maximum possible age, or just let anything happen since the partial differential equation is only first order with respect to A so that one boundary condition is sufficient.

It would be possible to introduce many refinements in a variety of different combinations. Each of the assumptions could be relaxed in a number of different ways. For example, two species could be considered and represented as two coupled partial differential equations. Or the distribution of food size might be made more realistic. Or earlier stages in the fish ontogeny, eggs and larvae, could be incorporated into the model each represented by a partial differential equation. Or other kinds of food such as insects or plankton might be incorporated either as a specified input to the system or as another population formulated in terms of its own partial differential equation. Through such ramifications it would be possible to lead outwardly from the model and gradually approach contact with other members of a hierarchy of models for the aquatic ecosystem. In addition, it would be possible to provide control points in the model, parameters representing possible manipulations of the real system such as changes in water quality, water temperature, fishing restrictions, stocking, etc.

One particular modification of the model that would seem to deserve early attention would be to relax the assumption of adequate food. Then $\rho(A, t)$ would have to be treated as another dependent variable and represented by a second partial differential equation expressing the growth of body mass as a net result of hindered predation and biomass losses accompanying the increased effort to search for and capture prey. $M(A)$ would then have to be treated as $M(A, t)$, a variable related to predation success and determined by the relative populations of prey and predators. Another possibility suggested by a growth model is to consider seasonal variation in metabolism as a function of water temperature which varies with the time of year, influencing body mass, food requirements, and level of activity. Another effect is that the fraction of food eaten which is assimilated increases with a decrease of $M(A, t)$.

VI. Mammalian Sociodemographic Ecomodels

This section considers some possibilities for a hierarchy of ecomodels in connection with mammals. The hierarchy suggested interrelates physiology, personality, behavior, sociometry, demography, development, and other relevant aspects of animal ecology. As will become evident, it is not feasible to incorporate more than a few of these aspects in any one model. Thus, a hierarchy is necessary. Such a hierarchy should be extendable to the ecosystem level, and also should enable possible control actions of man to be incorporated.

Some possibilities for dimensions of the space in which the model hierarchy is to be developed include the following:

(1) number of animals involved (1, 2, or larger natural groups),
(2) number of species involved (from 1 to possibly hundreds),
(3) trophic level (levels in a canonically drawn network of who eats whom),
(4) time scale (run duration, for characteristic frequency).

To make a two-dimensional space, these four parameters must be collapsed in order to allow mapping of model concepts. Since the first three are similar in representing different aspects of system complexity, the latter together with time scale represent two usable dimensions (Fig. 2). Each block in the diagram corresponds to one or more possible models, capsule descriptions of which can easily be imagined and written down since they are directly suggested by the combinations of row–column headings. That is, the processes that are relevant in each level of organization and on each time scale are fairly well suggested. There are many natural lateral and vertical dependencies between adjacent models in the diagram, which makes it easy to interface models and

	Diurnal	Annual	Multiyear
Populations containing at least two species	Daily predation by a population	Seasonal dietary shifts among prey species; annual control cycle by man	Animal succession; population evolution; pest abatement campaigns
Colony	Diurnal Pattern of encounters in a colony	Annual colony behavior pattern; reproductive behavior cycle	Population cycles; social structures; differentiation of behavior types
Interacting individuals	Interaction episodes	Annual behavior pattern of a mating pair	Long-term behavior shifts of interacting individuals
Individual animal	Daily behavior and allocation of one predator or herbivore; habitat suitability	Seasonal changes in metabolism or growth; annual cycle of herbivore diet	Lifetime growth; shifts or endocrine systems and of metabolism under social stress

FIG. 2. Map of some mammalian ecomodels.

thus develop a hierarchy. The use of a simple scheme such as this makes it possible to develop concepts systematically and of itself is an effective technique for generating novel and useful ideas.

It would now be possible to provide several pages of discussion for each of the blocks in Fig. 2 to illustrate how meaningful models might be developed in a hierarchical fashion. This would be inappropriate to the purposes of this chapter, however, and therefore the following selected comments will suffice. The models suggested by the row labeled "interacting individuals" might be approached from the viewpoint that will become evident in the next section, dealing with a model of human personality. At least some of the short- and long-term phenomena noted by Calhoun (1962) for a crowded rat colony seem approachable through a model designed around the variables discussed in Section VII. The lower left-hand block of Fig. 2 has been modeled by Holling (1966a, b). The column headed "multiyear" could be represented by a single model capable of explaining the population cycles of various wild mammalian species. The bottom three echelons at least seem essential to such an explanation (Clymer *et al.*, 1965).

VII. A Model of Human Personality and Interactions

One of the most neglected areas with relevance to public problems is human personality and its interaction with both the physical and social environment. In attempting to model personality factors the hierarchical approach would seem appropriate since various degrees of detail on a number of different time scales should be considered. The nature of a suitable hierarchy is uncertain at the present time, and perhaps to stimulate its development a tentative approach is described here at a level of concern to most people, namely, what occurs between two interacting personalities over a short period of time, say an hour. Considering the problem on this time scale will produce some ideas about how related phenomena of both shorter and longer duration might be approached. Beginning with a pairwise interaction also yields a basic building block for interactions in groups of any size.

The complex behavior of an individual interacting with another can be reduced to two basic time-varying parameters, "giving" and "taking." Each parameter can be scaled to vary between $\pm 100\%$, and if the parameters are regarded as axes of a coordinate system, then together they define a closed two-dimensional space, the "give-and-take plane." A person's attitude or behavior toward another person may be plotted instantaneously as a point in this square space. Two interacting people have two points as instantaneous loci. Any dynamic interactive process can be plotted as two trajectories.

The referents of giving and taking can be any resources at all available to the persons involved, such as knowledge, skill, love, status, power, etc. A detailed lower-echelon model would operate at this level, but the model concept under discussion would treat giving and taking in the abstract without concrete content or detail of communication. Different behavioral patterns or moods can be plotted quite reasonably as points in the give-and-take plane.

Figure 3 illustrates such a plane with *giving* as the ordinate and *taking* as the abscissa. The nine extremes are readily identifiable with familiar forms of behavior or stances, both of children and adults. Once the connotations of each region in the plane have become familiar, insights into interpersonal interactions can be obtained by observing behavioral trajectories of two simulated people as some form of computer-graphic output. Consider, for example, the "games people play" in which the two trajectories dart all over the plane as game after game is played out. All sorts of psychodynamic endeavors can be charted and later analyzed in terms of who did what to whom at each stage and what consequences resulted and why. On a longer time scale the progress of a developing personality disorder might be observed within some dynamic interpersonal context.

Only the behavioral aspect of interaction is depicted by the give-and-take plane. Important dependent variables which mediate this behavior are not apparent. A hypothetical model of personality which represents

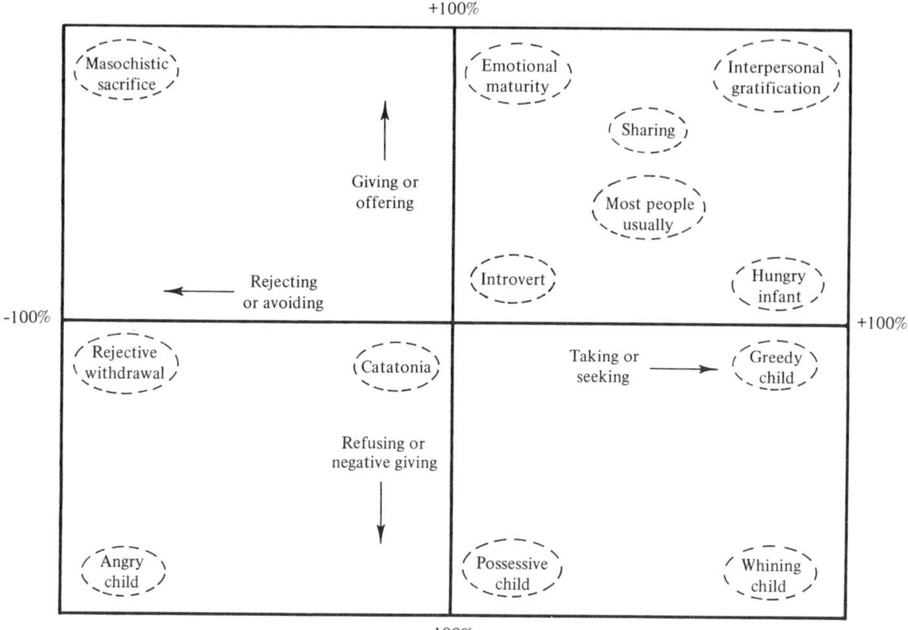

FIG. 3. The give-and-take plane.

some of these factors is illustrated in Fig. 4. The blocks contain processes and arrows convey process outputs to other blocks or to the outside world. Broken arrows represent relatively slow effects which can be considered constant on a short time scale. The solid-line arrows indicate rapid transfers, with the loops among the top four blocks being especially fast, occurring faster than real time in the imagination. The amount of planning and evaluation occurring in the upper blocks depends upon the tempo of the situation.

The contents of the blocks can only be described vaguely, although ultimately they must be represented by equations or algorithms. The value function is an algebraic equation each term of which is a product of a "value" (slowly changing parameter considered by a person representing the relative importance or desirability of the associated variable) and a variable denoting some aspect of the contemplated response. The value function allows a number to be given to each alternative response, the highest number representing the most desirable response.

The "response planning" block generates each contemplated response in fast time, outputs the relevant predicted consequences, and transfers them to the value function.

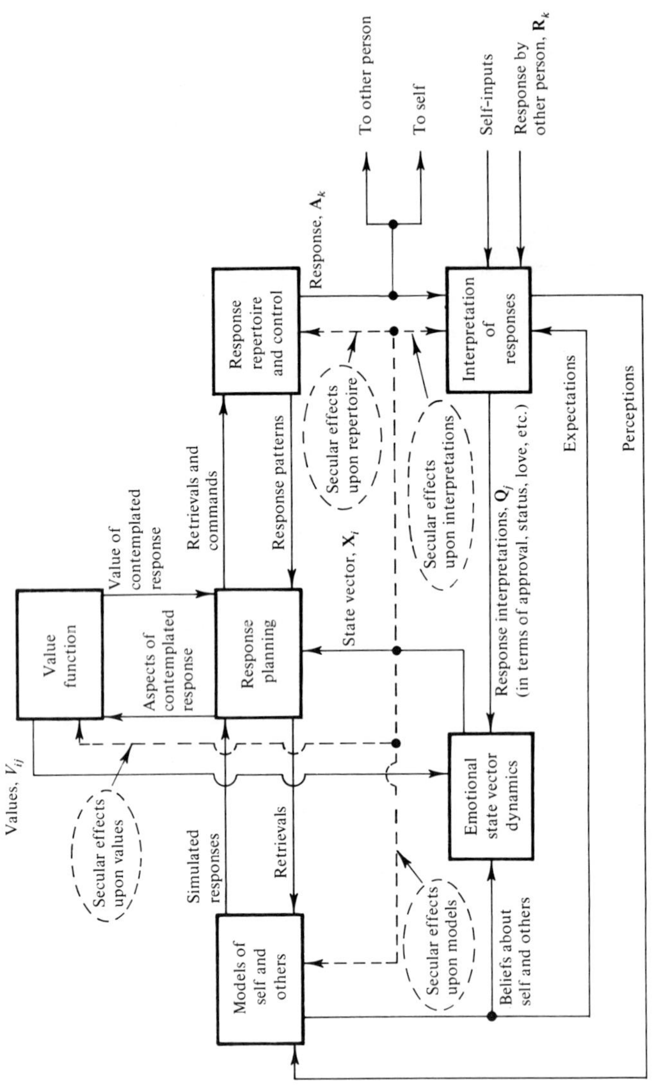

FIG. 4. A model of personality.

The "response repertoire" is a file that can be interrogated by an entry code determined by the emotional state vector so that the responses contemplated are restricted to those which are in some sense appropriate to need, which is a column vector. After a response is selected and initiated, the response planning block can proceed to work on the predicted and early perceived counter response and so begin planning the next response. Thus the response planning block may be in continuous use throughout a simulation trial.

The block for "emotional state vector dynamics" represents a set of ordinary differential equations whose dependent variables are the elements of the emotional state vector (such as needs, feelings, etc.). This vector may be influenced by stimuli from the other individual and also by self-interpreted responses, for example the enjoyment of one's own irony. There is also a loop with the lower centers of the central nervous system and the homeostatic control system of the body which is involved in psychosomatic phenomena. An engineer would regard the emotions as conscious concomitants of a column vector of error signals which the individual endeavors by his behavior to drive to zero. These error signals are differences between what is perceived and what it is desired to perceive, both in the state of one's own body (i.e., needs) and in one's concepts or images of oneself, other people, and the world in general.

The conception presented in Fig. 4 may not be very satisfactory, but it should at least provoke some thought. For example, a special case may be considered in which the number of rows in the emotional state vector and in the response matrix is restricted to correspond to some isolated phase of human life in order to reduce the number of variables. The aid of psychologists and other experts in human personality should be sought in the definition of appropriate dependent variables.

The secular effects represented in Fig. 4 by broken arrows may play an important role in the development or deterioration of personality. These effects operate as a result of long-term average emotional state variables playing upon slowly responsive parameters in the value function, response planning model parameters, response pattern parameters, etc., and urging them to change or not depending on the effect caused by the behavior feedback. The same mechanisms produce gradual shifts in human relationships, either toward or away from a stable state. Thus, the technical concept of stability and instability seems applicable to the phenomena of personality interactions. It should be possible with this long-term approach to modeling to track in time the development of conflicts, painful subjects, accumulated tension, inappropriate behavior patterns, etc., with or without increments associated with traumatic experiences. Further discussion of the non-mathematical aspects of this model appears in Carson (1969). The form of mathematical structure which such a model might have is suggested below.

The model might be driven by a column vector $\mathbf{R}_k$ denoting the response of the other person in each of the k degrees of freedom of his behaviors. Each element R_k of this response vector is taken to be a continuous variable representing the amount or intensity of the kth aspect of his behavior as a time series. In the case of a discrete simulation the elements R_k would represent the other individual's remark or action, but the formulation here will be developed in terms of continuous variables. The response vector $\mathbf{R}_k$ is perceived by the subject individual being modeled. This perception may be formulated as the product of the response vector and a coefficient matrix of proper dimensionality which serves the function of pattern recognition. The product is a column vector $\mathbf{Q}_j$ of the perceptions of resources being made available:

$$\mathbf{Q}_j = P_{jk}\mathbf{R}_k ,$$

where the elements of the matrix P_{jk} are subjective pattern recognition coefficients. The "resources" represented by the rows j of $\mathbf{Q}_j$ might be things such as:

(1) Love, Q_1, which ranges from intense love to intense hatred through intermediate values representing affection, approval, disapproval, etc.;

(2) Food, Q_2, which ranges in attractiveness from preferred foods to poison, and which also can range in amount;

(3) Sexual stimulation, Q_3 ;

(4) Help, Q_4 ;

(5) Material things, Q_5 ;

(6) Information, Q_6 ;

⋮

The matrix P_{jk} might be constant in a short-term model, but it would be slowly time varying in the long-term modeling of a changing personality, and might even include changes in the algebraic signs of some of the matrix elements which would tend to produce perversions of normal behaviors through distorted or inaccurate perceptions of resources.

The perceived resources $\mathbf{Q}_j$ affect a column vector of emotional state variables $\mathbf{X}_i$. The rows of this vector might be as follows:

(1) Self image, X_1, especially in terms of interpersonal interactions but also concerning other positive or negative attributes as perceived;

(2) Feeling, X_2, with respect to other people or things present or in mind, ranging from confidence to fear;

(3) Intensity, X_3, of some basic physiological need, e.g., food, sex, water, etc.;

(4) Sensed pain or pleasure, X_4, from all sensory modalities;

(5) Level, X_5, of alertness, arousal, or activity;

⋮

The dependent variables X_i depend not only upon resources $\mathbf{Q}_j$ received but also upon internal stimuli and upon each other (i.e., other X_i's). The following vector differential equation might represent these effects:

$$\dot{\mathbf{X}}_i = V_{ij}\mathbf{Q}_j - K_{ii'}\mathbf{X}_i - \mathbf{C}_i ,$$

where the matrix V_{ij} causes rates of change of emotional state variables due to received resources $\mathbf{Q}_j$, the matrix $K_{ii'}$ provides state variable interaction and also a means for letting the emotions lag or subside, and the vector $\mathbf{C}_i$ is the set of internal forcing functions deriving from the subject's constellation of images about himself and other people, his homeostatic error signals such as hunger, etc. The coefficients V_{ij} and P_{jk} probably vary slowly in accordance with first-order differential equations having large time constants and having the state variable vector as a forcing function. These coefficients would thus be "trained" to appropriate values as an indirect result of experience affecting one's emotional state. Conversely, unexpected, unpredictable, or unfamiliar inputs to a person may drive these coefficients to highly inappropriate values. The retraining of these coefficients is the task of therapy performed on the individual's input side. The value function V_{ij} may be a linear function of the components of the emotional state vector. These coefficients are capable of being slowly changed by exposure to long enduring and/or intense state variables. Behavior is designed to maximize this value function; for example, a hungry person seeks food.

Let A_k denote an element of a person's behavior in the kth degree of freedom. The column vector $\mathbf{A}_k$ is selected to maximize the predicted value function resulting from the other person's response vector $\mathbf{R}_k$ which in turn results from the vector $\mathbf{A}_k$. The design of the vector $\mathbf{A}_k$ might be treated as an optimization problem incorporated into the model, or it might be developed indirectly with the use of a few simplifying assumptions. For example, one person might attribute to the other person his own set of parameter values, and then generate actions $\mathbf{A}_k$ which in his estimation are most helpful in bolstering the other person's emotional state vector, whether or not they would do so in fact. A more realistic model would attempt to deal with empathy (the deduction of the other person's state vector and parameter values) as an approach to designing A_ks which would be more helpful to the other person. Still another approach would be for the first person to assume that the other person's state vector is some scalar times his own (or perhaps his inference of the other person's) V_{ij} and P_{jk} matrices, times the other person's response vector $\mathbf{R}_k$, assuming that the other individual is performing essentially a steepest-ascent solution on his own value function. To be sure, these approaches would not be able to incorporate defense mechanisms, concealed motives, indirect appeals, deceptive behaviors, "games," etc., but at least they provide a start on the problem.

Another type of model employing the previous concept as a building block would be to couple one person with all the other people with whom he comes in contact in a typical day, either one at a time or perhaps for certain purposes in small groups. Such a model would make it possible to assess the resources which the person receives daily in much the same way that the nutritional value of a person's diet would be assessed in terms of what he eats in a day.

Although this personality model delineated in concept would not be very potent in dealing with the endless variety and richness of human interactions, it might nevertheless be at the right level of realism to describe the dynamics of an individual rat in an overcrowded colony or some other situation of interest. Such a model should be able to manifest the personality changes and behavioral types which emerge in a stressed colony of rats (Calhoun, 1962).

A final potential application would be to the career of an alcoholic, who begins to drink and continues the habit because of some valued changes which drinking causes in his emotional state vector, but who, after the development of bizarre behavioral and severe physiological changes, tends to be trapped in a loop of compulsive behavior which harms him as well as others around him. If alcohol is the cause of half or more of motor vehicle fatalities and other innumerable forms of distress in society, it would seem desirable to pursue the applicability of the model concept to gain possibly an understanding of the regenerative processes which occur in the personality and in interpersonal behaviors, and in order to investigate various forms of therapy. These processes have a hierarchical organization (Lysloff, 1969) so that the hierarchical modeling approach appears appropriate.

VIII. A Hierarchy of Health Care System Models

Health care may not seem a suitable topic for a paper on the future of modeling ecology. On the other hand, health care is a crucial and increasingly important function of the human ecosystem, one that is concerned with disease organisms, epidemics, death rates, and other similar phenomena which are central in many ecological considerations. Presumably all plant and animal species suffer dynamics of health state in ways that are important in ecology, so that a model depicting human health care can be modified by deleting the care aspects to represent the dynamics of plant and animal health states, including selective predation, selective grazing, and other phenomena needed in ecomodels. In addition, health care is one function which can be provided by man on behalf of not only himself but also an ecosystem, such as in the application of pesticides, so that the principles of health care are certainly desirable and relevant in many aspects of public problems modeling.

Another purpose served by treating the subject here is that it provides a good illustration of the applicability of ordinary differential equations to systems which would be approached by most modelers from the standpoint of discrete events (e.g., hospital entry by a patient, as in Kennedy, 1969), queueing, operations research, statistics, etc. This will serve to enunciate the wider applicability of ordinary differential equations in big system modeling.

Figure 5 is an abbreviated block diagram of the dynamics of a community health care system. The two upper blocks represent the planning and management loop by which the system is gradually improved, with or without the aid of simulation. The system itself is represented inside the broken-line box by one typical care subsystem (j) and one typical population category (h, i). The system basically provides treatment in return for money, with the result that patients tend to flow from sicker categories back to good health states against the tide of infection, accident, degeneration, overstress, etc.

The entire system is hierarchical and in fact pyramidal in structure. The base is a rectangular array of blocks denoting numbers of people in health state h and population category i. The flux of people among these compartments occurs mostly as a result of the influence of the various health care subsystems, which constitute a hierarchy. For example, a particular hospital is a member of the local hospital federation, which in turn might have representatives on the local health care planning board. Upward from the base several echelons of health states and population categories might be aggregated. Health states, for example, could be lumped grossly along the spectrum from sick to well in the

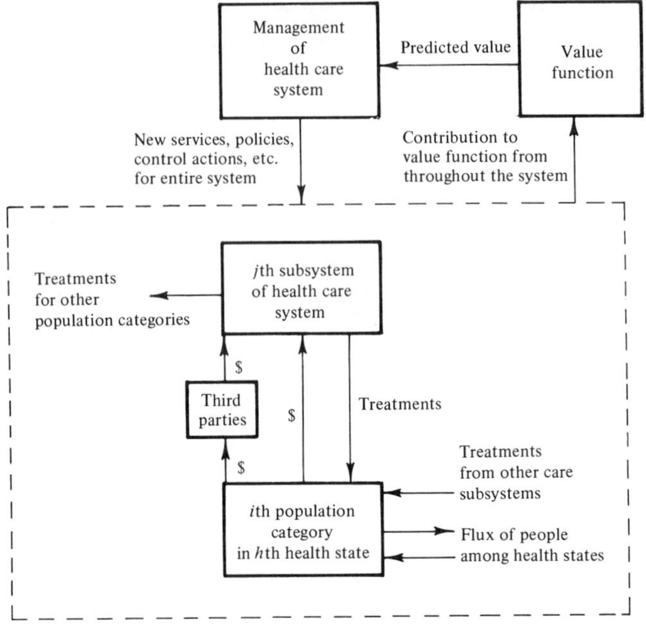

FIG. 5. Community health care system.

top echelon, while treated in smaller categories such as individual diseases, degrees of severity, and complications in lower echelons. Likewise, the population categories could be partitioned according to any degree of distinction with respect to geographical area, socioeconomic status, sex, age, education, marital status, ethnic and racial characteristics, etc. The health care subsystems could themselves be broken down to an arbitrary extent, such as recognizing individual hospitals from the set of hospitals, or by identifying particular care functions or facilities within a hospital, etc. Similarly, the set of physicians could be classified into an arbitrary set of echelons based on a variety of distinctions and so could all other health care subsystems such as nursing facilities, labor, pharmacies, emergency vehicles, etc.

The system itself being hierarchical makes it natural to organize its internal planning operations also hierarchically so that all echelons and blocks in the system benefit from good and mutually consistent planning. The planning efforts should be performed and funded by every block in the hierarchy of care subsystems, and all efforts should be closely coordinated to provide continuing consistency. The resulting plans which may undergo continual change with unforeseen conditions should be implemented by the hierarchy of care subsystems as often as possible consistent with the ability of the system to respond to management actions. Actions at upper echelons presumably would be less frequent than at lower echelons.

The hierarchy of plans should be generated according to a congruent hierarchy of value functions which need to be made progressively explicit and objective as well as mutually consistent. The design of value functions is a difficult undertaking, even individually, and the more so when they must be internally related within a hierarchy. Each value function should represent the quality of care in some relation to its costs and to the consequences of lowered quality measured in terms of a common denominator such as dollars per year equivalent net benefit to the community. The latter could be expressed in terms of some datum condition such as no care at all or the care in some particular reference year. Only by means of such a value function can the care provided by any subsystem of the total system be evaluated to make possible comparisons between the merits of alternative plans, leading to decisions involving payoffs, investments, and sacrifices in the short run for a higher or longer-term benefit. Only through having a value function is it possible to obtain benefit/cost ratios, returns on investments, or insights into weak or overly costly subsystems, consequences of omission or neglect, growing weaknesses of the overall system, and so forth.

Such a hierarchical system deserves to be rendered by a hierarchy of models so that each subsystem is represented by its own model and value function for use in plan evaluation and conception by computer simulation. The development of each model and concurrent simulation exercises can serve to clarify neighboring models in the hierarchy and to develop their mutual consistency. For example, the aggregation of outputs from individual models on one echelon should agree progressively with the output of the model above them in the next higher echelon since the latter model presumably subsumes the lower models in some more general or simplified form. The type of approach advocated is illustrated in the next section where some ideas toward a top-echelon model of the community health care system are presented.

Before proceeding to this, it would be good to call attention to a different kind of model of the same system wherein the patients are treated as discrete entities, discrete events befall them, and their meanderings through the system are treated stochastically, like balls in a pinball machine. A model in this form has some advantages in addressing detail but it would be expensive to exercise on a computer in competition with the proposed continuous model because of the large number of Monte Carlo trials that would have to be made for each plan to be evaluated. In a top echelon model it seems necessary to sacrifice attention to particular patients as human beings and to substitute treatment of "people flows." Better resolution would be appropriate, however, in lower echelon models, at least for some purposes.

IX. A General Model for a Community Total Health Care System

The model described in this section is under development by the Central Ohio Biomedical Engineering Community Council. Although it was derived for central Ohio, the model, or one like it, could be applied with little modification to any community, city, or region, and possibly to some entire countries. A block diagram appears in Fig. 6, the model itself being represented within the broken lines. At the top of Fig. 6 the loop closure is depicted whereby the model can be exercised repeatedly, incorporating different alternative plans in order to predict and evaluate the consequences of each plan. One has a means of choosing an optimum of the plans considered.

The blocks in the figure denote processes, and the arrows represent process inputs and outputs. The variables flowing out of the block are determined by equations representing the process dynamics, given the input variables and also a set of numerical values for both the inputs and the coefficients. The Fig. 6 diagram is abbreviated since only one type of health care is depicted explicitly. Also, only one category of people is considered, those in demographic category i and in health state h. All types of care are to be included across the top of the broken-line box, and all categories of people in the community are to be included across the bottom. In other words, every person and every existing health care service is implicitly represented.

The model is intended to portray realistically the essential processes of the system, introducing minimal detail and focusing mainly upon aspects of chief concern to the community. There are other models in the hierarchy discussed in Section VIII, which treat subsets of the total health care system. These other models incorporate fuller details. The present model is intended primarily as a tool for community-wide planning for the total health care system. Therefore, the categories of care and people to be included in any application of the model must be balanced so as not to bias the model in favor of or against any component of the system. The time scale for simulation is intended to be a few years in order to make it possible to neglect, at least in early development of the model, population changes due to net births and migration.

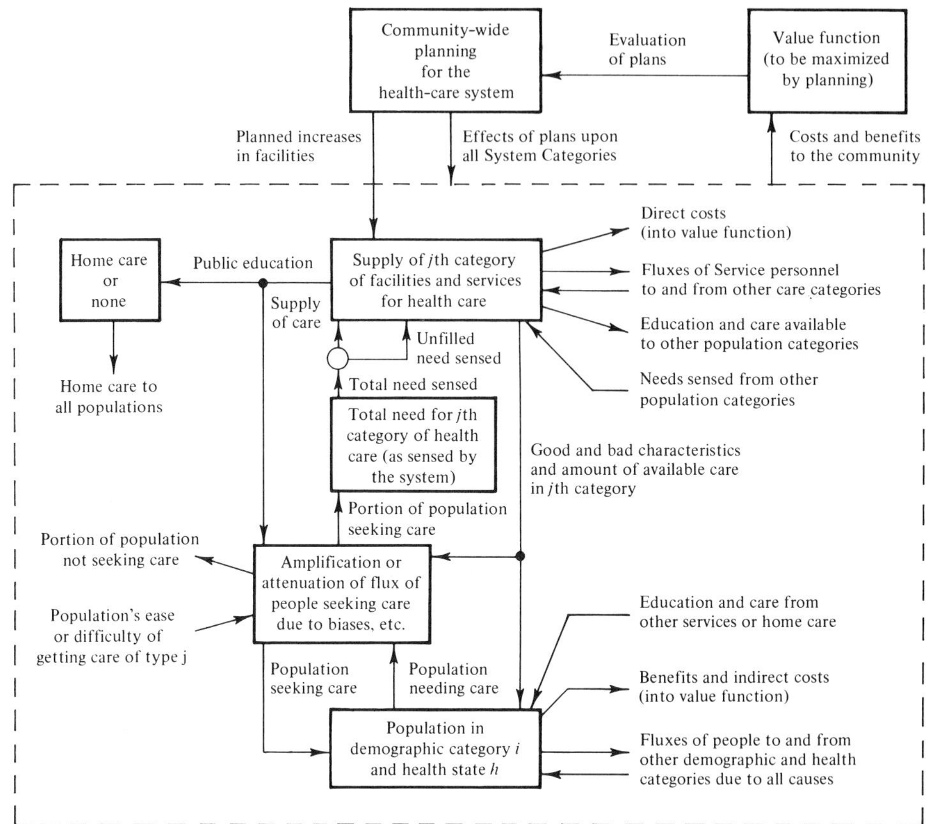

FIG. 6. Community health care system model in relation to value function and planning. The broken outline encloses the entire system, but only one care category (j) and population category (h, i) is shown within the system blocks.

In its gross features the model effectively transfers people from sick blocks to well blocks by the application of health care services. Therefore, the model must predict the flows of people per year between each pair of blocks in each direction, accounting for all of the causes of each flux. The kinds of causes considered include falling ill with a disease, suffering an accident, enjoying care of type j, spontaneous cure or cure from home care only, relapse or recurrence, death, environmental factors, referral of patients, etc. The sum of all the fluxes of people into a particular block minus the sum of fluxes out yields the instantaneous rate at which the number of people in the block changes:

$$dP_{hi}/dt = \sum_{\text{all } h'} F_{h'hi} - \sum_{\text{all } h'} F_{hh'i},$$

where P_{hi} is the instantaneous population in category (h, i), $F_{h'hi}$ is the flux into that category from another category (h', i), and $F_{hh'i}$ is the flux in the opposite direction. The model contains a differential equation of the above form for each category of people (h, i) included in the classification of the community and its health states.

The fluxes may be expressed as algebraic functions of the variables involved in the relevant mechanisms of people pumping. The simplest type of flux is one which occurs with known statistical incidence or probability per year per person, and essentially independently of any known factors other than demographic ones. Such a flux could be expressed mathematically as

$$F_{h'hi} = K_{h'hi} P_{h'i} ,$$

where $K_{h'hi}$ is a constant obtained from medical statistics, and $P_{h'i}$ is the population size of people category (h', i), where h' is a healthy state and h is a sick state into which the flux defined goes.

Another type of flux is one that depends upon the product of two populations, as in the case of the incidence of a communicable disease if the infected and uninfected populations were subjected to random encounters. Such a flux could be formulated as

$$F_{h'hi} = K_{h'hi} P_{hi} P_{h'i} ,$$

where h' is the uninfected state. The constant $K_{h'hi}$ might change as a result of implementing a plan, e.g., public education in areas such as venereal disease, smoking, or drug abuse.

The flux due to cures depends upon the product of the affected population and some function of the per capita amount of care actually rendered for the population. This kind of flux might be expressed as

$$F_{hh'i} = \sum_{j} K_{hh'i} P_{hi} f(q_{hij} / \mu_{hij} P_{hi}). \tag{1}$$

The summation is over all types of health care j, and the function f defines the applied care q_{hij}. This function could be of almost any form. For example, in a case like acute appendicitis, f would be a step function occurring at the level of care corresponding to an appendectomy and subsequent recuperatory care. In this case no less care can produce a cure and no more care can increase the flux. If the amount of applied care is not approximately uniform for all persons in P_{hi}, then it would be necessary to partition P_{hi} into two subsets having known ratio, calculate the fluxes separately for each, and add them together.

Consider now the care q_{hij} which is actually received by the people in category (h, i) from service j. Let q be measured in professional-level man-days per day, i.e., the number of professionals. It would not be necessary at this level to distinguish between the medical and para-medical professions. In a more detailed version of the model j could be disaggregated into services of different echelons of professional and subprofessional personnel.

Let Q denote available (but not necessarily rendered) care, also measured in man-days per day, and notated with subscripts representing the potential recipients and the type of care. In the case of inadequate supply of care type j a particular q is smaller than the corresponding Q because of a complex mutual selection process operating between people needing service and those potentially providing it. The details of this process can be finessed to some degree by treating the process as a gauntlet through which persons in need of service pass. In a digital computer program it might appear as a series of tests, the final output from which determines the fraction of people P_{hi} who receive care of type j and the complementary fraction who do not. Let μ_{hij} denote the fraction who do. Then the number of people of the category (h, i) who actually receive care of type j is $\mu_{hij}P_{hi}$. This number of individuals is in the argument of the function f in the Eq. (1). It can be determined by multiplying together the fractions representing probabilities of people passing the successive tests, such as: Does the person have sufficient biases (religious, ethnic, psychological, etc.) to avoid seeking care? Is the service rendered at a place or time which makes it essentially unavailable to the person? Can the person pay for the service directly or through an insurance program (if the facility j must have that kind of income)? And, is the service Q_j saturated by the services q_{hij} being rendered to all categories of people? The number of people $\mu_{hij}P_{hi}$ appears in the denominator of the argument of the cure function f because the effectiveness of the care depends upon the care per person actually receiving it.

In a short-time model the available amounts of care Q_j could be treated as known but manipulable constants. More realistically, each Q_j behaves according to a differential equation representing a process of dynamic population change of the persons available to offer services. The left-hand side of the equation would correspond to the rate of change of Q_j, and the right-hand side would include terms for death, net migration, specialization, introduction of new workers by recruiting, and so forth. With such an equation it would be possible to investigate, as an example, the dynamics of medical services available in a rural or urban area over a period of many years.

Between the extreme treatments of Q_j outlined above an intermediate approach would be to treat Q_j as an elastic variable, taking into account the possibility of a nonstandard work day. Then, Q_j could be used to measure the hours of work (service) required by each worker per day to render the health care needed and sought by and provided to all populations P_{hi}. In this approach the care rendered would be calculated by summing standard amounts of care over people served. There would be no distinction between Q_j and q_j. The work day would then be Q_j/N_j, where N_j is the number of people providing care of type j. If Q is man-days per year, then the quotient Q_j/N_j would be unitless (days per day: e.g., 1.5 would indicate a 12-hour day). This type of approach would not take realistic account of the saturation of services by patients, resulting in inadequate care, unless some allowable upper limit is set for each Q_j/N_j and unless q was decreased until Q_j/N_j returned to the limit whenever an overshoot occurred. Then the unsatisfactory service will be manifested in decreased effluxes by cures, due to decreases in the arguments of the cure functions. In this modification, $q = Q$ unless Q/N is saturated.

The model also must contain an algebraic formula for each element of cost or benefit that is of concern to the community, since these elements are needed inputs to the value function which planning is to maximize. The direct costs of providing services of type j can be taken as proportional to the services rendered q_j, and can be summed over all categories. An indirect cost could be assessed to represent the loss of work capacity by each person in a sick category, in recognition that such an externality is a very real cost to the community. Similarly, the prolonged productive life of individuals should be credited to the system when it increases the period that individuals spend in healthy states. Although the value function is peripheral to the system model, it is vital to utilization of the model so that great care and attention must be given to making it progressively more explicit and objective as the model development advances.

Total services and costs generated by the model are subject to validation against any available historical record for the actual system, as are the numbers and fluxes of people in and through the various health state categories. A further check on the input data can be made by comparing the turnover rates of patients as predicted by the model with rates based on statistics of dwell times of patients. The required input data are the numerical values of each K, μ, N, and the function f; initial conditions for each simulation trial (corresponding to each plan to be evaluated); cost and service data for each type of patient; and the coefficients of the value function to be maximized. The K's can be obtained for aggregated categories by determining weighted averages of disaggregated categories.

Development of the model would begin with gross categories *hij*. For example, the Central Ohio model has the following initial categories:

i: $i = 1$ is the central city of Columbus,
$i = 2$ is the rest of Central Ohio;

j: $j = 1$ is home care or none,
$j = 2$ is physician services for ambulatory patients,
$j = 3$ is hospital services,
$j = 4$ is other institutions, such as nursing homes;

h: $h = 1$ is relatively asymptomatic,
$h = 2$ is chronically physically ill,
$h = 3$ is acutely physically ill,
$h = 4$ is chronically mentally ill,
$h = 5$ is acutely mentally ill, and
$h = 6$ is dead.

Even with this small number of categories it is a difficult task to obtain the necessary input data with suitable accuracy. Low accuracy may be tolerated initially to permit early simulation trials as an aid to evaluating categories that most deserve to be rendered in greater detail in subsequent stages of development. The detail will increase the number of parameters accessible in the model for planning manipulation. As the elaboration process proceeds, the variables will become more meaningful and realistic, and the parameters can be expected to make better contact with available statistics. Also, the model outputs will be more easily compared with existing observations on the system. In addition, the models in lower echelons of the hierarchy of models will be increasingly interactive with the top-echelon model, facilitating cross-coupling within the hierarchy.

At some stage the model will be developed to an extent where community health care planners have sufficient confidence in its validity and sufficient interest in the manipulable parameters that they will begin to employ it as a planning tool. Allocation factors for public financing of various health care resources will be investigated. The impact of national health care insurance will be of interest. The loading of professionals and the utilization of their services will be of concern to the medical profession. Public health officials will be interested to see the model react to episodes of pollution, epidemics, and various disasters. Any significant wasteages of money revealed by the model will be of concern and investigated. Hospital planners will wish to consider effects of various alternative sites for contemplated new facilities. Educators may wish to determine future needs for personnel in all qualification categories. Social and welfare services may wish to identify categories of people in most need of improved care, and the impact of such care both upon the system and upon the total community. Operations researchers will wish to see where savings in time or changed procedures can increase the value function for the community at large. Researchers and administrators in medical technology will want to use the model to identify targets for research or to establish priorities for different projects. Insurance companies will wish to see how changes in their coverage policies will influence people fluxes and costs in the system. Thus, all interested public and private agencies and other groups will be able to use the model for their own purposes, and will have at their joint command a common tool by which they can try to arrive at compromised plans for the good of all.

This type of model is only one example of many different areas in which the same general approach can be applied. In the early stages of their development, such models may not represent accurately the mechanisms and interrelations of their subject systems, but at least they will serve as a heuristic tool for both their own perfection and also to stimulate holistic thinking about the systems and their possible interactions with other systems of concern to the community at large. Hierarchical models, then, appear to offer a reasonable approach to the handling of big public-problems systems and the gradual integration of the various facets of human activity into coordinated and well-planned patterns of both individual and collective life.

X. Closure

It has not been possible, really, in a broad-brush treatment such as this to express adequately the future prospects for modeling in the ecological context. Nor has it been possible to portray the enthusiasm and sense of good feelings which I share with all other workers in the general area of ecomodeling. One of the feelings is excitement because this work is on the frontier, and any contribution at all is helpful at this stage. The field is growing rapidly in scope and level of activity and in maturity.

Another feeling is satisfaction from reflection that one's relatively feeble and fumbling efforts are going to be built upon effectively to aid in at least partial cure of some of the world's ills, that one is a force for good, that one is helping to make the world a better place, that one is doing his share to sustain our environmental heritage and to enable mutual coexistence with our fellow organisms, that one is making one of the best possible applications of his technical skills, and in short that one is somehow in tune with what God would wish of us.

Another feeling is humility in the face of the huge task which we are entering upon so brashly, armed with a little technology and a little ecological data and knowledge, but faced with a host of unknowns and pitfalls. There is no way to predict the outcome of this endeavor, but also there is no alternative but to proceed in it with all dispatch, letting it evolve or run its course as chance would have it. This is the heritage of our evolution as a species, and the legacy that we now seek to build for future generations of humanity. What we learn may or may not shape a new world, reorder its ethical foundations, change the structures of outworn, archaic, and unresponsive institutions, or in general set our priorities for civilized life straight in terms of known truths of ecological existence. But if some of these things do occur then perhaps somehow we may have had a small hand in it. There is no way of knowing whether we sit at the terminus of a Golden Age of Science, or rather at the beginning of a new chapter—the pursuit of wholes. History will establish this and our role in determining it.

If we succeed in passing along a viable set of values, perspectives and methodologies to the future, then perhaps the next generation of ecologists, and generations after that may still be involved in some form of ecomodeling. And if they ever should be afforded the luxury of speculating about the future of their work, who in his wildest imagination would now hazard a guess about what they might contemplate as *their* "next-generation models in ecology"?

REFERENCES

Anon. (1970). *Simulation* **14**, 176–179.

Atkins, G. L. (1969). "Multicompartment Models for Biological Systems." Methuen, London.

Axelband, E. I., Grettenberg, T. L., Kalaba, R. E., and Sridhar, R. (1969). *IEEE Trans. Automatic Control* December 1969, 750.

Ayres, R. U. (1964). Special Aspects of Environment Resulting from Various Kinds of Nuclear Wars. Part II, Annex III, Application of Input-Output Analysis to a Homeostatic Ecosystem. Hudson Institute (HI-303-RR/A III), Harmon-On-Hudson, New York.

Bailey, N. T. J. (1967). "The Mathematical Approach to Biology and Medicine." Wiley, New York.

Barea, D. J. (1963). *Arch. Zootecnia* **12**, 252.

Bartholomew, G. A., and Hoel, P. G. (1953). *J. Mammal.* **34**, 417.
Bartlett, M. S. (1960). "Stochastic Population Models in Ecology and Epidemiology." Methuen, London.
Beverton, R. J. H., and Holt, S. J. (1957). *Min. Agr. Fish. Food (U.K.) Fish Investig. Ser. II* **19**.
Bledsoe, L. J., and Jameson, D. A. (1969). *In* "The Grasslands Ecosystem" (R. L. Dix and R. G. Beidleman, eds.), Range Sci. Dept., Sci. Ser., No. 2. Colorado State Univ., Fort Collins, Colorado.
Braun, K. N., and Clymer, A. B. (1963). *Proc. AIAA Conf. Simulat. Aerospace Flight,* 1963 244.
Brennan, R. D., de Wit, C. T., Williams, W. A., and Quattrin, E. V. (1970). *Oecologia* **4**, 113.
Calhoun, J. B. (1962). *Sci. Amer.* February 1962, 139.
Carson, R. C. (1969). "Interaction Concepts of Personality." Aldine, Chicago, Illinois.
Clymer, A. B. (1969a). *Proc. Conf. Simulat. Modeling, Midwestern Simulat. Council* April 1969, 107.
Clymer, A. B. (1969b). *Proc. Conf. Appl. Continuous Syst. Simulat. Languages, 1st* June 1969, 1.
Clymer, A. B. (1970). *Simulation* **15**, 50.
Clymer, A. B., and Bledsoe, L. J. (1969). *In* "Simulation and Analysis of Dynamics of a Semi-Desert Grassland." Range Sci. Dept., Colorado State Univ., Fort Collins, Colorado (Dec. 1970), pp. I-75 through I-99.
Clymer, A. B., Braun, K. N., and Davidson, R. S. (1965). The Desirability and Feasibility of Simulating Ecological Systems. Meeting, Midwestern and Central States Simulation Councils, January 1965, unpublished.
Cooper, C. F. (1969). *In* "The Ecosystem Concept in Natural Resource Management" (G. M. Van Dyne, ed.), Chapter 9. Academic Press, New York.
Cooper, C. F., and Jolly, W. C. (1969). Ecological Effects of Weather Modification. School of Natural Resources, University of Michigan, Ann Arbor, Michigan.
Cowser, K. E., Kaye, S. V., Rohwer, P. S., Snyder, W. S., and Struxness, E. G. (1967). Bioenvironmental and Radiological Safety Feasibility Studies, Atlantic-Pacific Interoceanic Canal. Phase I Final Report: Dose Estimation Studies. Battelle Memorial Institute, Columbus, Ohio.
Cushing, D. H. (1968). "Fisheries Biology. A Study in Population Dynamics." Univ. of Wisconsin Press, Madison, Wisconsin.
Dale, M. B. (1970). *Ecology* **51**, 2.
Davidson, R. S., and Clymer, A. B. (1966). *Ann. N. Y. Acad. Sci.* **128**, 790.
Duncan, W. G., Loomis, R. S., Williams, W. A., and Hanan, R. (1967). *Hilgardia* **38**, 181.
Forrester, J. W. (1969). "Urban Dynamics." MIT Press, Cambridge, Massachusetts.
Garfinkel, D. A. (1967). *J. Theoret. Biol.* **14**, 325.
Garfinkel, D. A., MacArthur, R. H., and Sack, R. (1964). *Ann. N. Y. Acad. Sci.* **115**, 943.
Garfinkel, D. A., and Sack, R. (1964). *Ecology* **45**, 502.
Goodall, D. W. (1969). *In* "Concepts and Models of Biomathematics: Simulation Techniques and Methods" (F. Heimmetz, ed.), pp. 211–236. Dekker, New York.
Holdridge, L. R. (1947). *Science* **105**, 367.
Holling, C. S. (1966a). *Mem. Entom. Soc. Canada* **48**, 1.
Holling, C. S. (1966b). *In* "Systems Analysis in Ecology" (K. E. F. Watt, ed.). Academic Press, New York.
Hufschmidt, M. M., and Fiering, M. B. (1966). "Simulation Techniques for Design of Water Resource Systems." Harvard Univ. Press, Cambridge, Massachusetts.

IBM Corporation (1966). "Bibliography on Simulation." IBM Corp., White Plains, New York.

Kadlec, J. (1971). "A Partial Annotated Bibliography of Mathematical Models in Ecology," Univ. of Michigan, School of Natural Resources, Ann Arbor, Michigan.

Kennedy, F. (1969). *IEEE Trans. Systems Science and Control* **5**, 199–207.

Krendel, E. S. (1970). *In Proc. 6th Annual Conf. on Manual Control, Dayton, Ohio, April 7–9, 1970*, 637–675.

Levins, R. (1966). *Amer. Sci.* **54**, 421.

Levins, R. (1968). "Evolution in Changing Environments." Princeton Univ. Press, Princeton, New Jersey.

Lysloff, G. O. (1969). *Yrbk. Soc. Gen. Syst. Res.* **14**, 17.

McLeod, J., and Defares, J. G. (1962). *AIEE Trans.* **62**, 8.

Neyman, J., and Scott, E. L. (1959). *Science* **130**, 303.

Olson, J. S. (1963a). *In* "Radioecology." *Proc. Nat. Symp. Radioecol.*, 1*st*, *Fort Collins, Colorado*, *September* 1961. Reinhold, New York.

Olson, J. S. (1963b). *Ecology* **44**, 322.

Olson, J. S. (1964). *Ecology* (suppl.) **52**, 99.

Olson, J. S. (1965). *Health Phys.* **11**, 1385.

Patten, B. C. (1965). Community Organization and Energy Relationships in Plankton. Oak Ridge Nat. Lab. Rep. 3634, Oak Ridge, Tennessee.

Patten, B. C. (ed.) (1971). "Systems Analysis and Simulation in Ecology," Vols. I and II. Academic Press, New York.

Pennycuick, C. J., Compton, R. M., and Beckingham, L. (1968). *J. Theoret. Biol.* **18**, 316.

Pielou, E. C. (1969). "An Introduction to Mathematical Ecology." Wiley (Interscience), New York.

Ricker, W. E. (1954). *J. Fish. Res. Bd. Canada* **11**, 559.

Robinson, J. (1967). *Nature* **215**, 33.

Schaefer, M. B., and Beverton, R. J. H. (1963). *In* "The Sea" (M. W. Hill, ed.). Wiley (Interscience), New York.

Schwartzman, G. (ed.) (1970). "Some Concepts of Modeling." Natural Resource Ecology Laboratory, Colorado State Univ., Fort Collins, Colorado.

Silliman, R. P. (1967). *Fish. Bull.* **66**, 31.

Simulation Councils, Inc. (1970). *Simulation* **14**, 176–179.

Smith, F. E. (1963). *Ecology* **44**, 651.

Taylor, N. W. (1967). *Ecology* **48**, 290.

Turner, F. B., and Jennrich, R. I. (1967). *In* "Radioecological Concentration Processes" (B. Aberg and F. P. Hungate, eds.). Pergamon, Oxford.

UNESCO (1969). "Human Engineering of the Planet." Impact of Science on Society **19**. UNESCO, Place de Fontenoy, 75 Paris-7e, France.

Van Dyne, G. M. (1969a). *In* "The Grassland Ecosystem" (R. L. Dix and R. G. Beidleman, eds.). Range Science Department, Science Series No. 2, Colorado State University, Fort Collins, Colorado.

Van Dyne, G. M. (ed.) (1969b). "The Ecosystem Concept in Natural Resource Management." Academic Press, New York.

Watt, K. E. F. (1956). *J. Fish. Res. Bd. Canada* **13**, 613.

Watt, K. E. F. (ed.) (1966). "Systems Analysis in Ecology." Academic Press, New York.

Watt, K. E. F. (1968). "Ecology and Resource Management." McGraw-Hill, New York.

Witherspoon, J. P., Auerbach, S. I., and Olson, J. S. (1964). *Ecol. Monogr.* **34**, 403.

Woo, K. B., Boersma, L., and Stone, L. N. (1966). *Water Resources Res.* **2**, 85.

Author Index

Numbers in italics refer to the pages on which the complete references are listed.

G

H

I

J

K

Y

Z

Subject Index

A

B

F

G

R